NEUROMETHODS ☐ 12

Drugs as Tools in Neurotransmitter Research

NEUROMETHODS

Program Editors: Alan A. Boulton and Glen B. Baker

1. **General Neurochemical Techniques**
   Edited by *Alan A. Boulton and Glen B. Baker,* 1985
2. **Amines and Their Metabolites**
   Edited by *Alan A. Boulton, Glen B. Baker, and Judith M. Baker,* 1985
3. **Amino Acids**
   Edited by *Alan A. Boulton, Glen B. Baker, and James D. Wood,* 1985
4. **Receptor Binding Techniques**
   Edited by *Alan A. Boulton, Glen B. Baker, and Pavel D. Hrdina,* 1986
5. **Neurotransmitter Enzymes**
   Edited by *Alan A. Boulton, Glen B. Baker, and Peter H. Yu,* 1986
6. **Peptides**
   Edited by *Alan A. Boulton, Glen B. Baker, and Quentin Pittman,* 1987
7. **Lipids and Related Compounds**
   Edited by *Alan A. Boulton, Glen B. Baker, and Lloyd A. Horrocks,* 1988
8. **Imaging and Correlative Physicochemical Techniques**
   Edited by *Alan A. Boulton, Glen B. Baker, and Donald P. Boisvert,* 1988
9. **The Neuronal Microenvironment**
   Edited by *Alan A. Boulton, Glen B. Baker, and Wolfgang Walz,* 1988
10. **Analysis of Psychiatric Drugs**
    Edited by *Alan A. Boulton, Glen B. Baker, and Ronald T. Coutts,* 1988
11. **Carbohydrates and Energy Metabolism**
    Edited by *Alan A. Boulton, Glen B. Baker, and Roger F. Butterworth,* 1989
12. **Drugs as Tools in Neurotransmitter Research**
    Edited by *Alan A. Boulton, Glen B. Baker, and Augusto V. Juorio,* 1989
13. **Psychopharmacology**
    Edited by *Alan A. Boulton, Glen B. Baker, and Andrew J. Greenshaw,* 1989
14. **Neurophysiology**
    Edited by *Alan A. Boulton, Glen B. Baker, and Case H. Vanderwolf*
15. **Neuropsychology**
    Edited by *Alan A. Boulton, Glen B. Baker, and Merril Hiscock*
16. **Molecular Neurobiology**
    Edited by *Alan A. Boulton, Glen B. Baker, and Anthony T. Campagnoni*

NEUROMETHODS

Program Editors: *Alan A. Boulton and Glen B. Baker*

NEUROMETHODS □ 12

Drugs as Tools in Neurotransmitter Research

Edited by

Alan A. Boulton

University of Saskatchewan, Saskatoon, Canada

Glen B. Baker

University of Alberta, Edmonton, Canada

and

Augusto V. Juorio

University of Saskatchewan, Saskatoon, Canada

Humana Press • Clifton, New Jersey

Library of Congress Cataloging in Publication Data

Main entry under title:

Drugs as tools in neurotransmitter research.

(Neuromethods ; 12)
Includes index.
1. Neurotransmitters—Research—Technique.
2. Neurotoxins. 3. Neurotransmitters—Agonists.
4. Neurotransmitters—Antagonists. 5. Nervous system—
Diseases—Animal models. I. Boulton, A. A. (Alan A.)
II. Baker, Glen B., Date . III. Juorio,
A. V. (Augusto V.) IV. Series. [DNLM: 1. Neural
Transmission—drug effects. 2. Neuroregulators—
pharmacology. 3. Synaptic Receptors—drug effects.
W1 NE337G v.12 / WL 102.8 D794]
QP364.7.D78 1988 612.8'072 88-32044
ISBN 0-89603-122-5

Crescent Manor
PO Box 2148
Clifton, NJ 07015

Printed in the United States of America

Preface to the Series

When the President of Humana Press first suggested that a series on methods in the neurosciences might be useful, one of us (AAB) was quite skeptical; only after discussions with GBB and some searching both of memory and library shelves did it seem that perhaps the publisher was right. Although some excellent methods books have recently appeared, notably in neuroanatomy, it is a fact that there is a dearth in this particular field, a fact attested to by the alacrity and enthusiasm with which most of the contributors to this series accepted our invitations and suggested additional topics and areas. After a somewhat hesitant start, essentially in the neurochemistry section, the series has grown and will encompass neurochemistry, neuropsychiatry, neurology, neuropathology, neurogenetics, neuroethology, molecular neurobiology, animal models of nervous disease, and no doubt many more "neuros." Although we have tried to include adequate methodological detail and in many cases detailed protocols, we have also tried to include wherever possible a short introductory review of the methods and/or related substances, comparisons with other methods, and the relationship of the substances being analyzed to neurological and psychiatric disorders. Recognizing our own limitations, we have invited a guest editor to join with us on most volumes in order to ensure complete coverage of the field. These editors will add their specialized knowledge and competencies. We anticipate that this series will fill a gap; we can only hope that it will be filled appropriately and with the right amount of expertise with respect to each method, substance or group of substances, and area treated.

Alan A. Boulton
Glen B. Baker

Preface

Serendipitous discovery has played an important role in the identification of physiological mechanisms in central and peripheral neurotransmission. In many cases, the subsequent followup of these discoveries has been substantially aided by previously identified natural substances that interfered with these mechanisms. The site of neuromuscular blockade produced by curare extracts, for example, was shown in the 1850s by Claude Bernard, and Sir Henry H. Dale in 1906 discovered the reversal of the pressor effects of adrenal extracts by ergot alkaloids. These findings were the first clues that led eventually to the definition of nicotinic cholinergic receptors and the present classification of adrenergic receptors.

In the 1950s many discoveries, including the identification of the catecholamines and 5-hydroxytryptamine in the central nervous system, followed by the development of specific sensitive biochemical and histochemical methods and the introduction of modern psychotropic drugs led to our increased understanding of chemical transmission and the formulation of theories that attempt to explain the causes of some mental disorders. The finding that hydroxylated derivatives of noradrenaline and 5-hydroxytryptamine are readily taken up by catecholaminergic or serotonergic neurons, and that this is followed by poisoning and death of the cell, encouraged the use of these compounds for the mapping of central and peripheral catecholamine or 5-hydroxytryptamine systems, and the investigation of the relevance of these pathways to receptor regulation or behavioral responses. As early as 1906, the effects of nicotine and curare on muscle contractions were interpreted by J. N. Langley in terms of their effects on the same "receptive substance," but it took a further 70 years and the development of radioligand-binding techniques before substantial advances in receptor pharmacology occurred. Using these techniques, investigators have been able to study drugs and neurotransmitter receptors as discrete molecular entities, and they have described changes in receptor numbers and affinities in a variety of situations.

The multidisciplinary nature of neurotransmitter research coupled with the diversity of techniques that are needed and the rapid growth of this area prompted us to assemble this volume, which deals with some of the pharmacological tools that have been used throughout the years in the study of neurotransmission. Some subjects are discussed in more than one chapter; each author, however, has viewed the situation from a different perspective and has presented material from different utilitarian points of view. The first three chapters deal with the use of neurotoxins for the study of catecholamine, 5-hydrotryptamine, and cholinergic pathways. Subsequent chapters examine the interactions between dopamine, 5-hydroxytryptamine, and GABA and their receptor sites, with special emphasis being placed on the conditions required for the assays. The final chapters examine the effect of drugs that interfere with monoamine storage, the various techniques that are available for the estimation of the metabolism of monoamines and acetylcholine, as well as behavioral and biochemical models of disease.

A. V. Juorio

Contents

NEUROTOXINS THAT AFFECT CENTRAL INDOLEAMINE NEURONS

Tomás A. Reader

TOXINS AFFECTING CHOLINERGIC NEURONS
Vincent A. Chiappinelli

GABA ANTAGONISTS: THEIR USE AND MECHANISMS OF ACTION
M. Farrant and R. A. Webster

DRUG-INDUCED BEHAVIORAL MODELS OF CENTRAL DISORDERS
David M. Jackson

DETERMINATION OF ACETYLCHOLINE DYNAMICS
Darwin L. Cheney, John Lehmann, Cristina Cosi, and Paul L. Wood

Contributors

GLEN B. BAKER • *Neurochemical Research Unit, Department of Psychology, University of Alberta, Edmonton, Alberta, Canada*

ALAN A. BOULTON • *Neuropsychiatric Research Unit, University of Saskatchewan, Saskatoon, Saskatchewan, Canada*

DARWIN L. CHENEY • *Neuroscience Research, CIBA-GEIGY Corporation, Summit, New Jersey*

VINCENT A. CHIAPPINELLI • *Department of Pharmacology, St. Louis University School of Medicine, St. Louis, Missouri*

CRISTINA COSI • *NIH, Bethesda, Maryland*

R. C. DUVOISIN • *Department of Neurology, University of Medicine and Dentistry of New Jersey, Robert Wood Johnson Medical School, Piscataway, New Jersey*

MARK FARRANT • *Department of Pharmacology, University College of London, London, England*

SILVIO GARATTINI • *Instituto di Ricerche Farmacologiche "Mario Negri," Milan, Italy*

MARIA HADJICONSTANTINOU • *Department of Pharmacology, The Ohio State College of Medicine, Columbus, Ohio*

RICHARD E. HEIKKILA • *Department of Neurology, University of Medicine and Dentistry of New Jersey, Robert Wood Johnson Medical School, Piscataway, New Jersey*

DAVID M. JACKSON • *Department of Pharmacology, University of Sydney, New South Wales, Australia*

AUGUSTO V. JUORIO • *University of Saskatchewan, Saskatoon, Saskatchewan, Canada*

RICHARD M. KOSTRZEWA • *Department of Pharmacology, East Tennessee State University, Quillen-Dishner College of Medicine, Johnson City, Tennessee*

PIERRE M. LADURON • *Department of Biochemical Pharmacology, Janssen Pharmaceutica, Beerse, Belgium*

JOHN LEHMANN • *Division of Neurobiology, Fondax Groupe de Recherches Servier, Puteaux, France*

J. E. LEYSEN • *Department of Biochemical Pharmacology, Janssen Pharmaceutica, Beerse, Belgium*

T. MENNINI • *Instituto di Ricerche Farmacologiche "Mario Negri," Milan, Italy*

NORTON N. NEFF • *Department of Pharmacology, The Ohio State College of Medicine, Columbus, Ohio*

THOMÁS A. READER • *Centre de Recherche en Sciences Neurologiques, Department de Physiologic, Université de Montréal, Montreal, Quebec, Canada*

P. K. SONSALLA • *Department of Neurology, University of Medicine and Dentistry of New Jersey, Robert Wood Johnson Medical School, Piscataway, New Jersey*

R. A. WEBSTER • *Department of Pharmacology, University College of London, London, England*

PAUL L. WOOD • *G. D. Searle & Co., St. Louis, Missouri*

Neurotoxins That Affect Central and Peripheral Catecholamine Neurons

Richard M. Kostrzewa

1. Scope, Objective, and Format of the Review on Neurotoxins

The catecholamine (CA) neurotoxins are so extensively used in neuroscience research today, for chemical lesioning of noradrenergic and/or dopaminergic neurons, that a comprehensive review on all aspects of neurotoxin use is virtually an insurmountable task. The focus of this review article on CA neurotoxins is on the general utility of these substances, regarding expected pharmacological actions for given agents. Because of the relative specificity of the widely used 6-hydroxydopamine (6-OHDA) as a neurotoxin for CA neurons, the major portion of the review is devoted to this agent. It is expected, after reading this review, that the naive scientist will be able to reasonably develop a suitable experimental protocol for chemical lesioning of noradrenergic and/or dopaminergic nerves for particular studies. For the scientist who is experienced in the use of chemical neurotoxins, this review article should serve to highlight the critical aspects of neurotoxic drugs, while summarizing the mechanisms by which each of the various agents acts and indicating the practical uses of each agent.

Other reviews have been written on the topic of CA neurotoxins, and the reader is referred to these for

(a) a perspective on how the field of CA neurotoxins has developed
(b) greater detail on dose–effect relationships
(c) a more global review of how the thrust of research activity in this area has changed over time and
(d) reference to topics that are not included in this review.

(*See* Levi-Montalcini and Angeletti, 1966; Thoenen and Tranzer, 1973; Kostrzewa and Jacobowitz, 1974; Breese, 1975; Jonsson, 1980; Kostrzewa, 1988.)

The format of the present review consists of a description of the various CA neurotoxins that are commonly used today. The general utility of each agent is indicated, including the limitations on use, the selectivity of action, potency, advantages, and disadvantages.

2. History of the Development of Catecholamine Neurotoxins

Historically, a principal means of studying a particular nerve function has been to surgically ablate the axonal process of that nerve or to surgically remove the ganglion from which the nerve emanates, and then determine what function has been lost. This general application was most prominently espoused for the sympathetic nervous system by W. B. Cannon (*see* Cannon and Rosenblueth, 1937). Obvious limitations of this procedure are: ganglia are often difficult to isolate, a nerve tract contains only a portion of a specific neurochemical input to a structure, or the axon contains several different types of neurochemical inputs to a structure.

An improvement over this methodology was achieved by Levi-Montalcini, who, together with Cohen and Booker, demonstrated that an antibody to "nerve growth factor"—administered to mice during the first several days after birth—prevented the development of a major portion of the sympathetic nervous system (Cohen, 1960; Levi-Montalcini and Booker, 1960). This technique is known today as "immunosympathectomy."

Several years after this discovery, Thoenen and Tranzer, who were investigating phenethylamines that would form dense aggregates at the ultrastructural level in sympathetic nerves, found that the substance 2,4,5-trihydroxyphenethylamine produced an overt destruction of sympathetic nerve terminals, and that this action was a selective one (Tranzer and Thoenen, 1967, 1968; Thoenen and Tranzer, 1968). Other investigators demonstrated that intracerebral injection of this compound, which is generally known as 6-hydroxydopamine (6-OHDA), resulted in the destruction of central noradrenergic and dopaminergic neurons (Ungerstedt, 1968; Bloom et al., 1969), and a permanent reduction of both norepinephrine (noradrenaline, NE) and dopamine (DA) in the central nervous system (CNS) (Breese and Traylor, 1970; Uretsky and Iversen, 1970).

A variety of analogs of 6-OHDA have been synthesized and have been shown to produce similar effects. This includes 6-hydroxydopa (6-OHDOPA), a substance that crosses the blood–brain barrier and is metabolized to 6-OHDA (Ong et al., 1969; Berkowitz et al., 1970).

N-(2-Chloroethyl)-*N*-ethyl-bromobenzylamine (DSP-4), a compound unrelated to 6-OHDA, also is capable of crossing the blood–brain barrier, but produces selective effects on noradrenergic neurons, leaving dopaminergic neurons intact (Jonsson et al., 1979). Guanethidine, a sympathetic neuronal blocking agent, is capable of producing destruction of sympathetic ganglia (Burnstock et al., 1971; Eränkö and Eränkö, 1971).

The most recently discovered CA neurotoxin is 1-methyl-4-phenyl-1,2,3,6-tetrahydropyridine (MPTP), which was found as a contaminant in a drug of abuse and shown to be responsible for production of clinical irreversible Parkinsonism in humans (Langston et al., 1983). This substance has since been used experimentally to produce relatively selective destruction of the pars compacta of the substantia nigra of primates (Burns et al., 1983; Langston et al., 1984a) and of a more convenient animal model, the mouse (Heikkila et al., 1984a). Each of the above aspects of catecholamine neurotoxins is addressed in greater detail in the appropriate section of this review.

3. Anti-NGF

Levi-Montalcini and Cohen, recipients of the 1986 Nobel Prize in Medicine, played an instrumental role in enriching our understanding of neuronal development and growth, and demonstrating the concept of growth factors in general and "nerve growth factor" (NGF) in particular. Levi-Montalcini first showed that a mouse sarcoma, implanted onto chick embryos, promoted the growth of fibers from paravertebral ganglia toward the implant (Levi-Montalcini and Hamburger, 1951; Levi-Montalcini, 1952). The factor responsible for this effect, NGF, was characterized by Cohen (Cohen et al., 1954), who subsequently developed an antiserum, which is often termed anti-NGF (Cohen, 1960).

Use of this anti-NGF represented the beginning of a chemical approach to alter the integrity of CA neurons. In essence, this antiserum was the first selective neurotoxin for CA neurons. When injected into neonatal chicks, rats, or mice, anti-NGF produces an "immunosympathectomy" characterized by a 90–95% loss of paravertebral ganglion cells and a moderate loss of prevertebral ganglion cells. Associated with the reduction in sympathetic cells is a marked reduction in sympathetic noradrenergic innervation to the various peripheral tissues (Cohen, 1960; Levi-Montalcini and Booker, 1960; Levi-Montalcini and Angeletti, 1962, 1966). It has now been shown that mouse NGF, injected into rats, promotes the production by the rat of an anti-NGF that cross-reacts with its own NGF. During pregnancy, this anti-NGF inactivates fetal NGF, resulting in an early immunosympathectomy and providing confirmation that peripheral noradrenergic neuronal development is impaired, whereas central CA fibers are unimpaired (Dolkart-Gorin and Johnson, 1979).

Mechanistically, anti-NGF appears to inactivate endogenous NGF, which itself is normally incorporated into sympathetic terminals and perikarya and transported to the perikaryon (Hendry et al., 1974; Yanker and Shooter, 1979). Acting at sites on the nuclear membrane or inside the nucleus, NGF is considered to mediate changes in gene expression that lead to posttranscriptional molecular events that promote axonal growth (*see* Levi-Montalcini, 1987).

Although NGF and anti-NGF have dramatic neuroteratologic actions on sympathetic ganglia, the central effects on CA-containing neurons are negligible (Konkol et al., 1978b; Olson et al., 1979). The advantage of anti-NGF is that it permanently removes a large portion of sympathetic innervation throughout life. This animal model has been of value in studying the development and function of sympathetic neurons. Disadvantages of immunosympathectomy include the fact that a substantial portion of the peripheral sympathetic system remains intact—virtually all terminal ganglia, along with a fair to moderate number of prevertebral ganglia and 5–10% of paraganglia. Also, anti-NGF must be administered to developing animals, since actions on mature sympathetic neurons are not substantial. Since peripheral tissue function may be regulated by only a small portion of the normal complement of sympathetic fibers, conclusions about the functional role of groups of fibers are not easily drawn.

4. 6-OHDA

4.1. Molecular Mechanism of Action

The agent 6-OHDA, an isomer of NE that differs in the location of an hydroxyl moiety, is a chemically unstable substance with a low oxidation potential that undergoes spontaneous autooxidation to a proposed series of intermediates. Saner and Thoenen (1971), on the basis of ultraviolet spectroscopy, suggested the intermediate was a *p*-quinone, trihydroxyindole, and/or trihydroxyindoline. Blank and coworkers (1972a), on the basis of NMR, mass spectrometry, infrared, and fluorescence spectroscopy, proposed a dihydroxyindole. This latter group of investigators, moreover, have concluded that, upon oxidation and rearrangement, all of the neurotoxins similar in structure to 6-OHDA (i.e., see Tranzer and Thoenen, 1973) could generate a dihydroxy or aminohydroxyindole (in the case of 6-aminodopamine or α-methyl-6-aminodopamine). The dihydroxyindoles also represent intermediates of 5,6- and 5,7-dihydroxytryptamine, serotoninergic neurotoxins (Baumgarten et al., 1972), indicating that this species is a common feature of different classes of neurotoxic agents.

Because radiolabel is detectable in tissue weeks after administration of [^{14}C] or [^{3}H]6-OHDA and is associated with the particulate fraction of a tissue homogenate, it is apparent that 6-OHDA or one of its byproducts binds to membrane components (Porter et al., 1965; Thoenen and Tranzer, 1968). It is thought that nerve destruction is brought about by a loss of membrane integrity (Saner and Thoenen, 1971) or by interruption of the respiratory transport chain in mitochondria and interference with energy poroduction (Wagner and Trendelenburg, 1971), or by interruption of some other neuronal function.

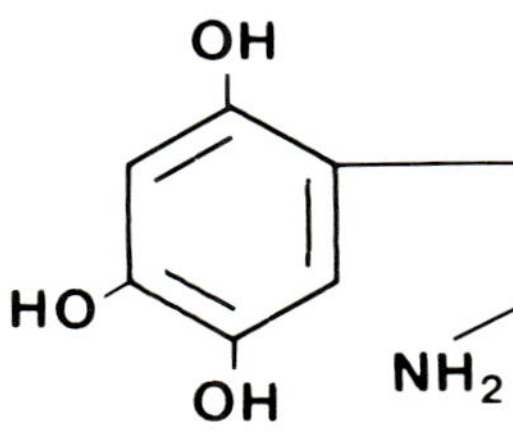

6-OHDA

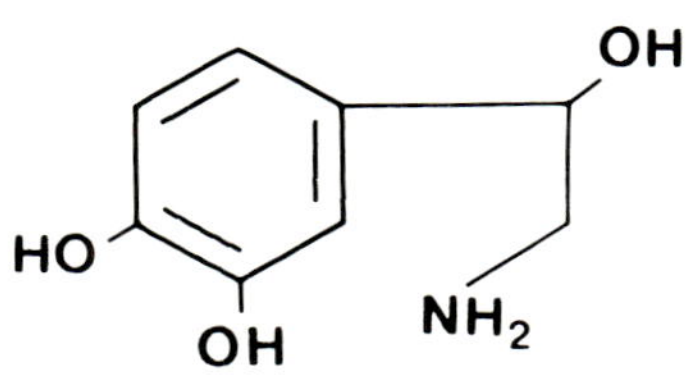

NE

An alternative but compatible view on the mechanism of 6-OHDA and related neurotoxins is that these substances auto-oxidize and liberate peroxides, superoxides, singlet oxygen, and hydroxy radicals, and that this series of oxidative intermediates reacts with neuronal elements, thereby interfering with neuronal function to cause nerve destruction (Heikkila and Cohen, 1971, 1973; Cohen and Heikkila, 1974; Heikkila and Cabbat, 1977). The importance of liberated free radicals is demonstrated by the protection against nerve damage that is afforded by catalase and free radical scavengers.

The generation of dihydroxyindoles and other intermediates and the production of free radicals are not mutually exclusive hypotheses for the mechanism of nerve damage by neurotoxins. Both processes may be involved in nerve damage, with auto-oxidation products promoting peroxides, free radicals, and so on. Alternatively, it is possible for the CA neurotoxins to produce damage by generation of peroxides and related species without significant formation of quinone-related compounds, although the toxic potential may be greater when both groups of species are formed (Borchardt et al., 1977).

4.2. Specificity of Action

The auto-oxidation of 6-OHDA will occur in a test tube, in the absence of biological material. Therefore, any tissue that accumulates 6-OHDA is subject to its destructive potential. The classification of 6-OHDA as a CA-neurotoxin arises because 6-OHDA is selectively accumulated by this particular type of nerve. Selective uptake into CA neurons is via the same active uptake site by which NE is taken into noradrenergic neurons and by which DA is taken into dopaminergic neurons. Blockade of this site by uptake inhibitors such as desmethylimipramine, metaraminol, 5-hydroxydopamine, and cocaine effectively protects the neuron from 6-OHDA damage (Stone et al., 1964, Breese and Traylor, 1971; Jonsson and Sachs, 1971).

The proposed neuronal actions of 6-OHDA are illustrated in Fig. 1. Once accumulated by a neuron, 6-OHDA may undergo oxidative metabolism, via monoamine oxidase (MAO), to inactive products. A portion of the 6-OHDA can be taken into the granule storage pool for NE and act as a false neurotransmitter. However, the majority of 6-OHDA is believed to auto-oxidize to substances

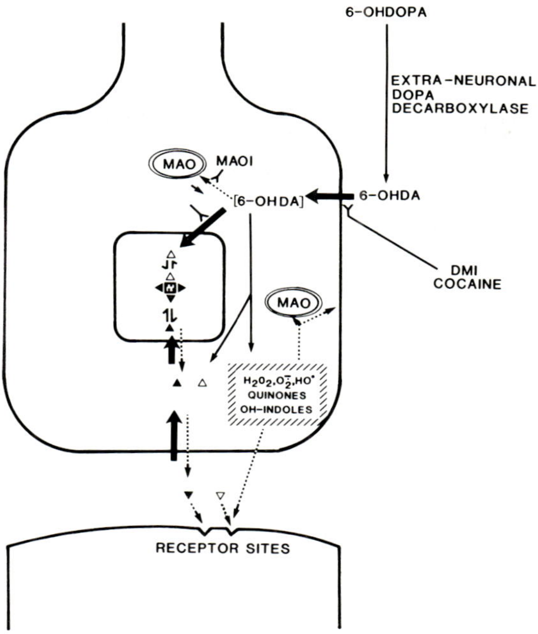

Fig. 1. Schematic illustration of the intraneuronal actions of the neurotoxin 6-OHDOPA and 6-OHDA. The diagram illustrates the extraneuronal decarboxylation of 6-OHDOPA to 6-OHDA, a conversion that can be prevented by decarboxylase inhibitors. The neurotoxin 6-OHDA (administered as 6-OHDA or generated from administered 6-OHDOPA) is taken up by an active energy-dependent process on the limiting membrane of catecholamine-containing neurons. Inside the neuron, 6-OHDA can undergo several transformations. First, it can be deaminated and inactivated by monoamine oxidases (MAOs). Second, it can be taken up into the granule storage pool, whence it can be released as a false neurotransmitter, upon nerve stimulation. Third, 6-OHDA undergoes spontaneous oxidation and rearrangement to form products or postulated products such as quinones, hydroxyindoles, peroxides, superoxides, and hydroxyl radicals. These products are believed to act in a nonspecific manner (a) with mitochondrial cytochromes and enzymes, resulting in a disruption of oxidative phosphorylation, and impairment of energy production by the neuron; (b) with the neuronal membrane and ATPases, resulting in loss of the ionic gradient across the membrane. These processes are thought to cause the observed degeneration of catecholamine-containing neurons. It has also been considered that 6-OHDA or its metabolites may interact with the receptor sites for neurotransmitters, and possibly result in the loss of some of the receptor sites.

that act on various components of the neuron or receptor site and produce the damaging effects already described.

4.3. Factors Affecting Neurotoxicity of 6-OHDA in the PNS

In the iris, where there is a dense interdigitating network of both cholinergic and noradrenergic fibers, [^{3}H]6-OHDA has been found to accumulate along the course of histofluorescent noradrenergic fibers (Ljungdahl et al., 1971). In an electron microscopic field, apparently intact cholinergic fibers, myelinated axons, smooth muscle cells, Schwann cells, and endothelial cells are seen when there is marked damage to noradrenergic terminals (Lever et al., 1971). Therefore, 6-OHDA may be considered as a highly selective neurotoxin for peripheral sympathetic nerves. Several factors affecting the selectivity and potency of 6-OHDA in the peripheral nervous system (PNS) are considered in this section.

4.3.1. Dose.

The greater the dose of 6-OHDA, the greater is the extent of neurotoxicity. Higher doses of 6-OHDA increase the time needed for recovery from damage. Divided doses are often used to decrease acute adverse effects and to decrease nonspecific damage.

4.3.2. Route of Administration.

The general order of decreasing effectiveness for 6-OHDA by various routes is intravenous (iv), intraperitoneal (ip), and subcutaneous (sc). Special applications of 6-OHDA may be desirable. For example, simply painting a solution of 6-OHDA on the mucosal surface was found to be a highly effective means for destroying sympathetic periarteriolar nerves (Siggins and Bloom, 1970). In other procedures, a slow iv drip or an intra-arterial perfusion may be of value.

4.3.3. Age.

Contrasted with the complete regeneration of damaged sympathetic fibers in adult species is the permanent partial sympathectomy after neonatal treatment with high doses of 6-OHDA, as described subsequently. Therefore, susceptibility of sympathetic fibers to damage by 6-OHDA is greater in immature animals.

4.3.4. Blood Supply.

Tissues with a rich vascular supply will receive a proportionately greater amount of 6-OHDA. Therefore, sympathetic fibers in atria, which are highly vascularized, will be more susceptible to 6-OHDA than will those fibers innervating a tissue such as the vas deferens, which is poorly vascularized. In the vas deferens, blood vessels are situated in such a manner that nutrients and 6-OHDA have to diffuse considerable distances to the sympathetic terminals. Connective tissue sheaths, in tissues such as the intestines, may also prevent access of 6-OHDA to sympathetic terminals.

4.3.5. Length of Axonal Process.

Sympathetic fibers originating in paravertebral ganglia have axonal processes that are longer than those originating in prevertebral ganglia. The latter, in turn, have longer axons than those sympathetic perikarya located in the so-called terminal ganglia. After damage, those sympathetic neurons with short axonal processes can regenerate more rapidly and can more quickly reinnervate their respective end-organs. Thus, recovery of sympathetic innervation to the vasa deferentia after 6-OHDA is quite rapid, whereas recovery in the heart occurs more slowly.

4.3.6. Density of Innervation.

In order to produce destruction, a critical intraneuronal concentration of 6-OHDA must be attained (Jonsson and Sachs, 1971). In a densely innervated tissue, such as the atrium, the amount of 6-OHDA taken up by any single terminal may be below that needed for the critical intraneuronal level associated with production of damage. In a sense, adjacent fibers serve to protect each other, since each may accumulate a subthreshold amount of 6-OHDA. With a tissue having a low density of innervation, such as the cardiac ventricles, 6-OHDA would be taken up by fewer terminals, so that levels of 6-OHDA in any single terminal would be relatively high. The usual finding after 6-OHDA is less damage of the sympathetic terminals in atria compared to ventricles.

4.3.7. Other Factors.

It is not known if cytoplasmic mechanisms of inactivation of 6-OHDA are greater in terminals of specific tissues. Undefined factors could also affect 6-OHDA neurotoxicity.

4.4. Factors Affecting Neurotoxicity of 6-OHDA in the CNS

4.4.1. Blood–Brain Barrier.

In mature animals, a blood–brain barrier for monoamines prevents entry of 6-OHDA to the brain. Therefore, except in specialized regions where this blood–brain barrier is imperfectly formed or is absent, as in the circumventricular organs (Klara et al., 1976), damage by peripherally administered 6-OHDA is essentially limited to the sympathetic nervous system in mature animals. The entry of small amounts of 6-OHDA into the CNS, nevertheless, may occur and is facilitated by the elevation in blood pressure caused by the neurotoxin (Johansson and Henning, 1980).

In developing and immature animals, the blood–brain barrier is absent or incomplete. For at least a major portion of the brain, the blood–brain barrier in the rat does not form until about 9 d after birth (Sachs, 1973) and is still not impervious for several weeks to the passage of high doses of 6-OHDA (Sachs and Jonsson, 1973). Therefore, in adult animals, peripherally administered 6-OHDA will not produce actions on central CA-neurons, whereas in neonatal animals, 6-OHDA destroys CA nerve fibers in the CNS.

In considering the selectivity of action of 6-OHDA for CA neurons in the CNS, all of the factors addressed for peripheral sympathetic neurons are of importance. However, because 6-OHDA crosses the blood–brain barrier of immature animals only, a separate classification of factors is required for the adult and neonatal groups.

4.4.2. Dose

4.4.2.1. Adult. The lower the dose of 6-OHDA by an intraventricular (ivtr) or intracisternal (ic) route, the greater is the selectivity for noradrenergic neurons. By administering several lower doses of 6-OHDA (ca. 50 μg), greater destructive effects are exerted on noradrenergic neurons, in the absence of effects on dopaminergic neurons (Breese and Traylor, 1971). Greater selectivity of 6-OHDA for dopaminergic fibers is attained by pretreatment of animals with an NE uptake inhibitor, such as desmethylimipramine or protriptyline (Evetts and Iversen, 1970; Breese and Traylor, 1971). Greater destruction of dopaminergic neurons is also promoted by treatment with an MAO-inhibitor, such as pargyline, prior to 6-OHDA (Breese and Traylor, 1970, 1971). Selective DA uptake inhibitors, which would enhance 6-OHDA selectivity for noradrenergic neurons, have recently become available (Heikkila and Manzino, 1984).

4.4.2.2. Neonate. As in adults, a series of low doses of 6-OHDA will help in retaining selectivity of 6-OHDA for neonate noradrenergic neurons, whereas pretreatment with desmethylimipramine will produce a more selective action of 6-OHDA on dopaminergic neurons (Smith et al., 1973).

4.4.3. Route

In order to bypass the blood–brain barrier of adults, 6-OHDA must be administered directly into the cerebrospinal fluid or into the parenchyma of the brain or spinal cord. When injected directly into the parenchyma of the brain, only small amounts of 6-OHDA (2–8 μg) are required for destruction of CA neurons. All cellular elements that are bathed in high concentrations of 6-OHDA at the tip of the cannula are subject to the destructive effects of 6-OHDA (Ungerstedt, 1971b; Poirier et al., 1972; Hökfelt and Ungerstedt, 1973). Nonspecific damage is reduced by lowering the amount of 6-OHDA that is injected and by infusing low volumes of fluid at a slow rate.

When injected into one of the cerebral ventricles or cisternae, a high dose of 6-OHDA (250 μg in 20 μ1 in the rat) produces destruction of the ependyma and underlying nervous tissue of the injected ventricle. With a very high dose, this damage extends to other ventricles (Poirier et al., 1972). With lower doses of 6-OHDA, damage is limited.

When neonatal rats are given a peripherally administered dose of 6-OHDA, the effect on cellular elements, other than CA-neurons in the brain, is thought to be minimal. Perikarya of locus ceruleus neurons are reportedly unaffected by such treatment (Jonsson et al., 1974; McBride et al., 1985). Non-CA cells that lack a specific uptake mechanism for CA would accumulate even less 6-OHDA.

4.4.4. Age

The younger the animal, the greater is the sensitivity of central CA neurons to 6-OHDA destruction (Lytle et al., 1972). In neonates, satellite cells encompass the neuron to a much lesser extent than in the adult, and this may affect accessibility of 6-OHDA to the neuron. Also, the dendritic elaboration is much less in the undeveloped neuron, as is the nerve terminal network. Intraneuronal levels of MAO are less in the neonate, and less inactivation of 6-OHDA would occur. Differences in the pharmacokinetics of 6-OHDA in the neonate might also represent a factor in enhanced CA neuronal damage by 6-OHDA. These and other differences between adult and neonate may render the developing CA neurons more susceptible to 6-OHDA.

4.4.5. Length of Axonal Process

As with sympathetic neurons, those noradrenergic terminals associated with long axonal processes are more readily destroyed by 6-OHDA. Fibers in the neocortex, hippocampus, cerebellum, and spinal cord that are distal to their locus ceruleus nucleus of origin are largely destroyed by 6-OHDA, whereas noradrenergic fibers in more proximal brain regions are only modestly affected.

4.4.6. Segment of Neuron

Different components of CA neurons are not affected equally by 6-OHDA. Nerve terminals are most sensitive to 6-OHDA, the axonal segments are partially resistant to damage, and the cell bodies are most resistant. This variation in sensitivity is probably dependent primarily on the relative number of uptake-1 sites on the outer membrane and the relative cytoplasmic volume for each component. Thus, nerve terminals are able to accumulate 6-OHDA in high amounts, but the volume of each nerve terminal is small, resulting in the attainment of high cytoplasmic concentrations of 6-OHDA. In the cell body, 6-OHDA is accumulated to a far lesser degree, and the volume of the perikaryon is relatively large, resulting in the attainment of much lower cytoplasmic concentrations of 6-OHDA. In the axon, density of uptake-1 sites and intraneuronal volume are intermediate to those extremes found in the terminals and cell bodies. Therefore, the critical (threshold) cytoplasmic concentration of 6-OHDA needed to produce destruction is first attained in the nerve terminal, then in the axon, and finally in the cell body (*see* Kostrzewa and Jacobowitz, 1974).

4.4.7. Locus Ceruleus vs Lateral Tegmental Noradrenergic Neurons

It is noteworthy that locus ceruleus neurons are more susceptible to 6-OHDA action than ventral tegmental noradrenergic neurons (*see* Kostrzewa and Jacobowitz, 1974; Levitt and Moore, 1980; Versteeg et al., 1980). After high or repeated doses of 6-OHDA, a near total destruction of locus ceruleus noradrenergic fibers in the neocortex, hippocampus, cerebellum, and spinal cord is obtained. Noradrenergic fibers to the hypothalamus and brain stem, which are derived in large part from ventrotegmental nuclei, are partially resistant to 6-OHDA action. The reasons for this may be related to some of the factors already discussed in this portion of the chapter.

4.5. Chemical Sympathectomy with 6-OHDA

4.5.1. Acute Changes after 6-OHDA in Adults

Postganglionic sympathetic noradrenergic nerve endings actively accumulate 6-OHDA via uptake sites for NE, and then are destroyed by the noncatalytic auto-oxidation products of 6-OHDA. As a consequence, end organs lose their associated sympathetic input. There is a decrease in tissue content of NE, their rate of in vitro accumulation of ^{3}H-NE, and activities of tyrosine hydroxylase (T-OH) and dopamine-β-hydroxylase (DBH). The above events occur within the first 30 min after moderate to high doses of 6-OHDA (about 30 mg/kg or more) in the rat (Thoenen and Tranzer, 1968).

The spontaneous release of NE from damaged sympathetic fibers produces an immediate sympathomimetic effect. This is associated with an elevation in blood pressure, an inotropic and chronotropic action with possible production of arrhythmias, and possible changes in smooth muscle activity and glandular secretion (DeChamplain, 1971; DeChamplain and Nadeau, 1971; Laverty et al., 1965). In large animals such as cats and dogs, the total sympathetic storage pool is larger, and the potential for production of arrhythmias, ventricular fibrillation, and lethal effects is much greater. Chronotropic responses in the dog can be observed up to 2 h after a dose of 6-OHDA of only 1 mg/kg (iv) (Gauthier et al., 1972).

4.5.2. Long-Term Changes after 6-OHDA in Adults

Several days to weeks after 6-OHDA, a widespread loss of sympathetic noradrenergic fibers is reflected by a reduction in plasma DBH (Weinshilboum and Axelrod, 1971). Loss of sympathetic input to the different peripheral tissues and organs is nonuniform. The decreasing order of tissue sensitivity for destruction of sympathetic input by 6-OHDA is heart, salivary gland, spleen, vas deferens, and vascular arteries (*see* Kostrzewa and Jacobowitz, 1974).

Both a presynaptic and postsynaptic supersensitivity to NE may develop (Haeusler et al., 1969; Haeusler, 1971). The former occurs because of the loss of the noradrenergic terminal network and attendant loss of uptake processes for removing synaptic NE; the latter represents a proliferation of NE receptor sites on the postsynaptic membrane. Destruction of receptor sites for NE is felt to be an insignificant effect, since 6-OHDA is initially removed from the synapse by surviving presynaptic noradrenergeric fibers, and 6-OHDA-released stores of NE would compete at the receptor site for NE. Potentially, uptake-2 sites could accumulate 6-OHDA and either inactivate the compound or be damaged by it.

As a consequence of the loss of sympathetic terminals after 6-OHDA, a series of compensatory changes occurs within the sympathetic ganglia. Activity of T-OH is not induced by reserpine treatment, as it normally would be, and activities of T-OH, DBH, and MAO become reduced. Content of NE increases, possibly because of the reduction of MAO activity and disruption of NE transport down the axon. These changes persist for a period of 2–3 wk (Brimijoin, 1972; Brimijoin and Molinoff, 1971; Cheah et al., 1971).

The critical facet of 6-OHDA-induced destruction of sympathetic fibers is that recovery typically occurs. Surviving cell bodies and undamaged axons are able to send new sprouts to the sympathetically denervated tissue over a period of time. Therefore, a permanent sympathetic denervation is unlikely to occur. Treatment with 6-OHDA may, nonetheless, be of value for studying the physiological role of sympathetic input to a tissue prior to damage, immediately after damage and during recovery from damage.

4.5.3. Compensatory Response of Adrenal Medulla

In the adrenal medulla, total CA content is unaltered or elevated after 6-OHDA (Thoenen and Tranzer, 1968; Thoenen et al., 1969; Lew and Quay, 1971; Lytle et al., 1972). This tissue appears to be resistant to the direct action of 6-OHDA. However, activities of T-OH and phenylethanolamine-*N*-methyl-transferase increase. CA turnover likewise increases as a consequence of reduced sympathetic tone. These changes are mediated by enhanced splanchnic nerve activity. As the sympathetic nervous system recovers from 6-OHDA damage, the compensatory changes in the adrenal medulla abate (Mueller et al., 1969; Thoenen et al., 1969).

4.5.4. Acute Changes after 6-OHDA in Neonates

Following administration of high doses of 6-OHDA (100 mg/kg or higher, ip or sc) to mice or rats, over the first several days after birth, there is a marked destruction of sympathetic nerve terminals as in adults. However, in the neonate, the destructive action of 6-OHDA extends to both pre- and paravertebral sympathetic ganglion cells, resulting in the development of a much smaller ganglion (Angeletti, 1971; Angeletti and Levi-Montalcini, 1972). Ganglion cell death is thought to occur as a consequence to the loss of retrograde axonal transport of NGF to the cell body from the peripherally innervated target tissue. Lacking this trophic factor, the sympathetic cell is unable to survive (Hendry et al., 1974). It is known that exogenous NGF will prevent the death of sympathetic cells after neonatal 6-OHDA (Aloe et al., 1975; Levi-Montalcini et al., 1975). Initially, sympathetic input to the various peripheral smooth muscles and exocrine glands is lost, but as the animals mature, regeneration of damaged sympathetic fibers occurs. In time, certain tissues, such as the vas deferens, retain a full complement of fibers. In other tissues, such as the heart and spleen, there may be a permanent reduction in sympathetic innervation (Jaim-Etcheverry and Zieher, 1971; Lew and Quay, 1971; Clark et al., 1972; Sachs and Jonsson, 1972b; Jacobowitz, 1975). Nevertheless, a complete permanent chemical sympathectomy is not obtained by neonatal 6-OHDA treatment.

4.5.5. Chemical Sympathectomy with Other CA-Neurotoxins

In attempting to produce a chemical sympathectomy with the various CA-neurotoxins, it seems that all have the same limitations as 6-OHDA. Some, such as 6-OHDOPA, are less potent, whereas others, like DSP-4 or α-methyl-6-aminodopamine, are of equal or greater potency. However, in all cases, the degree of permanent sympathectomy is not substantially greater than that obtained with 6-OHDA. Therefore, although study of recovery/regeneration of a markedly impaired sympathetic system may be of value, one is unable to obtain a permanent complete sympathectomy by chemical means.

4.6. CNS Effects of 6-OHDA in Adults

The approximate dose–effect relationship between 6-OHDA and CA depletion in rat brain is summarized in Fig. 2. A single 50 μg ic dose of 6-OHDA produces a moderate depletion of whole brain NE, whereas a dose of approximately 100 μg is needed for initial DA depletion. Higher doses of 6-OHDA produce proportionately greater depletion of both NE and DA. However, after a total dose of approximately 500 μg, increasing amounts of 6-OHDA produces little further depletion of CA, although lethality increases dramatically (Breese and Traylor, 1970; Hedreen and Chalmers, 1972; Uretsky and Iversen, 1970). This latter effect is

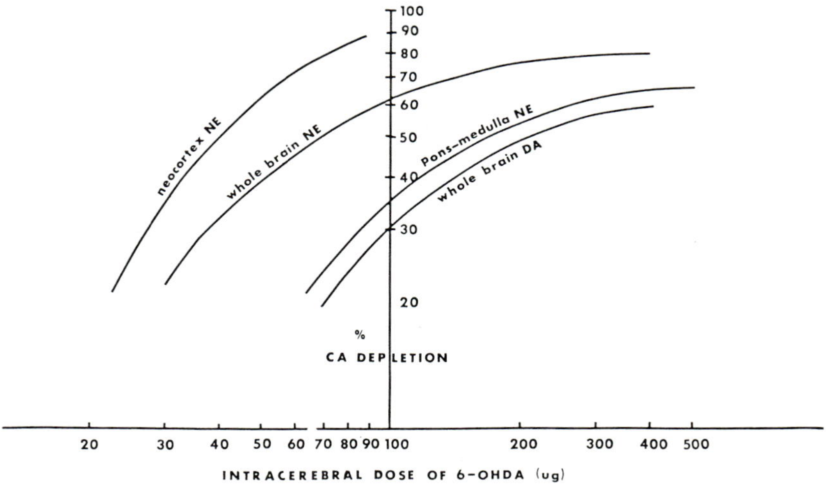

Fig. 2. Approximate long-term depletion of the catecholamines (CA) norepinephrine (NE) and dopamine (DA) in rat brain after different doses of 6-hydroxydopamine (6-OHDA), administered intracisternally or intraventricularly.

circumvented to some degree by the use of divided doses of 6-OHDA.

Generally, the decreasing order of sensitivity for NE depletion by 6-OHDA is hippocampus=spinal cord=cerebellum, neocortex, hypothalamus=thalamus=midbrain=striatum, pons, and medulla (*see* Kostrzewa and Jacobowitz, 1974). The approximate degree of CA depletion in several brain regions after various doses of 6-OHDA can be seen in Fig. 2.

Depletion of DA in the brain after 6-OHDA follows a different time course than that for NE. With NE, depletion occurs within 30 min of 6-OHDA treatment. With DA there may be an elevation above control levels during the first 24 h (Bell et al., 1970).

A variety of other parameters, reflecting the loss of catecholaminergic fibers, virtually parallel the changes in monoamine content: T-OH activity (Breese and Traylor, 1970; Uretsky and Iversen, 1970), DBH activity (Reis and Molinoff, 1972), ^{3}H-monoamine uptake rates (Breese and Traylor, 1970; Jacks et al., 1972; Jonsson and Sachs, 1970), and 3-methoxy-4-hydroxyphenylglycol (MHPG) excretion level (Breese et al., 1972). Actual ultrastructural damage to noradrenergic and dopaminergic neurons is evident by electron microscopy (Bloom et al., 1969). A decrease in the areolar fiber density is notable by histofluorescence microscopy (Malmfors and Sachs, 1968; Ungerstedt, 1968, 1971d).

If one were to select a single parameter to assess alterations in CA fiber integrity after 6-OHDA, the most reliable indicators would generally be in vitro uptake of ^{3}H-CA or actual observation of the fibers with an immunocytochemical method. Uptake of ^{3}H-CA is a measure of functional uptake sites on the neuronal membrane. Since the number of uptake sites is correlated with numbers of fibers, the V_{max} for uptake is an indicator of the number of CA nerve terminals. Observation of histofluorescent fibers would indicate numbers of fibers in a particular brain region. With a formaldehyde condensation method or glyoxylic acid method, the intensity of fluorescence is dependent on CA content. This limitation poses the potential problem that more fibers become visible when CA content is increased, and this may be interpreted as an increase in numbers of fibers in a region. An FITC-immunocytochemical procedure would be dependent on the T-OH or DBH enzymes. In principle, increased enzyme levels in a swollen axon or preterminal axon could present the same problem as with the other histochemical methods. However, because the sensitivity of the procedure is so great, the problem with interpretation of results is usually not considered to be a major one.

A major caution with administration of 6-OHDA into the parenchyma or by an ic or ivtr route is that nonspecific damage will still occur. Precautions previously described should be utilized.

In brain regions that are hypoinnervated by noradrenergic fibers, T-OH activity and NE synthesis rate increase several-fold in the residual fibers (Acheson et al., 1980). This change in enzyme activity is associated with an activation of T-OH molecules, signaled by an increase in the V_{max} and a shift in the pH optimum for the enzyme (Acheson and Zigmond, 1981). The turnover rate of NE is the same as in controls (Jonsson et al., 1974, 1979). Therefore, it appears that the residual fibers synthesize NE at a higher rate and release NE at a greater rate than individual fibers in intact controls. This compensatory increase in activity is correlated with an increase in the number of α_1, α_2, and β adrenoceptors for NE (Harden et al., 1977, 1979; Jonsson and Hallman, 1978; U'Prichard et al., 1980) and an increase in isoproterenol-stimulated cyclic AMP accumulation (Harden et al., 1977). In brain regions that have become hyperinnervated as a consequence to collateral fiber sprouting, NE turnover rate is also the same as in controls (Jonsson et al., 1974, 1979), and noradrenergic receptor number is reduced

(Jonsson and Hallman, 1978; Harden et al., 1979; U'Prichard et al., 1980). Thus, both pre- and postsynaptic adaptive mechanisms occur after noradrenergic fiber reorganization.

Loss of catecholaminergic fibers after 6-OHDA persists indefinitely in the CNS of adult rats, and collateral sprouting presumably persists indefinitely. High doses of 6-OHDA destroy locus ceruleus perikarya *per se* (Descarries and Saucier, 1972; Poirier et al., 1972).

4.7. CNS Effects of 6-OHDA in Neonates

4.7.1. ic or ivtr Administration

When introduced into the cerebrospinal fluid of neonatal rats, 6-OHDA produces marked effects on CA-containing neurons (Lew and Quay, 1971; Breese and Traylor, 1972; Lytle et al., 1972). Doses of 6-OHDA in the range of 20–80 μg are associated with the destruction of noradrenergic terminals in virtually all regions of the CNS (Schmidt et al., 1980).

As the animal develops, a small degree of regeneration or proliferation of undamaged fibers occurs, but brain regions such as the hippocampus, neocortex, and spinal cord remain permanently hypoinnervated by noradrenergic fibers. In the brain stem and cerebellum, a collateral sprouting of noradrenergic fibers occurs, and these regions become permanently hyperinnervated if the neonatal 6-OHDA dose is reasonably low (Jaim-Etcheverry and Zieher, 1975; Pappas et al., 1976; Zieher and Jaim-Etcheverry, 1979; Schmidt and Bhatnagar, 1980; Schmidt et al., 1980). With high neonatal doses of 6-OHDA, the brain stem and cerebellum remain permanently hypoinnervated by noradrenergic fibers (Schmidt et al., 1980).

In addition to selection of a critical 6-OHDA dose for the above effects, the age at the time of treatment is also critical. The collateral sprouting and resultant hyperinnervation of the hindbrain do not occur if the 6-OHDA is given after the first week after birth (Schmidt et al., 1980).

Although low to moderate i.c. or i.vtr. doses of 6-OHDA (≤40 μg) do not produce any apparent destructive effects on locus ceruleus perikarya, there is a discrepancy in the effects of higher doses. In assessing the appearance of locus ceruleus cells by histofluorescence microscopy, it was concluded that cells remained intact after high doses (80 or 100 μg) of 6-OHDA (Konkol et al., 1978a; Schmidt et al., 1980). By means of classical histological techniques, destruction of cells in the mid to caudal regions of the nucleus was reported (Sievers et al., 1980).

High ic or ivtr doses of 6-OHDA in neonatal rats destroy dopaminergic fibers in the brain, resulting in a permanent dopaminergic hypoinnervation of the striatum, and an enhanced reactivity to DA agonists, which produce a variety of stereotypic behaviors, including self-biting and self-mutilation as the rats mature (Breese et al., 1984). As a consequence of the reduction of dopaminergic input to the striatum, serotoninergic fibers hyperinnervate this region (Berger et al., 1985: Snyder et al., 1986).

The selectivity of 6-OHDA for noradrenergic or dopaminergic neurons is a dose-dependent one that is modified by pretreatment with several agenys (e.g., desmethylimipramine or protriptyline). The regeneration and collateral fiber sprouting is highly dependent on the dose of 6-OHDA and the age of the animal at the time of treatment. Consideration must be given to the possibility that other nerve fibers may sprout into brain regions in which the CA fibers have been reorganized.

4.7.2. ip or sc Administration

When administered ip or sc to neonates, 6-OHDA produces effects that are much the same as those found after ic or ivtr routes. Permanent noradrenergic hypoinnervation of neocortex, hippocampus, and spinal cord occurs. Collateral fiber sprouting into the brain stem ensues after a 100 μg/g dose given several times during the first week after birth (Jonsson et al., 1974). However, if the first dose of 6-OHDA is administered more than 9 d after birth, a noradrenergic hyperinnervation of the brain stem does not occur (Sachs, 1973).

In the cerebellum, noradrenergic hyperinnervation does not develop if the dose of 6-OHDA is too high or if treatment with 6-OHDA is 3 d or more after birth (Kostrzewa and Jacobowitz, 1974; Kostrzewa and Harper, 1975). Locus ceruleus perikarya are not destroyed by a moderately high neonatal sc dose of 6-OHDA (100 μg/g × 2, 24h interval), but a doubling of this dose destroys 18% of the cells (McBride et al., 1985).

Selectivity of 6-OHDA for CA neurons in the neonate is indicated by the lack of observed effects on brain content of serotonin, acetylcholine, GABA, and several amino acids (Jacks et al., 1972; Sachs and Jonsson, 1972b). There are two major distinctions between neonatal 6-OHDA being administered peripherally vs centrally. First, dopaminergic fibers are particularly resistant to

6-OHDA by the peripheral route. Second, the development of noradrenergic hyperinnervation in the cerebellum can occur up to 7 d after birth following ic or ivtr 6-OHDA, but only for the first 2 d after birth following ip or sc 6-OHDA.

4.8. Locus Ceruleus Lesion

Efferents from the locus ceruleus provide noradrenergic innervation of the dorsal brain structures (neocortex, hippocampus), frontal brain regions (olfactory bulbs, septum), the brain stem (midbrain, pons, medulla), cerebellum, and spinal cord (Ungerstedt, 1971d). When total ablation of the noradrenergic innervation to any or all of these specific brain regions is desired a convenient means involves destruction of the ceruleus. Frequently, electrolytic lesions are made, because this type of lesion is considered to be more complete than a chemical lesion and damage to the surrounding tissue is not felt to be a confounding factor. Numerous groups have employed this method in rats, and some of the common features are presented. After anesthetization, mounting in a stereotaxic apparatus, and suitable preparation of the animal, electrodes that are insulated to within 0.5–1.0 mm of the tip are positioned in the locus ceruleus. Occasionally, two different placements are made in the locus ceruleus, in order to more extensively damage the nucleus, while minimizing damage to surrounding tissue (Heydorn et al., 1986). In some cases, the electrode is advanced in a caudal to rostral position at a slight angle, in order to avoid the transverse sinus and to better position the tip of the electrode in the locus ceruleus (Schmidt and Bhatnagar, 1979a, 1979b). The lesion is made by passing an anodal current of about 2 mA for 3–10 s (Savaki et al., 1984). Sham controls consist of placement of the electrode in a shallower position in the brain and the absence of a current.

When 6-OHDA lesions of the locus ceruleus are made, animals are often first pretreated with the MAO inhibitor pargyline (50 mg/kg ip) several minutes before 6-OHDA. The 6-OHDA (4–8 μg), in a solution containing ascorbic acid (a.a.), is administered in a small volume (ca. 2 μL) and infused over an interval of several minutes (Koda et al., 1978; Velley et al., 1983). In neonatal rats, 4 d after birth, 6-OHDA lesions of the locus ceruleus have been made by administering the 6-OHDA at three different cannula placements in small volumes (0.15 μL ea; 4 μg of 6-OHDA•HBr/0.15 μL) at a slow rate of administration (Lanfumey et al., 1981).

Both the electrolytic and chemical lesioning procedures produce a marked destruction of locus ceruleus cells. The nonspecific damage to surrounding tissue is minimized by reducing the current, or by using two or more electrode/cannula placements, and by slow infusion of small amounts of 6-OHDA. The particular research study under consideration is a factor in selection of the lesion method.

4.9. Dorsal Bundle Lesions

4.9.1. Adults

Dorsal bundle lesions are commonly made by electrothermal means or by injection of 6-OHDA into the bundle. When using 6-OHDA, 4–8 μg are frequently administered in 2–4 μL over several minutes (Hökfelt and Ungerstedt, 1973; U'Prichard et al., 1979, 1980; Britton et al., 1983; Dahlgren et al., 1981; Velley et al., 1983). Noradrenergic fibers anterograde to the lesion site degenerate, and forebrain structures become hypoinnervated. Retrograde to the lesions, the fibers become swollen and engorged with NE, DBH, and T-OH. The brain stem and locus ceruleus demonstrate elevated T-OH and DBH for 2 d after the lesion, but exhibit reduced levels of those enzymes for the 7–21-d interval following injury. At 4 wk, T-OH and DBH activities return to control levels in the brain stem and ceruleus, even when forebrain regions have markedly reduced noradrenergic fiber input (Reis and Ross, 1973; Ross et al., 1975). In the cerebellum, DBH activity does not initially increase, as in the ceruleus, but is reduced 7–14 d after injury, returning to control levels at 3 wk (Reis and Ross, 1973). Because of prolonged elevation of NE in the cerebellum after a dorsal bundle lesion, it is considered that collateral fiber sprouting occurs after injury. This change in the cerebellum is associated with a reduction in the number of α_1 and β sites, whereas in the cortex, an elevation in the number of α_1, α_2, and β sites is correlated with the noradrenergic hypoinnervation (U'Prichard et al., 1980). The initial depression of T-OH and DBH enzyme activities is thought to reflect a shift in priorities of the neuron—sacrificing neurotransmitter synthesis so that energy is expended toward repair of the injury (Reis and Ross, 1973), as was suggested in the peripheral nervous system (Cheah et al., 1971).

Hökfelt and Ungerstedt (1973) observed that nonspecific damage after 6-OHDA was minimal, and that undamaged myelinated (nonaminergic) fibers could be seen in the same bundles as damaged unmyelinated (primarily noradrenergic) fibers. The selectivity of 6-OHDA is supported by studies of Hedreen and Chalmers (1972). It is felt that the nonspecific damage with 6-OHDA is far less than that seen after electrothermal lesions (Willis et al., 1987).

4.9.2. Neonates

Although ascending tract lesions in the neonate are commonly made by surgical transection of the tract, the changes after injury are in some ways similar to those observed after lesion of the same tract in adults. Immediately after injury, noradrenergic fibers anterograde to the lesion site degenerate, whereas the same fibers retrograde to the injury site become swollen and engorged with T-OH, DBH, and NE. These changes are analogous to those seen in adults. In the following weeks, noradrenergic collateral fiber sprouting occurs in both the brain stem and the cerebellum. The ensuing noradrenergic hyperinnervation of the hindbrain regions is permanent (Jonsson and Sachs, 1982; Klisans-Fuenmayor et al., 1986), even though the locus ceruleus cell number may be reduced (Kostrzewa et al., 1988). The reactivity of noradrenergic neurons in both developmentally immature and mature animals is an indication of the high degree of plasticity of this neuronal type in the CNS.

4.10. Nigro-Neostriatal Tract Lesion

Ungerstedt studied the unique sensitivity of nigral dopaminergic cell bodies to 6-OHDA destruction. When the tip of the injection cannula was positioned slightly rostral to the nigra, to avoid mechanical injury, 6-OHDA (8 μg) destroyed virtually all cell bodies considered to be dopaminergic (Ungerstedt, 1971a,b). Nonspecific injury at the injection site was minimal (Hökfelt and Ungerstedt, 1973).

Associated with the destruction of these cells is the loss of dopaminergic terminals in the striatum, along with development of striatal supersensitivity to the DA receptor agonist, apomorphine, and the occurrence of aphagia and adipsia with bilateral lesions. The positioning of the cannula, and the amount and rate of 6-OHDA administration is critical, lest injury to nearby mesolimbic dopaminergic fibers and dorsal bundle noradrenergic fibers occurs (Ungerstedt, 1971a, b, and c).

The removal of dopaminergic innervation of the striatum by an intrastriatal injection of 6-OHDA is a less desirable procedure than intranigral 6-OHDA administration. A high dose of 6-OHDA (8 μg) in the striatum does effectively destroy dopaminergic terminals, but the potential for destruction of nondopaminergic fibers is great (Ungerstedt, 1971a, b, and c). It is noted that destruction of nigrostriatal axonal terminals and preterminals may result in the degeneration of the nigral cell bodies (Andén et al., 1965).

Specific lesions of substantia nigral cell bodies by 6-OHDA injection is a common procedure that is contrasted with MPTP destruction of these same groups of neurons. This is discussed in a separate section of this review.

5. 6-OHDOPA

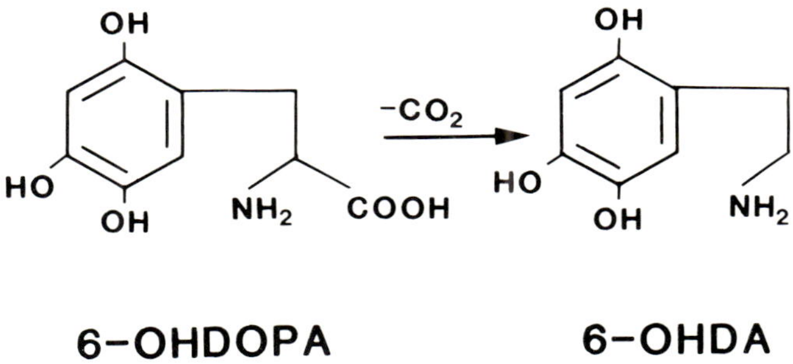

5.1. *Effects on CA Neurons in Adults*

The agent 6-OHDOPA was synthesized specifically because of its predicted ability to (a) gain entry to the CNS following a noncentral parenteral route of administration and (b) be converted through decarboxylation to 6-OHDA to effect the destruction of central monoaminergic neurons (Ong et al., 1969). When injected noncentrally, 6-OHDOPA also produces a destruction of peripheral sympathetic noradrenergic neurons. The greatest effects are seen in the heart, where NE content and the number of histochemically demonstrable noradrenergic fibers are correspondingly reduced. Lesser effects of 6-OHDOPA are seen in other peripheral tissues. Levels of NE in the vas deferens, a tissue with an abundance of sympathetic innervation, are not altered by even a high dose of 6-OHDOPA (Kostrzewa and Jacobowitz, 1972; Sachs and Jonsson, 1972a,c).

Comparison of these findings with those after 6-OHDA illustrates that 6-OHDOPA is far less potent than 6-OHDA. Related to this is the shorter recovery time of noradrenergic fibers damaged after 6-OHDOPA. The lesser potency of 6-OHDOPA can be the result of a number of different processes.

(a) A portion of 6-OHDOPA gains entry to the CNS, accounting for a wider distribution than 6-OHDA
(b) 6-OHDOPA may also accumulate in a greater variety of neurons and extraneuronal tissues, so that 6-OHDA generated in this manner would also have a wider distribution than 6-OHDA
(c) The rate of conversion of 6-OHDOPA to 6-OHDA may be sufficiently slow to allow partial inactivation of generated 6-OHDA
(d) A portion of 6-OHDOPA *per se* may be metabolized or excreted, thereby reducing the amount convertible to 6-OHDA or
(e) Metabolites that lack neurotoxic properties may be generated from 6-OHDOPA.

The postulated metabolites and wider distribution of 6-OHDOPA are likely mechanisms, since the lethality of 6-OHDOPA is so much greater than that of 6-OHDA (Corrodi et al., 1971). This comparison also holds true in neonates, where both 6-OHDOPA and 6-OHDA will enter the brain after noncentral injection (Kostrzewa and Harper, 1974).

The central NE-depleting action of 6-OHDOPA was initially described by Ong and coworkers (1969) and later confirmed by others (Jacobowitz and Kostrzewa, 1971; Sachs and Jonsson, 1972a,c). In adult mice, a moderately large dose of 6-OHDOPA (100–300 mg/kg ip) reduces whole brain NE content by only about 50%. This effect persists for more than 2 mo. Accompanying the NE depletion is a reduction in ^{3}H-NE and ^{3}H-metaraminol uptake by cortical slices, and a loss of histofluorescent fibers throughout the brain (Jacobowitz and Kostrzewa, 1971; Sachs and Jonsson, 1972a,c; Jonsson and Sachs, 1973b). Later electron microscopic studies provided the definitive evidence for the neurotoxic action of 6-OHDOPA (Kostrzewa and Harper, 1974).

Because of ablation of the terminal portion of noradrenergic fibers, axonal segments of the neuron accumulate NE in a retrograde manner and become highly visible after 6-OHDOPA. This effect has been used as a means of mapping the nonradrenergic network of the brain (Jacobowitz and Kostrzewa, 1971; Sachs et al., 1973). The actions of 6-OHDOPA are relatively selective for noradrenergic neurons. Levels of DA are only slightly altered (Jacobowitz and Kostrzewa, 1971), and activities of MAO, catechol-*O*-methyltransferase, and choline acetyltransferase are not changed after 6-OHDOPA. Cholinesterase activity is slightly reduced at 2 wk in the neocortex, but is unchanged in the brain stem and cerebellum (Kostrzewa and Jacobowitz, 1973). The serotonin content of the brain is unaltered after 6-OHDOPA alone (Berkowitz et al., 1970; Corrodi et al., 1971) and only slightly reduced after nialamid pretreatment (Sachs and Jonsson, 1972c).

Despite this apparent selectivity of 6-OHDOPA, there are several disadvantages to its use.

(a) There is a broad spectrum of effects on noradrenergic neurons. With noncentral administration, both peripheral and central effects are produced. Pretreatment with a peripherally acting dopa decarboxylase inhibitor, such as carbidopa, would afford protection of the peripheral noradrenergic fibers, while potentiating central destructive actions of 6-OHDOPA (Ong et al., 1969; Kostrzewa and Jacobowitz, 1972, 1973; Sachs and Jonsson, 1972c).

(b) However, a disadvantage, not so easily overcome, is that a major portion of the central noradrenergic system is still intact after 6-OHDOPA, and this remaining component may be able to carry out functions with minimal impairment.

(c) The potency of 6-OHDOPA by an ivtr route is still far less than that of 6-OHDA (Thoa et al., 1972; Richardson and Jacobowitz, 1973), thereby obviating the use of 6-OHDOPA in place of 6-OHDA.

(d) Even though the destructive action of 6-OHDOPA is the result of generated 6-OHDA, there is still a much greater lethal action of 6-OHDOPA. Perhaps 6-OHDOPA or its metabolites interact with nonadrenergic elements, making it more nonselective than would appear and, therefore, more toxic.

These limitations on the effects of 6-OHDOPA relegate its use in adult animals to a secondary role. Its greatest utility is in the study of regenerative sprouting in the brains of animals treated as neonates.

5.2 Effects on CA Neurons in Neonates

Because of the development of a blood–brain barrier to CAs in rats by 9 d after birth (Sachs, 1973), there are limitations on the use of 6-OHDA in the ontogenic study of CA neurons in the CNS. Being a CA analog, 6-OHDA cannot cross this barrier. However, 6-OHDOPA is able to permeate the blood–brain barrier and, in fact, is able to cross the placental barrier to alter the ontogeny of both central and peripheral noradrenergic neurons in fetal rats and mice (Jaim-Etcheverry and Zieher, 1975; Zieher and Jaim-Etcheverry, 1973, 1975; Kostrzewa et al., 1978). It has been shown that the development of noradrenergic innervation into the telencephalon is permanently impaired when rodents are treated with 6-OHDOPA at 16 or 17 d after conception, whereas noradrenergic innervation to the brain stem and cerebellum is permanently enhanced when rodents are treated as early as 13 d after conception. The plastic response of noradrenergic fibers to damage changes at different developmental times for each of the brain regions (Jaim-Etcheverry and Zieher, 1975; Kostrzewa, 1975; Zieher and Jaim-Etcheverry, 1975; Kostrzewa and Garey, 1976, 1977). Therefore, 6-OHDOPA is a particularly suitable agent for probing the ontogenic recovery potential of noradrenergic fibers, since the blood–brain barrier does not limit the usefulness of this agent. These changes in the reorganization of central noradrenergic fiber distribution are qualitatively the same as when rats or mice are treated shortly after birth with 6-OHDA: Forebrain regions become permanently hypoinnervated, whereas hindbrain regions become permanently hyperinnervated by noradrenergic fibers.

As demonstrated with silver degeneration stains and by electron microscopy, locus ceruleus cells are destroyed by neonatal 6-OHDOPA treatment (Kostrzewa and Harper, 1974, 1975). Those cells in the mid to caudal portion of the nucleus are most susceptible to the destructive action (Clark et al., 1979). When rats are cotreated with opioids, such as morphine, enkephalins or β-endorphin, the destructive action of 6-OHDOPA on locus ceruleus cells is enhanced (unpublished), whereas the extent of noradrenergic collateral fiber sprouting in the brain stem and cerebellum is similarly increased (Harston et al., 1980, 1981, 1982). This action of opioids develops into one that is receptor-mediated, since naloxone is able to antagonize the effects of opioids severals days after birth (Kostrzewa and Klisans-Fuenmayor, 1984).

Substance P reduces development of the hindbrain sprouting response and promotes recovery of forebrain noradrenergic fibers from 6-OHDA. Opioids and substance P are thought to modulate the recovery of fibers from damage, and not simply modify the toxicity of 6-OHDA or 6-OHDOPA (Jonsson and Hallman, 1982a, b). The neonatal actions of neurotoxins are analogous to the changes in noradrenergic fiber reorganization that occur after neonatal transections of the dorsal bundle or medial forebrain bundle (Jonsson and Sachs, 1982; Klisans-Fuenmayor et al., 1986), and this includes damage to locus ceruleus cells (Kostrzewa et al., 1988).

The ability of 6-OHDOPA to affect central noradrenergic neurons at virtually any stage of pre- or postnatal development makes this agent an indispensible chemical tool for studying noradrenergic neuronal ontogeny, regeneration, and collateral sprouting. Not being limited pre- or postnatally by barriers, 6-OHDOPA has advantages that cannot be reproduced by 6-OHDA. Therefore, 6-OHDOPA is commonly used in studies on developing noradrenergic fibers.

6. Analogs of 6-OHDA

Aside from 6-OHDOPA, a number of other compounds chemically related to 6-OHDA have been found to produce degeneration of noradrenergic fibers. These include 6-aminodopamine (6-ADA), α-methyl-6-ADA, α-methyl-6-OHDA, 4-amino-2,5-dihydroxyphenethylamine, 5-amino-2,5-dihydroxyphenethylamine, and 2,3,5-trihydroxyphenethylamine (Tranzer and Thoenen, 1973) as well as 2,3,4,5-tetrahydroxyphenethylamine and 2,3,5,6-tetrahydroxyphenethylamine (Lundstrom et al., 1973). Another compound, 3-*O*-methyl-6-OHDA, was reported to be more selective for dopaminergic fibers, but this has not been confirmed (Ho et al., 1973).

The 6-ADA has been studied to some degree, and appears to be more selective than 6-OHDA for noradrenergic vs dopaminergic neurons (Blank et al., 1972b; Jonsson and Sachs, 1973a; Lundstrom et al., 1973; Oke et al., 1974). A similar selectivity of action is observed with α-methyl-6-ADA (Kostrzewa et al., 1980).

Although 5,6-dihydroxytryptamine is generally considered to be a serotoninergic neurotoxin, the compound is sufficiently nonspecific that it is able to destroy CA neurons (Baumgarten et al., 1972). This substance is considered to have little or no utility as a CA neurotoxin. Despite certain advantages in selectivity of the ADA compounds (6-ADA and α-methyl-6-ADA) over 6-OHDA, they have not come into common use, nor have any of the other agents listed in this section.

7. DSP-4

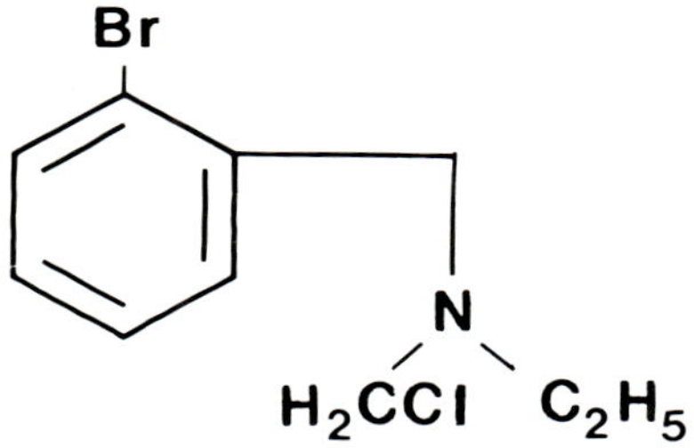

DSP–4

A bretylium-like compound, *N*-(2-chloroethyl)-*N*-ethyl-2-bromobenzylamine (DSP-4), that produces a long-term inhibition of ^{3}H-NE uptake by tissues and reduces DBH activity (Ross et al., 1973; Ross, 1976; Ross and Renyi, 1976) is now considered to be a neurotoxic substance with selectivity for noradrenergic neurons. DSP-4 is accumulated by noradrenergic neurons via an uptake process that is blocked by desmethylimpramine, which also inhibits the neurotoxic actions of DSP-4 (Jonsson et al., 1982). Destruction is thought to be brought about by alkylating reactions, which would interfere with vital neuronal functions. When given to adult rats, DSP-4 destroys sympathetic terminals, as does 6-OHDA. However, unlike 6-OHDA, DSP-4 crosses the blood–brain barrier and also destroys central noradrenergic fibers. The spectrum of action of DSP-4 in adult animals is much like that observed after ivtr 6-OHDA treatment of adults, except that less destruction of hypothalamic fibers occurs with DSP-4 (Lookingland et al., 1986).

In neonates, DSP-4 does not destroy sympathetic ganglion cells as does 6-OHDA (Jaim-Etcheverry and Zieher, 1980). Centrally, the same destructive action on noradrenergic fibers occurs as with 6-OHDA. Thus, virtually all noradrenergic terminals are initially subject to damage, and this is followed by a collateral sprouting of noradrenergic fibers in hindbrain regions as the animal develops. One distinction between DSP-4 and 6-OHDA in neonates is that noradrenergic hyperinnervation that develops in the hindbrain is greater in the cerebellum after DSP-4, but greater in the brain stem after 6-OHDA (Jaim-Etcheverry and Zieher, 1980; Jonsson et al., 1982).

This substance represents an alternative to 6-OHDA in the study of peripheral and central noradrenergic neurons. DSP-4 is relatively selective for noradrenergic neurons. Levels of DA, epinephrine, acetylcholine, GABA, glycine, aspartic acid, and glutamic acid are little affected by this neurotoxin (Ross, 1976; Dudley et al., 1981; Jonsson et al., 1982; Dooley et al., 1983). Although serotoninergic neurons are slightly affected by DSP-4 (Ross, 1976; Jonsson et al., 1982), this lack of selectivity can be reduced by pretreatment of animals with a 5-HT uptake inhibitor (Archer et al., 1986).

8. MPTP

The previous discussion on 6-OHDA and chemical analogs highlights the fact that none of these agents has selectivity for dopaminergic neurons. Recently, a relatively selective neurotoxin for dopaminergic neurons was discovered. This compound, 1-methyl-4-phenyl-1,2,3,6-tetrahydropyridine (MPTP), was initially found as a contaminant of 1-methyl-4-phenyl-4-propionoxypiperidine, the active ingredient in "synthetic heroin," a street drug (Langston et al., 1983).

Extensive studies have demonstrated that MPTP is a highly selective toxin for dopaminergic pars compacta cells in the substantia nigra of primates (Burns et al., 1983; Langston et al., 1983, 1984a) and mice (Heikkila et al., 1984a), but not rats (Chiueh et al., 1984). The MPTP *per se* was initially considered as an inhibitor of MAO, but is now recognized as a substrate, specifically for MAO-B (Chiba et al., 1984). Pargyline, an inhibitor of MAO-B, prevents the toxicity of MPTP in mice (Heikkila et al., 1984b) and monkeys (Langston et al., 1984b).

At this time, it is not known which species of metabolic products are responsible for the toxicity of MPTP or how these metabolites produce neurotoxicity. Oxidation of the pyridine ring by MAO-B results in the formation of a positively charged pyridinium ion, MPP^+ (Markey et al., 1984; Langston et al., 1984c), which has a half-life of days vs the half-life of hours for MPTP. The concentration of MPP^+ in monkey brain regions and peripheral tissues is markedly reduced after pargyline pretreatment (Langston et al., 1984a), whereas MPTP levels are markedly elevated.

MPTP has been shown by autoradiography to bind with high affinity to areas of the brain that are destroyed, but high affinity binding is also found in regions of the brain that are not grossly affected by MPTP (Javitch et al., 1984). In vivo, MPP^+ is found in far greater concentrations than MPTP in the neocortex, cerebellum, and basal ganglia (Langston et al., 1984c). Chiba and coworkers (1984) have suggested a dihydropyridinium species as the principal intermediate in the formation of MPP^+ from MPTP. It has been postulated that a mixture of dopamine oxidation products, formed by a catalytic action of the intermediary dihydropyridinium ion, accounts for the neurotoxicity of MPTP.

In both monkeys and rats, uptake of MPTP into dopaminergic neurons is similar. However, in monkeys MPP^+ persists, whereas in rats MPP^+ disappears rapidly. These differences may be related to the distribution of MAO-B, which metabolizes MPTP to MPP^+ (Chiueh et al., 1984). The relatively selective loss of substantia nigra cells by in vivo MPTP (Burns et al., 1983; Langston et al., 1984a,b) may be associated with high DA content or the presence of neuromelanin pigment. However, it is noteworthy that MPTP is effective in the mouse, which does not have a pigmented substantia nigra (Heikkila et al., 1984a), and MPTP destroys the unpigmented rat substantia nigra in culture (Mytilineou and Cohen, 1984). Therefore, neuromelanin pigment *per se* may be unimportant (Javitch et al., 1984). No evidence directly vindicates high DA levels as having an association with MPTP toxicity.

It would seem that the selectivity of MPTP for substantia nigral cell bodies represents another dimension to the conventional CA neurotoxins, which have a greater selectivity for nerve terminal regions. However, in mice MPTP may act primarily on the dopaminergic terminals (Hallman et al., 1984). Also, although MPTP exhibits relative selectivity for substantia nigral cells, those dopaminergic cells in the ventral tegmental area, as well as locus ceruleus noradrenergic cells, may be damaged by MPTP in primates (Mitchell et al., 1985). An in-depth discussion of the actions and mechanisms of MPTP is included in a later chapter of this volume by Heikkila.

9. Guanethidine

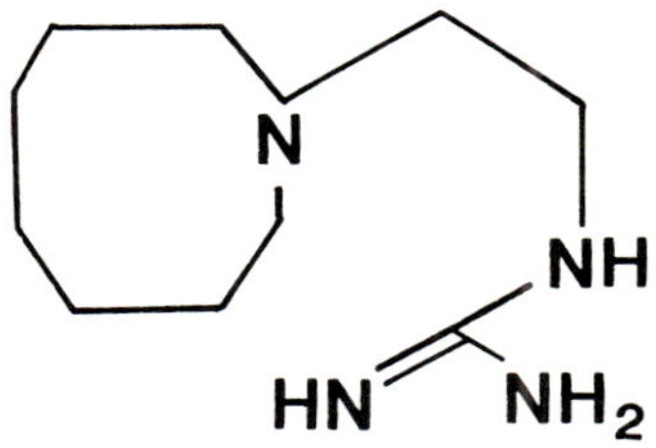

Guanethidine

Guanethidine is a neuronal blocking agent used in the treatment of human hypertension. In certain species, including the rat, guanethidine is able to destroy immature sympathetic ganglia and produce a chemical sympathectomy (Burnstock et al., 1971; Eränkö and Eränkö, 1971; Angeletti and Levi-Montalcini, 1972). Compared to anti-NGF, 6-OHDA, or 6-OHDA analogs, which can be used for this purpose, guanethidine has several advantages.

First, guanethidine (50 mg/kg/d sc × 20, from birth) destroys prevertebral and terminal ganglia, along with paravertebral ganglia. Consequently, NE content of the mesentery and vas deferens is reduced for at least 10 wk after birth, and stimulation of sympathetic vasomotor and periarteriolar nerves produces little physiological response (Johnson et al., 1975). Additionally, brain CA neurons are affected little or not at all by neonatal guanethidine. Finally, this chemical sympathectomy can also be produced by guanethidine treatment of adult rats (Johnson and Hunter, 1979).

Destruction of sympathetic ganglia occurs by an immune-mediated process, characterized by small cell infiltration of the ganglia. Immunosuppressive agents (γ-radiation, cyclophosphamide) prevent guanethidine toxicity (Manning et al., 1982), whereas immune reconstitution of these animals via transfer of syngeneic splenic and bone marrow cells restores guanethidine toxicity (Manning et al., 1983). NGF protects from guanethidine neurotoxicity

(Johnson et al., 1975), and it has been established that NGF interferes with the immune response (Manning et al., 1983). These studies indicate that guanethidine sympathectomy is quite different from that produced by other CA neurotoxins, and can occur in both immature and adult rats without altering CNS noradrenergic neurons.

10. Anti-DBH

Unlike the other CA neurotoxins, which destroy the neuron by formation of reactive products and through generation of free radicals, treatment of animals with an antibody to DBH sets in motion a complement-mediated immunodegenerative process (Furness et al., 1977). Peripheral administration of anti-DBH to mice or guinea pigs results in the destruction of sympathetic fibers to the heart and intestines. Not all neurons are affected, since sympathetic innervation to the iris and vas deferens remains intact (Costa et al., 1976; Rush et al., 1976). It may be of significance that the two latter tissues are innervated by fibers that emanate from prevertebral and terminal ganglia vs the paravertebral origin of fibers to the heart and intestines.

Although anti-DBH produces selective destruction of noradrenergic fibers in the CNS, the extent of nerve terminal destruction is less than that which occurs after 6-OHDA, and cell bodies appear to remain intact. It is also of importance that complement often must be added with anti-DBH in the rat, even to produce destruction of noradrenergic fibers in the CNS (Blessing et al., 1977; Geffen et al., 1978; Costa et al., 1979). These findings suggest that anti-DBH may not be able to totally destroy a defined system, but may still be of value in studies on regeneration and sprouting of noradrenergic neurons.

11. Summary

A number of neurotoxins are available as chemical tools to probe the function and plasticity of CA-containing neurons. The description of actions and mechanisms of action of commonly used agents have been presented and should serve as an aid toward the experimental approach of studying these neurons. By selection of the proper agent and with proper administration, one can achieve the most suitable experimental model, while limiting nonspecific damage of non-CA cellular elements and other neurochemical systems.

The CA neurotoxins have broad applications that are tempered by the inherent limitations of each agent. When used properly, the CA neurotoxins are indispensible tools in neuropharmacology.

Acknowledgment

The author is grateful to Ms. Betty Hughes for typing the manuscript.

References

Acheson A. L. and Zigmond M. J. (1981) Short and long term changes in tyrosine hydroxylase activity in rat brain after subtotal destruction of central noradrenergic neurons. *J. Neuroscience* **1,** 493–504.

Acheson A. L., Zigmond M. J., and Stricker E. M. (1980) Compensatory increase in tyrosine hydroxylase activity in rat brain after intraventricular injections of 6-hydroxydopamine. *Science* **207,** 537–540.

Aloe L., Mugnaini E., and Levi-Montalcini, R. (1975) Light and electron microscope studies on the excessive growth of sympathetic ganglia in rats injected with 6-OHDA and NGF. *Arch. Ital. Biol.,* **113,** 326–353.

Andén N.-E., Dahlström A., Fuxe K., and Larsson K. (1965) Further evidence for the presence of nigro-neostriatal dopamine neurons in the rat. *Am. J. Anat.* **116,** 329–334.

Angeletti P. U. (1971) Chemical sympathectomy in newborn animals. *Neuropharmacol.* **10,** 55–59.

Angeletti P. U. and Levi-Montalcini R. (1972) Growth inhibition of sympathetic cells by some adrenergic blocking agents. *Proc. Nat. Acad. Sci. USA* **69,** 86–88.

Archer T., Jonsson G., Minor B. G., and Post C. (1986) Noradrenergic-serotonergic interactions and nociception in the rat. *Eur. J. Pharmacol.* **120,** 295–308.

Baumgarten H. G., Evetts K. D., Holman R. B., Iversen L. L., Vogt M., and Wilson Y. (1972) Effects of 5,6-dihydroxytryptamine on monoaminergic neurons in the central nervous system of the rat. *J. Neurochem.* **19,** 1587–1597.

Bell L. J., Iversen L. L., and Uretsky N. J. (1970) Time course of the effects of 6-hydroxydopamine on catecholamine-containing neurons in rat hypothalamus and striatum. *Br. J. Pharmacol.* **40,** 790–799.

Berger T. W., Kaul S., Stricker E. M., and Zigmond M. J. (1985) Hyperinnervation of the striatum by dorsal raphe afferents after dopamine-depleting brain lesions in neonatal rats. *Brain Res.* **336,** 354–358.

Berkowitz B. A., Spector S., Brossi A., Focella A., and Teitel S. (1970) Preparation and biological properties of (–) and (+)-6-hydroxydopa. *Experientia* **26,** 982–983.

Blank C. L., Kissinger P. T., and Adams R. N. (1972a) 5,6-Dihydroxyindole formation from oxidized 6-hydroxydopamine. *Eur. J. Pharmacol.* **19,** 391–394.

Blank C. L., Murrill E., and Adams R. N. (1972b) Central nervous system effects of 6-aminodopamine and 6-hydroxydopamine. *Brain Res.* **45,** 635–637.

Blessing W. W., Costa M., and Geffen L. B. (1977) Immune lesions of noradrenergic neurons in rat central nervous system produced by antibodies to dopamine-β-hydroxylase. *Nature* **267,** 368–369.

Bloom F. E., Algeri S., Gropetti A., Revuelta A., and Costa E. (1969) Lesions of central norepinephrine terminals with 6-OH-dopamine: Biochemistry and fine structure. *Science* **166,** 1284–1286.

Borchardt R. T., Burgess S. K., Reid J. R., Liang Y. O., and Adams R. N. (1977) Effects of 2- and/or 5-methylated analogues of 6-hydroxydopamine on norepinephrine and dopamine containing neurons. *Mol. Pharmacol.* **13,** 805–818.

Breese G. R. (1975) Chemical and immunochemical lesions by specific neurotoxic substances and antisera, in *Handbook of Psychopharmacology.* (Iversen, L. L., Iversen, S. D., and Snyder S. H. eds.), Vol. 1, Plenum, New York, pp. 137–189.

Breese G. R., Baumeister A. A., McCown T. J., Emerick S. G., Frye G. D., Crotty K., and Mueller R. A. (1984) Behavioral differences between neonatal and adult 6-hydroxydopamine-treated rats to dopamine agonists: Relevance to neurological symptoms in clinical syndromes with reduced brain dopamine. *J. Pharmacol Exp. Ther.* **231,** 343–354.

Breese G. R., Prange A. J., Howard J. L., Lipton M. A., McKinney W. T., Bowman R. E., and Bushnell P. (1972) 3-Methoxy-4-hydroxyphenylglycol excretion and behavioural changes in rat and monkey after central sympathectomy with 6-hydroxydopamine. *Nature (New Biol.)* **240,** 286–287.

Breese G. R. and Traylor T. D. (1970) Effect of 6-hydroxydopamine on brain norepinephrine and dopamine: Evidence for selective degeneration of catecholamine neurons. *J. Pharmacol. Exp. Ther.* **174,** 413–420.

Breese G. R. and Traylor T. D. (1971) Depletion of brain noradrenaline and dopamine by 6-hydroxydopamine. *Br. J. Pharmacol.* **42,** 88–99.

Breese G. R. and Traylor T. D. (1972) Developmental characteristics of brain catecholamines and tyrosine hydroxylase in the rat: effects of 6-hydroxydopamine. *Br. J. Pharmacol.* **44,** 210–222.

Brimijoin S. (1972) Evidence that 6-hydroxydopamine decreases synthesis of dopamine-β-hydroxylase in sympathetic ganglia. *J. Pharmacol. Exp. Ther.* **183,** 298–306.

Brimijoin S. and Molinoff P. B. (1971) Effects of 6-hydroxydopamine on the activity of tyrosine hydroxylase and dopamine-β-hydroxylase in sympathetic ganglia of the rat. *J. Pharmacol. Exp. Ther.* **178,** 417–424.

Britton K. T., Svensson T., Schwartz J., Bloom F. E., and Koob G. F. (1983) Dorsal noradrenergic bundle lesions fail to alter opiate withdrawal or suppression of opiate withdrawal by clonidine. *Life Sci.* **34,** 133–139.

Burns R. S., Chiueh C. C., Markey S. P., Ebert M. H., Jacobowitz D. M., and Kopin I. J. (1983) A primate model of parkinsonism: Selective destruction of dopamineric neurons in the pars compacta of the substantia nigra by *N*-methyl-4-phenyl-1,2,3,6-tetrahydropyridine. *Proc. Nat. Acad. Sci. USA* **80,** 4546–4550.

Burnstock G., Doyle A. E., Gannon B. J., Gerkens J. F., Iwayama T., and Mashford M. L. (1971) Prolonged hypotension and ultrastructural changes in sympathetic neurons following guanacline treatment. *Eur. J. Pharmacol.* **13,** 175–187.

Cannon W. B. and Rosenblueth A. (1937) *Autonomic Neuroeffector Systems.* Macmillan Co., New York.

Cheah T. B., Geffen L. B., Jarrott B., and Ostberg A. (1971) Action of 6-hydroxydopamine on lamb sympathetic ganglia, vas deferens and adrenal medulla: a combined histochemical, ultrastructural and biochemical comparison with the effects of reserpine. *Br. J. Pharmacol.* **42,** 543–557.

Chiba K., Trevor A., and Castagnoli N. (1984) Metabolism of the neurotoxic tertiary amine, MPTP, by brain monoamine oxidase. *Biochem. Biophys. Res. Commun.* **120,** 574–578.

Chiueh C. C., Markey S. P., Burns R. S., Johnannessen J. N., Pert A., and Kopin I. J. (1984) Neurochemical and behavioral effects of systemic and intranigral administration of *N*-methyl-4-phenyl-1,2,3,6-tetrahydropyridine in the rat. *Eur. J. Pharmacol.* **100,** 189–194.

Clark D. W. J., Laverty R., and Phelan E. L. (1972) Long-lasting peripheral and central effects of 6-hydroxydopamine in rats. *Br. J. Pharmacol.* **44,** 233–243.

Clark M. B., King J. C., and Kostrzewa R. M. (1979) Loss of nerve cell bodies in caudal locus coeruleus following treatment of neonates with 6-hydroxydopa. *Neurosci. Lett.* **13,** 331–336.

Cohen G. and Heikkila R. E. (1974) The generation of hydrogen peroxide, superoxide radical and hydroxyl radical by 6-hydroxydopamine, dialuric acid and related cytotoxic agents. *J. Biol. Chem.* **249,** 2447–2452.

Cohen S. (1960) Purification of a nerve growth promoting protein from the mouse salivary gland and its neurocytotoxic antiserum. *Proc. Nat. Acad. Sci. U.S.A.* **46,** 302–311.

Cohen S., Levi-Montalcini R., and Hamburger V. (1954) A nerve growth stimulating factor isolated from sarcoma 37 and 180. *Proc. Nat. Acad. Sci. U.S.A.* **40,** 1014–1018.

Corrodi H., Clark W. G., and Masuoka D. T. (1971) The synthesis and effects of D,L-6-hydroxydopa, in *6-Hydroxydopamine and Catecholamine Neurons* (Malmfors T. and Thoenen H., eds.), North-Holland, Amsterdam, pp. 187–192.

Costa M., Geffen L. B., Rush R. A., Bridges D., Blessing W. W., and Heath J. (1979) Immune lesions of central noradrenergic neurons produced by antibodies to dopamine-β-hydroxylase. *Brain Res.* **173,** 65–78.

Costa M., Rush R. A., Furness J. B., and Geffen L. B. (1976) Histochemical evidence for the degeneration of peripheral noradrenergic axons following intravenous injection of antibodies to dopamine-β-hydroxylase. *Neurosci. Lett.* **3,** 201–207.

Dahlgren N., Lindvall T., Sakabe T., Stenevi U., and Siesjo B. K. (1981) Cerebral blood flow and oxygen consumption in the rat brain after lesions of the noradrenergic locus coeruleus system. *Brain Res.* **209,** 11–23.

De Champlain J. (1971) Degeneration and regrowth of adrenergic nerve fibers in the rat peripheral tissues after 6-hydroxydopamine. *Can. J. Physiol. Pharmacol.* **49,** 345–355.

De Champlain J. and Nadeau R. (1971) 6-Hydroxydopamine, 6-hydroxydopa and degeneration of adrenergic nerves. *Fed. Proc.* **30,** 877–885.

Descarries L. and Saucier G. (1972) Disappearance of the locus coeruleus in the rat after intraventricular 6-hydroxydopamine. *Brain Res.* **37,** 310–316.

Dolkart-Gorin P. and Johnson E. M. (1979) Experimental autoimmune model of nerve growth factor deprivation: Effects on developing peripheral sympathetic and sensory neurons. *Proc. Natl. Acad. Sci. U.S.A.* **76,** 5382–5386.

Dooley D. J., Bittiger H., Hauser K. L., Bischoff S. F., and Waldmeier P. C. (1983) Alteration of central alpha 2 and beta-adrenergic receptors in the rat after DSP-4, a selective noradrenergic neurotoxin. *Neuroscience* **9,** 889–898.

Dudley M. W., Butcher L. L., Kammerer R. C., and Cho A. K. (1981) The actions of xylamine on central noradrenergic neurons. *J. Pharmacol. Exp. Ther.* **217,** 834–840.

Eränkö L. and Eränkö O. (1971) Effect of guanethidine on nerve cells and small intensely fluorescent cells in sympathetic ganglia of newborn and adult rats. *Acta. Pharmacol Toxicol.* **30,** 403–416.

Evetts K. D. and Iversen L. L. (1970) Effects of protriptyline on the depletion of catecholamines induced by 6-hydroxydopamine in the brain of the rat. *J. Pharm. Pharmacol.* **22,** 540–543.

Furness J. B., Lewis S. Y., Rush R. A., Costa M., and Geffen L. B. (1977) Involvement of complement in degeneration of sympathetic nerves after administration of antiserum to dopamine-β-hydroxylase. *Brain Res.* **136,** 67–75.

Gauthier P., Nadeau R., and DeChamplain J. (1972) Acute and chronic cardiovascular effects of 6-hydroxydopamine in dogs. *Circ. Res.* **31,** 207–217.

Geffen L. B., Costa M., Rush R. A., and Bridges D. (1978) Immune lesions of central noradrenergic neurons produced by dopamine-β-hydroxylase (DBH) antibodies, in *Catecholamines: Basic and Clinical Frontiers,* (Usdin E., Kopin I. J., and Barchas J., eds.) **Vol. 2,** Pergamon, New York, pp. 1652–1654.

Haeusler G. (1971) Early pre-postjunctional effects of 6-hydroxydopamine. *J. Pharmacol. Exp. Ther.* **178,** 49–62.

Haeusler G., Haefely W., and Thoenen H. (1969) Chemical sympathectomy of the cat with 6-hydroxydopamine. *J. Pharmacol. Exp. Ther.* **170,** 50–61.

Hallman H., Olson L., and Jonsson G. (1984) Neurotoxicity of the meperidine analogue *N*-methyl-4-phenyl-1,2,3,6-tetrahydropyridine on brain catecholamine neurons in the mouse. *Eur. J. Pharmacol.* **97,** 133–136.

Harden T. K., Mailman R. B., Mueller R. A., and Breese, G. R. (1979) Noradrenergic hyperinnervation reduces the density of β-adrenergic receptors in rat cerebellum. *Brain Res.* **166,** 194–198.

Harden T. K., Wolfe B. B., Sporn J. R., Poulos B. K., and Molinoff, P. B. (1977) Effects of 6-hydroxydopamine on the development of the β-adrenergic receptor adenylate cyclase system in rat cerebral cortex *J. Pharmacol. Exp. Ther.* **203,** 132–143.

Harston C. T., Clark M. B., Hardin J. C., and Kostrzewa R. M. (1981) Opiate-enhanced toxicity and noradrenergic sprouting in rats treated with 6-hydroxydopa. *Eur. J. Pharmacol.* **71,** 365–373.

Harston C. T., Clark M. B., Hardin J. C., and Kostrzewa R. M. (1982) Developmental localization of noradrenergic innervation to the rat cerebellum following neonatal 6-hydroxydopa and morphine treatment. *Dev. Neuroscience* **5,** 252–262.

Harston C. T., Morrow A., and Kostrzewa R. M. (1980) Enhancement of sprouting and putative regeneration of central noradrenergic fibers by morphine. *Brain Res. Bull.* **5,** 421–424.

Hedreen J. C. and Chalmers J. P. (1972) Neuronal degeneration in rat brain induced by 6-hydroxydopamine, a histological and biochemical study. *Brain Res.* **47,** 1–36.

Heikkila R. E. and Cabbat F. S. (1977) Chemiluminescence from 6-hydroxydopamine. Involvement of hydrogen peroxide, the superoxide and the hydroxyl radical: A potential role for singlet oxygen. *Res. Commun. Chem. Pathol. Pharmacol.* **17,** 649–662.

Heikkila R. E. and Cohen G. (1971) Inhibition of biogenic amine uptake by hydrogen peroxide: A mechanism for toxic effects of 6-hydroxydopamine. *Science* **172,** 1257–1258.

Heikkila R. E. and Cohen G. (1973) 6-Hydroxydopamine: Evidence for superoxide radical as an oxidative intermediate. *Science* **181,** 456–457.

Heikkila R. E., Hess A., and Duvoisin R. C. (1984a) Dopaminergic neurotoxicity of 1-methyl-4-phenyl-1,2,5,6-tetrahydropyridine in mice. *Science* **224,** 1451–1453.

Heikkila R. E., Manzino L., Cabbat F. S., and Duvoisin R. C. (1984b) Protection against the dopaminergic neurotoxicity of 1-methyl-4-phenyl-1,2,3,6-tetrahydropyridine by monoamine oxidase inhibitors. *Nature* **311,** 467–469.

Heikkila R. E. and Manzino L. (1984) Behavioral properties of GBR 12909, GBR 13069 and GBR 13098: Specific inhibitors of dopamine uptake. *Eur. J. Pharmacol.* **103,** 241–248.

Hendry I. A., Stach R., and Herrup K. (1974) Characteristics of the retrograde axonal transport for nerve growth factor in the sympathetic nervous system. *Brain Res.* **82,** 117–128.

Heydorn W. E., Nguyen K. Q., Creed G. J., Kostrzewa R. M., and Jacobowitz D. M. (1986) Effects of bilateral lesion of the locus coeruleus and of neonatal administration of 6-hydroxydopamine on the concentration of individual proteins in rat brain. *Brain Res.* **367,** 31–38.

Ho B. T., Meyer A. L., and Taylor D. (1973) Selective depletion of dopamine following *O*-methylation of 6-hydroxydopamine. *Res. Commun. Chem. Pathol.* **6,** 47–56.

Hökfelt T. and Ungerstedt U. (1973) Specificity of 6-hydroxydopamine-induced degeneration of central monoamine neurons: An electron and fluorescence microscopic study with special reference to intracerebral injection on the nigro-striatal dopamine system. *Brain Res.* **60,** 269–297.

Jacks B. R., De Champlain J., and Cordeau J. P. (1972) Effects of 6-hydroxydopamine on putative transmitter substances in the central nervous system. *Eur. J. Pharmacol.* **18,** 353–360.

Jacobowitz D. M. (1975) Long-term effects on peripheral adrenergic nerves of 6-hydroxydopamine injected into newborn rats, in *Chemical Tools in Catecholamine Research,* **Vol. 1,** (Jonsson, G., Malmfors T., and Sachs Ch., eds.) North Holland, Amsterdam, pp. 153–162.

Jacobowitz D. and Kostrzewa R. (1971) Selective action of 6-hydroxydopa on noradrenergic terminals: Mapping of preterminal axons of the brain. *Life Sci.* **10,** 1329–1341.

Jaim-Etcheverry G. and Zieher L. M. (1971) Permanent depletion of peripheral norepinephrine in rats treated at birth with 6-hydroxydopamine. *Eur. J. Pharmacol.* **13,** 272–276.

Jaim-Etcheverry G. and Zieher L. M. (1975) Alterations in the development of central adrenergic neurons produced by 6-hydroxydopa, in *Chemical Tools in Catecholamine Research,* **Vol. 1** (Jonsson G, Malmfors T., and Sachs Ch. eds.) North-Holland, Amsterdam, pp. 173–180.

Jaim-Etcheverry, G. and Zieher, L. M. (1980) DSP-4: A novel compound with neurotoxic effects on noradrenergic neurons of adult and developing rats. *Brain Res.* **188,** 513–523.

Javitch J. A., D'Amato R. J., Strittmatter S. M., and Snyder S. H. (1984) Parkinsonism-inducing neurotoxin, 1-methyl-4-phenyl-1,2,3,6-tetrahydropyridine: uptake of the metabolite *N*-methyl-4-phenylpyridine by dopamine neurons explains selective toxicity. *Proc. Nat. Acad. Sci. U.S.A.* **82,** 2173–2177.

Johansson B. B. and Henning M. (1980) 6-Hydroxydopamine and the blood-brain barrier in adult conscious rats. *Acta Physiol. Scand.* **110,** 1–4.

Johnson E. M., Cantor E., and Douglas J. R. (1975) Biochemical and functional evaluation of the sympathectomy produced by the administration of guanethidine to newborn rats. *J. Pharmacol. Exp. Ther.* **193,** 503–512.

Johnson E. M. and Hunter F. E. (1979) Chemical sympathectomy by guanidinium adrenergic neuron blocking agents. *Biochem. Pharmacol.* **28,** 1525–1531.

Jonsson G. (1980) Chemical neurotoxins as denervating tools in neurobiology. *Ann. Rev. Neurosci.* **3,** 169–187.

Jonsson G. and Hallman H. (1978) Changes in β-receptor binding sites in rat brain after neonatal 6-hydroxydopamine treatment. *Neurosci. Lett.* **9,** 27–32.

Jonsson G. and Hallman H. (1982a) Substance P counteracts neurotoxin damage on norepinephrine neurons in rat brain during ontogeny. *Science* **215,** 75–77.

Jonsson G. and Hallman H. (1982b) Substance P modifies the 6-hydroxydopamine-induced alteration of postnatal development of central noradrenaline neurons. *Neuroscience* **7,** 2909–2918.

Jonsson G., Hallman H., and Sundstrom E. (1982) Effects of the noradrenaline neurotoxin DSP4 on the postnatal development of central noradrenaline neurons in the rat. *Neuroscience* **7,** 2895–2907.

Jonsson G., Pycock Ch., Fuxe K., and Sachs Ch. (1974) Changes in the development of central noradrenaline neurons following neonatal administration of 6-hydroxydopamine. *J. Neurochem.* **22,** 419–426.

Jonsson G. and Sachs Ch. (1970) Effects of 6-hydroxydopamine on the uptake and storage of noradrenaline in sympathetic adrenergic neurons. *Eur. J. Pharmacol.* **9,** 141–155.

Jonsson G. and Sachs Ch. (1971) Uptake and accumulation of ^{3}H-6-hydroxydopamine in adrenergic nerves. *Eur. J. Pharmacol.* **16,** 55–62.

Jonsson G. and Sachs Ch. (1973a) 6-Aminodopamine-induced degeneration of catecholamine neurons. *J. Neurochem.* **21,** 117–124.

Jonsson G. and Sachs Ch. (1973b) Pharmacological modifications of the 6-hydroxy-DOPA induced degeneration of central noradrenaline neurons. *Biochem. Pharmacol.* **22,** 1709–1716.

Jonsson G. and Sachs Ch. (1982) Changes in the development of central noradrenaline neurons after neonatal axon lesions. *Brain Res. Bull.* **9,** 641–650.

Jonsson G., Wiesel F.-A., and Hallman, H. (1979) Developmental plasticity of central noradrenaline neurons after damage—changes in transmitter functions. *J. Neurobiol.* **10,** 337–353.

Klara P. M., Kostrzewa R. M., and Brizzee K. R. (1976) Ultrastructural alterations in rat area postrema following parenteral 6-hydroxydopamine. *Brain Res.* **104,** 187–192.

Klisans-Fuenmayor D., Harston C. T., and Kostrzewa R. M. (1986) Alterations in noradrenergic innervation of the brain following dorsal bundle lesions in neonatal rats. *Brain Res. Bull.* **16,** 47–54.

Koda L. Y., Schulman J. A., and Bloom F. E. (1978) Ultrastructural identification of noradrenergic terminals in rat hippocampus: unilateral destruction of the locus coeruleus with 6-hydroxydopamine. *Brain Res.* **145,** 190–195.

Konkol R. J., Bendeich E. G., and Breese G. R. (1978a) A biochemical and morphological study of altered growth pattern of central catecholamine neurons following 6-hydroxydopamine. *Brain Res.* **140,** 125–135.

Konkol R. J., Mailman R. B., Bendeich E.-G. Garrison A. M., Mueller R. A., and Breese, G. R. (1978b) Evaluation of the effects of nerve growth factor and anti-nerve growth factor on the development of central catecholamine-containing neurons. *Brain Res.* **144,** 277–285.

Kostrzewa R. M. (1975) Effects of neonatal 6-hydroxydopa on monamine content of rat brain and peripheral tissues. *Res. Commun. Chem. Pathol. Pharmacol.* **11,** 567–579.

Kostrzewa R. M. (1988) Reorganization of noradrenergic neuronal systems following neonatal chemical and surgical injury. *Prog. Brain Res.* **78,** 405–423.

Kostrzewa R. M., Fukushima H., Morrow A., Blank C. L., and Lehr R. (1980) *alpha*-Methyl-6-aminodopamine: Depletion of norepinephrine in mouse brain and peripheral tissues. *Life Sci.* **27,** 2245–2250.

Kostrzewa R. M. and Garey R. E. (1976) Effects of 6-hydroxydopa on noradrenergic neurons in developing rat brain. *J. Pharmacol. Exp. Ther.* **197,** 105–118.

Kostrzewa R. M. and Garey R. E. (1977) Sprouting of noradrenergic terminals in rat cerebellum following neonatal treatment with 6-hydroxydopa. *Brain Res.* **124,** 385–391.

Kostrzewa R. M., Hardin J. C., and Jacobowitz D. M. (1988) Destruction of cells in the midportion of the locus coeruleus by a dorsal bundle lesion in neonatal rats. *Brain Res.* **442,** 321–328.

Kostrzewa R. M. and Harper J. W. (1974) Effect of 6-hydroxydopa on catecholamine-containing neurons in brains of newborn rats. *Brain Res.* **69,** 174–181.

Kostrzewa R. M. and Harper J. W. (1975) Comparison of the neonatal effects of 6-hydroxydopa and 6-hydroxydopamine on growth and development of noradrenergic neurons in the central nervous system, in *Chemical Tools in Catecholamine Research,* **Vol. 1** (Malmfors T., Jonsson G., and Sachs Ch., eds.) North Holland, Amsterdam, pp. 181–189.

Kostrzewa R. and Jacobowitz D. (1972) The effect of 6-hydroxydopa on peripheral adrenergic neurons. *J. Pharmacol. Exp. Ther.* **183,** 284–296.

Kostrzewa R. and Jacobowitz D. (1973) Acute effects of 6-hydroxydopa on central monoaminergic neurons *Eur. J. Pharmacol.* **21,** 70–80.

Kostrzewa R. M. and Jacobowitz D.M. (1974) Pharmacological actions of 6-hydroxydopamine. *Pharmacol. Rev.* **26,** 199–188.

Kostrzewa R. M., Klara J. W., Robertson J., and Walker L. C. (1978) Studies on the mechanism of sprouting of noradrenergic terminals in rat and mouse cerebellum after neonatal 6-hydroxydopa. *Brain Res. Bull.* **3,** 525–531.

Kostrzewa R. M. and Klisans-Fuenmayor D. (1984) Development of an opioid-specific action of morphine in modifying recovery of neonatally-damaged noradrenergic fibers in rat brain. *Res. Commun. Chem. Pathol. Pharmacol.* **46,** 3–11.

Lanfumey, L., Arluison M., and Adrien J. (1981) Destruction of noradrenergic cell bodies by intracerebral 6-hydroxydopamine in the newborn rat. *Brain Res.* **214,** 445–450.

Langston J. W., Ballard P., Tetrud J. W., and Irwin I. (1983) Chronic parkinsonism in humans due to a product of meperidine analog synthesis. *Science* **219,** 979–980.

Langston J. W., Forno L. S., Rebert C., and Irwin I. (1984a) Selective nigral toxicity after systemic administration of 1-methyl-4-phenyl-1,2,5,6-tetrahydropyrine (MPTP) in the squirrel monkey. *Brain Res.* **292,** 390–394.

Langston J. W., Irwin I., Langston E. B., and Forno L. S. (1984b) Pargyline prevents MPTP-induced parkinsonism in primates. *Science* **225,** 1480–1482.

Langston J. W., Irwin I., Langston E. B., and Forno L. S. (1984c) The importance of the '4–5' double bond for neurotoxicity in primates of the pyridine derivative MPTP. *Neurosci. Lett.* **50,** 289–294.

Laverty R., Sharman D. E., and Vogt M. (1965) Action of 2,4,5-trihydroxyphenylethylamine on the storage and release of noradrenaline. *Br. J. Pharmacol.* **24,** 549–560.

Lever J. D., Spriggs T. L. B., Ivens C., Joo F., Mottram D. R., and Graham J. D. P. (1971) Vasomotor innervation in the hind limb muscles of the normal and 6-hydroxydopamine-treated cat. *Z. Zellforsch. Mikroskop. Anat.* **122,** 389–396.

Levi-Montalcini R. (1952) Effects of mouse tumor transplantation on the nervous system. *Ann. N.Y. Acad. Sci.* **55,** 330–343.

Levi-Montalcini R. (1987) The nerve growth factor 35 years later. *Science* **237,** 1154–1162.

Levi-Montalcini R., Aloe L., Mugnaini E., Oesch F., and Thoenen H. (1975) Nerve growth factor induces volume increase and enhances tyrosine hydroxylase in sympathetic ganglia of newborn rats. *Proc. Nat. Acad. Sci. U.S.A.* **72,** 595–599.

Levi-Montalcini R. and Angeletti P. U. (1962) Noradrenaline and monoaminoxidase content in immunosympathectomized animals. *Int. J. Neuropharmacol.* **1,** 161–164.

Levi-Montalcini R. and Angeletti P. U. (1966) Immunosympathectomy. *Pharmacol. Rev.* **18,** 619–628.

Levi-Montalcini R. and Booker B. (1960) Destruction of the sympathetic ganglia in mammals by an antiserum to the nerve growth promoting factor. *Proc. Nat. Acad. Sci. U.S.A.* **46,** 384–391.

Levi-Montalcini R. and Hamburger V. (1951) Selective growth stimulating effects of mouse sarcoma on the sensory and sympathetic nervous system of the chick embryo. *J. Exp. Zool.* **116,** 321–361.

Levitt P. and Moore R. Y. (1980) Organization of brain stem noradrenaline hyperinnervation following neonatal 6-hydroxydopamine treatment in rat. *Anat. Embryol.* **158,** 133–150.

Lew G. M. and Quay W. B. (1971) Noradrenaline contents of hypothalamus and adrenal gland increased by postnatal administration of 6-hydroxydopamine. *Res. Commun. Chem. Pathol. Pharmacol.* **2,** 807–812.

Ljungdahl A., Hökfelt T., Jonsson G., and Sachs Ch. (1971) Autoradiographic demonstration of uptake and accumulation of ^{3}H-6-hydroxydopamine in adrenergic nerves. *Experientia* **27,** 297–299.

Lookingland K. J., Chapin D. S., McKay D. W., and Moore K. E. (1986) Comparative effects of the neurotoxins *N*-chloroethyl-*N*-ethyl-*N*-2-bromobenzylamine hydrochloride (DSP4) and 6-hydroxydopamine on hypothalamic noradrenergic, dopaminergic and 5-hydroxytryptaminergic neurons in the male rat. *Brain Res.* 365, 228–234.

Lundstrom J., Ong H., Daly J., and Creveling C. R. (1973) Isomers of 2,4,5-trihydroxyphenethylamine (6-hydroxydopamine): Long term effects on the accumulation of [^{3}H]-norepinephrine in mouse heart in vivo. *Mol. Pharmacol.* **9,** 505–513.

Lytle L. D., Shoemaker W. J., Cottman K., and Wurtman R. J. (1972) Long-term effects of postnatal 6-hydroxydopamine treatment on tissue catecholamine levels. *J. Pharmacol. Exp. Ther.* **183,** 56–64.

Malmfors T. and Sachs Ch. (1968) Degeneration of adrenergic nerves produced by 6-hydroxydopamine. *Eur. J. Pharmacol.* **3,** 89–92.

Manning P. T., Powers C. W., Schmidt R. E., and Johnson E. M. (1983) Guanethidine-induced destruction of peripheral sympathetic neurons occurs by an immune-mediated mechanism. *J. Neuroscience* **3,** 714–724.

Manning P. T., Russell J. H., and Johnson E. M. (1982) Immunosuppressive agents prevent guanethidine-induced destruction of rat sympathetic neurons. *Brain Res.* **241,** 131–143.

Markey F. P., Johannessen J. N., Chiueh C. C., Burns R. S., and Herkenham M. A. (1984) Intraneuronal generation of a pyridinium metabolite may cause drug-induced parkinsonism. *Nature* **311,** 464–466.

McBride R. L., Ozment R. V., and Sutin J. (1985) Neonatal 6-hydroxydopamine destroys spinal cord noradrenergic axons from the locus coeruleus, but not those from lateral tegmental cell groups. *J. Comp. Neurol.* **235,** 375–383.

Mitchell I. J., Cross A. J., Sambrook M. A., and Crossman A. R. (1985) Sites of the neurotoxic action of 1-methyl-4-phenyl-1,2,3,6-tetrahydropyridine in the macaque monkey include the ventral tegmental area and the locus coeruleus. *Neurosci. Lett.* **61,** 195–200.

Mueller R. A., Thoenen H., and Axelrod J. (1969) Adrenal tyrosine hydroxylase compensatory increase in activity after chemical sympathectomy. *Science* **163,** 468–469.

Mytilineou C. and Cohen G. (1984) 1-Methyl-4-phenyl-1,2,3,6-tetrahydropyridine destroys dopamine neurons in explants of rat embryo mesencephalon. *Science* **225,** 529–531.

Oke A., Freeman R., and Adams R. N. (1974) Comparison of central effects of 6-aminodopamine and 6-hydroxydopamine on catecholamine levels. *Eur. J. Pharmacol.* **26,** 125–127.

Olson L., Ebendal T., and Seiger Å. (1979) NGF and anti-NGF: Evidence against effects on fiber growth in locus coeruleus from cultures of perinatal CNS tissues. *Dev. Neurosci.* **2,** 160–176.

Ong H. H., Creveling C. R., and Daly J. W. (1969) The synthesis of 2,4,5-trihydroxyphenylalanine (6-hydroxydopa). A centrally active norepinephrine-depleting agent. *J. Med. Chem.* **12,** 458–469.

Pappas B. A., Saari M., and Peters D. A. V. (1976) Regional brain catecholamine levels after intraventricular 6-hydroxydopamine in the neonatal rat. *Res. Commun. Chem. Pathol. Pharmacol.* **14,** 751–754.

Poirier L. J., Langelier P., Roberge A., Boucher R., and Kitsikis A. (1972) Non-specific histopathological changes induced by the intracerebral injection of 6-hydroxy-dopamine (6-OH-DA). *J. Neurol. Sci.* **16,** 401–416.

Porter C.C., Totaro J. A., and Burcin A. (1965) The relationship between radioactivity and norepinephrine concentrations in the brains and hearts of mice following administration of labelled methyldopa or 6-hydroxydopamine. *J. Pharmacol. Exp. Ther.* **150,** 17–22.

Reis D. J. and Molinoff P. B. (1972) Brain dopamine-β-hydroxylase: Regional distribution and effects of lesions and 6-hydroxydopamine on activity. *J. Neurochem.* **19,** 195–204.

Reis D. J. and Ross R. A. (1973) Dynamic changes in brain dopamine-β-hydroxylase activity during anterograde and retrograde reactions to injury of central noradrenergic axons. *Brain Res.* **57,** 307–326.

Richardson J. S. and Jacobowitz D. M. (1973) Depletion of brain norepinephrine by intraventricular injection of 6-hydroxydopa: A biochemical, histochemical and behavioral study in rats. *Brain Res.* **58,** 117–133.

Ross, R. A., Joh T. H., and Reis D. J. (1975) Reversible changes in the accumulation and activities of tyrosine hydroxylase and dopamine-β-hydroxylase in neurons of nucleus locus coeruleus during the retrograde reaction. *Brain Res.* **92,** 57–72.

Ross, S. B. (1976) Long-term effects of *N*-2-chloroethyl-*N*-ethyl-2-bromobenzylamine hydrochloride on noradrenergic neurons in the rat brain and heart. *Br. J. Pharmacol.* **58,** 521–527.

Ross S. B., Johansson J. G., Lindborg B., and Dahlbom R. (1973) Cyclizing compounds. I. Tertiary *N*-(2-bromobenzyl)-*N*-haloalkylamines with adrenergic blocking actions. *Acta. Pharm. Suec.* **10,** 29–42.

Ross S. B. and Renyi A. L. (1976) On the long-lasting inhibitory effect of *N*-(2-chloroethyl)-*N*-ethyl-2-bromobenzylamine (DSP-4) on the active uptake of noradrenaline. *J. Pharm. Pharmacol.* **28,** 458–459.

Rush R. A., Costa M., Furness J. B., and Geffen L. B. (1976) Changes in tyrosine hydroxylase and dopamine-β-hydroxylase activities during degeneration of noradrenergic axons produced by antibodies to dopamine-β-hydroxylase. *Neurosci. Lett.* **3,** 209–213.

Sachs Ch. (1973) Development of the blood-brain barrier for 6-hydroxydopamine. *J. Neurochem.* **20,** 1753–1760.

Sachs, Ch. and Jonsson, G. (1972a) Degeneration of central and peripheral noradrenergic neurons produced by 6-hydroxydopa. *J. Neurochem.* **19,** 1561–1575.

Sachs Ch. and Jonsson G. (1972b) Degeneration of central noradrenaline neurons after 6-hydroxydopamine in newborn animals. *Res. Commun. Chem. Pathol. Pharmacol.* **4,** 203–220.

Sachs Ch. and Jonsson G. (1972c) Selective 6-hydroxy-DOPA induced degeneration of central and peripheral noradrenaline neurons. *Brain Res.* **40,** 563–568.

Sachs Ch. and Jonsson G. (1973) Changes in central noradrenaline neurons after systemic 6-hydroxydopamine administration. *J. Neurochem.* **21,** 1517–1524.

Sachs Ch., Jonsson G., and Fuxe K. (1973) Mapping of central noradrenaline pathways with 6-hydroxy-dopa. *Brain Res.* **63,** 249–261.

Saner A. and Thoenen H. (1971) Model experiments on the molecular mechanism of action of 6-hydroxydopamine. *Mol. Pharmacol.* **7,** 147–154.

Savaki H. E., Graham D. I., Grome J. J., and McCulloch, J. (1984) Functional consequences of unilateral lesion of the locus coeruleus: A quantitative [^{14}C]2-deoxyglucose investigation. *Brain Res.* **292,** 239–249.

Schmidt R. H. and Bhatnagar R. K. (1979a) Assessment of the effects of neonatal subcutaneous 6-hydroxydopamine on noradrenergic and dopaminergic innervation of the cerebral cortex. *Brain Res.* **166,** 309–319.

Schmidt R. H. and Bhatnagar R. K. (1979b) Distribution of hypertrophied locus coeruleus projection to adult cerebellum after neonatal 6-hydroxydopamine. *Brain Res.* **172,** 23–33.

Schmidt R. H. and Bhatnagar R. K. (1980) Intracisternal dose-response analysis of 6-hydroxydopamine-induced noradrenergic sprouting in the neonatal rat cerebellum. *J. Pharmacol. Exp. Ther.* **212,** 456–461.

Schmidt R. H., Kasik S. A., and Bhatnagar R. K. (1980) Regenerative critical periods for locus coeruleus in postnatal rat pups following intracisternal 6-hydroxydopamine: a model of noradrenergic development. *Brain Res.* **191,** 173–190.

Sievers J., Klemm H. P., Jenner S., Baumgarten H. G., and Berry M. (1980) Neuronal and extraneuronal effects of intracisternally administered 6-hydroxydopamine on the developing rat brain. *J. Neurochem.* **34:** 765–771.

Siggins G. R. and Bloom F. E. (1970) Cytochemical and physiological effects of 6-hydroxydopamine on periateriolar nerves of frogs. *Circ. Res.* **27,** 23–38.

Smith R. D., Cooper B. R., and Breese G. R. (1973) Growth and behavioral changes in developing rats treated intracisternally with 6-hydroxydopamine: Evidence for involvement of brain dopamine. *J. Pharmacol. Exp. Ther.* **185,** 609–619.

Snyder A. M., Zigmond M. J., and Lund R. D. (1986) Sprouting of serotoninergic afferents into striatum after dopamine-depleting lesions in infant rats: A retrograde transport and immunocytochemical study. *J. Comp. Neurol.* **245,** 274–281.

Stone C. A., Porter C. C., Stavorski J. M., Ludden C. T., and Totaro J. A. (1964) Antagonism of certain effects of catecholamine-depleting agents by antidepressant and related drugs. *J. Pharmacol. Exp. Ther.* **144,** 196–204.

Thoa N. B., Eichelman B., Richardson J. S., and Jacobowitz D. (1972) 6-Hydroxydopa depletion of brain norepinephrine and the facilitation of aggressive behavior. *Science* **178,** 75–77.

Thoenen H., Mueller R. A., and Axelrod J. (1969) Transynaptic induction of adrenal tyrosine hydroxylase. *J. Pharmacol. Exp. Ther.* **169,** 249–254.

Thoenen H. and Tranzer J. P. (1968) Chemical sympathectomy by selective destruction of adrenergic nerve endings with 6-hydroxydopamine. *Naunyn-Schmiedeberg's Arch. Pharmacol.* **261,** 271–288.

Thoenen H. and Tranzer J. P. (1973) The pharmacology of 6-hydroxydopamine. *Ann. Rev. Pharmacol.* **13,** 169–180.

Tranzer J. P. and Thoenen H. (1967) Ultra-morphologische Veranderungen der sympatischen Nervendigunden der Katze nach Vorbehandlung mit 5- und 6-hydroxy-dopamin. *Naunyn-Schmiedeberg's Arch. Pharmacol.* **257,** 343–344.

Tranzer J. P. and Thoenen H. (1968) An electron microscopic study of selective acute degeneration of sympathetic nerve terminals after administration of 6-hydroxydopamine. *Experientia* **24,** 155–156.

Tranzer J. P. and Thoenen H. (1973) Selective destruction of adrenergic nerve terminals by chemical analogues of 6-hydroxydopamine. *Experientia* **29,** 314–315.

Ungerstedt U. (1968) 6-Hydroxydopamine induced degeneration of central monoamine neurons. *Eur. J. Pharmacol.* **5,** 107–110.

Ungerstedt U. (1971a) Adipsia and aphagia after 6-hydroxydopamine induced degeneration of the nigro-striatal dopamine system. *Acta. Physiol. Scand. Suppl.* **367,** 95–122.

Ungerstedt U. (1971b) Histochemical studies on the effect of intracerebral and intraventricular injections of 6-hydroxydopamine on monoamine neurons in the rat brain, in *6-Hydroxydopamine and Catecholamine Neurons* (Malmfors T. and Thoenen H., eds.) North-Holland, Amsterdam, pp. 101–128.

Ungerstedt U. (1971c) Postsynaptic supersensitivity after 6-hydroxydopamine induced degeneration of the nigro-striatal dopamine system. *Acta. Physiol. Scand. Suppl.* **367,** 69–73.

Ungerstedt U. (1971d) Stereotaxic mapping of the monoamine pathways in the rat brain. *Acta. Physiol. Scand. Suppl.* **367,** 1–48.

U'Prichard D. C., Bechtel W. D., Rouot B. R., and Snyder S. H. (1979) Multiple apparent alpha-noradrenergic receptor binding sites in rat brain: effect of 6-hydroxydopamine. *Mol. Pharmacol.* **16,** 47–60.

U'Prichard D. C., Reisine T. D., Mason S. T., Fibiger H. C., and Yamamura H. I. (1980) Modulation of rat brain α- and β-adrenergic receptor populations by lesion of the dorsal noradrenergic bundle. *Brain Res.* **187,** 143–154.

Uretsky N. J. and Iversen L. L. (1970) Effects of 6-hydroxydopamine on catecholamine containing neurones in the rat brain. *J. Neurochem.* **17,** 269–278.

Velley L., Nassif S., Kempf E., and Cardo B. (1983) Enhancement of learning four weeks after stimulation of the nucleus locus coeruleus in the rat: Differential effects of dorsal noradrenergic bundle lesion and lesion of the locus coeruleus proper. *Brain Res.* **265,** 273–282.

Versteeg D. H. G., Wijnen H. J. L. M., DeKloet E. R., and DeJong W. (1980) Differential effects of neonatal 6-hydroxydopamine treatment on the catecholamine content of hypothalamic nuclei and brain stem regions. *Neurosci. Lett.* **7,** 341–346.

Wagner K. and Trendelenburg U. (1971) Effect of 6-hydroxydopamine on oxidative phosphorylation and on monoamine oxidase activity. *Naunyn-Schmiedeberg's Arch. Pharmacol.* **269,** 110–116.

Weinshilboum R. and Axelrod J. (1971) Serum dopamine-β-hydroxylase: Decrease after chemical sympathectomy. *Science* **173,** 931–934.

Willis G. L., Sleeman M., Pavey G. M., and Smith G. C. (1987) Further studies on the neurochemical specificity of 6-hydroxydopamine as compared to radiofrequency lesions. *Brain Res.* **403,** 15–21.

Yanker B. A. and Shooter E. M. (1979) Nerve growth factor in the nucleus: Interaction with receptors on the nuclear membrane. *Proc. Nat. Acad. Sci. U.S.A.* **76,** 1269–1273.

Zieher L. M., and Jaim-Etcheverry G. (1973) Regional differences in the long-term effect of neonatal 6-hydroxydopa treatment on rat brain noradrenaline. *Brain Res.* **60,** 199–207.

Zieher L. M. and Jaim-Etcheverry G. (1975) 6-Hydroxydopa during development of central adrenergic neurons produces different long-term changes in rat brain noradrenaline. *Brain Res.* **86,** 271–281.

Zieher L. M. and Jaim-Etcheverry G. (1979) 6-Hydroxydopamine during development: Relation between opposite regional changes in brain noradrenaline. *Eur. J. Pharmacol.* **58,** 217–223.

Neurotoxins That Affect Central Indoleamine Neurons

Tomás A. Reader

1. Introduction

1.1 Serotoninergic and Catecholaminergic Systems

The indoleamine serotonin (5-hydroxytryptamine, 5-HT) was described in the central nervous system (CNS) more than 30 years ago (Amin et al., 1954; Bogdanski et al., 1956; De Robertis, 1964; Twarog and Page, 1953; Zieher and De Robertis, 1963) based on biochemical determinations of this amine in different brain regions. These studies were rapidly followed by the use of histofluorescent microscopy (Anden et al., 1965, 1966, 1967; Dahlstrom and Fuxe, 1964; Fuxe, 1965; Moore et al., 1978; Ungerstedt, 1971) and measurements of the activity of tryptophan hydroxylase and 5-HT uptake (Kuhar et al., 1972). This work was later complemented by anatomical descriptions of the now chemically identified neurotransmitter systems using radioautographic tracing techniques that take advantage of the uptake properties of the 5-HT system (Azmitia, 1978; Azmitia and Segal, 1978; Beaudet and Descarries, 1981; Parent et al., 1981) as well as by the use of immunocytochemical methods with antibodies against 5-HT itself (Steinbusch, 1981; Steinbusch et al., 1978; Steinbusch and Nieuwenhuys, 1983).

Similarly, the presence of monoamine (MA)-containing neurons in the mammalian CNS was first demonstrated in correlative biochemical and histochemical investigations. The catecholamine (CA) neurons and their projections were the first to be described as chemically identified neurotransmitter systems related to the development of histofluorescence microscopy (Falck et al., 1962). Very soon, and combining such histofluorescence methods with biochemical information on endogenous levels and/or enzymes of synthesis, three CAs were described in the CNS; i.e., noradrenaline (NA; Carlsson et al., 1962; De Robertis, 1964; De Robertis et al., 1965; Fuxe, 1965; Pellegrino de Iraldi et al., 1963; Ungerstedt, 1971; Versteeg et al., 1976; Vogt, 1954), dopamine

(DA; Carlsson et al., 1972; Hokfelt et al., 1974; Kehr et al., 1976; Lindvall et al., 1974; Lindvall and Bjorklund, 1974, 1983, 1984; Versteeg et al., 1976), and adrenaline (AD; Beart et al., 1979; Hokfelt et al., 1973; Koslow and Schlumpf, 1974; Reader, 1982, 1983; Van der Gugten et al., 1976).

1.2. Use of Neurotoxins

The research on the MA systems has benefited from the use of different drugs and agents used as chemical tools to biochemically dissect out the different CA and 5-HT projections, while at the same time providing interesting models for the study of changes in sensitivity, uptake capacities, behavioral modifications, as well as sprouting and/or plastic properties of the CNS. Reserpine, among the first drugs used, was found to decrease the levels of both NA and 5-HT, and therefore showed little selectivity. Another potent compound reported to decrease 5-HT levels more selectively in brain was *p*-chlorophenylalanine (PCPA; Koe and Weissman, 1966); this drug was considered to block 5-HT synthesis at the rate-limiting step. With the information from histofluorescence microscopy, it soon became an obvious step to perform lesions in discrete CNS areas to specifically denervate the brain from one or another MA. Such deafferentations could be produced either electrolytically or by the use of so-called specific neurotoxins, such as 6-hydroxydopamine (6-OHDA) for the CA systems (Kostrzewa and Jacobowitz, 1974; Lidbrink and Jonsson, 1974; Tranzer and Thoenen, 1967; Uretsky and Iversen, 1970) or 5,6- and 5,7-dihydroxytryptamine for the 5-HT systems (Baumgarten et al., 1971; Baumgarten and Lachenmayer; 1972; Baumgarten et al., 1974, 1982a; Costa et al., 1972; Daly et al., 1973, 1974; Schlossberger and Kuch, 1960).

2. Dihydroxytryptamines

2.1. Mechanisms of Action of Dihydroxytryptamine Neurotoxins

The neurotoxic dihydroxytryptamines (DHTs) are preferentially taken up by 5-HT transport mechanisms (uptake sites) of MA neurons, and this constitutes a requisite for their effects. The transport mechanism does not recognize oxidized DHTs as an adequate substrate for uptake; therefore the compounds to be used as 5-HT-neurotoxins must be protected from premature oxidation, usually by adding ascorbic acid as antioxidant. In fact, lack of affinity of a compound toward the 5-HT transport mechanism coincides with a lack of neurotoxicity for 5-HT neurons. Once taken up, all the toxic DHTs form quinone-like metabolites (indolequinones; *see* Fig. 1). This intraneuronal oxidation (autoxidation) of the DHTs is the initial *intracellular* step in the chemical chain of events leading to the destruction of the 5-HT terminals, and several hypothesis have been advanced (Baumgarten and Bjorklund, 1976) to explain their mode of action at a cellular and/or molecular level. The most important of these are (1) the oxidation and formation of hydrogen peroxide and (2) covalent binding to essential proteins.

The intracellular oxidation of the hydroxylated tryptamines to quinones is an essential step in the mode of action of these neurotoxic agents. Only the indoleamines (IAs) that become oxidized (or autoxidized) at biological pH and at a moderate rate are potential neurotoxins. If the autoxidation of the IA is very high (i.e., 5,6,7-trihydroxytryptamine), the compound will very rapidly be transformed in the cerebrospinal fluid (CSF) and/or extracellular brain fluid (interneuronal compartment) into a quinone that cannot be used as a substrate for the high-affinity uptake mechanism of 5-HT neurons (Baumgarten et al., 1982a). The first neurotoxic DHT used was 5,6-DHT (Baumgarten et al., 1971, 1972; Schlossberger and Kuch, 1960), and it exibits a somewhat lower toxicity than 5,7-DHT. This different neurotoxic potency can be explained because, although 5,6-DHT is taken up with a relatively high affinity, its oxidation velocity (determined by the oxygen consumption rate) is slower than that of 5,7-DHT. On the other hand, 5,7-DHT is a meta-substituted dihydroxytryptamine, and an aqueous a solution of this compound will inevitably be composed of a mixture (at equilibrium) of different tautomeric keto-enol forms, which retain their affinity for the 5-HT uptake mechanism. A fraction of 5,7-DHT may, however, be oxidized quite rapidly, but there is still a conservation of its substrate properties because one of the hydroxyl groups remains functional. In addition, it has been proposed that brain tissue provides an environment that protects 5,7-DHT more efficiently than 5,6-DHT from too rapid oxidation, whereas the autoxidation of 5,6-DHT is catalytically maintained and/or promoted by the oxygen peroxide it generates. In contrast, 5,7-DHT lacks this capacity of autocatalytic oxidation. Other biochemical

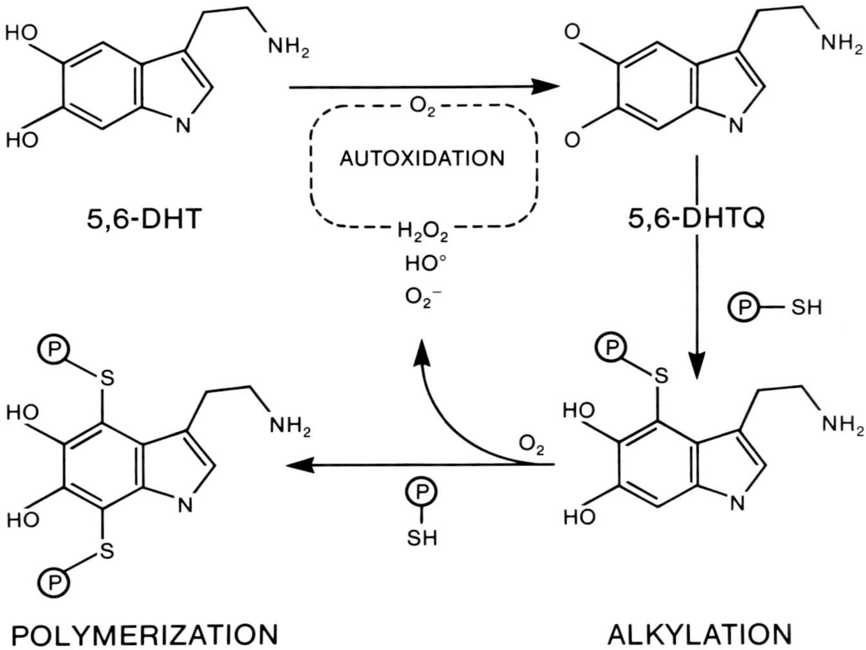

Fig. 1. Mechanism of action of 5,6-dihydroxytryptamine (5,6-DHT). The compound is transported into 5-HT neurons by a high-affinity uptake. Once inside the cells, 5,6-DHT is oxidized (autoxidized) into a quinone (5,6-DHTQ). This process is autocatalytic, favored and maintained by the hydrogen peroxide formed (autoxidation). The quinones produced in this first intraneuronal step can alkylate sulfhydryl (—SH) groups of proteins (P—SH), which are extremely reactive. The alkylation of these —SH bonds can cause protein denaturation and thus inactivate enzymes essential for the regulation of the cell i.e., enzymes and transporters involved in membrane pump mechanisms or energy-producing enzymes of the respiratory chains. A further step involves the alkylation of a second —SH group, thus leading to polymerization by crosslinking of homologous or heterologous proteins and/or peptides (modified from Sinhababu et al., 1985).

properties that may explain the different potencies of these compounds are related to their binding to proteins and peptides.

In a series of biochemical studies (Klemm et al., 1980), it has been shown that the initial oxidation of 5,6-DHT is slow, but this is followed by an accelerated reation with O_2; i.e., the autocatalytic oxidation of this compound is caused by the formation of hydrogen peroxide. Regarding oxidation by enzymes, it has also been documented that the DHTs can be metabolized by monoamine oxidase (MAO) and the aldehyde dehydrogenase systems of mitochondria (Klemm et al., 1978, 1979). The resulting metabolites (dihydroxyindoleacetic acids) as well as nondeaminated DHTs may oxidize further, in the presence of cytochrome *c* oxidase.

The other proposed mechanism of action of the neurotoxic DHTs is covalent binding to nucleophilic groups of protein, particularly to their sulfhydryls, leading to protein denaturation. The 5,6-substituted IA reacts readily and rapidly with nucleophiles and can cross-link protein and peptides, thus leading to the formation of melanoid polymers of high molecular weight that can be found in combination with nonoxidized 5,6-DHT (Fig. 1). The compound 5,7-DHT does not seem to form high-molecular-weight polymers easily. Among the different proteins to which the neurotoxic DHTs may bind, MAO seems to play an important role. It has been shown that alpha-methylated analogs of neurotoxic tryptamines, which are also MAO-resistant, have only slight effects on NA and 5-HT neurons (Baumgarten et al., 1975a; Bjorklund et al., 1975c). Interestingly, MAO inhibitors may antagonize the effects of 5,7-DHT on NA neurons (Baumgarten and Bjorklund, 1976). It has also been proposed that metabolism by MAO may be a prerequisite for 5,7-DHT to develop neurotoxicity (Creveling et al., 1975). Contrary to these findings, metabolism by MAO seems to play no critical role in the in vivo neurotoxicity of 5,6-DHT and only a minor role in the in vivo effects of 5,7-DHT (Baumgarten et al., 1978, 1981; Klemm et al., 1979). The most important intracellular mechanism documented to date that increases the toxicity of the IAs is their oxidation by cytochrome *c* oxidase (Baumgarten et al., 1982b). Finally, the neurotoxic DHTs and/or their degradation products may uncouple oxidation and phosphorylation mechanisms, with a reduction in ATP formation, or the indolequinones may replace the ubiquinones in the respiratory enzymatic chain, with a consequent blockade in ATP production.

2.2 Modes and Routes of Administration

The most widely used route of administration of the neurotoxic DHTs to produce 5-HT deafferentations in the CNS is the intraventricular (ivt; lateral ventricle route) injection (Noble et al., 1967). Usually the ivt administration is performed using stereotaxic coordinates (Fig. 2) and the DHT given in the smallest volume

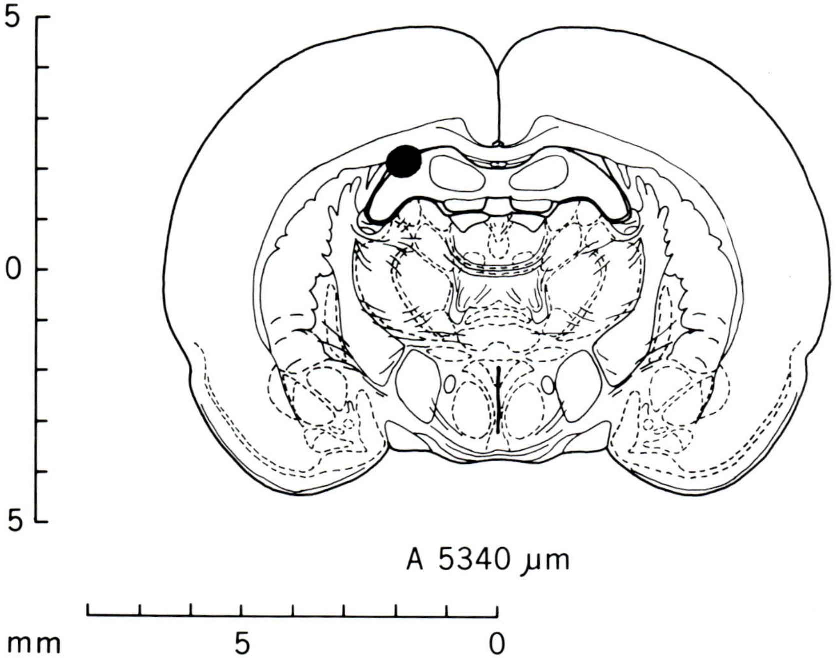

Fig. 2. The schematic transversal drawing of the rat brain is based upon the stereotaxic atlas of Konig and Klippel (1963) and shows the intraventricular (ivt) injection site for neurotoxic agents; coordinates are anterior = 5.40, lateral = 2.20, and horizontal = +2.20 mm. The ivt injections can also be made in the frontal horn of the lateral ventricle at anterior 8.3 mm, lateral = 1.0 mm, and horizontal = +1.8 mm (Gershanik et al., 1979). The ivt injection can also be performed using cranial bone landmarks, i.e., 1.2 mm posterior to bregma, 3.0 mm lateral to midline, and 3.0 mm below the surface of the cortex (Ross et al., 1976; Trulson et al., 1976). Another injection site used (Ferron et al., 1982) is at anterior = 5.4 mm, lateral 2.4 mm, and horizontal = +2.75 mm, according to the stereotaxic coordinates of De Groot (1959).

possible (15–20 μL). The vehicle is usually sterile saline (NaCl, 0.9%). Ascorbic acid (0.01–0.5%) is added prior to dissolving the DHT, thus avoiding the premature oxidation of the compound, which would hinder its uptake (*see also* Tables 1–3). We have usually performed the injections of 200 μg of 5,7-DHT (free base) in 18–20 μL of sterile saline containing 0.1% of ascorbic acid; the injection is conducted slowly over a 4–5-min period (speed of infusion 4–5 μL/min) using a microinjection unit (Model 5000,

Table 1
Doses and Modes of Administration of 5,6-dihydroxytryptamine[a]

Dose, μg	Volume, μL	Vehicle	Ascorbic acid, %	Reference
Lateral ventricle				
25–75	20	Diluted Ringer	0.01	Baumgarten et al., 1971
50	20	Saline	0.01	Baumgarten et al., 1972
75	20	Saline	0.02	Baumgarten et al., 1974
75	20	Diluted Ringer	0.01	Bjorklund et al., 1973b
75	20	Diluted Ringer	0.02	Bjorklund et al., 1980
80	10	Diluted Ringer	0.01	Donelson et al., 1982
150[b]	15		0.02	Gershanik et al., 1979
75	10	Saline	0.5	Honma et al., 1979
75	20	Saline	0.02	Wiklund and Bjorklund, 1980
50	5	Saline	0.01	Vogt, 1974
Cisterna magna				
25	20	Saline	0.02	Nygren and Olson, 1977
75	20	Saline	0.02	Wiklund et al., 1978
Medial forebrain bundle				
10	4	Ringer	0.1	Lorez et al., 1975

[a]The dose of 5,6-dihydroxytryptamine represents the free base or [b] as salt.

David Kopf Instruments, Tujunga, CA) and a gage 31 injection needle.

It has been claimed that when using the ivt route the magnitude and symmetry (the injections are performed either in the right or left lateral ventricle) of the 5-HT deafferentations may vary according to the speed of infusion (Gershanik et al., 1979). These authors have documented that slow infusions (1 μL/min) resulted in a more extensive and symmetric deafferentation of periventricular regions (i.e., right and left striatum) than did faster infusions (3 μL/min). In a very careful analysis of the effects of infusion speed, however, Baumgarten and colleagues (Baumgarten et al., 1982b) disclaimed these observations. They compared a

Table 2
Doses and Modes of Administration of 5,7-Dihydroxytryptamine into the Lateral Ventricle (IVT) and Cisterna Magna (IC)[a]

Doses, μg	Volume, μL	Vehicle	Ascorbic acid, %	Reference
Lateral ventricle				
200	30	Saline	0.01	Baez et al., 1980
200	20	Saline	0.02	Barbeau and Bedard, 1981a
10–200	20	Saline	0.1	Baumgarten et al., 1973a, 1974
150	20	NaCl 0.15*M*	0.1	Baumgarten et al., 1973b
150	20	Saline	0.1	Bjorklund et al., 1975a, c,
150	20	Saline	0.02	Bjorklund and Lindvall, 1979
200	20	Saline	0.1	De Montigny et al., 1980
200	20	Saline	0.1	Ferron et al., 1981, 1982
200	5	Saline	0.5	Fukata et al., 1984
200	20	Saline	0.1	Gauthier and Reader, 1982
200	25	CSF	0.5	Gerson and Baldessarini, 1975
200	10	Saline	0.01	Lyness et al., 1980
200	10	Saline	0.1	McCall and Aghajanian, 1979
100	30	Saline	0.1	Morrow et al., 1985
200	20	Saline	0.1	Reader and Gauthier, 1984
200	20	Saline	0.1	Ross et al., 1976
Cisterna magna				
200	20	Saline	0.5	Breese and Cooper, 1975

[a]The dose of 5,7-dihydroxytryptamine represents the free base.

slow infusion (2 μL/min) to a rapid pulse injection, and by the use of [^{14}C]5,7-DHT found a fairly even distribution of the radioactivity in the forebrain regions (cerebral cortex, hippocampus, and striatum) 1 h after the pulse injection. In the slowly infused animals, the radioactivity retained 1 h later had an asymmetric dis-

Table 3
Doses and Modes of Localized Administration of 5,7-Dihydroxytryptamine[a]

Dose, μg	Volume, μL	Vehicle	Ascorbic acid, %	Reference
Hypothalamus				
5	0.4	Saline	0.02	Frankfurt et al., 1985 Luine et al., 1983
Medial forebrain bundle				
4	4	Saline	0.02	Fuxe et al., 1978
Ventromedial tegmentum				
20	10	Water	0.01	Wang et al., 1979
Raphe nuclei				
8[b]	2	Saline		Blackburn et al., 1980
10[b]	5	Saline	0.2	Lorens et al., 1976
5[b]	1	Saline		Segal, 1979
4.4[b]	2	Saline	0.1	Vergnes et al., 1977
Intrathecal				
200	10	CSF	0.5	Vasko et al., 1984

[a]The dose of 5,7-dihydroxytryptamine represents the free base or [b] as salt.

tribution, i.e., much greater levels were measured in the striatum and hippocampus homolateral to the injected ventricle. These initial differences seem to level off; the asymmetries in the distribution of radioactivity are only minor when measured 24 h after the injections. The degree of 5-HT depletion, which is greater at day 1 for the homolateral periventricular regions than for the corresponding contralateral regions of slowly infused animals, becomes more uniform when the depletions are estimated at day 8. Furthermore, differences in the extent of the 5-HT deafferentations between slowly infused and pulse-injected animals disappear when the assessments of the depletions are performed after 7–8 d.

Several representative rat CNS locations where the DHTs have been injected are given in Tables 1–3. Young (and newborn) rats can also be treated with ivt injections; 10–50 μg free base dissolved in 10μL of 0.9% saline, containing 1 mg/μL of ascorbic acid (Lytle et al., 1974). For ivt injections in cat brain, higher doses of the DHTs are used, i.e., 125 μg of 5,6-DHT dissolved in 100–200 μL of saline containing 0.01% ascorbic acid (Beleslin and Samardzic, 1979) or 1 mg of 5,6-DHT dissolved in 200 μL of saline containing 0.01% ascorbic and infused at a rate of 10 μL/min on 7 consecutive d for a total dose of 7.0 mg (Steeves et al., 1980).

2.3. Specificity of the Compounds

The specificity of effects induced by the neurotoxic DHTs is determined by several factors, and has been very carefully and extensively examined by Baumgarten and colleagues (Baumgarten et al., 1982a,). These include: (1) the site of injection; (2) the injection volume; (3) the amounts and type of antioxidant used to protect the DHTs; (4) the type of anesthetic used and its dose; and (5) the use of a CA uptake blocking agent to impair the transport of the DHTs into noradrenergic, dopaminergic, or adrenergic neurons.

The administration of 100 μg (free base) of 5,7-DHT into the lateral ventricle (ivt) of ether-anesthetized rats pretreated with desipramine (DMI; 25 mg/kg; intraperitoneally, ip) reduces 5-HT levels in the forebrain (maximum at 5 d), and this reduction shows no recovery up to 25 d after the ivt injection. The same dose of 5,7-DHT administered into the fourth ventricle (intracisternal; ic) leads to a less extensive 5-HT depletion in the forebrain, with a maximum at 12 d and a slight recovery within 2 mo for the forebrain and a more pronounced recovery for the brainstem. Although the NA fibers seem to be protected against damage, there is a loss of DA fibers in the caudate, accumbens, bed nucleus of the stria terminalis, and medial part of the olfactory tubercles after the ivt administration, and only in the olfactory tubercles after the ic injection. Higher doses of 5,7-DHT (150 μg free base) administered either by the ivt or the ic route cause damage to the NA systems, in spite of the DMI pretreatment, which fails to completely protect the NA pathways in the brainstem (Baumgarten et al., 1978; Bjorklund et al., 1975a). Still higher doses of 5,7-DHT (200 μg free base) injected by the ivt route, even under the protection of DMI, reduce NA levels in the forebrain. In the hippocampus, such a treatment reduced 5-HT levels by 89% 1–2 mo after ivt administration, but the NA reduction was found to be 65% (Table 4). It could be argued that since the hippocampus is readily accessed by the CSF of the ventricles, it is an extremely vulnerable target to the neurotoxic affects of 5,7-DHT. In fact, the same treatment does not markedly effect NA levels in more distant regions, such as the cerebral

Table 4
Endogenous Serotonin (5-HT), Dopamine (DA), and Noradrenaline (NA) Contents in the Rat Hippocampus 1–2 mo after the Intraventricular Injection of Saline, 5,7-Dihydroxytryptamine,[a] and 6-hydroxydopamine[b,c]

		5-HT, ng/g	DA, ng/g	NA, ng/g
Saline	(n = 9)	255 ± 45	187 ± 24	444 ± 34
5,7-DHT,	(n = 9)	27 ± 14[d]	93 ± 13[e]	315 ± 46[e]
% change		–89	–50	–29
6-OHDA,	(n = 7)	94 ± 37[e]	19 ± 8[f]	105 ± 20[f]
% change		–63	–90	–76

[a]5,7-DHT; 200 μg, free base; pretreatment with desipramine, DMI= 25 mg/kg ip, 1 h before).

[b]6-OHDA; 200 μg, free base.

[c]The assays were performed radioenzymatically (i.e., 5-HT by the method of Saavedra et al., 1973, whereas the CAs were measured by the technique of Coyle and Henry (1973). The values are the means ± SEM, in ng/g wet tissue weight. The number of animals in every group is indicated in parentheses (n). Statistical significance was calculated by unpaired t-Student's test. The 5,7-DHT and 6-OHDA injections were made under chloral hydrate anesthesia (400 mg/kg, ip). The 5,7-DHT-injected animals were pretreated with desipramine (DMI; 25 mg/kg, ip) 1 h prior to the ivt injection of the neurotoxin.

[d]$p < 0.001$.

[e]$p < 0.05$.

[f]$p < 0.01$.

cortex. In this region, however, the DA levels were significantly reduced (Table 5), thus confirming a certain degree of sensitivity of DA projections to the DHTs.

In order to gain specificity, the most important single factor is probably the choice of a NA and/or DA uptake blocking agent to limit the toxic effects of the DHTs to the serotoninergic systems (Bjorklund et al., 1975c; Breese and Cooper, 1975; Breese and Mueller, 1978; Gerson and Baldessarini, 1975). In our own work, to avoid the nonspecific effects of 5,7-DHT on central CA systems, the animals were pretreated with DMI (25 mg/kg, ip) 1 h before the ivt administration of the neurotoxin (*see* Tables 6–8). In fact, in most of the regions examined in which 5-HT deafferentations were as-

Table 5
Endogenous Serotonin (5-HT), Dopamine (DA), and Noradrenaline (NA) Contents in Rat Frontoparietal Cortex, 15 d after the Intraventricular Injection of 5,7-Dihydroxytryptamine[a] or Electrolytic Lesions of the Raphe Nuclei Complex, and 2 d after the Administration of *p*-Chlorophenylalanine[b,c]

		5-HT ng/g	DA, ng/g	NA, ng/g
Controls	(n = 10)	360 ± 62	160 ± 32	330 ± 21
5,7-DHT,	(n = 10)	32 ± 8[d]	52 ± 4[d]	309 ± 17
%change		–91	–67	–6
Electrolytic lesion,	(n = 10)	55 ± 19[d]	47 ± 4[e]	327 ± 18
% change		–85	–71	–1
PCPA,	(n = 10)	32 ± 2[d]	138 ± 45	207 ± 30[f]
% change		–91	–14	–37

[a]5,6-DHT; 200 μg free base; pretreatment with desipramine, DMI: 25 mg/kg ip, 1 h before.

[b]PCPA, 2 × 400 mg/kg, ip.

[c]The assays were performed radioenzymatically, by the method of Saavedra et al. (1973) for 5-HT and by the method of Coyle and Henry (1973) for NA and DA. The values are the means ± SEM, in ng/g wet weight tissue weight. The number of animals in every group is indicated in parentheses (n). Statistical significance was calculated by unpaired t-Student's test; The 5,7-DHT injections and the electrolytic lesions of the raphe complex were made under choral hydrate anesthesia (400 mg/kg, ip). For the electrolytic lesions a monopolar stainless-steel electrode (tip diameter, 0.3 mm and length of exposed tip, 2.5 mm) was implanted in the region of the raphe, following stereotaxic coordinates. A cathodal current of 2.5 nA was passed for 10 s, with the reference electrode in the anus of the animal (*see also* Lorens et al., 1976; Lorez et al., 1975; Quick et al., 1977; Jacobs et al., 1979; Juvancz 1980. The 5,7-DHT-injected animals were pretreated with desipramine (DMI; 25 mg/kg, ip 1 h prior to the ivt injection of the neurotoxin.

[d]$p < 0.001$.

[e]$p < 0.05$.

[f]$p < 0.01$.

sessed biochemically, no significant changes in NA were measured, thus confirming that inhibiting the uptake mechanisms of NA fibers prevented their nonselective destruction by 5,7-DHT. Interestingly, there were significant reductions in NA levels in the samples from the pons-medulla and cerebellum (Table 8), for which we have no explanation. A possible interpretation may reside in the greater susceptibility to 5,7-DHT of the NA neurons of the locus coeruleus (pons-medulla sample) and of the neighboring NA fibers projecting to the cerebellum. Most likely, this nonspecific effect can be related to the relatively high dose of 5,7-DHT we used (200 μg free base) versus the more conservative amounts (150 μg free base) classically employed and recommended (Baumgarten et al., 1982a,b). Two regions, the hippocampus and the hypothalamus (Table 7), showed nonsignificant increases in NA content; this is perhaps a reflection of the activation at days 5–10 of tyrosine hydroxylase, reported to be under the control of a tonic 5-HT inhibition (Lewis et al., 1976; McRae-Degueurce et al., 1982; Pujol et al., 1978; Renaud et al., 1975). Since the samples were all assayed at 15 d, the activity of tyrosine hydroxylase was probably returning to basal levels. The more generalized effect of 5,7-DHT on the CA was a reduction in endogenous content, however. The protection of DMI did not extend to the DA and AD systems, since reductions in these CAs were also observed. A further degree of protection could have been obtained by a combined pretreatment with nomifensine (15 mg/kg, ip) and DMI (15 mg/kg, ip), as proposed by Baumgarten and colleagues (Baumgarten et al., 1981, 1982a,b), or by a pretreatment with DMI and anfenolic acid (Liston et al., 1982) or DMI (25 mg/kg, ip) plus pargyline (40 mg/kg, ip), as proposed by Lyness and coworkers (Lyness et al., 1980).

The DHTs may alter other putative neurotransmitters. Substance P, which coexists in some of the 5-HT neurons (Chan-Palay et al., 1978; Hokfelt et al., 1978), is reduced together with 5-HT, especially in the descending fiber tracts of the ventral spinal cord (Gilbert et al., 1982; Singer et al., 1979). This is very important to take into account, especially in behavioral and/or physiological studies of pain and nociception (Messing and Lytle, 1977; Segal, 1979; Sugrue, 1979; Vasko et al., 1984; Vogt, 1974; Yaksh et al., 1977). In this particular case of coexistence of neurotransmitters, the removal of the 5-HT terminals by the neurotoxic DHT will inevitably eliminate and reduce the levels of the coexisting substance P. In other situations, also involving analgesia and nociception, some of these interactions may be of a different nature. For example, 5,7-DHT (as well as PCPA) have recently been shown to reduce the brain levels of β-endorphin, α-melanotropin, and adrecorticotropin (Fukata et al., 1984), whereas 5,7-DHT-induced deafferentations may affect luteinizing hormone levels (van de Kar et al., 1980).

Table 6
Endogenous Serotonin (5-HT), Dopamine (DA), Adrenaline (AD), and Noradrenaline (NA) Contents in the Rat Forebrain, Brain Stem, and Spinal Cord 15 d after the Intraventricular Injection of 5,7-Dihydroxytryptamine[a] or 2 d after *p*-Chlorophenylalanine[b,c]

		5-HT, ng/mg P	DA, ng/mg P	AD, ng/mg P	NA, ng/mg P
Forebrain					
Controls	(n = 6)	3.111 ± 0.211	1.491 ± 0.176	0.334 ± 0.080	3.911 ± 0.176
5,7-DHT,	(n = 6)	0.634 ± 0.148	0.901 ± 0.073	0.223 ± 0.076	4.311 ± 0.402
% change		–80[d]	–40[e]	–33 NS	+10 NS
PCPA,	(n = 6)	0.564 ± 0.188	0.538 ± 0.084	0.116 ± 0.018	2.113 ± 0.215
% change		–82[d]	–64[d]	–65[d]	–46[e]
Brain stem					
Controls	(n = 6)	4.466 ± 0.265	1.489 ± 0.240	0.584 ± 0.123	3.877 ± 0.875
5,7-DHT,	(n = 6)	1.973 ± 0.488	1.015 ± 0.124	0.460 ± 0.224	2.636 ± 0.206
% change		–56[e]	–32 NS	–21 NS	%–32 NS
PCPA,	(n = 6)	0.959 ± 0.342	0.826 ± 0.152	0.284 ± 0.064	3.139 ± 0.534
% change		–79[e]	–45[f]	–51[f]	19 NS

Spinal cord					
Controls	(n = 9)	2.886 ± 0.252	0.427 ± 0.103	0.748 ± 0.191	2.567 ± 0.353
5,7-DHT,	(n = 6)	0.616 ± 0.269	0.381 ± 0.077	0.520 ± 0.065	2.267 ± 0.227
% change		–79[d]	–11 NS	–30 NS	–12NS
PCPA,	(n = 6)	0.404 ± 0.167	0.208 ± 0.047	0.157 ± 0.021	1.476 ± 0.366
% change		–86[d]	–51[f]	–79[e]	–42

[a] 5,7-DHT; 200 μg free base.

[b] PCPA; 2 ×400 mg/kg ip.

[c] The assays were performed radioenzymatically by the methods of Saavedra et al. (1973) for 5-HT and that of Cuello (1979) for the CAs. The values represent the mean content in ng/mg protein ± SEM. Protein contents was measured by the method of Lowry et al. (1951). The number of samples assayed in every group is indicated in parentheses (n). The percent changes in every treated group were compared to the control group. Statistical significance was calculated by unpaired t-Student's test. [f] $p < 0.05$; [e] $p < 0.01$ and [d] $p < 0.001$. The 5,7-DHT injected animals were pretreated with desipramine (DMI; 25 mg/kg ip) 1 h prior to the ivt injection of the neurotoxin (data from Gauthier and Reader, 1982).

[d] $p < 0.001$.

[e] $p < 0.01$.

[f] $p < 0.05$.

Table 7
Endogenous Serotonin (5-HT), Dopamine (DA), Adrenaline (AD), and Noradrenaline (NA) Contents in the Rat Hypothalamus, Hippocampus, and Cerebral Cortex 15 d after the Intraventricular Injection of 5,7-Dihydroxytryptamine[a] or 2 d after *p*-Chlorophenylalanine[b,c]

		5-HT, ng/mg P	DA, ng/mg P	AD, ng/mg P	NA, ng/mg P
Hypothalamus					
Controls	(n = 6)	4.324 ± 0.399	3.002 ± 0.304	0.334 ± 0.080	7.327 ± 0.956
5,7-DHT,	(n = 6)	0.823 ± 0.296	1.797 ± 0.238	0.223 ± 0.076	8.964 ± 0.819
% change		–82[d]	–40[e]	–33 NS	+22 NS
PCPA,	(n = 6)	0.426 ± 0.131	1.336 ± 0.225	0.116 ± 0.018	3.702 ± 0.346
% change		–90[d]	–55[d]	–65[f]	–49[e]
Hippocampus					
Controls	(n = 6)	2.718 ± 0.119	0.689 ± 0.147	0.584 ± 0.123	2.142 ± 0.362
5,7-DHT,	(n = 6)	0.477 ± 0.129	1.483 ± 0.144	0.460 ± 0.224	2.279 ± 0.251
% change		–82[e]	–30 NS	–21 NS	+ 6 NS

PCPA,	($n = 6$)	0.568 ± 0.223	0.211 ± 0.050	0.284 ± 0.064	1.468 ± 0.204
% change		–79[d]	–69[f]	–51[f]	–31 NS
Cerebal cortex					
Controls	($n = 6$)	2.291 ± 0.395	0.781 ± 0.245	0.249 ± 0.049	2.264 ± 0.328
5,7-DHT,	($n = 6$)	0.600 ± 0.103	0.424 ± 0.121	0.065 ± 0.018	1.688 ± 0.284
% change		–74[d]	–46 NS	–74[e]	–25 NS
PCPA,	($n = 6$)	0.362 ± 0.183	0.326 ± 0.133	0.104 ± 0.043	1.171 ± 0.220
% change		–84[d]	–58[f]	–58; $p = 0.052$	–48[f]

[a] 5,7-DHT; 200 μg free base.

[b] PCPA; 2 × 400 mg/kg ip.

[c] The assays were performed radioenzymatically by the methods of Saavedra et al. (1973) for 5-HT and that of Cuello (1979) for the CAs. The values represent the mean content in ng/mg protein ± SEM. Protein contents was determined by the method of Lowry et al. (1951). The number of samples assayed in every group is indicated in parentheses (n). The percent changes in every treated group were compared to the control group. Statistical significance was calculated by unpaired t-Student's test; The 5,7-DHT injected animals were pretreated with desipramine (DMI; 25 mg/kg ip) 1 h prior to the ivt injection of the neurotoxin (data from Reader and Gauthier, 1984).

[d] $p < 0.001$.

[e] $p < 0.01$.

[f] $p < 0.05$.

Table 8

Endogenous Serotonin (5-HT), Dopamine (DA), Adrenaline (AD), and Noradrenaline (NA) Contents in the Rat Mesencephalon, Pons-Medulla, and Cerebellum 15 d after the Intraventricular Injection of 5,7-Dihydroxytryptamine[a] or 2 d after *p*-Chlorophenylalanine[b,c]

		5-HT, ng/mg P	DA, ng/mg P	AD, ng/mg P	NA, ng/mg P
Mesencephalon					
Controls	(n = 6)	3.232 ± 0.496	2.207 ± 0.299	0.371 ± 0.101	2.961 ± 0.662
5,7-DHT,	(n = 6)	1.387 ± 0.276	1.623 ± 0.259	0.138 ± 0.041	2.395 ± 0.475
% change		–57[d]	–26 NS	–62; p = 0.057	–19 NS
PCPA,	(n = 6)	0.548 ± 0.180	1.673 ± 0.124	0.270 ± 0.089	2.933 ± 0.390
% change		–83[e]	–24 NS	–27 NS	–1 NS
Pons-medulla					
Controls	(n = 6)	5.698 ± 0.306	0.771 ± 0.222	0.644 ± 0.220	5.595 ± 1.068
5,7-DHT,	(n = 6)	2.058 ± 0.595	0.405 ± 0.101	0.371 ± 0.185	2.876 ± 1.249
% change		–64[e]	–47 NS	–42 NS	–48[d]
PCPA,	(n = 6)	0.548 ± 0.180	0.310 ± 0.075	0.298 ± 0.119	3.344 ± 0.724
% change		–83[e]	–60; p = 0.076	–53 NS	–40 NS

Cerebellum					
Controls	($n = 6$)	1.947 ± 0.667	0.206 ± 0.072	0.191 ± 0.042	1.690 ± 0.322
5,7-DHT,	($n = 6$)	0.502 ± 0.265	0.274 ± 0.080	0.113 ± 0.017	0.800 ± 0.179
% change		–74[f]	–33 NS	–40 NS	–53[f]
PCPA,	($n = 6$)	0.276 ± 0.207	0.122 ± 0.058	0.146 ± 0.072	1.492 ± 0.395
% change		–86[f]	–41 NS	–23 NS	–11 NS

[a] 5,7-DHT; 200 μg free base.

[b] PCPA; 2 × 400 mg/kg ip.

[c] The assays were performed radioenzymatically by the method of Saavedra et al. (1973) for 5-HT and that of Cuello (1979) for the CAs. The values represent the mean content in ng/mg protein ± SEM. Protein content was determined by the method of Lowry et al. (1951). The number of samples assayed in every group is indicated in parentheses (*n*). The percent changes in every treated group were compared to the control group. Statistical significance was calculated by unpaired *t*-Student's test; The 5,7-DHT injected animals were pretreated with desipramine (DMI; 25 mg/kg ip) 1 h prior to the ivt injection of the neurotoxin (data from Reader and Gauthier, 1984).

[d] $p < 0.01$.

[e] $p < 0.001$.

[f] $p < 0.05$.

2.4. Anesthetic Procedures

The anesthesia recommended by Baumgarten and colleagues is a combination of pentobarbitone (30 mg/kg, ip) and ketamine (50 mg/kg, ip). We have usually used a gaseous anesthetic (halothane, 2.3% in a mixture of 95% O_2 and 5% CO_2), but, after awakening, the animals had convulsive episodes. In some experimental situations we have also employed chloral hydrate (400 mg/kg, ip). Short-acting barbiturates can also be used; these include Brietal (Lilly; 40 mg/kg, ip), as proposed by Bjorklund and coworkers (Bjorklund et al., 1975a).

A list of the most widely used anesthetic procedures employed for the administration of the DHTs in rat brain is given in Table 9. Other less frequently used anesthetics are Equithesin (2 ml/kg, ip; Lyness et al., 1980), as well as a combination of Ketalar (Ketamine hydrochloride; Parke Davis; ip) and Rompun (Xylazine, Haver Lockhardt; im), as employed by Luine and coworkers (Luine et al., 1983). For the administration in cats, the anesthetic used is usually a barbiturate, such as pentobarbital sodium (35–40 mg/kg, ip), as employed by Beleslin and Samardzic (1979) or Steeves et al. (1980).

2.5. Assessment of the Deafferentations

2.5.1 Biochemical Methods

The assessment of the degree, magnitude, and extent of the DHT-induced deafferentations is an extremely important methodological aspect in the use of these neurotoxins. The most generalized and widely employed procedure is to rely on biochemical assays of endogenous 5-HT levels, 5-HT uptake sites, or tryptophan hydroxylase (the rate-limiting enzyme in 5-HT synthesis). The fluorometric determinations of 5-HT (Ahtee et al., 1970; Bertler, 1961; Bogdanski et al., 1956; Curzon and Green, 1970; Maickel et al., 1968; Snyder et al., 1965; Snyder and Taylor, 1972; Vogt and Wilson, 1972) were among the first biochemical assay tools successfully used to assess deafferentations, but for the past 10–12 yr they have been replaced by radioenzymatic (Boireau et al., 1976; Hammel et al., 1978; Palkovits et al., 1978; Reader, 1980; Saavedra et al., 1973) or high-performance liquid chromatography (HPLC) methods (Forchetti et al., 1982; Lackovic et al., 1981; Ponzio and Jonsson, 1979; Sasa et al., 1978; Semerdjian-Rouquier et al., 1981; Vasko et al., 1984; Warsh et al., 1979) that have proven to be more

Table 9
Anesthetic Procedures for the Intracerebral Administration of the Neurotoxic Dihydroxytryptamines

Anesthetic/s	Dose and route	Reference
Chloral hydrate	400 mg/kg, ip	De Montigny et al., 1980 McCall and Aghajanian, 1979 Wang et al., 1979
Ether	Alone	Baumgarten et al., 1971 Morrow et al., 1985 Vogt, 1974
Ether Plus pentobarbital sodium,	15 mg/kg, ip	Baez et al., 1980a Baumgarten et al., 1973
	20–30 mg/kg, ip	Breese and Cooper, 1975
Plus phenobarbital sodium	30 mg/kg, ip	Breese and Cooper, 1975
Halothane		Blackburn et al., 1980 Nygren and Olson, 1977 Gauthier and Reader, 1982 Reader and Gauthier, 1984
Methohexital sodium (Brietal; Lilly)	40 mg/kg, ip	Bjorklund et al., 1975a Bjorklund and Lindvall, 1979 Bjorklund et al., 1973a,b Bjorklund and Wiklund, 1980 Wiklund and Bjorklund, 1980
Pentobarbital sodium (Nembutal; Abbot)	40 mg/kg, ip	Fukata et al., 1984 Honma et al., 1979 Sugrue et al., 1979 Trulson et al., 1976 Trulson and Jacobs, 1978 Vergnes et al., 1977
	45 mg/kg, ip	Lorez et al., 1975
	50 mg/kg, ip	Ross et al., 1976

sensitive and specific, especially when dealing with relatively small and restricted regions. Details of a radioenzymatic method and an HPLC procedure for the measurement of tissue levels of 5-HT are given below.

In order to verify the specificity of effects, and since other MA systems besides 5-HT may be affected, most investigators have also relied on CA assays by fluorometric (Anton and Sayre, 1964; Bertler et al., 1958; Chang, 1964; Haggendal, 1963; Laverty and Taylor, 1968; Snyder and Taylor, 1972), radioenzymatic (Coyle and Henry, 1973; Cuello, 1979; Cuello et al., 1973; Peuler and Johnson, 1977; Reader, 1981, 1982), or HPLC methodologies (Adams, 1979; Ehrenstrom and Johansson, 1985; Felice et al., 1978; Hallman et al., 1978; Keller et al., 1976; Moyer and Jiang, 1978; Refshauge et al., 1974). Finally, there are also available several fluorometric assays (Holman et al., 1976; Jacobowitz and Richardson, 1978; Miller et al., 1970; Schellenberger and Gordon, 1971) and HPLC methodologies (Lakhdar-Ghazal et al., 1986; Mefford, 1981; Niwa et al., 1984, Reader et al., 1986; Renner and Luine, 1984; Saller and Salama, 1984) that allow for the simultaneous measurement of 5-HT and the CAs.

The destruction of the serotoninergic terminals by the DHTs also implies a loss of the 5-HT uptake sites, and this is the rationale behind the use of determinations of uptake of radioactive amines to assess the deafferentations. Originally based on the work of Hamberger (1967) and that of Sachs and Jonsson (1972), they not only constitute an excellent index of 5-HT terminal loss, but also have contributed to our knowledge on the mechanisms of action of these neurotoxins (Baumgarten et al., 1973a; Bjorklund et al., 1973b, 1975b; Bjorklund and Wiklund, 1980; Coyle and Snyder, 1969; Horn et al., 1973; Luine et al., 1983; Wang et al., 1979). Similarly, the determination of [^{3}H]imipramine binding has recently been considered a good index of 5,7-DHT-induced deafferentations (Luine et al., 1983). Finally, the assay of tryptophan hydroxylase can also be taken as a biochemical measurement of the loss of 5-HT terminals (Baumgarten et al., 1973b, 1975b; Carlsson et al., 1972; Palkovits et al., 1978).

2.5.2. Anatomical Methods

The histofluorescence technique of Falck-Hillarp (Falck et al., 1962) gave a considerable impetus to the studies of the MA systems in the CNS, and soon the original method was modified and improved to permit the visualization of IA cell bodies and their nerve terminals and thus be able to assess morphologically a DHT-induced deafferentation (Aghajanian et al., 1973; Baumgarten et al., 1975a; Bjorklund et al., 1972, 1973a, 1980; Bjorklund and Skagerberg, 1979; Falck et al., 1982; Fuxe and Jonsson, 1967, 1974; Loren et al., 1976, 1980, 1982; Nobin et al., 1973). A most interesting aspect of 5,6- and 5,7-DHT and some of their related mono- and dihydroxylated tryptamines is that they are accumulated in 5-HT neurons during the first 12–24 h and will react with formaldehyde to give an intense fluorophore (probably a β-carboline). If the uptake of the hydroxylated tryptamines is confined to the 5-HT system by preventing the uptake by NA and DA neurons and fibers with the appropriate uptake-blocking agents, they become selective markers for the 5-HT neurons and nerve fibers (Baumgarten et al., 1975a, 1982a,b; Bjorklund et al., 1973a; Nobin et al., 1973). Another related interesting anatomical application of the DHTs is to use [^{14}C]-5,6-DHT or [^{14}C]-5,7-DHT as tracers of 5-HT pathways, provided their uptake by CA neurons is prevented (Baumgarten et al., 1982a,b). The DHT-induced disappearance of 5-HT terminals can also eventually be documented using radioautographic (Beaudet and Descarries, 1981; Parent et al., 1981) or immunocytochemical methods (Steinbusch et al., 1978; Steinbusch, 1981; Wallace et al., 1982).

2.5.3. Radioenzymatic Assays of Monoamines

2.5.3.1. SEROTONIN ASSAY. The 5-HT contents can be determined using the method developed by Saavedra and colleagues, based on the enzymatic acetylation of 5-HT, and then by a methylation (Saavedra et al., 1973, 1974). In the first reaction, the acetylation is by the enzyme *N*-acetyltransferase (NAT, EC 2.3.1.5) in the presence of acetylcoenzyme A (AcCoA). The product of this first reaction is *N*-acetylserotonin, which is the substrate for methylation by hydroxyindole-*O*-methyltransferase (HIOMT, EC 2.1.1.4), an enzyme that will produce [^{3}H]melatonin, using as a donor of [^{3}H]methyl groups [^{3}H]methyl-*S*-adenosyl-L-methionine ([^{3}H]SAM). The end product, i.e., [^{3}H]melatonin (or [^{3}H]methyl-acetylserotonin) can thereafter be purified either by solvent extractions (Saavedra et al. 1973, Reader et al., 1979) or by thin layer chromatography (TLC), as previously reported by Boireau et al., (1976). The chromatographic separation of [^{3}H]melatonin (Fig. 3) was found to be necessary, since some of the batches of [^{3}H]SAM may give high blank values when the end-product is purified by simple organic extraction (Reader, 1980, 1981). Other mod-

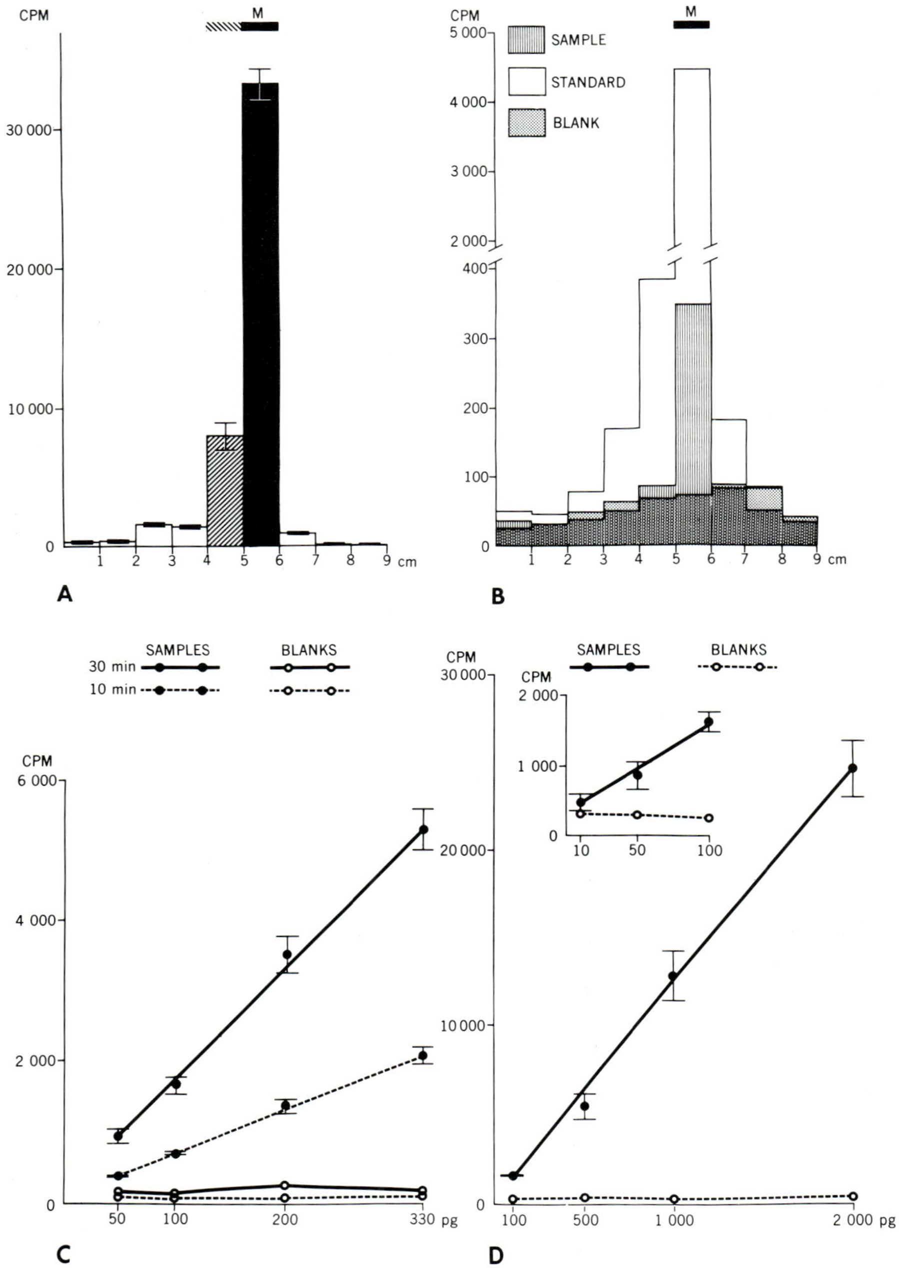

CPM
M
30 000
20 000
10 000
0
1 2 3 4 5 6 7 8 9 cm
A
CPM
5 000
4 000
3 000
2 000
400
300
200
100
0
SAMPLE
STANDARD
BLANK
M
1 2 3 4 5 6 7 8 9 cm
B
SAMPLES
BLANKS
30 min
10 min
CPM
6 000
4 000
2 000
0
50 100 200 330 pg
C
CPM
30 000
20 000
10 000
0
SAMPLES
BLANKS
CPM
2 000
1 000
0
10 50 100
100 500 1 000 2 000 pg
D

ifications of the original radioenzymatic assay can be used, such as the simplified procedure of Hammel et al. (1978). NAT is usually prepared from rat livers (Weissbach et al., 1961) and HIOMT from bovine pineal glands (Axelrod and Weissbach, 1961). The assays are always run with internal standards for each and every sample, prepared by adding a known amount (usually 1.0 ng) of authentic 5-HT to aliquots of the supernatants. In addition, every sample should be done in duplicate and with a blank obtained by replacing AcCoA in the first incubation by an equal volume of 0.1 m*M* HCl. The sensitivity of the method (twice the blank value) allows for the measurement of amounts above 25 pg.

For the radioenzymatic assays, brain tissue samples are homogeneized in 50 vol (wt/vol) of 0.1*N* cold perchloric acid ($HClO_4$) containing 4 m*M* of EGTA and 70 m*M* of $MgCl_2$ (Reader et al., 1976, 1979; Reader, 1981). The homogenates are centrifuged at 12,500 rpm for 30–45 min in a refrigerated centrifuge and the supernatants decanted for the radioenzymatic assays. The pellets can be resuspended in 1 mL of 1*N* NaOH for protein assay (Lowry et al., 1951).

Aliquots of 10 μL of the $HClO_4$ extract are transferred to 15-mL glass stoppered tubes containing 10 μL of a solution made with 10

←

Fig. 3. (A) Chromatogram of authentic melatonin to determine the separation and recovery. Twenty-four picograms of ^{3}H-melatonin plus 10 μg of cold melatonin (carrier) were spotted on the plate, and after developing the channels were scraped. The black bar indicates the melatonin (M) spot visualized under UV light (recovery from 1 cm), and the hatched one the tailing (2nd cm in 2 cm recoveries). The values represented are the average cpm of three chromatograms. (B) Chromatogram of an actual 5-HT assay in a sample of the rat frontoparietal cortex. The samples were run in triplicate, with a blank (no AcCoA) and an internal standard made of 1.0 ng of authentic 5-HT added to the homogenate. Methylation time, 10 min. Recovery was exclusively from the melatonin spot (M; 1 cm). (C) Samples of authentic 5-HT (50–330 pg) were acetylated and then incubated with HIOMT in the presence of ^{3}H SAM for 10 or 30 min. ^{3}H Melatonin was then purified by TLC. Each point is the mean ± SEM of three determinations. Blanks were not substracted from these curves, but plotted separately, and recoveries were from 1 cm. (D) Standard curve for authentic 5-HT, between 10 and 2,000 pg. Methylation time, 30 min. Each point is the mean ± SEM of three determinations (from Reader, 1980).

parts of 0.2*M* sodium phosphate buffer pH 7.9 and 1.1 parts of 1*N* NaOH (final pH 7.9). The first enzymatic reaction is carried out at 37°C and is started by the addition of 5 μL of NAT and AcCoA (4 mg/mL of AcCoA in 0.1 mM HCl) in equal proportions; 1:1 (vol/vol). After incubating for 30 min, the tubes are placed in an ice bath, and 5 μL of a mixture containing three parts of HIOMT, one part of [^{3}H]SAM (Amersham; specific activity 5–15 Ci/mmol), and one part of 0.2*M* sodium phosphate buffer, pH 7.9 are added. The reaction is allowed to proceed for 30 min at 37°C and is stopped by placing the tubes in an ice bath with 500 μL of 0.5*M* borate buffer at pH 10.0. As a cold carrier, 50 μL of a 1-mg/mL solution of melatonin in 25% ethanol is added. [^{3}H]Melatonin is extracted into 5 mL of toluene and then phase separated by a low-speed centrifugation (3000 rpm; 5 min). Four milliliters of the organic phase are transferred to glass counting vials, already containing 2 mL of toluene, and dried in a chromatography oven at 80°C overnight. The dry residue is resuspended in 100 μL of ethanol, and spotted on predried 250-μm-mesh Whatman silica-gel plates (LKD5F). The chromatograms are developed in toluene:acetic acid:ethyl acetate:water (80:40:20:5), and the melatonin spot is identified under UV light and scraped into liquid scintillations vials containing 1 mL of ethanol. Nine milliliters of phosphor cocktail are added and the radioactivity is counted by liquid scintillation spectroscopy.

2.5.3.2. Catecholamine Assays. The determination of endogenous CAs in DHT-treated animals is important since it constitutes: (1) an *internal* control of the specificity; (2) it allows one to follow regeneration of the NA systems in response to a 5-HT deafferentation, and (3) DHT-treated animals can be utilized as a model to investigate the functional and/or the compensatory mechanisms of the CA systems in response to 5-HT deafferentations.

The CA content of the samples in most of our studies was determined by radioenzymatic methods based on the procedures developed by Cuello et al. (1973) and Coyle and Henry (1973). Basically, the CAs are methylated by the enzyme catechol-*O*-methyltransferase (COMT, EC 2.1.1.11) in the presence of [^{3}H]SAM as the methyl donor. The ^{3}H-derivatives are thereafter isolated by solvent extraction (Coyle and Henry, 1973; Reader et al., 1976, 1979) or by chromatography on paper (Cuello, 1979) or TLC (Peuler and Johnson, 1977). COMT is usually prepared and partially purified from rat liver (Axelrod and Tomchick, 1958). In our hands, and using the TLC procedure, the sensitivity of the CA assay (twice the blank value) was 15 pg for NA and AD and 8–10 pg for DA.

For the CA assays, 50 μL of the acidified supernatants are transferred to 15-mL glass stoppered tubes containing 10 μL of 0.1*N* $HClO_4$. The enzymatic reaction is carried out at 37°C and started by the addition of 25 μL of incubation medium containing 0.45 mg of dithiothreitol, 11.4 μL of 3*M* Tris base; 1 μL of 300 m*M* $MgCl_2$, 11.4 μL of partly purified COMT, and 1 μL of [^{3}H]SAM (Amersham, specific activity 5–15 Ci/mmol). After 60 min the reaction is stopped by placing the tubes in an ice bath and adding 150 μL of 0.5*M* borate buffer, pH 10, containing 10 mg/mL each of 3-methoxytyramine, normetanephrine, and metanephrine as cold carriers. Fifty microliters of 1.5% tetraphenylboron is added, and then the methylated derivatives are extracted into 2.5 mL of toluene:isoamyl alcohol (3:2 vol/vol). The aqueous and organic phases are separated by low-speed centrifugation (3000 rpm for 5 min). Two milliliters of the organic phase are transferred to another set of tubes containing 150 μL of 0.1*N* HCl, and the methylated derivatives back-extracted into the acid phase. After washing the acid phase twice with toluene:isoamyl alcohol, 100 μL of the acid phase are spotted on predried 250-μm-Whatman silica-gel plates (LKD5F). The chromatograms are developed in tertiary-amyl alcohol:toluene:methylamine 40% (6:2:3 vol/vol) for 90 min. The spots are identified under UV light and scraped into liquid scintillation vials containing 1 mL of 0.05*M* ammonium hydroxide. [^{3}H]Methoxytyramine is counted by liquid scintillation spectroscopy after adding 10 mL of toluene:isoamyl alcohol:Liquifluor (70:30:5 vol/vol). [^{3}H]Normetanephrine and [^{3}H]metanephrine are oxidized by adding 50 μL of 4% sodium metaperiodate. After 5 min, the reaction is stopped with 50 μL of 10% glycerol, and 100 μL of glacial acetic acid is added to each vial. The derivatives are then extracted into 10 mL of toluene:Liquifluor (1000:50 vol/vol) and counted by liquid scintillation spectroscopy.

2.5.4. HPLC Assays of Monoamines

2.5.4.1. Serotonin Assays. There are several simple procedures to assay 5-HT in samples of brain tissue using HPLC and electrochemical detection of the eluated peaks (Lackovic et al., 1981; Ponzio and Jonsson, 1979; Sasa and Blank, 1977; Sasa et al., 1978; Warsh et al., 1979). We have used chromatographic conditions that allow for the measurement of 5-HT and CAs simultaneously in the same chromatographic run (see below). Therefore, using the same conditions as described for the HPLC assay of monoamines, but with an increased flow-rate (1.5–2.0 mL/min), the assay of 5-HT can be performed in 20–25 min (Fig. 4B). The first peaks could fuse and be difficult to separate in tissue samples, but 5-HT will elute within a reasonably short time (20 min). Further decreases in the retention time can be obtained by increasing the methanol concentration of the mobile phase to 15%, as well as by heating the column to 40–45°C.

2.5.4.2. Catecholamine Assays. There are several procedures to assay the CAs in brain tissue using HPLC methodology, and usually they involve a purification step of the brain extracts using alumina (Anton and Sayre, 1962, 1964) before the actual chromatography is performed (Fig. 4A). The sensitivity of the present electrochemical detectors is sufficiently high to enable the measurement of the CA in small samples of brain tissue (1–10 mg/wet weight). The disadvantage of relying on a separate CA assay is that two chromatograms have to be performed to obtain the 5-HT and CA levels, and since very often the conditions of the mobile phase are not the same, the two assays cannot be performed in succession unless two chromatographs are available.

The assay of only the CAs by HPLC involves a sample preparation, using an alumina clean-up step, based on the procedure developed by Anton and Sayre (1962). Briefly, 250 μL of the perchloric acid (0.1*N*) supernatant of brain tissue homogenates are pipeted into tubes already containing 10–20 mg of acid-washed alumina (A103; BAS, Bioanalytical Systems, West Lafayette, IN). An internal standard of isoproterenol HCl (5 ng) is added in 50 μL of $HClO_4$ to all samples as well as to the standards to calculate the recovery from the alumina clean-up procedure. The pH is changed by adding a buffer made out of Tris base, 1.5*M*, titrated to pH 8.6 with 5*N* HCl and containing 25 mg/mL of Na_2EDTA (ethylenediaminetetraacetic acid, disodium salt; Sigma Chemical Co., St. Louis, MO). The tubes are shaken several times (vortex mixer) and, after 15 min, once the alumina settles, the buffer is aspirated. The alumina *slurry* is washed twice with 1 mL of distilled water, and is then pipeted into a microfiltration unit (MF-1 microfilter holders and RC-58 filter membranes, 0.2 μm pore size, BAS). The microfiltration tubes are centrifuged (5 min at 2000 rpm), a new receiver tube (MF-2; BAS) is inserted into the microfiltration unit, and the CAs eluted with 500 μL of 0.2*N* $HClO_4$ (Fig. 4A). To the acid eluate, 500 μL of HPLC buffer are added prior to the injection into the column.

2.5.4.3. MONOAMINE ASSAYS. It is obviously advantageous to perform the determinations of 5-HT and CA in the same chromatographic run. In this way there are no losses caused by two different chromatographic conditions (i.e., the time involved in equilibrating the column to a different mobile phase). An additional advantage of an HPLC assay would be the ability to inject the MAs extract directly into the column, with few preparative steps. As mentioned above, there are several HPLC methods by which to determine content of biogenic amines, including all the CAs and 5-HT, as well as some of their major metabolites (for review, *see* Mefford, 1981). This constitutes an added bonus, since sometimes the levels of metabolites are a good indication of MA turnover.

At present we are using an HPLC system composed of a Bioanalytical Systems (BAS) PM30A double-piston pump, a AS-48 automatic sampler (Biorad; Richmond, CA) equipped with a model 7010 sample injector (Rheodyne Co.; Cotati, CA) and a column (250 × 4.6 mm) packed with RP18, 5 μm particle size (BAS). The electrochemical detection system is a model LC4B amperometic detector, a TL5 glassy carbon, and an RE-1 Ag/AgCl reference electrode (BAS). The working electrode is maintained at a potential of +850 mV relative to the reference electrode. Peaks caused by the oxidation of the compounds in the column eluates are recorded on a strip-chart recorder (Model 355, Linear Instruments Corp., Irvine, CA) and integrated with a Minigrator (Spectra-Physics: Santa Clara, CA). The mobile phase is made of 0.1*M* monochloroacetic acid adjusted with sodium hydroxide to pH 3.30–3.35, containing 700–800 mg/L of ethylenediaminetetraacetic acid (EDTA, disodium salt; Sigma Chemical Co., St. Louis, MO), 200–300 ml/L of octyl sodium sulfate (OSS; Kodak; Rochester, NY), and 10% (vol/vol) of HPLC-grade methanol, filtered through 0.22 μm pores (Types GS, Millipore Corporation; Bedford, MA) and degassed under vacuum prior to use. This mobile phase is pumped at a rate of 0.7–1.5 mL/min, and the temperature is maintained at 32–36°C (column temperature controller LC22A; BAS). By varying the flow, the amount of OSS and the temperature of the column (Kempf and Mandel, 1981; Kontur et al., 1984; Reader et al., 1986), separation of the compounds 4-hydroxy-3-methoxyphenylglycol (MHPG), NA, AD, 3,4-dihydroxyphenylacetic acid (DOPAC), 5-hydroxyindole-3-acetic acid (5-HIAA), DA, 4-hydroxy-3-

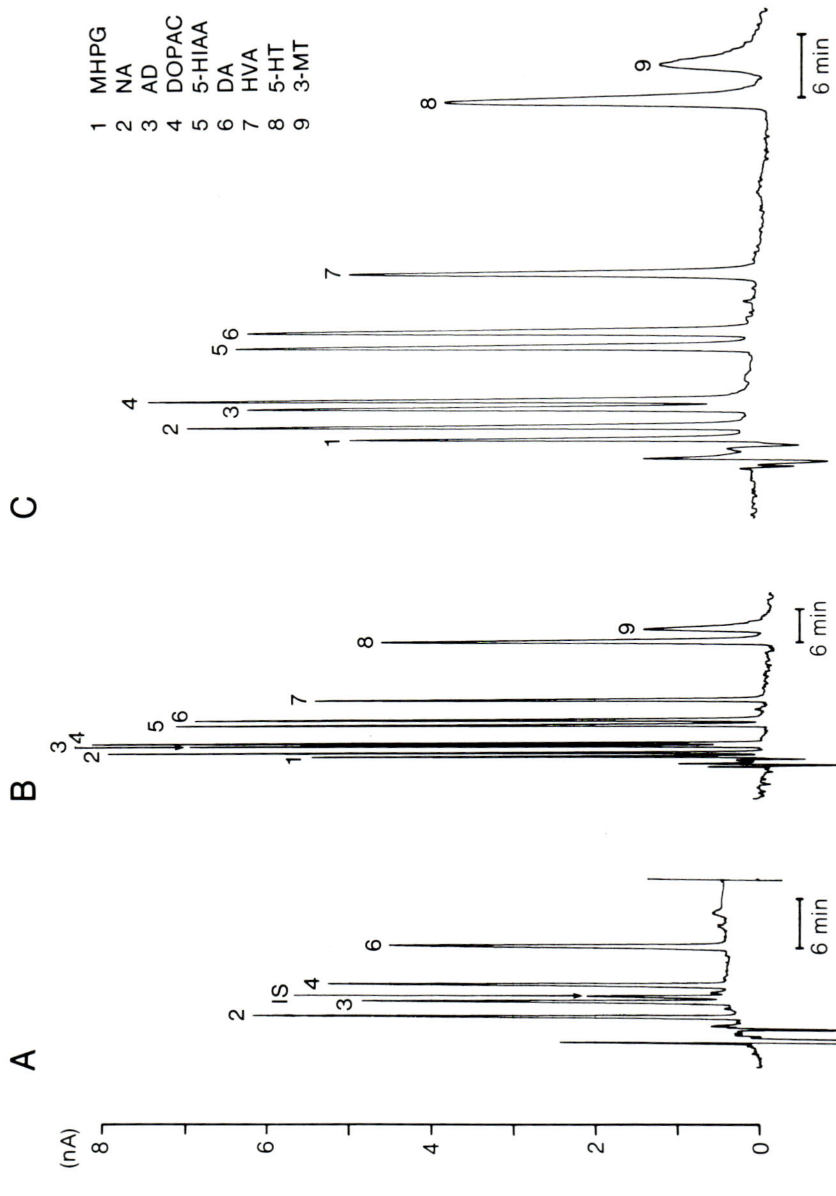
A
B
C
(nA)
8
6
4
2
0
IS
6 min
1 MHPG
2 NA
3 AD
4 DOPAC
5 5-HIAA
6 DA
7 HVA
8 5-HT
9 3-MT

Fig. 4. Chromatograms of external standards of monoamines, using electrochemical detection to determine the peaks of the eluate. The mobile phase was the same for the three chromatograms, i.e., a monoachloroacetic buffer, 0.1*M*, at pH 3.35 containing 800 mg/L of Na_2EDTA, 280 mg/L of sodium octyl phosphate, and 10% methanol (vol/vol). (A) After alumina clean-up, the HPLC allows the separation of the catecholamines noradrenaline (NA), adrenaline (AD), 3,4-dihydroxyphenylacetic acid (DOPAC), and dopamine (DA) (3.25 ng of each). IS is the internal standard (isoproterenol HCl; 2.5 ng) used to calculate recovery. Flow rate, 0.8 mL/min; column temperature, 34°C. (B) Chromatogram of external standards (3.25 ng each, in 200 µL of injection volume) of NA, AD, DOPAC, DA, 4-hydroxy-3-methoxyphenylglycol (MHPG), 4-hydroxy-3-methoxyphenylacetic acid (homovanillic acid, HVA), 5-hydroxyindole-3-acetic acid (5-HIAA), serotonin (5-HT), and 3-methoxytyramine (3-MT). Flow rate, 1.5 mL/min; column temperature, 36°C. (C) The same standard was chromatographed at a flow of 1.0 mL/min, and now the CAs, their metabolites, and 5-HT show a better separation. The procedure was found to be linear between 250 pg and 250 ng.

methoxyphenylacetic acid (homovanillic acid; HVA), 5-HT, and 3-methoxytyramine (3-MT) can be obtained within 30–40 min (Figs. 4B and C). For these compounds the detection limit is 10–25 pg, varying according to the noise levels of the electrochemical detector. The linearity is extremely good, i.e., standard curves made with amounts between 250 pg and 250 ng gave regression coefficients (least-square linear regression analysis) never below 0.998. Other compounds that can be resolved and separated using these chromatographic conditions are 5-hydroxytryptophan, metanephrine, and normetanephrine (Lakhdar-Ghazal et al., 1986).

3. Synthesis Inhibitors and Related Compounds

3.1. Use of *p*-Chlorophenylalanine (PCPA)

This compound was originally reported to selectively decrease 5-HT in brain (Koe and Weissman, 1966) by inhibiting tryptophan hydroxylase, the rate-limiting enzyme in 5-HT synthesis. In this pioneer study, it was shown that in the rat brain PCPA significantly lowers 5-HT and only slightly reduces NA content. However, the data on dog brain regions show marked reductions in both 5-HT and NA measured in the caudate, hypothalamus, and thalamus. Even a single dose of PCPA (400 mg/kg, ip) has a significant effect on rat brain NA, and the NA-depleting effects last for several days, according to the region examined (Miller et al., 1970). We have used slightly higher doses (500 mg/kg, ip) and for 2 d, in order to compare its depleting effects to those obtained by 5,7-DHT (Tables 5–8). The increase in the dose and the fact the effects of PCPA are long lasting clearly stand out in view of the important 5-HT depletions that can be obtained; usually reductions as great or even greater that those documented for 5,7-DHT deafferentations are achieved. The specificity of PCPA can, however, be questioned, since the contents of CAs are also affected. In fact, NA depletions were observed in several regions and were important in frontal cerebral cortex (Tables 5 and 7), hypothalamus (Table 7), and spinal cord (Table 6). In other regions, the NA levels were only slightly decreased and the depletions determined varied and were nonsignificant. Most regions showed reductions in DA levels, but the DA depletions were significant only in hippocampus and hypothalamus (Table 7). Similarly, there were reductions in the AD levels, especially in the hypothalamus (Table 7) and spinal cord (Table 6), but since this cryptic amine only represents a small percentage of total CA content and adrenergic systems (neurons using AD as their neurotransmitter) have not yet been fully described, this does not warrant further speculation. The lack of specificity of PCPA has been interpreted as a direct effect on tyrosine hydroxylase or as a result of interacting neuronal mechanisms (Ferron et al., 1982; Fuller, 1980; Reader and Jasper, 1984).

3.2. Mechanisms of Action and Effects of PCPA

The mechanism of action proposed for PCPA is the direct inhibition of tryptophan hydroxylase (Koe and Weissman, 1966; Jequier et al., 1967), although a blockade of uptake of the amino acid precursor could also contribute to the effect of decreasing 5-HT synthesis (Koe and Weissman, 1966). It has also been demonstrated, using histofluorescence and biochemical methods, that the 5-HT depletion produced by PCPA is of greater magnitude in the terminal fields of innervation than in the 5-HT neurons of the raphe complex (Aghajanian et al., 1973). In fact, the reduction in the raphe nucleus amounted to about 50%, whereas in the forebrain the 5-HT depletions were greater than 80%. Because of its mode of action, the 5-HT depletions induced by PCPA tend to be more generalized than those caused by the neurotoxic DHTs. For example, a 80–90% decrease in 5-HT content induced by PCPA and measured in the fields of innervation implies that the synthesis is probably equally blocked by 80–90% in all 5-HT terminals that are still present. In contrast, a 80–90% depletion of 5-HT caused by a DHT is the result of the loss by destruction of 80–90% of the terminals, and only 10–20% of them are present and may actually contain normal amounts of 5-HT. These considerations are important when dealing with hypersensitivity phenomena (see below) and regeneration of 5-HT systems after deafferentation with the DHTs.

3.3. Other Compounds

Besides PCPA, other compounds that have been used to reduce brain 5-HT levels are *p*-chloroamphetamine and *p*-chloro-*N*-methylamphetamine (Fuller et al., 1965, 1975; Lippmann and Wishnik, 1965; Miller et al., 1970; Ogren et al., 1976; Pletscher et al., 1963; Sanders-Bush et al., 1972), *N*-hydroxy-4-chloroamphetamine (Ross et al., 1974), and fenfluramine (Fuller, 1980; Trulson and

Jacobs, 1976). These agents have an initial reserpine-like effect, i.e., after releasing 5-HT, they further decrease the endogenous 5-HT pools, and this more sustained effect may involve the inhibition of tryptophan hydroxylase.

3.4. Assessment of the Depletions

Most of the studies using these 5-HT synthesis inhibitors and/or depleting agents rely on biochemical assays to determine the degree and extent of reduction in endogenous 5-HT levels. Usually only 5-HT is measured, but sometimes its main metabolite (5-HIAA) is also determined. The assay of the CAs is also extremely important when these drugs are used, since it indicates the degree of specificity of the 5-HT depletions. Very few studies have combined different methods of estimating the 5-HT reductions obtained by these agents, and the work on the differential effects of PCPA on neurons and terminals by Aghajanian and coworkers (Aghajanian et al., 1973) remains a very fine example of how such assessments should be done.

4. Neurobiology of Serotonin Deafferentation and Depletion

4.1. Neurotoxic Deafferentation *vs* Synthesis Inhibition

The neurotoxic DHTs have been employed as pharmacological tools to study properties of 5-HT receptors in the CNS. By suppressing the 5-HT afferents to a given region in the brain, the remaining functional capacity of MA projections can be thus analyzed in the absence of 5-HT nerve terminals. The use of an inhibitor of 5-HT synthesis, such as PCPA, can also provide useful information and help decide if changes in responsiveness of target cells are caused by a loss in 5-HT innervation (DHTs) or by a temporary blockade of 5-HT synthesis with conservation of the serotoninergic elements (PCPA). In numerous neuropharmacological, electrophysiological, and behavioral studies, both PCPA as well as the DHTs have been used, and in some cases their different effects compared (*see* Beleslin and Samardzic, 1979; Ferron et al., 1981, 1982; Fukata et al., 1984; Gauthier and Reader, 1982; Honma et al., 1979; Lyness et al., 1980; Quick et al., 1977; Reader, 1982; Reader and Gauthier, 1984; Steeves et al., 1980;

Sugrue et al., 1979; Tremblay et al., 1985; Trulson et al., 1976; Trulson and Jacobs, 1978; Vogt, 1974; Wang et al., 1979). An added impetus to this research with the neurotoxic DHTs came when it was documented that after such deafferentation lesions, there is a plastic capacity of the 5-HT as well as the CA systems to sprout and reinnervate the deafferented regions (Baumgarten et al., 1974; Bjorlund and Lindvall, 1979; Bjorklund and Wiklund, 1980; Bjorklund et al., 1981; Nygren et al., 1974; Stenevi et al., 1973; Wiklund et al., 1978).

4.2. Functional Changes to Serotoninergic Deafferentation

After deafferenting the CNS with the neurotoxic DHTs, one of the first phenomena that appears is an increased responsiveness to 5-HT. This denervation supersensitivity (Cannon, 1939) has been documented for the 5-HT systems in the amygdala (Wang et al., 1979), facial nucleus (McCall and Aghajanian, 1979), and cerebral cortex (Ferron et al., 1981, 1982). It does not develop in the 5,7-DHT-deafferented hippocampus (De Montigny et al., 1980) because of regional differences in either the pattern of 5-HT innervation or the mode of action of 5-HT in this particular region. In the neocortex, the denervation supersensitivity to 5-HT is caused by the absence of 5-HT afferents, rather than that of 5-HT itself, since no supersensitivity could be detected following equally severe depletions of endogenous 5-HT stores with PCPA (Ferron et al., 1981, 1982; Reader and Jasper, 1984). These results are consistent with the lack of 5-HT supersensitivity observed after PCPA in behavioral experiments (Trulson et al., 1976) as well as in spinal cord responses (Barbeau and Bedard, 1981a,b; Tremblay et al., 1985). Alternatively, it could be envisaged that following a 5-HT deafferentation with the neurotoxic DHTs, the important lowering of 5-HT represents a total loss of the majority of 5-HT nerve endings at a certain target site, whereas after PCPA no target cells were completely deprived of a certain degree of 5-HT input. The possibility cannot be excluded that after PCPA there still is a residual minimal release of 5-HT, with or without an increased turnover, and that this suffices to maintain a normal sensitivity of the target cells.

4.3. Interactions with Other Neurotransmitter Systems

Specific functional interactions between the CAs and 5-HT have not yet been clearly demonstrated, except for the reciprocal effects observed after deafferenting the 5-HT systems, i.e., enhancement of responses to 5-HT after 5,7-DHT (pre- as well as postsynaptic types of supersensitivity), with a concomitant reduction in the responses to the CAs (Ferron et al., 1982; Reader and Jasper, 1984). Temporary increases in the activity of tyrosine hydroxylase have been measured following 5-HT deafferentations of the locus ceruleus with 5,6-DHT (Lewis et al., 1976; Pujol et al., 1978; McRae-Degueurce et al., 1982; Renaud et al., 1975) or after depletions of its 5-HT content with PCPA (Crespi et al., 1980). It has also been documented that descending spinal 5-HT pathways can regulate adrenomedullary tyrosine hydroxylase (Gauthier et al., 1979; Quick et al., 1977).

4.4. Biochemical Aspects

Whether triggered by the absence of 5-HT fibers or by that of 5-HT itself, two mechanisms could account for the denervation supersensitivity: a suppression of 5-HT uptake into the nerve terminals (a *presynaptic* component) and a modification of 5-HT receptors on the target (a *postsynaptic* component). It has been shown that these two mechanisms may account for the increased sensitivity to 5-HT of postsynaptic neurons after 5,7-DHT deafferentation (De Montigny et al., 1980; Ferron et al., 1982; Reader and Jasper, 1984). The presynaptic component can, however, be obtained in nondeafferented animals by blocking the uptake of 5-HT with fluoxetine (Ferron et al., 1982; Wong et al., 1974). It remains to be determined whether the postsynaptic component of denervation supersensitivity to 5-HT is the consequence of an up-regulation (increased number) of 5-HT receptors and/or modifications in their affinities and transducing mechanisms. It has also been shown that 14 d after lesioning the ascending 5-HT pathways with 5,7-HDT there is an increase in the cerebral cortex of the density as well as the affinity of LSD binding sites (Fuxe et al., 1978; *see also* Morrow et al., 1985; Nelson et al., 1978).

Acknowledgments

The personal studies referred to in this chapter were supported by grants from the Medical Research Council of Canada (MT-6967), the University of Montreal, and the Quebec Heart Foundation. The author belongs to the Equipe de Recherche sur les

Neurotransmetteurs du Cerveau, Fonds de la recherche en sante du Quebec, from which he is currently a Senior Scholar (Chercheur-boursier Senior). The expert assistance of Dr. Nico M. van Gelder with the HPLC system is acknowledged. The excellent technical aid of Miss L. Grondin is greatly appreciated, as well as the help of Miss Helene Dussault in typing the manuscript.

References

Adams R. N. (1979) Electrochemical Methodology, in *The Neurobiology of Dopamine* (Horn A. L., Korf J., and Westerink B. H. C., eds.) Academic, London.

Aghajanian G. K., Kuhar M. J., and Roth R. H. (1973) Serotonin-containing neuronal perikarya and terminals: Differential effects of *p*-chlorophenylalanine. *Brain Res.* **54,** 85–101.

Ahtee L., Sharman D. F., and Vogt M. (1970) Acid metabolism of monoamines in avian brain: Effects of probenecid and reserpine. *Br. J. Pharmacol.* **38,** 72–85.

Amin A. H., Crawford T. B. B., and Gaddum J. H. (1954) The distribution of substance P and 5-hydroxytryptamine in the central nervous system of the dog. *J. Physiol.* (Lond.) **126,** 596–618.

Anden N.-E., Dahlstrom A., Fuxe K., and Larsson K. (1965) Mapping out of catecholamine and 5-hydroxytryptamine neurons innervating the telencephalon and diencephalon. *Life Sci.* **4,** 1275–1279.

Anden N.-E., Dahlstrom A., Fuxe K., Larsson K., Olson L., and Ungerstedt U. (1966) Ascending monoamine neurons to the telencephalon and diencephalon. *Acta Physiol. Scand.* **67,** 313–326.

Anden N.-E., Fuxe K., and Ungerstedt U. (1967) Monoamine pathways to cerebellum and cerebral cortex. *Experientia* **23,** 838–842.

Anton A. H. and Sayre D. F. (1962) A study of the factors affecting the aluminum oxide-trihydroxyindole procedure for the analysis of catecholamines. *J. Pharmacol. Exp. Ther.* **138,** 360–375.

Anton A. H. and Sayre D. F. (1964) The distribution of dopamine and dopa in various animals and a method for their determination in diverse biological material. *J. Pharmacol. Exp. Ther.* **45,** 326–336.

Axelrod J. and Tomchick R. (1958) Enzymatic O-methylation of epinephrine and other catechols. *J. Biol. Chem.* **233,** 702–705.

Axelrod J. and Weissbach H. (1961) Purification and properties of hydroxyindole-O-methyltransferase. *J. Biol. Chem.* **236,** 211–213.

Azmitia E. C. (1978) The Serotonin-Producing Neurons of the Midbrain and Dorsal Raphe Nuclei, in *Handbook of Psychopharmacology, Chemical*

Pathways in the Brain (Iversen L. L., Iversen S. D., and Snyder S. H., eds.) vol. 9, Plenum, New York.

Azmitia E. F. and Segal M. (1978) An autoradiographic analysis of the differential projections of the dorsal and median raphe nuclei in the rat. *J. Comp. Neurol.* **179,** 641–668.

Baez L. A., Browning R. A., and Cusatis M. (1980) Evaluation of body weight changes after selective serotonin depletion with 5,7-dihydroxytryptamine. *Prog. Neuro-Psychopharmacol.* **4,** 123–127.

Barbeau H. and Bedard P. (1981a) Similar motor effects of 5-HT and TRH in rats following chronic spinal transection and 5,7-dihydroxytryptamine injection. *Neuropharmacology* **20,** 477–481.

Barbeau H. and Bedard P. (1981b) Denervation supersensitivity to 5-hydroxytryptophan in rats following spinal transection and 5,7-dihydroxytryptamine injection. *Neuropharmacology* **20,** 611–616.

Baumgarten H. G. and Bjorklund A. (1976) Neurotoxic indoleamines and monoamine neurons. *Ann. Rev. Pharmacol. Toxicol.* **16,** 101–111.

Baumgarten H. G. and Lachenmayer L. (1972) 5,7-Dihydroxytryptamine: Improvements in chemical lesioning of indoleamine neurons in the mammalian brain. *Z. Zellfrosch.* **135,** 399–414.

Baumgarten H. G., Bjorklund A., Lachenmayer L., Nobin A., and Stenevi U. (1971) Long-lasting selective depletion of brain serotonin by 5,6-dihydroxytryptamine. *Acta Physiol. Scand. 373* (suppl.) 1–16.

Baumgarten H. G., Bjorklund A., Lachenmmayer L., Rensch A., and Rosengren E. (1974) De- and regeneration of the bulbospinal serotonin neurons in the rat following 5,6- or 5,7-dihydroxytryptamine treatment. *Cell Tiss. Res* **152,** 271–281.

Baumgarten H. G., Bjorklund A., Lachenmayer L., and Nobin A. (1973a) Evaluation of the effects of 5,7-dihydroxytryptamine on serotonin and catecholamine neurons in the rat CNS. *Acta Physiol. Scand.* 391 (suppl.), 1–19.

Baumgarten H. G., Victor S. J., and Lovenberg W. (1973b) Effect of intraventricular injection of 5,7-dihydroxytryptamine on regional tryptophan hydroxylase of rat brain. *J. Neurochem.* **21,** 251–253,

Baumgarten H. G., Bjorklund A., Nobin A., Rosengren E., and Schlossberger H. G. (1975a) Neurotoxicity of hydroxylated tryptamines: Structure–activity relationships. I. Long-term effects on monoamine content and fluorescence morphology of central monoamine neurons. *Acta Physiol. Scand.* 429 (suppl.), 1–27.

Baumgarten H. G., Victor S. J., and Lovenberg W. (1975b) A developmental study of the effect of 5,7-dihydroxytryptamine on regional tryptophan hydroxylase in rat brain. *Psychopharmacol. Commun.* **1,** 75–88.

Baumgarten H. G., Jenner S., Bjorklund A., Klemm H. P., and Schlossberger H. G. (1982a) Serotonin Neurotoxins, in *Biology of*

Serotonergic Transmission (Osborne N. N., ed.) John Wiley, Chichester-New York.

Baumgarten H. G., Klemm H. P., Sievers J., and Schlossberger H. G. (1982b) Dihydroxytryptamines as tools to study the neurobiology of serotonin. *Brain Res. Bull.* **9,** 131–150.

Baumgarten H. G., Evetts K. D., Holman R. B., Iversen L. L., Vogt M., and Wilson G. (1972) Effects of 5,6-dihydroxytryptamine on monoaminergic neurones in the central nervous system of the rat. *J. Neurochem.* **9,** 1587–1597.

Baumgarten H. G., Jenner S., and Klemm H. P. (1981) Serotonin neurotoxins: Recent advances in the mode of administration and molecular mechanism of action. *J. Physiol.* (Paris) **77,** 309–314.

Baumgarten H. G., Klemm H. P., Lachenmayer L., Bjorklund A., Lovenberg W., and Schlossberger H. G. (1978) Mode and mechanism of action of neurotoxic indoleamines: A review and progress report. *Ann. NY Acad. Sci.* **305,** 3–24.

Beart P. M., Prosser D., and Louis W. J. (1979) Adrenaline and phenylethanolamine-*N*-methyltransferase in rat medullary and anterior hypothalamic nuclei. *J. Neurochem.* **33,** 947–950.

Beaudet A. and Descarries L. (1981) The fine structure of central serotonin neurons. *J. Physiol.* (Paris) **77,** 193–203.

Beleslin D. B. and Samardzic R. (1979) Effects of *para* chlorophenylalanine and 5,6-dihydroxytryptamine on aggressive behavior evoked by cholinominetics and anticholinesterases injected into the cerebral ventricles of conscious cats. *Neuropharmacology* **18,** 251–257.

Bertler A. (1961) Effect of reserpine on the storage of catecholamines in brain and other tissues. *Acta Physiol. Scand.* **51,** 75–83.

Bertler A., Carlsson A., Rosengren E., and Waldeck A. (1958) A method for the fluorimetric determinations of adrenaline, noradrenaline and dopamine in tissues. *Kungl, Fysiogr. Sallsk. Lund Forh.* **28,** 121–123.

Bjorklund A. and Lindvall O. (1979) Regeneration of normal terminal innervation patterns by central noradrenergic neurons after 5,7-dihydroxytryptamine-induced axotomy in the adult rat. *Brain Res.* **171,** 271–293.

Bjorklund A. and Skagerberg G. (1979) Simultaneous use of retrograde fluorescent tracers and fluorescence histochemistry for convenient and precise mapping of monoaminergic projections and collateral arrangements in the CNS. *J. Neurosci. Meth.* **1,** 261–277.

Bjorklund A. and Wiklund L. (1980) Mechanisms of regrowth of the bulbospinal serotonin system following 5,6-dihydroxytryptamine induced axotomy. I. Biochemical correlates. *Brain Res.* **191,** 109–127.

Bjorklund A., Wiklund L., and Descarries L. (1981) Regeneration and plasticity of central serotoninergic neurons: A review. *J. Physiol.* (Paris) **77,** 247–255.

Bjorklund A., Nobin A., and Stenevi U. (1973a) The use of neurotoxic dihydroxytryptamines as tools for morphological studies and localized lesioning of central indoleamine neurons. *Z. Zellforsch.* **145,** 479–501.

Bjorklund A., Nobin A., and Stenevi U. (1973b) Effects of 5,6-dihydroxytryptamine on nerve terminal serotonin and serotonin uptake in the rat brain. *Brain Res.* **53,** 117–127.

Bjorklund A., Baumgarten H. G., Lachenmayer L., and Rosengren E. (1975a) Recovery of brain noradrenaline after 5,7-dihydroxytryptamine-induced axonal lesions in the rat. *Cell Tiss. Res.* **161,** 145–155.

Bjorklund A., Baumgarten H. G., Horn A. S., Nobin A., and Schlossberger H. G. (1975b) Neurotoxicity of hydroxylated tryptamines: Structure–activity relationships. 2. In vitro studies on monoamine uptake inhibition and uptake impairment. *Acta Physiol. Scand.* **429** (suppl.), 31–61.

Bjorklund A., Baumgarten H. G., and Rensch A. (1975c) 5,7-dihydroxytryptamine: Improvement of its selectivity for serotonin neurons in the CNS by pretreatment with desipramine. *J. Neurochem.* **24,** 833–835.

Bjorklund A., Falck B., Lindvall O., and Loren I. (1980) The aluminium-formaldehyde (ALFA) histofluorescence method for improved visualization of catecholamines and indoleamines. 2. Model experiments. *J. Neurosci. Meth.* **2,** 301–318.

Bjorklund A., Falck B., and Owman Ch. (1972) Fluorescence Microscopic and Microspectrofluorometric Techniques for the Cellular Localization of Biogenic Amines, in *Methods in Investigative and Diagnostic Endocrinology* vol. I *The Thyroid and Biogenic Amines* (Rall J. E. and Kopin I. J. eds.) North-Holland, Amsterdam.

Blackburn T. P., Foster G. A., Heapy C. G., and Kemp J. D. (1980) Unilateral 5,7-dihydroxytryptamine lesions of the dorsal raphe nucleus (DRN) and rat rotational behavior. *Eur. J. Pharmacol.* **67,** 427–438.

Bogdanski D. F., Pletscher A., Brodie B. B., and Udenfriend S. (1956) Identification and assay of serotonin in brain. *J. Pharmacol. Exp. Ther.* **117,** 82–88.

Boireau A., Ternaux J. P., Bourgoin S., Hery F., Glowinski J., and Hamon M. (1976). The determination of picogram levels of 5-HT in biological fluids. *J. Neurochem.* **26,** 201–204.

Breese G. R. and Cooper B. R. (1975) Behavioral and biochemical interactions of 5,7-dihydroxytryptamine with various drugs when administered intracisternally to adult and developing rats. *Brain Res.* **98,** 517–527.

Breese G. R. and Mueller R. (1978) Alterations in the neurotoxicity of 5,7-dihydroxytryptamine by pharmacologic agents in adult and developing rats. *Ann. NY Acad. Sci.* **305,** 160–174.

Cannon W. B. (1939) A law of denervation. *Am. J. Med. Sci.* **198,** 737–750.

Carlsson A., Davis J. N., Kehr W., Lindvist M., and Atack C. V. (1972) Simultaneous measurements of tyrosine hydroxylase and tryptophan hydroxylase activities in brain in vivo using an inhibitor of the aromatic amino acid decarboxylase. *Naunyn Schmiedebergs Arch. Pharmacol.* **272,** 153–168.

Carlsson A., Falck B., and Hillarp N.-A. (1962) Cellular localization of brain monoamines. *Acta Physiol. Scand.* **56,** (suppl. 196), 1–28.

Chang C. C. (1964) A sensitive method for the spectrofluorometric assay of catecholamines. *Int. J. Neuropharmacol.* **3,** 643–649.

Chan-Palay V., Jonsson G., and Palay S. L. (1978) Serotonin and substance P coexist in neurons of the rat's central nervous system. *Proc. Natl. Acad. Sci. USA* **75,** 1582–1586.

Costa E., Daly J., Lefevre H., Meck J., Revuelta A., Sparro F., Strada S., and Daly J. (1972) Serotonin and catecholamine concentrations in brains of rats injected intracerebrally with 5,6-dihydroxytryptamine. *Brain Res.* **44,** 304–308.

Coyle J. T. and Henry D. (1973) Catecholamines in fetal and newborn rat. *J. Neurochem.* **21,** 61–67.

Coyle J. T. and Snyder S. H. (1969) Catecholamine uptake by synaptosomes in homogenates of rat brain: Stereospecificity in different areas. *J. Pharmacol. Exp. Ther.* **170,** 221–231.

Crespi F., Buda M., McRae-Degueurce A., and Pujol J.-F. (1980) Alteration of tyrosine hydroxylase activity in the locus coeruleus after administration of p-chlorophenylalanine. *Brain Res.* **191,** 501–509.

Creveling C. R., Lundstrom J., McNeal E. T., Tice L., and Daly J. W. (1975) Dihydroxytryptamines: Effects on noradrenergic function in mouse heart in vivo. *Mol. Pharmacol.* **11,** 211–222.

Cuello A. C. (1979) The Radioenzymatic Assay of Catecholamines, in *The Neurobiology of Dopamine* (Horn A. S., Korf J., and Westerink B. H. C., eds.) Academic, London.

Cuello A. C., Hiley R., and Iversen L. L. (1973) Use of catechol-O-methyltransferase for the enzyme radiochemical assay of dopamine. *J. Neurochem.* **21,** 1337–1340.

Curzon G. and Green A. R. (1970) Rapid method for determination of 5-hydroxytryptamine and 5-hydroxyindoleacetic acid in small regions of rat brain. *Br. J. Pharmacol.* **39,** 653–655.

Dahlstrom A. and Fuxe K. (1964) Evidence for the existence of monoamine containing neurons in the central nervous system. I. Demonstration of monoamines in the cell bodies of brainstem neurons. *Acta Physiol. Scand.* **62** (suppl. 232), 1–55.

Daly J., Fuxe K., and Jonsson G. (1973) Effects of intracerebral injection of 5,6-dihydroxytryptamine on cerebral monoamine neurons: Evidence for selective degeneration of central 5-hydroxytryptamine neurons. *Brain Res.* **49,** 476–482.

Daly J., Fuxe K., and Jonsson G. (1974) 5,7-Dihydroxytryptamine as a tool for the morphological and functional analysis of central 5-hydroxytryptamine neurons. *Res. Commun. Chem. Path. Pharmacol.* **7,** 175–187.

De Groot J. (1959) The rat forebrain in stereotaxic coordinates. *Verk. K. Ned. Akad. Wet., B. Natuwrkd.* **2,** 1–40.

De Montigny C., Wang R. Y., Reader T. A., and Aghajanian G. K. (1980) Monoaminergic denervation of the rat hippocampus: Microiontophoretic studies on pre- and post-synaptic supersensitivity to norepinephrine and serotonin. *Brain Res.* **200,** 363–376.

De Robertis E. D. P. (1964). *Histophysiology of Synapses and Neurosecretion* Pergamon, New York.

De Robertis E., Pellegrino de Iraldi A., Rodriguez de Lores Arnaiz G., and Zieher L. M. (1965) Synaptic vesicles from the rat hypothalamus. Isolation and norepinephrine content. *Life Sci.* **4,** 193–201.

Donelson A. C., Manian A. A., and Maickel R. P. (1982) Effects of compounds structurally related to 5,6-dihydroxytryptamine on behavior and brain biogenic amines in the rat. *Res. Commun. Chem. Pathol. Pharmacol.* **38,** 517–520.

Ehrenstrom F. and Johansson P. (1985) A method for very rapid determinations of catechols using ion-pairing reverse phase HPLC with electrochemical detection: Effects of L-DOPA treatment on the catechol content in various rat brain structures. *Life Sci.* **36,** 867–879.

Falck B., Bjorklund A., and Lindvall O. (1982) Recent progress in aldehyde fluorescence histochemistry. *Brain Res. Bull.* **9,** 3–10.

Falck B., Hillarp H.-A., Thieme G., and Thorp A. (1962) Fluorescence of catechol amines and related compounds condensed with formaldehyde. *J. Histochem. Cytochem.* **10,** 348–354.

Felice L. J., Felice J. D., and Kissinger P. T. (1978) Determination of catecholamines in rat brain parts by reverse-phase ion-pair liquid chromatography. *J. Neurochem.* **31,** 1461–1465.

Ferron A., Descarries L., and Reader T. A. (1982) Altered neuronal responsiveness to biogenic amines in rat cerebral cortex after serotonin denervation or depletion. *Brain Res.* **231,** 93–108.

Ferron A., Reader T. A., and Descarries L. (1981) Responsiveness of cortical neurons to serotonin after 5,7-DHT denervation or PCPA depletion. *J. Physiol.* (Paris) **77,** 381–384.

Forchetti C. M., Marco E. J., and Meek J. L. (1982) Serotonin and γ-aminobutyric acid turnover after injection into the median raphe of substance P and D-Ala-Met-enkephalin amide. *J. Neurochem.* **38,** 1336–1341.

Frankfurt M., Renner K., Azmitia E., and Luine V. (1985) Intrahypothalamic 5,7-dihydroxytryptamine: Temporal analysis of effects on 5-hydroxytryptamine content in brain nuclei and on facilitated lordosis behavior. *Brain Res.* **340,** 127–133.

Fukata J., Nakai Y., Imura H., and Takeuchi J. (1984) Effect of serotonin depleters on the contents of β-endorphin, α-melanotropin and adrenocorticotrophin in rat brain and pituitary. *Brain Res.* **324,** 289–293.

Fuller R. W. (1980) Pharmacology of central serotonin neurons. *Ann. Rev. Pharmacol. Toxicol.* **20,** 111–127.

Fuller R. W., Hines C. W., and Mills J. (1965) Lowering of brain serotonin by chloroamphetamines. *Biochem. Pharmacol.* **14,** 483–488.

Fuller R. W., Perry K. W., and Molloy B. B. (1975) Effect of 3-(*p*-trifluoromethylphenoxy)-N-methyl-3-phenylpropylamine on the depletion of brain serotonin by 4-chloroamphetamine. *J. Pharmacol. Exp. Ther.* **193,** 796–803.

Fuxe K. (1965) Evidence for the existence of monoamine neurons in the central nervous system. IV. Distribution of monoamine nerve terminals in the central nervous system. *Acta Physiol. Scand.* **64** (suppl. 247), 37–85.

Fuxe K. and Jonsson G. (1967) A modification of the histochemical fluorescence method for the improved localization of 5-hydroxytryptamine. *Histochemie* **11,** 161–166.

Fuxe K. and Jonsson G. (1974) Further mapping of central 5-hydroxytryptamine neurons. Studies with the neurotoxic dihydroxytryptamines. *Adv. Biochem. Psychopharmacol.* **10,** 1–12.

Fuxe K., Ogren S.-O., Agnati L. F., Jonsson G., and Gustafsson J.-A. (1978) 5,7-Dihydroxytryptamine as a tool to study the functional role of central 5-hydroxytryptamine neurons. *Ann. NY Acad. Sci.* **305,** 346–369.

Gauthier P. and Reader T. A. (1982) Adrenomedullary secretory response to midbrain stimulation in rat: Effects of depletion of brain catecholamines or serotonin. *Can. J. Physiol. Pharmacol.* **60,** 1464–1474.

Gauthier S., Gagner J.-P., and Sourkes T. L. (1979) Role of descending spinal pathways in the regulation of adrenal tyrosine hydroxylase. *Exp. Neurol.* **66,** 42–54.

Gershanik O. S., Heikkila R. E., and Duvoisin R. C. (1979) Asymmetric action of intraventricular monoamine neurotoxins. *Brain Res.* **174,** 345–350.

Gerson S. and Baldessarini R. J. (1975) Selective destruction of serotonin terminals in rat forebrain by high doses of 5,7-dihydroxytryptamine. *Brain Res.* **85,** 140–145.

Gilbert R. F. T., Emsom P. C., Hunt S. P., Bennett G. W., Marsden C. A., Sandberg B. E. B., Steinbusch H. W. M., and Verhofstad A. A. J. (1982) The effects of monoamine neurotoxins on peptides in rat spinal cord. *Neuroscience* **7,** 69–87.

Haggendal J. (1963) An improved method for fluorimetric determination of small amounts of adrenaline and noradrenaline in plasma and tissues. *Acta Physiol. Scand.* **59,** 242–254.

Hallman H., Farnebo L.-O., Hamberger B., and Jonsson G. (1978) A sensitive method for the determination of plasma catecholamines using liquid chromatography with electrochemical detection. *Life Sci.* **23,** 1049–1052.

Hamberger B. (1967). Reserpine-resistant uptake of catecholamines in isolated tissues of the rat. *Acta Physiol. Scand.* **295** (suppl.), 1–47.

Hammel I., Naot Y., Ben-David E., and Ginsburg H. (1978) A simplified microassay for serotonin: Modification of the enzymatic isotopic assay. *Anal. Biochem.* **90,** 840–843.

Hokfelt T., Fuxe K., Goldstein M., and Johansson O. (1973) Evidence for adrenaline neurons in the rat brain. *Acta Physiol. Scand.* **89,** 286–288.

Hokfelt T., Fuxe K., Johansson O., and Ljungdhal A. (1974) Pharmacohistochemical evidence for the existence of dopamine nerve terminals in the limbic cortex. *Eur. J. Pharmacol.* **25,** 108–112.

Hokfelt T., Ljungdhal A., Steinbusch H. W. M., Verhofstad A., Nilsson G., Brodin E., Pernow B., and Goldstein M. (1978) Immunohistochemical evidence of substance P-like immunoreactivity in some 5-hydroxytryptamine-containing neurons in the rat central nervous system. *Neuroscience* **3,** 517–538.

Holman R. B., Angwin P., and Barchas J. D. (1976) Simultaneous determination of indole- and catecholamines in small brain regions in the rat using a weak cation exchange resin. *Neuroscience* **1,** 147–150.

Honma K.-I., Watanabe K., and Hiroshige T. (1979) Effects of parachlorophenylalanine and 5,6-dihydroxytryptamine on the free-running rhythms of locomotor activity and plasma corticosterone in the rat exposed to continuous light. *Brain Res.* **169,** 531–544.

Horn A. S., Baumgarten H., and Schlossberger H. G. (1973) Inhibition of the uptake of 5-hydroxytryptamine, noradrenaline and dopamine into rat homogenates by various hydroxylated tryptamines. *J. Neurochem.* **21,** 233–236.

Jacobowitz D. M. and Richardson J. S. (1978) Method for the rapid determination of norepinephrine, dopamine and serotonin in the same brain region. *Pharmacol. Biochem. Behav.* **8,** 515–519.

Jacobs B. L., Wise W. D., and Taylor K. M. (1979) Differential behavioral and neurochemical effects following lesions of the dorsal or median raphe nuclei in rats. *Brain Res.* **79,** 353–361.

Jequier E., Lovenberg W., and Sjoerdsma A. (1967) Tryptophan hydroxylase inhibition: The mechanism by which p-chlorophenylalanine depletes rat brain serotonin. *Mol. Pharmacol.* **3,** 274–278.

Juvancz P. (1980) The effect of raphe lesion on sleep in the rat. *Brain Res.* **194,** 371–376.

Kehr W., Lindqvist M., and Carlsson A. (1976) Distribution of dopamine in the rat cerebral cortex. *J. Neural Trans.* **38,** 173–180.

Keller R., Oke A., Mefford I., and Adams R. N. (1976) Liquid chromatographic analysis of catecholamines. Routine assay for regional brain mapping. *Life Sci.* **19,** 995–1004.

Kempf E. and Mandel P. (1981) Reverse-phase high-performance liquid chromatographic separation and electrochemical detection of norepinephrine, dopamine, serotonin, and related major metabolites. *Anal. Biochem.* **112,** 223–231.

Klemm H. P., Baumgarten H. G., and Schlossberger H. G. (1978) Interaction of 5,6-and 5,7-dihydroxytryptamine with tissue monoamine oxidase. *Ann. NY Acad. Sci.* **305,** 36–56.

Klemm H. P., Baumgarten H. G., and Schlossberger H. G. (1979) In vitro studies on the interaction of brain monoamine oxydase with 5,6- and 5,7-dihydroxytryptamine. *J. Neurochem.* **32,** 111–119.

Klemm H. P., Baumgarten H. G., and Schlossberger H. G. (1980) Polarographic measurements of spontaneous and mitochondria-promoted oxydation of 5,6- and 5,7-dihydroxytryptamine. *J. Neurochem.* **35,** 1400–1408.

Koe B. K. and Weissman A. (1966) p-Chlorophenylalanine: A specific depletor of brain serotonin. *J. Pharmacol. Exp. Ther.* **154,** 499–516.

Konig J. F. R. and Klippel R. A. (1963) *The Rat Brain, a Stereotaxic Atlas of the Forebrain and Lower Parts of the Brainstem* Williams and Wilkins, Baltimore, Maryland.

Kontur P., Dawson R., and Monjan A. (1984) Manipulation of mobile phase parameters for the HPLC separation of endogenous monoamines in rat brain tissue. *J. Neurosci. Meth.* **11,** 5–18.

Koslow S. H. and Schlumpf M. (1974) Quantitation of adrenaline in rat brain nuclei by mass fragmetography. *Nature* **252,** 530–531.

Kostrzewa R. M. and Jacobowitz D. M. (1974) Pharmacological actions of 6-hydroxydopamine. *Pharmacol. Rev.* **26,** 199–288.

Kuhar M. J., Aghajanian G. K., and Roth R. H. (1972) Tryptophan hydroxylase activity and synaptosomal uptake of serotonin in discrete brain regions after midbrain raphe lesions: Correlations with serotonin levels and histochemical fluorescence. *Brain Res.* **44,** 165–176.

Lackovic Z., Parenti M., and Neff N. H. (1981) Simultaneous determination of femtomole quantities of 5-hydroxytryptophan, serotonin and 5-hydroxyindoleacetic acid in brain using HPLC with electrochemical detection. *Eur. J. Pharmacol.* **69,** 347–352.

Lakhdar-Ghazal N., Grondin L., Bengelloun W. A., and Reader T. A. (1986) Alpha-adrenoceptors and monoamine contents in the cerebral cortex of the rodent *Jaculus orientalis:* Effects of acute cold exposure. *Pharmacol. Biochem. Behav.* **25,** 903–911.

Laverty R. and Taylor K. M. (1968) The fluorometric assay of catecholamines and related compounds. Improvements and extension to the hydroxyindole technique. *Anal. Biochem.* **22,** 269–279.

Lewis B. D., Renaud B., Buda M., and Pujol J.-F. (1976) Time course variations in tyrosine hydroxylase activity in the rat locus coeruleus after electrolytic destruction of the nuclei raphe dorsalis or raphe centralis. *Brain Res.* **108,** 339–349.

Lidbrink P. and Jonsson G. (1974) Noradrenaline nerve terminals in the cerebral cortex: Effects on noradrenaline uptake and storage following axonal lesion with 6-hydroxydopamine. *J. Neurochem.* **22,** 617–626.

Lindvall O. and Bjorklund A. (1974) The organization of the ascending catecholamine neuron systems in the rat brain as revealed by the glyoxylic acid fluorescence method. *Acta Physiol. Scand.* **412,** (Suppl.), 1–48.

Lindvall O. and Bjorklund A. (1983) Dopamine and Norepinephrine-Containing Neuron Systems: Their Anatomy in the Rat Brain, in *Chemical Neuroanatomy* (Emson P. C., ed.) Raven, New York.

Lindvall O. and Bjorklund A. (1984) General Organization of Cortical Monoamine Systems, in *Monoamine Innervation of Cerebral Cortex* (Descarries L., Reader T. A., and Jasper H. H., eds.) Alan R. Liss, New York.

Lindvall O., Bjorklund A., Moore R. Y., and Stenevi U. (1974) Mesencephalic dopamine neurons projecting to neocortex. *Brain Res.* **81,** 325–331.

Lippmann W. and Wishnik M. (1965) Effect of dl-p-chloro-N-methylamphetamine on the concentrations of monoaminers in the cat and rat brain and rat heart. *Life Sci.* **4,** 849–857.

Liston D. R., Franz D. N., and Gibb J. W. (1982) Biochemical evidence for alteration of neostriatal dopaminergic functions by 5,7-dihy-droxytryptamine. *J. Neurochem.* **38,** 1329–1335.

Loren W. H., Bjorklund A., Falck B., and Lindvall O. (1976) An improved histofluorescence procedure for freeze-dried paraffin-embedded tissue based on combined formaldehyde-glyoxylic acid perfusion with high magnesium content and acid pH. *Histochemistry* **49,** 177–192.

Loren I., Bjorklund A., Falck B., and Lindvall O. (1980) The aluminium-formaldehyde (ALFA) histofluorescence method for improved visualization of catecholamines and indoleamines. 1. A detailed account of the methodology for central nervous tissue using paraffin, cryostat or vibratome sections. *J. Neurosci. Meth.* **2,** 277–300.

Loren I., Bjorklund A., Lindvall O., and Schmidt R. H. (1982) Improved catecholamine histofluorescence in the developing brain based on the magnesium and aluminium (ALFA) perfusion techniques: Methodology and anatomical observations. *Brain Res. Bull.* **9,** 11–26.

Lorens S. A., Guldberg H. C., Hole K., Kohler C., and Srebro B. (1976) Activity, avoidance learning and regional 5-hydroxytryptamine following intra-brain stem 5,7-dihydroxytryptamine and electrolytic midbrain raphe lesions in the rat. *Brain Res.* **108,** 97–113.

Lorez H. P., Pieri L., and Richards J. G. (1975) Dissappearance of supra-ependymal 5-HT axons in the rat forebrain after electrolytic and 5,6-DHT-induced lesions of the medial forebrain bundle. *Brain Res.* **100,** 1–12.

Lowry O. H., Rosebrough N. J., Farr A. L., and Randall R. J. (1951) Protein measurements with Folin phenol reagent. *J. Biol. Chem.* **193,** 265–275.

Luine V. N., Frankfurt M., Rainbow T. C., Biegon A., and Azmitia E. (1983) Intrahypothalamic 5,7-dihydroxytryptamine facilitates sexual behavior and decreases [^{3}H]imipramine binding and 5-HT uptake. *Brain Res.* **264,** 344–348.

Lyness W. H., Demarest K. T., and Moore K. E. (1980) Effects of d-amphetamine and disruption of 5-hydroxytryptaminergic neuronal systems on the synthesis of dopamine in selected regions of the rat brain. *Neuropharmacology* **19,** 883–889.

Lytle L. D., Jacoby J. H., Nelson M. F., and Baumgarten H. G. (1974) Long term effects of 5,7-dihydroxytryptamine administered at birth on the development of brain monoamines. *Life Sci.* **15,** 1203–1217.

Maickel R. P., Cox R. H., Jr., Saillant J., and Miller F. P. (1968) A method for the determination of serotonin and norepinephrine in discrete areas of rat brain. *Int. J. Neuropharmacol.* **7,** 275–281.

McCall R. B. and Aghajanian G. K. (1979) Denervation supersensitivity to serotonin in the facial nucleus. *Neuroscience* **4,** 1501–1510.

McRae-Degueurce A., Berod A., Keller A., Chouvet G., Joh T. H., and Pujol J.-F. (1982) Alterations in tyrosine hydroxylase activity elicited by raphe nuclei lesions in the rat locus coeruleus: Evidence for the involvement of serotonin afferents. *Brain Res.* **235,** 285–301.

Mefford I. N. (1981) Application of high performance liquid chromatography with electrochemical detection to neurochemical analysis: Measurement of catecholamines, serotonin and metabolites in rat brain. *J. Neurosci. Meth.* **3,** 207–224.

Messing R. B. and Lytle L. D. (1977) Serotonin-containing neurons: Their possible role in pain and analgesia. *Pain* **4,** 1–21.

Miller F. P., Cox R. H., Jr., Snodgrass W. R., and Maickel R. P. (1970) Comparative effects of *p*-chlorophenylalanine, *p*-chloroamphetamine and *p*-chloro-*N*-methylamphetamine on rat brain norepinephrine, serotonin and 5-hydroxyindole-3-acetic acid. *Biochem. Pharmacol.* **19,** 435–442.

Moore R. Y., Halaris A. E., and Jones B. E. (1978) Serotonin neurons of the midbrain raphe: Ascending projections. *J. Comp. Neurol.* **180,** 417–438.

Morrow A. L., Norman A. B., Battaglia G., Loy R., and Creese I. (1985) Up-regulation of serotonergic binding sites labeled by [^{3}H]WB4101 following fimbrial transection and 5,7-dihydroxytryptamine-induced lesions. *Life Sci.* **37,** 1913–1922.

Moyer T. P. and Jiang N.-S. (1978) Optimized isocratic conditions for analysis of catecholamines by high-performance reversed-phase paired-ion chromatography with amperometric detection. *J. Chromatog.* **153,** 365–372.

Nelson D. L., Herbert A., Bourgoin S., Glowinski J., and Hamon M. (1978) Characteristics of central 5-HT receptors and their adaptive changes following intracerebral 5,7-dihydroxytryptamine administration in the rat. *Mol. Pharmacol.* **14,** 983–995.

Niwa M., Kunisada K., Himeno A., and Ozaki M. (1984) Serotonin in the rat sympathetic ganglion: Microdeterminations of monoamines and their metabolites by high-performance liquid chromatographic electrochemical detection. *Jpn. J. Pharmacol.* **35,** 237–245.

Nobin A., Baumgarten H. G., Bjorklund A., Lachenmayer L., and Stenevi U. (1973) Axonal degeneration and regeneration of the bulbospinal indoleamine neurons after 5,6-dihydroxytryptamine. *Brain Res.* **56,** 1–24.

Noble E. P., Wurtman R. J., and Axelrod J. (1967) A simple and rapid method for injecting H^3-norepinephrine into the lateral ventricle of the rat brain. *Life Sci.* **6,** 281–291.

Nygren L.-G. and Olson L. (1977) Intracisternal neurotoxins and monoamine neurons innervating the spinal cord: Acute and chronic effects on cell and axon counts and nerve terminal densities. *Histochemistry* **52,** 281–306.

Nygren L.-G., Fuxe K., Jonsson G., and Olson L. (1974) Functional regeneration of 5-hydroxytryptamine nerve terminals in the rat spinal cord following 5,6-dihydroxytryptamine induced degeneration. *Brain Res.* **78,** 377–394.

Ogren S. O., Kohler C. H., Ross S. B., and Srebro B. (1976) 5-Hydroxytryptamine depletion and avoidance acquisition in the rat. Antagonism of the long-term effects of *p*-chloroamphetamine with a selective inhibitor of 5-hydroxytryptamine uptake. *Neurosci. Lett.* **3,** 341–347.

Palkovits M., Saavedra J. M., Jacobowitz D. M., Kizer J. S., and Brownstein M. J. (1978) Serotoninergic innervation of the forebrain: Effects of lesions on serotonin and tryptophan hydroxylase levels. *Brain Res.* **130,** 121–134.

Parent A., Descarries L., and Beaudet A. (1981) Organization of ascending serotonin systems in the adult rat brain. A radioautographic study after intraventricular administration of [^{3}H]-5-hydroxytryptamine. *Neuroscience* **6,** 115–138.

Pellegrino de Iraldi A., Farini-Duggan H., and De Robertis E. (1963) Adrenergic synaptic vesicles in the anterior hypothalamus of the rat. *Anat. Rec.* **145,** 521–535.

Peuler J. D. and Johnson G. A. (1977) Simultaneous single isotope radioenzymatic assay of plasma norepinephrine, epinephrine and dopamine. *Life Sci.* **21,** 625–636.

Pletscher A., Burkard W. P., Bruderer H., and Gey K. F. (1963) Decrease of cerebral 5-hydroxytryptamine and 5-hydroxyindoleacetic acid by an arylalkylamine. *Life Sci.* **2,** 828–833.

Ponzio F. and Jonsson G. (1979) A rapid and simple method for the determination of picogram levels of serotonin in brain tissue using liquid chromatography with electrochemical detection. *J. Neurochem.* **32,** 129–132.

Pujol J.-F., Keane P., McRae A., Lewis B. D., and Renaud B. (1978) Biochemical Evidence for Serotonergic Control of the Locus Coeruleus, in *Interactions Between Putative Neurotransmitters in the Brain* (Garattini S., Pujol J.-F., and Samanin R., eds.) Raven, New York.

Quick M., Sourkes T. L., Dubrovsky B. O., and Gauthier S. (1977) Role of the raphe nuclei in the regulation of adrenal tyrosine hydroxylase. *Brain Res.* **122,** 183–190.

Reader T. A. (1980) Serotonin distribution in rat cerebral cortex: Radioenzymatic assays with thin-layer chromatography. *Brain Res. Bull.* **5,** 609–613.

Reader T. A. (1981) Distribution of catecholamines and serotonin in the rat cerebral cortex: Absolute levels and relative proportions. *J. Neural Trans.* **50,** 13–27.

Reader T. A. (1982) Catecholamines and serotonin in rat frontal cortex after PCPA and 6-OHDA: Absolute amounts and ratios. *Brain Res. Bull.* **8,** 527–534.

Reader T. A. (1983) The Role of Catecholamines in Neuronal Excitability, in *Basic Mechanisms on Neuronal Hyperexcitability* (Jasper H. H. and Van Gelder N. M., eds.) Alan R. Liss, New York.

Reader T. A. and Gauthier P. (1984) Catecholamines and serotonin in the rat central nervous system after 6-OHDA, 5,7-DHT and p-CPA. *J. Neural Trans.* **59,** 207–227.

Reader T. A. and Jasper H. H. (1984) Interactions Between Monoamines and Other Transmitters in Cerebral Cortex, in *Monoamine Innervation of Cerebral Cortex* (Descarries L., Reader T. A., and Jasper H. H., eds.) Alan R. Liss, New York.

Reader T. A., Briere R., Grondin L., and Ferron A. (1986) Effects of p-chlorophenylalanine on cortical monoamines and on the activity of noradrenergic neurons. *Neurochem. Res.* **11,** 1025–1035.

Reader T. A., De Champlain J., and Jasper H. H. (1976) Catecholamines released from cerebral cortex in the cat: Decrease during sensory stimulation. *Brain Res.* **111,** 95–108.

Reader T. A., Masse P., and De Champlain J. (1979) The intracortical distribution of norepinephrine, dopamine and serotonin in the cerebral cortex of the cat. *Brain Res.* **177,** 499–513.

Refshauge C., Kissinger P. T., Dreiling R., Blank L. R., Freeman R., and Adams R. N. (1974) New high performance liquid chromatographic analysis of brain catecholamines. *Life Sci.* **14,** 311–322.

Renaud B., Buda M., Lewis B. D., and Pujol J.-F. (1975) Effects of 5,6-dihydroxytryptamine on tyrosine-hydroxylase activity in central catecholaminergic neurons of the rat. *Biochem. Pharmacol.* **24,** 1739–1742.

Renner K. and Luine V. (1984) Determination of monoamines in brain nuclei by high performance liquid chromatography with electrochemical detection: Young vs middle age rats. *Life Sci.* **34,** 2077–2083.

Ross C. A., Trulson M. E., and Jacobs B. L. (1976) Depletion of brain serotonin following intraventricular 5,7-dihydroxytryptamine fails to disrupt sleep in the rat. *Brain Res.* **114,** 517–523.

Ross S. B., Gosztonyi T., and Renyi A. L. (1974) Long-term effects of N-hydroxy-4-chloroamphetamine on the level and accumulation of 5-hydroxytryptamine in the rat brain. *Eur. J. Pharmacol.* **28,** 222–224.

Saavedra J. M., Brownstein M. J., and Axelrod J. (1973) A specific and sensitive enzymatic-isotopic microassay for serotonin in tissues. *J. Pharmacol. Exp. Ther.* **186,** 505–515.

Saavedra J. M., Palkovits M., Brownstein M. J., and Axelrod J. (1974) Serotonin distribution in the nuclei of the rat hypothalamus and preoptic region. *Brain Res.* **77,** 157–165.

Sachs Ch. and Jonsson G. (1972) Degeneration of central and peripheral noradrenaline neurons produced by 6-hydroxy-DOPA. *J. Neurochem.* **19,** 1561–1575.

Saller C. F. and Salama A. I. (1984) Rapid automated analysis of biogenic amines and their metabolites using reversed-phase high-performance liquid chromatography with electrochemical detection. *J. Chromatog.* **309,** 287–298.

Sanders-Bush E., Bushing J. A., and Sulser F. (1972) Long-term effects of p-chloroamphetamine on tryptophan hydroxylase activity and on the levels of 5-hydroxytryptophan and 5-hydroxyindoleacetic acid in brain. *Eur. J. Pharmacol.* **20,** 385–390.

Sasa W. and Blank C. L. R. (1977) Determination of serotonin and dopamine in mouse brain tissue by high performance liquid chromatography with electrochemical detection. *Anal. Chem.* **49,** 354–359.

Sasa S., Blank C. L. R., Wenke D. C., and Sczupak C. A. (1978) Liquid-chromatographic determination of serotonin in serum and plasma. *Clin. Chem.* **24,** 1509–1514.

Schellenberger M. K. and Gordon J. H. (1971) A rapid simplified procedure for simultaneous assay of norepinephrine, dopamine, and 5-hydroxytryptamine from discrete brain areas. *Anal. Biochem.* **39,** 356–372.

Schlossberger H. G. and Kuch H. (1960) Synthese des 5,6-dihydroxytryptamins. *Chem. Ber.* **93,** 1318–1323.

Segal M. (1979) Serotonergic innervation of the locus coeruleus from the dorsal raphe and its action on responses to noxious stimuli. *J. Physiol.* (Lond.) **286,** 401–415.

Semerdjian-Rouquier L., Bossi L., and Scatton B. (1981) Determination of 5-hydroxytryptophan, serotonin and 5-hydroxyindoleacetic acid in rat and human brain and biological fluids by reversed-phase high-performance liquid chromatography with electrochemical detection. *J. Chromatog.* **218,** 663–670.

Singer E., Sperk G., Placheta P., and Leeman S. E. (1979) Reduction of substance P levels in the ventral cervical spinal cord of the rat after intracisternal 5,7-dihydroxytryptamine injection. *Brain Res.* **174,** 362–365.

Sinhababu A. K., Ghosh A. K., and Borchardt R. T. (1985) Molecular mechanism of action of 5,6-dihydroxytryptamine. Synthesis and biological evaluation of 4-methyl-, 7-methyl-, and 4,7-dimethyl-5,6-dihydroxytryptamines. *J. Med. Chem.* **28,** 1273–1279.

Snyder S. H. and Taylor K. M. (1972) Assay of Biogenic Amines and Their Deaminating Enzymes in Animal Tissues, in *Research Methods in*

Neurochemistry vol. 1 (Marks N. and Rodnight R., eds.) Plenum, New York.

Snyder S. H., Axelrod J., and Zweig M. (1965) A sensitive and specific fluorescence assay for tissue serotonin. *Biochem. Pharmacol.* **14,** 831–835.

Steeves J. D., Schmidt B. J., Skovgaard B. J., and Jordan L. M. (1980) Effect of noradrenaline and 5-hydroxytryptamine depletion on locomotion in the cat. *Brain Res.* **185,** 349–362.

Steinbusch H. W. M. (1981) Distribution of serotonin-immunoreactivity in the central nervous system of the rat: Cell bodies and terminals. *Neuroscience* **6,** 557–618.

Steinbusch W. M. and Nieuwenhuys R. (1983) The Raphe Nuclei of the Rat Brainstem: A Cytoarchitectonic and Immunohistochemical Study, in *Chemical Neuroanatomy* (Emson P. C., ed.) Raven, New York.

Steinbusch H. V. M., Verhofstad A. J., and Joosten W. J. (1978) Localization of serotonin in the central nervous system by immunohistochemistry: Description of a specific and sensitive technique and some applications. *Neuroscience* **3,** 811–819.

Stenevi U., Bjorklund A., and Moore R. Y. (1973) Morphological plasticity of central adrenergic neurons. *Brain Behav. Evol.* **8,** 110–134.

Sugrue M. F. (1979). Effect of depletion of rat brain 5-hydroxytryptamine on morphine-induced antinociception. *J. Pharm. Pharmacol.* **31,** 253–255.

Tranzer J. P. and Thoenen, H. (1967) Ultra-morphologische Veranderungen der sympatischen Nervendigungen der Katze nach Vorbehandlung mit 5- und 6-hydroxydopamin. *Naunyn Schmiedebergs Arch. Exp. Path. Pharmak.* **257,** 343–344.

Tremblay L. E., Bedard P. J., Maheux R., and Di Paolo T. (1985) Denervation supersensitivity to 5-hydroxytryptamine in the rat spinal cord is not due to the absence of 5-hydroxytryptamine. *Brain Res.* **330,** 174–177.

Trulson M. E. and Jacobs B. L. (1976) Behavioral evidence for the rapid release of CNS serotonin by PCA and fenfluramine. *Eur. J. Pharmacol.* **36,** 149–154.

Trulson M. E. and Jacobs B. L. (1978) Behavioral evidence for denervation supersensitivity after destruction of central serotonergic nerve terminals. *Ann. NY Acad. Sci.* **305,** 497–509.

Trulson M. E., Eubanks E. E., and Jacobs B. L. (1976) Behavioral evidence for supersensitivity following destruction of central serotonergic nerve terminals by 5,7-dihydroxytryptamine. *J. Pharmacol. Exp. Ther.* **198,** 23–32.

Twarog B. M. and Page I. H. (1953) Serotonin contents of some mammalian tissues and urine and a method for its determination. *J. Physiol.* (Lond.) **175,** 157–161.

Ungerstedt U. (1971) Stereotaxic mapping of the monoamine pathways in the rat brain. *Acta Physiol. Scand.* **82** (Suppl. 367), 1–48.

Uretsky N. J. and Iversen L. L. (1970) Effects of 6-hydroxydopamine on catecholamine containing neurons in the rat brain. *J. Neurochem.* **17,** 269–278.

van de Kar L. D., Lorens S. A., Vodraska A., Allers G., Green M., van Orden D. E., and van Orden L. S. (1980) Effect of selective midbrain and diencephalic 5,7-dihydroxytryptamine lesions on serotonin content in individual preopticohypothalamic nuclei and on serum luteinizing hormone level. *Neuroendocrinology* **31,** 309–315.

Van der Gugten J., Palkowitz M., Wijnen H. L. J. M., and Versteeg D. H. G. (1976) Regional distribution of adrenaline in rat brain. *Brain Res.* **107,** 171–175.

Vasko M. R., Pang I.-H., and Vogt M. (1984) Involvement of 5-hydroxytryptamine neurons in antinociception produced by injection of morphine into nucleus raphe magnus or onto spinal cord. *Brain Res.* **306,** 341–348.

Vergnes M., Penot C., Kempf E., and Mack G. (1977) Lesion selective des neurones serotoninergiques du raphe par la 5,7-dihydroxytryptamine: effets sur le comportement d'agression interspecifique du Rat. *Brain Res.* **133,** 167–171.

Versteeg D. H. G., der Gugten J. V., de Jong W., and Palkowitz M. (1976) Regional concentrations of noradrenaline and dopamine in rat brain. *Brain Res.* **12,** 563–574.

Vogt M. (1954) The concentration of sympathin in different parts of the central nervous system under normal conditions and after the administration of drugs. *J. Physiol.* (Lond.) **123,** 451–481.

Vogt M. (1974) The effect of lowering the 5-hydroxytryptamine content of the rat spinal cord on analgesia produced by morphine. *J. Physiol.* (Lond.) **236,** 483–498.

Vogt M. and Wilson G. (1972) Concentrations of 5-hydroxytryptamine and its acid metabolites in ventricle-near regions of the rat brain. *J. Neurochem.* **19,** 1599–1600.

Wallace J. A., Petrusz P., and Lauder J. M. (1982) Serotonin immunocytochemistry in the adult and developing rat brain: Methodological and pharmacological considerations. *Brain Res. Bull.* **9,** 117–129.

Wang R. Y., De Montigny C., Gold B. I., Roth R. H., and Aghajanian G. K. (1979) Denervation supersensitivity to serotonin in rat forebrain: Single cell studies. *Brain Res.* **178,** 479–497.

Warsh J. J., Chiu A., Godse D. D., and Coscina D. V. (1979) Determination of picogram levels of brain serotonin by a simplified liquid chromatographic electrochemical detection assay. *Brain Res. Bull.* **4,** 567–570.

Weissbach H., Redfield B. G., and Axelrod J. (1961) The enzymic acetylation of serotonin and other naturally occurring amines. *Biochem. Biophys. Acta* **54,** 190–192.

Wiklund L. and Bjorklund A. (1980) Mechanisms of regrowth in the bulbospinal serotonin system following 5,6-dihydroxytryptamine induced axotomy. II. Fluorescence histochemical observations. *Brain Res.* **191,** 129–160.

Wiklund L., Bjorklund A., and Nobin A. (1978) Regeneration of serotonin neurons in the rat brain after 5,6-dihydroxytryptame-induced axotomy. *Ann. NY Acad. Sci.* **305,** 370–384.

Wong D. T., Horng J. S., Bymaster F. P., Hausser K. L., and Molloy B. B. (1974) A selective inhibitor of serotonin uptake: Lilly-110140, 3-(p-trifluoromethylphenoxy)-N-methyl-3-phenyl-propylamine. *Life Sci.* **15,** 471–479.

Yaksh T. L., Plant R. L., and Rudy T. A. (1977) Studies on the antagonism by raphe lesions of the antinociceptive action of systemic morphine. *Eur. J. Pharmacol.* **41,** 399–408.

Zieher L. M. and De Robertis E. (1963) Subcellular localization of 5-hydroxy-tryptamine in rat brain. *Biochem. Pharmacol.* **12,** 596–598.

Toxins Affecting Cholinergic Neurons

Vincent A. Chiappinelli

1. Introduction

Acetylcholine (ACh) is known to be a neurotransmitter in the nervous systems of vertebrates and invertebrates. The fact that numerous plants, animals, and bacteria have evolved toxins that selectively interfere with cholinergic transmission is an indication of the relative importance of ACh as a neurotransmitter substance. In order for a toxin to be biologically effective, it must produce a sustained action at low concentrations. The high potency and selectivity of these toxins have allowed them to play a crucial role in understanding ACh's physiological effects and in identifying the properties of the enzymes and receptors subserving cholinergic neurotransmission. For example, snake venom α-neurotoxins have been valuable tools in identifying and purifying nicotinic acetylcholine receptors. In fact, the availability of the α-neurotoxins is a major reason the vertebrate muscle nicotinic receptor is currently the best characterized of all neurotransmitter receptors.

1.1. Location of Cholinergic Neurons

1.1.1. Central Nervous System

One particularly accessible cholinergic synapse in the vertebrate central nervous system is that between axon collaterals of spinal motoneurons and Renshaw cells. The effects of several neurotoxins on this synapse have been examined electrophysiologically (Miledi and Szczepaniak, 1975). Supraspinal cholinergic pathways also exist, although they are generally less amenable to experimental manipulations. Choline acetyltransferase (EC 2.3.1.6), the synthetic enzyme for acetylcholine, is found only in cholinergic neurons. Detectable levels of choline acetyltransferase are measured in essentially all brain regions, indicating a wide distribution of cholinergic cell bodies and terminals. Regions with very high enzyme activity include the accumbens, the caudate putamen, the olfactory tubercule, and the habenula (Hoover et al., 1978). Much lower levels are found in such areas as the cerebellum, the globus pallidus, and the hippocampus. The identification of central cholinergic neurons has been greatly facilitated by the availability of monoclonal antibodies to choline acetyltransferase, which can be used in light and electron micrographic localization studies (Ross et al., 1983).

The invertebrate nervous system also contains ACh. *Aplysia* neurons respond to ACh via at least three distinct receptors (Kehoe et al., 1976). In insects, cholinergic responses are also detected (Sattelle et al., 1983). Presumably as protection against insect predation, numerous plants manufacture insecticidal toxins. Nicotine, a cholinomimetic found in the leaves of the tobacco plant, produces an overexcitation in the insect nervous system that can lead to death. Nicotine exhibits similar stimulatory effects in the vertebrate central nervous system.

1.1.2. Peripheral Nervous System

Cholinergic neurons and pathways are difficult to study in the vertebrate central nervous system. Fortunately, two peripheral systems, autonomic ganglia and the neuromuscular junction, not only contain ACh, but are accessible, simple model systems in which the physiology and biochemistry of cholinergic neurotransmission can be readily examined.

1.1.2.1. AUTONOMIC GANGLIA. In both sympathetic and parasympathetic ganglia, the vast majority of presynaptic terminals contain ACh. Postsynaptic parasympathetic ganglion neurons are also cholinergic, and use ACh as a neurotransmitter at their end-organs, which consist largely of smooth muscle and glandular cells. Some sensory neurons also appear to be cholinergic (Crowcroft et al., 1971).

1.1.2.2. NEUROMUSCULAR JUNCTION. The best characterized of all cholinergic systems is the vertebrate neuromuscular junction. Here, cholinergic motoneurons innervate skeletal muscle, and the release of ACh produces the contraction that is the *sine qua non* of voluntary muscle. The electrophysiology, biochemistry, and pharmacology of the vertebrate neuromuscular junction have been the subjects of innumerable studies spanning several centuries (reviewed in Zaimis, 1976; Popot and Changeux, 1984).

2. Methods

2.1. Purification of Neurotoxins from Snake Venom

The venom glands of poisonous snakes are derived from salivary glands that originally contained a number of digestive enzymes (Kochva et al., 1983). Over millions of years of evolution, the RNAses and phospholipases in these glands have been modified into a variety of potent and selective toxins (Strydom, 1979). Although viper and rattlesnake venom contains toxins acting largely on the circulatory system, the venom of kraits, cobras, mambas, and sea snakes contains mostly neurotoxic components. These elapid and hydrophid snakes inject their neurotoxins in an effort to paralyze or otherwise incapacitate their prey prior to feeding. The toxins must exhibit a potent and prolonged action to be effective, properties that make them ideal tools for studying neurotransmission.

The venom of *Bungarus multicinctus,* the banded krait of China and Southeast Asia, contains three distinct types of toxins that affect cholinergic neurons. Two of these, α-bungarotoxin and κ-bungarotoxin, bind to nicotinic acetylcholine receptors, whereas the third, β-bungarotoxin, affects presynaptic release of ACh. The purification of these three neurotoxins is described below.

2.1.1. Sources of Crude Venom

Although some snakes are easily identified and cared for, many elapids do not survive well in captivity and can be incorrectly identified. Obviously, collecting crude venom from sick or poorly characterized snakes can lead to problems during toxin purification. A very reliable source for a wide variety of crude snake venom is the Miami Serpentarium (University of Utah Research Park, Salt Lake City, Utah 84108). Other suppliers include Sigma Chemical Company (St. Louis, Missouri) and Latoxan (Rosans 05150, France). Crude venom should be purchased lyophilized and stored at –20 to –100°C until used. The venom is stable for at least 5 yr under these conditions.

Although "purified" toxins are available from a number of suppliers, these are unreliable, since they have been found to contain numerous impurities (Chiappinelli et al., 1981; Toldi et al., 1983; Saiani et al., 1984). It is therefore best to begin with crude venom and to carefully assess the quality of any toxin isolated.

Table 1
Conditions for Running Cation Exchange Columns[a]

	Column A	Column B	Column C
Size, cm	30 × 1.5	19 × 1.0	50 × 1.0
Cation exchange resin	CM-Cellulose (Whatman CM52)	CM52	CM-Sephadex (Sigma C-50-120)
Flow rate, mL/h	25	20	8
Initial buffer			
pH	5.0	5.4	5.4
Conc. ammonium acetate, m*M*	75	150	150
Final buffer			
pH	6.9	6.1	6.1
Conc. ammonium acetate, m*M*	400	400	400
Total volume of gradient, mL	800	500	500

[a]Modified from Chiappinelli, 1983.

2.1.2. Cation Exchange Chromatography

Snake venom neurotoxins are basic (p*I* > 8.0) polypeptides of 6000–25,000 dalton. An excellent procedure for initial purifications of these toxins is low-pressure cation-exchange liquid chromatography. Glass columns should be siliconized (Sigmacote, Sigma Chemical Company) prior to use. Because of the long periods of time required to run the columns, the use of a cold room or chromatography refrigerator set at 4°C is strongly recommended.

Three sequential cation-exchange columns are required for complete purification of κ-bungarotoxin. The conditions for running these columns are listed in Table 1. A linear gradient of ammonium acetate is used to differentially elute the toxins. Fractions (3 mL) are collected and the elution is monitored by examining optical density at 280 nm.

The crude venom (0.5–0.7 g) is dissolved in 10–20 mL of 75 mM ammonium acetate, pH 5.0. The venom solution, which is slightly yellow but not cloudy, is layered onto the first column. Initial buffer is then run through the column for 2–3 h until the VOID peak (Fig. 1A) appears, after which the gradient is begun. A number of peaks (about 12–14) can be distinguished. The largest peak (peak 2 in Fig. 1A) elutes soon after the gradient is begun and contains α-bungarotoxin and possibly small amounts of other toxins. Peak 4 (Fig. 1A) contains κ-bungarotoxin and other toxins, and peak 10 (Fig. 1A) contains β-bungarotoxins.

Subsequent columns are used to remove the contaminating toxins. This process is shown for κ-bungarotoxin (peak 4A in Fig. 1B and KBgT in Fig. 1C) because there is so little of this toxin recovered (0.1–0.2% of crude venom protein), that the conditions described should be carefully followed in order to maximize yield. To remove contaminating toxins from α-bungarotoxin, similar procedures are used. In addition, molecular sizing chromatography on Sephadex G-50 can be helpful in removing high molecular weight (15,000 dalton) neurotoxins from α-bungarotoxin-containing fractions (Quik and Lamarca, 1982). The β-bungarotoxin peak (peak 10 in Fig. 1A) probably contains several very closely related β-bungarotoxins (Kondo et al., 1978, 1982a,b).

2.1.3. Concentration of Column Fractions

It is necessary to concentrate the collected fractions after the various columns. This can be done by lyophilization, since the only salt present is ammonium acetate. This technique can be time-consuming and troublesome, however. At high initial salt concentrations, melting can occur before lyophilization is completed.

A more reliable method of concentration is ultrafiltration under pressure in a stirred cell device (Model 8050, Amicon, Danvers, Massachusetts) using a nominal 1000-dalton cutoff membrane filter (Amicon YM-2). The toxins are retained in the solution above the filters and the recovery is typically 90–100%. This system allows for some control of the final concentration of toxin in the solution, which should be between 0.5 and 5 mg/mL. For unknown toxins, assume that a solution having 1.00 OD at 280 nm is 1 mg/mL in toxin. For α-bungarotoxin, this concentration is actually reached at 1.32 OD (Hanley et al., 1977), and the comparable value for κ-bungarotoxin is 0.816 OD (Chiappinelli and Lee, 1985). Toxins should be stored in 75–150 mM ammonium acetate, pH 5.0, at –100°C. Repeated freezing and thawing should be avoided by aliquoting the toxin solutions.

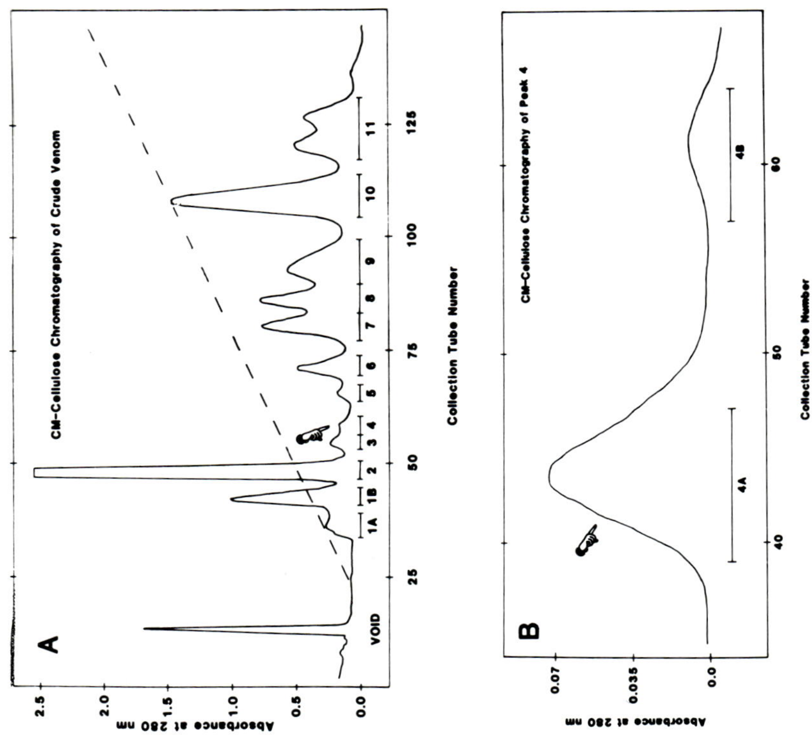
A
CM-Cellulose Chromatography of Crude Venom
Absorbance at 280 nm
0.0
0.5
1.0
1.5
2.0
2.5
VOID
1A
1B
2
3
4
5
6
7
8
9
10
11
25
50
75
100
125
Collection Tube Number
B
CM-Cellulose Chromatography of Peak 4
Absorbance at 280 nm
0.0
0.035
0.07
4A
4B
40
50
60
Collection Tube Number

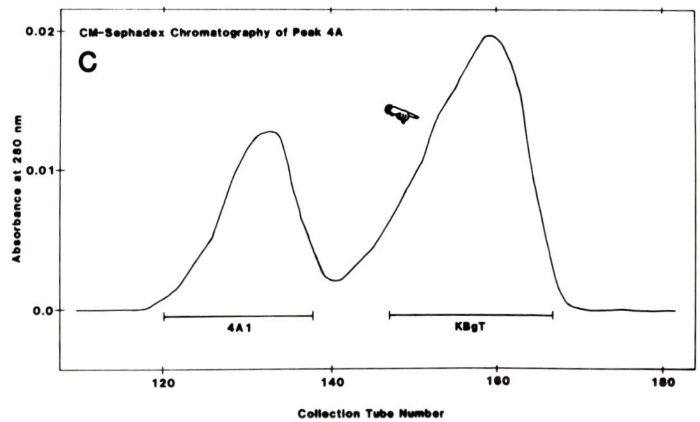

Fig. 1. Purification of three toxins from *Bungarus multicinctus* crude venom. (A) Crude venom is separated into 12 peaks by the first CM-cellulose column. Dashed line represents linear gradient of ammonium acetate. Peak 2 is α-bungarotoxin. Peak 10 consists of β-bungarotoxins. Peak 4, which is further purified in the next two columns, contains κ-bungarotoxin and several other toxins. (B) Peak 4 from panel A is further purified on CM-cellulose. The pointing finger highlights the peak containing κ-bungarotoxin in all three panels. (C) Peak 4A from panel B is separated into two peaks by CM-Sephadex. κ-Bungarotoxin is eluted near the end of the gradient. Details of column conditions are in Table 1 (from Chiappinelli, 1983).

2.1.4. High-Pressure Liquid Chromatography

High-pressure liquid chromatography (HPLC) can also be used to further purify venom components. A cation-exchange column (Bio-Rad TSK-SP-5-PW; 75 × 7.5 mm) is attached to an HPLC pump that has an automatic gradient maker. The column is pre-equilibrated in start buffer [90% (vol/vol) 25 m*M* sodium phosphate, pH 6.0; 10% (vol/vol) acetonitrile]. The sample is loaded in start buffer, followed by 10 min of start buffer pumped at 1 mL/min. A linear gradient is then run between start buffer and end buffer [90% (vol/vol) 500 m*M* sodium phosphate, pH 6.0; 10% (vol/vol) acetonitrile]. Total volume of the gradient is 120 mL and running time is 120 min at 1 mL/min. One milliliter fractions are collected, and columns can be run at room temperature. The column has a maximum capacity of 5 mg per venom component. For analytical runs, as little as 1–2 μg of toxin can be detected by monitoring optical density at 210 nm. For purification runs, optical density is monitored at 280 nm.

An example illustrates the separating power of the HPLC technique. A sample containing 2 μg of α-bungarotoxin and 1.5 μg of κ-bungarotoxin is applied to the HPLC column (Fig. 2). The two toxins are easily separated, and the position at which each toxin

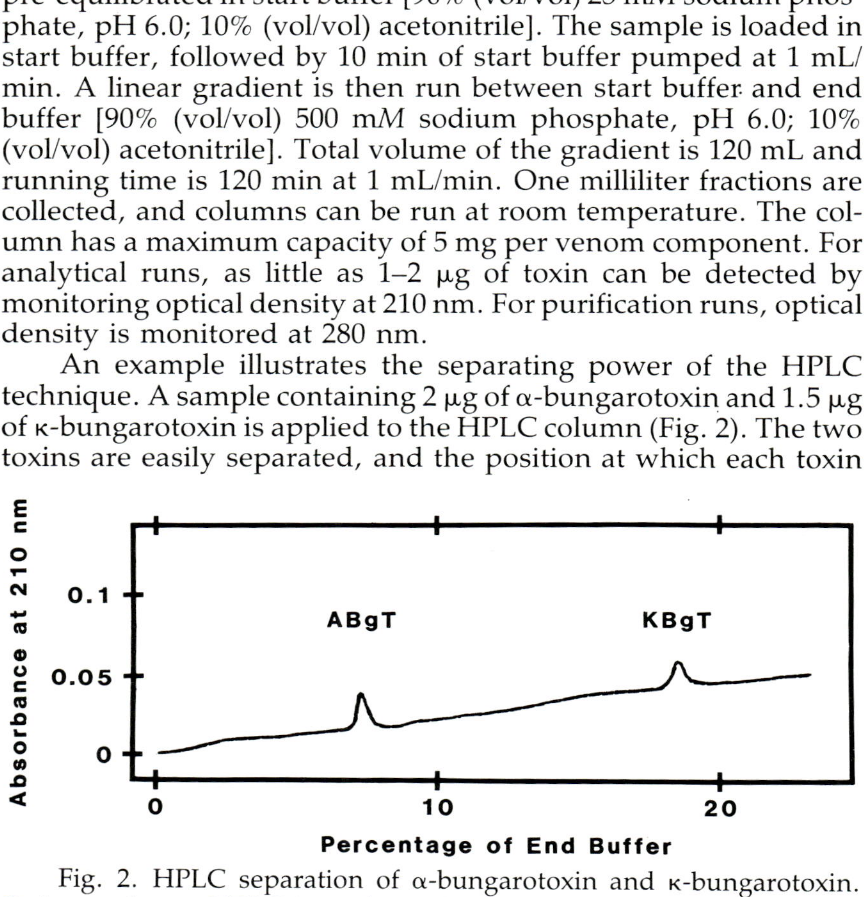

Fig. 2. HPLC separation of α-bungarotoxin and κ-bungarotoxin. Cation exchange HPLC is performed as described in section 2.1.4 on a sample containing 2.0 μg of α-bungarotoxin (ABgT) and 1.5 μg of κ-bungarotoxin (KBgT). The two peptides are readily separated by the linear gradient, which is indicated on the abscissa by the percentage of end buffer. The small amounts of toxin used for this HPLC separation would not be detected by the low-pressure chromatography method, thus pointing out the usefulness of HPLC as an analytical tool.

elutes (i.e., the percentage of end buffer required) is very reproducible from one run to the next. It is still possible, however, for several toxins to elute at identical positions, as is the case with the low-pressure columns. The HPLC method provides a further indication of the purity and identity of isolated venom components.

2.2. Biochemical Characterization of Neurotoxins

Because even minor contamination of snake toxins with other venom components can have drastic effects on the apparent physiological properties of the toxins (Chiappinelli et al., 1981; Quik and Lamarca, 1982), the purity of the toxin preparation should always be carefully tested. A number of methods are useful for this purpose.

2.2.1. Polyacrylamide Gel Electrophoresis

Electrophoresis in the presence of sodium dodecyl sulfate (Weber and Osborn, 1969) is a relatively rapid method for examining the purity of venom components. Slab gels (18 × 32 cm; 0.75 mm thick) are prepared using a linear gradient (5–39% wt/vol) of polyacrylamide. The ingredients of the solutions used are listed in Table 2. There is no separate stacking gel. Slabs are run in the vertical position (Hoefer Model SE600, San Francisco, California) with a constant current of 15 mA, temperature of 4°C, and running distance of about 20 cm. Total running time is 8.5 h. Samples (10–30 μg total protein) are pretreated for 10 min at 95°C in denaturing buffer of the following composition: 2% (wt/vol) sodium dodecyl sulfate; 10% (vol/vol) glycerol; 5% (vol/vol) β-mercaptoethanol; 70 m*M* TRIS, pH 6.8. Low molecular weight standards are from Bethesda Research Labs (Gaithersburg, Maryland). Gels are post-stained with Coomassie Blue R250 (Fairbanks et al., 1971).

α-Bungarotoxin runs as a single band of protein at 8000 dalton. Impurities may appear at 6500 (Chiappinelli et al., 1981) or 15,000 dalton (Quik and Lamarca, 1982). κ-Bungarotoxin runs as a single band of protein at 6500 dalton, and potential contaminating toxins generally appear at 8000 dalton. β-Bungarotoxins run as two bands of protein at 14,000 and 9000 dalton. Nondenaturing gel electrophoresis can also be used to characterize the toxins (Chiappinelli and Lee, 1985).

2.2.2. Isoelectric Focusing

Isoelectric focusing in a polyacrylamide slab gel provides an additional test of toxin purity. The method has been described in

Table 2
Solutions for 5–39% Gradient Acrylamide Gel[a]

5% gel solution		
3.7	mL	Low conc. acrylamide stock [30% (wt/vol) Acrylamide; 0.8% (wt/vol) BIS]
12.5	mL	H_2O (double-distilled)
5.5	mL	1.5*M* TRIS, pH 8.8
0.22	mL	10% (wt/vol) Sodium dodecylsulfate
48	μL	10% (wt/vol) Ammonium persulfate
8	μL	TEMED (add just before use)
39% Gel solution		
14.7	mL	High conc. acrylamide stock in glycerol [58% (wt/vol) Acrylamide; 1.6% (wt/vol) BIS; 50% (vol/vol) glycerol]
1.5	mL	H_2O
5.5	mL	1.5*M* TRIS, pH 8.8
0.22	mL	10% (wt/vol) Sodium dodecylsulfate
48	μL	10% (wt/vol) Ammonium persulfate
8	μL	TEMED (add just before use)

[a]Abbreviations: BIS, *N,N'*-methylene-*bis*-acrylamide; TEMED, *N,N,N'*,N'-tetramethylethylenediamine; TRIS, Tris(hydroxymethyl)aminomethane.

detail (Chiappinelli, 1983). The basic nature of the toxins makes it difficult to obtain precise and reproducible p*I* values. α-Bungarotoxin is always more basic than κ-bungarotoxin and runs as a single band of protein at p*I* = 9.5–9.8. κ-Bungarotoxin runs with a p*I* = 8.9–9.2.

Purity of radiolabeled toxins is readily assessed by isoelectric focusing (Chiappinelli, 1983). The migration of the radioactive toxin sample (0.1–1.0 μ Ci) is examined by slicing the gel into 2-mm segments and counting the radioactivity in each slice in a gamma radiation counter. Pure radiolabeled toxins will exhibit a single radioactive peak. This peak will either appear in the same slice (i.e., at the same p*I*) as the native toxin, or will be one to two slices more acidic than the native toxin.

2.2.3. Molecular Sizing Columns

Sephacryl S-200 (Sigma Chemical Company) is an excellent medium for examining the apparent molecular size of snake neurotoxins (Chiappinelli and Lee, 1985). The column can be used to examine purity of toxin preparations as well as to determine whether a particular toxin exhibits any self-association. κ-Bungarotoxin, which exists entirely as a dimer of identical subunits in physiological solutions, runs with an apparent molecular weight of 12,000 dalton on nondenaturing Sephacryl S-200 columns. When sodium dodecyl sulfate or urea are added to the eluting solution, however, the dimer dissociates and a single 6500-dalton protein is observed (Chiappinelli and Lee, 1985). Columns should be run at 4°C, except for urea-containing columns, which are run at room temperature.

2.2.4. Amino Acid Sequencing and Compositional Analysis

The definitive test of identity of a snake neurotoxin is complete amino acid sequencing of the protein and comparison with the known sequences of other toxins. Complete sequences are available for α-bungarotoxin (Mebs et al., 1971), κ-bungarotoxin (Grant and Chiappinelli, 1985), and the β-bungarotoxins (Kondo et al., 1982a,b). Complete amino acid sequencing is an expensive procedure and requires considerable amounts of toxin. An alternative approach is to perform automated Edman degradation on the intact protein to determine the first 10–12 amino acid residues on the amino-terminal end of the toxin (Grant and Chiappinelli, 1985). This procedure can readily be done with 0.5–1.0 nmol of toxin, and should allow for positive identification if the snake toxin has been previously sequenced. Any impurities may also be detected by this method. Amino acid compositional analysis is considerably less expensive and requires even less toxin than Edman degradation. The technique has only limited usefulness in confirming the identity of a toxin, however, because of the similar amino acid compositions of many snake neurotoxins.

2.3. Radiolabeling of Neurotoxins

Radioiodination is a relatively gentle procedure for obtaining labeled neurotoxins with very high specific activities that retain the biological actions of the native toxins. These radiolabeled neurotoxins are used in binding assays and localization studies.

Snake neurotoxins are iodinated in glass tubes previously coated with 4.3 μg Iodogen reagent (Pierce Chemical Company). The toxin (1.25 nmol) is combined with 1.0 mCi sodium [^{125}I]iodide (Amersham) in 85 μL of 0.25*M* sodium phosphate, pH 7.4. After a 15-min reaction at 22°C, the labeled toxin is separated from unreacted ^{125}I by Sephadex G10 filtration in a disposable polypropylene column (8 × 1.5 cm). Concentrations of radiolabeled toxin solutions are confirmed in binding assays by dilution with the appropriate nonradiolabeled toxin. Initial specific activities are typically between 700 and 1100 Ci/mmol.

Radiolabeled toxins should be stored at 4°C in 70–150 m*M* ammonium acetate, pH 7.4, containing 10 mg/mL bovine serum albumin and 0.02% (wt/vol) sodium azide. The relatively short half-life (60 d) of ^{125}I places a limit of 30–60 d on the useful lifetime of the radiolabeled toxins.

2.4. Electrophysiology in Autonomic Ganglia

Autonomic ganglia provide an excellent model system for studying cholinergic neurotransmission and for determining the effects of toxins on cholinergic function. The choice of the species and particular ganglion used will vary depending on the interests of the investigator. Some of the most frequently utilized preparations include the rat superior cervical ganglion (Perri et al., 1970), the guinea pig inferior mesenteric ganglion (Crowcroft and Szurszewski, 1971), the rat submandibular ganglion (Kawa and Roper, 1984), and the ciliary ganglion of the chick. The methods for electrophysiological recording described below are applicable to any of these preparations. The chick ciliary ganglion will be used for illustrative purposes in the following treatment.

2.4.1. Dissection of Ganglia

After decapitation, the head of a newly hatched chick is cleaned of skin and feathers, and a fine scissors is used to bisect the head down the midline. One half of the head is pinned down in a Petri dish whose bottom contains hardened silicone resin (RTV615; General Electric). The head is pinned with the cut surface (the brain) facing down. The remainder of the dissection, and the electrophysiological recording, must be done using a high-quality stereomicroscope, such as the Nikon SMZ-10, providing a zoom magnification of up to 80 ×. By clearing connective tissue between the eye and the head, the eye becomes loosely attached and can then be retracted away from the head and pinned down. The small white ciliary nerves can be seen where they exit the iris. They travel down the eyeball and join to form a single nerve at a position that is at the midpoint of a line drawn between the iris and the external auditory meatus (the earhole). The ciliary nerve is followed back to the ciliary ganglion, a fusiform structure that abuts the skull.

Using a fine iridectomy scissors, the skull around the preganglionic oculomotor nerve is removed. The oculomotor nerve is cut several millimeters away from the ganglion, and the ciliary nerve is cut as it enters the eyeball. The ciliary ganglion with attached pre- and postganglionic nerves is removed and placed in a small Petri dish filled with physiological solution for final cleaning. Connective tissue is removed from the ganglion and attached nerves. For intracellular recording, it is particularly important to remove all of the fine connective tissue sheath surrounding the ganglion. This is done with sharp no. 5 jeweler's forceps. The ganglion is now ready to be placed in the recording chamber.

2.4.2. Electrophysiological Chamber

The electrophysiological chamber, which is made of plexiglass, is shown in Fig. 3. It consists of two chambers connected by three drilled holes. The bottom of the recording chamber is filled with silicone resin, except where the inlet and connecting holes are drilled. The back chamber need not be filled with resin. Care should be taken that the holes connecting the two chambers are completely submerged during recording. Solution level is regulated by raising or lowering a 22-gage hypodermic needle (connected to a vacuum source) into the back chamber. This arrangement prevents wave artifacts produced by the suction needle from interfering with the electrical recording.

The ganglion is placed in the recording chamber. If extracellular recording is to be performed, glass-tipped suction electrodes are used to secure both pre- and post-ganglionic nerves. The seal between electrode and nerve is regulated (by melting the glass tip) so that a tight fit results, and the electrodes then hold the entire preparation securely suspended above the bottom of the chamber. Once the sizes of the glass tips required are established, the same electrodes can be used repeatedly for numerous preparations.

For intracellular recording, the ganglion must be pinned securely to the bottom of the recording chamber. This is done by cutting fine tungsten wire (0.025 mm diameter, Aesar, Seabrook New Hampshire) into very small segments, which are then stiff enough to be used as micropins. Six to eight micropins are used to secure the ganglion, and the preganglionic oculomotor nerve is then taken up into a suction electrode. The postsynaptic nerve need only be attached to a suction electrode if it is desired to record from a single ganglion both intracellularly and extracellularly.

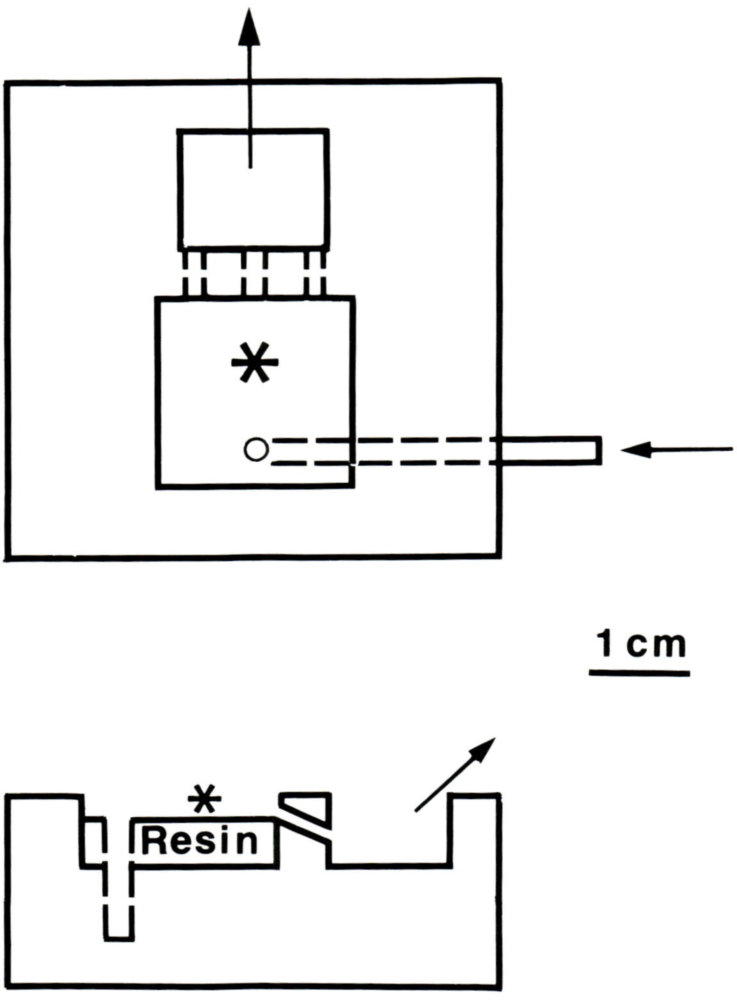

Fig. 3. Chamber for electrophysiological experiments. A top view of the electrophysiological chamber is shown (upper diagram) as well as a cross-sectional side view (lower diagram). The position of the ganglion in the recording chamber is indicated (asterisk). Arrows indicate direction of flow of perfusion fluid, which enters from below the recording chamber and travels across the ganglion and through the connecting holes into the back chamber, where the suction needle controls the fluid levels in both chambers. The entire chamber is made from a single block of plexiglass.

A gravity-feed system is used for flowing solution through the chamber. Perfusion bottles are raised 4 ft above the recording chamber and are attached to a five-gang switching valve (MX431-5FM; Medex, Hilliard, Ohio) for selecting the perfusion solution. The single small (1.2 mm, inside diameter) polyethylene tube exiting the switching valve is coiled inside a small water bath (Haake E15) kept between 22 and 60°C. The tube exits the water bath and shortly (within 2 ft) attaches to the inflow tube of the recording chamber. The temperature in the chamber is constantly monitored by a hypodermic temperature probe attached to a battery-operated temperature detection unit (Model 43TD, YSI, Yellow Springs, Ohio).

Temperature in the recording chamber is regulated by changing either the temperature of the water bath or the flow rate of the solution. The flow rate should be kept between 2–5 mL/min. For extracellular recording, recording chamber temperatures between 22 and 37°C can be used. For intracellular recording, temperatures between 34 and 37°C provide the most stable impalements.

The physiological solution used has the following composition: 150 m*M* NaCl; 3 m*M* KCl; 5m*M* $CaCl_2$; 2 m*M* $MgCl_2$; 10 m*M* HEPES; 17 m*M* glucose; pH 7.4 at 37°C and bubbled with 100% O_2. The high Ca^{2+} concentration improves the stability of intracellular impalements.

2.4.3. Extracellular Recording

For extracellular recording, the two leads (one inside and one outside the glass tip) of the suction electrode attached to the postsynaptic ciliary nerve are connected to a differential ac amplifier (Tektronix 5A22N), and the resulting signal is displayed on a storage oscilloscope (Tektronix 5111). The suction electrode attached to the presynaptic oculomotor nerve is connected to an electrical stimulator (Grass S88) via a constant current stimulus isolation unit (Grass PSIU6). Repeated pulses (1 Hz; 0.01–0.1 ms duration) of current are delivered to the presynaptic nerve while observing the ciliary nerve response. Maximum stimulation generally requires an indicated current of 0.3–8 mA.

2.4.4. Intracellular Recording

Intracellular recording requires several additional pieces of equipment. An isolation table (Micro-G; TMC, Peabody, Massachusetts) is essential to maintain impalements. A high-quality micromanipulator (Aus Jena or Leitz) is used to control the movements of the intracellular electrode. Glass microelectrodes (1.0 mm OD, 0.5 mm ID, Omega Dot; FHC, Brunswick, Maine) are pulled on a horizontal microelectrode puller (Industrial Science Associates, Ridgewood, New York or Model P-80 Brown-Flaming, Sutter Instrument, San Francisco, California), and have resistances of 50–120 MΩ when filled with 3*M* KCl. A WPI (New Haven, Connecticut) M707 electrometer is used as the intracellular amplifier. A Grass S88 electrical stimulator is used to pass constant dc current and current pulses through the intracellular electrode. A second, digital oscilloscope (Model 3091; Nicolet) is useful for simultaneously monitoring fast and slow synaptic events and for processing of data. Because of the increased amounts of data obtained with intracellular recording, a four- or eight-channel analog instrumentation tape recorder (Vetter, Rebersburg, Pennsylvania) is used to capture current, voltage, trigger, and voice signals for later analysis.

The intracellular electrode is advanced slowly to the surface of the ganglion. The best way to determine when the electrode reaches the ganglion surface is to feed the electrode voltage signal into a Grass Audio Monitor. A distinct sound can then be heard when the electrode reaches the surface. By alternately applying negative capacity compensation and advancing the microelectrode in steps, cells can be impaled. Upon impalement, hyperpolarizing dc current is injected through the electrode to maintain a steady –70 mV membrane potential. After rebalancing the electrode and waiting several minutes to allow the membrane to seal around the electrode, this dc current is removed and a true resting membrane potential of –40 to –70 mV is revealed. Healthy cells have overshooting action potentials 70–100 mV in amplitude.

Three different types of intracellular units are observed in the chick ciliary ganglion. Postsynaptic ciliary and choroid neurons are distinguished by their responses to presynaptic nerve stimulation. Ciliary neurons are singly innervated and exhibit electrical as well as nicotinic transmission, whereas choroid cells are multiply innervated and do not exhibit electrical responses (Dryer and Chiappinelli, 1985c). Presynaptic nerve terminals synapsing on ciliary neurons can also be impaled, and these units completely lack chemically mediated responses to presynaptic nerve stimulation (Dryer and Chiappinelli, 1985a).

2.5. Detecting Physiological Effects of Neurotoxins on Cholinergic Neurons

In the case of the parasympathetic ciliary ganglion, both the presynaptic and postsynaptic elements are cholinergic. This can be useful for some experiments, particularly with neurotoxins of an unknown character that may exhibit either pre- or postsynaptic effects.

2.5.1. α-Bungarotoxin

α-Bungarotoxin is a potent antagonist of nicotinic receptors in vertebrate muscle. It is frequently ineffective, however, in blocking nicotinic receptors found in autonomic ganglia and in the vertebrate central nervous system. This is true in the chick ciliary ganglion, as shown in Fig. 4. The compound action potential recorded extracellularly from the ciliary nerve is unchanged after exposure to a high concentration (1.2 μ*M*) of α-bungarotoxin for 1 h. During toxin exposures, the flow through the chamber is halted. This is done to conserve the toxins. Control experiments show little or no effect on extracellular or intracellular recordings following stopped flow of normal physiological solution for up to 90 min.

2.5.2. κ-Bungarotoxin

Although structurally related to α-bungarotoxin, κ-bungarotoxin exhibits a pharmacological spectrum of activity opposite to that of α-bungarotoxin. κ-Bungarotoxin is a weak antagonist of muscle nicotinic receptors (Chiappinelli et al., 1985), but is a potent inhibitor of ganglionic nicotinic receptors (Chiappinelli, 1983). This is illustrated in Fig. 4, in which 75 n*M* κ-bungarotoxin rapidly and completely blocks nicotinic transmission in the chick ciliary ganglion. Since the high dose of α-bungarotoxin was maintained throughout the κ-bungarotoxin exposure, κ-bungarotoxin blocks ganglionic transmission at a site not recognized by α-bungarotoxin. Considerable evidence suggests that this site is associated with the physiologically detected nicotinic receptor (Chiappinelli, 1983; Dryer and Chiappinelli, 1983).

After 60–90 min exposure of the ganglion to a toxin, the flow through the chamber is begun again. Toxins such as κ-bungarotoxin (Fig. 4) and κ-flavitoxin (Chiappinelli et al., 1986) have a prolonged effect on the ganglion, and complete recovery is not seen until 5 h after the washing out of the toxin is begun. It is useful to establish the reversible nature of postsynaptic agents

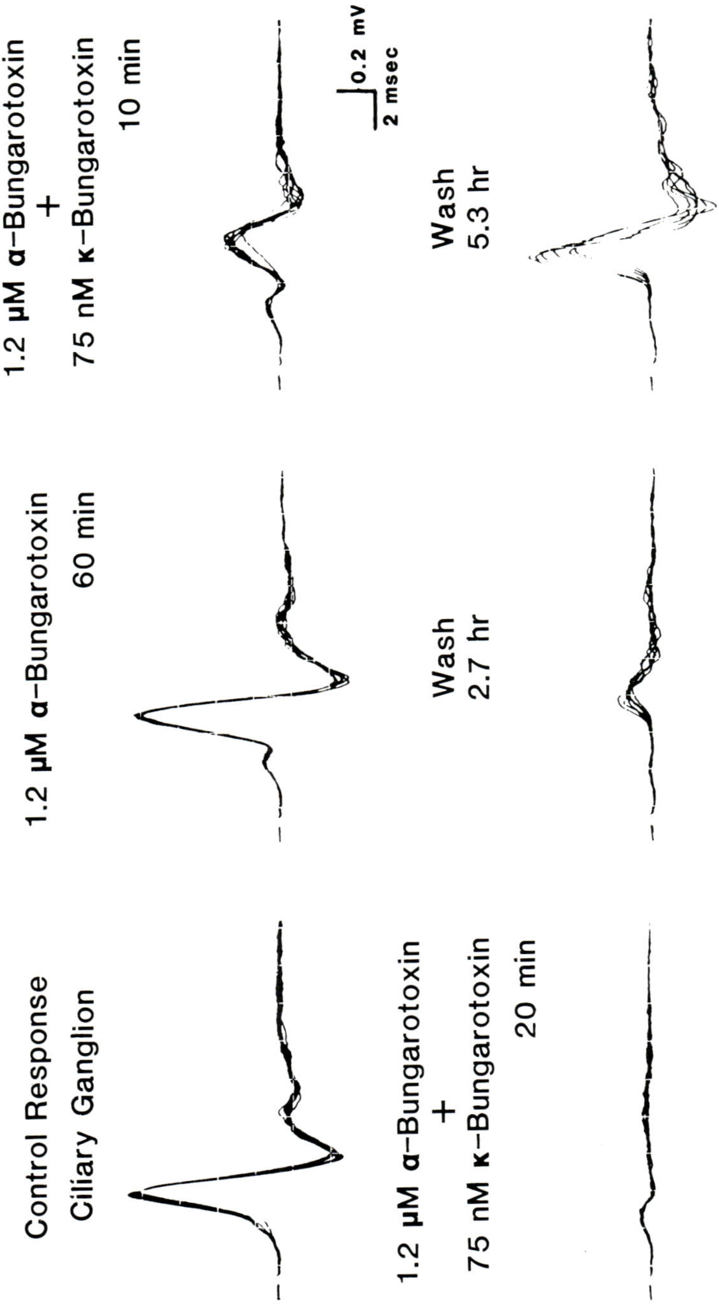
Control Response
Ciliary Ganglion
1.2 μM α–Bungarotoxin
60 min
1.2 μM α–Bungarotoxin
+
75 nM κ–Bungarotoxin
10 min
0.2 mV
2 msec
1.2 μM α–Bungarotoxin
+
75 nM κ–Bungarotoxin
20 min
Wash
2.7 hr
Wash
5.3 hr

Fig. 4. α-Bungarotoxin does not protect against action of κ-bungarotoxin. This sequence of photos illustrates an electrophysiological experiment on a single chick ciliary ganglion. Responses to preganglionic nerve stimulation at 1 Hz are recorded extracellularly from the postganglionic ciliary nerve. Stimulation occurs at break in record and is followed by a compound action potential mediated by neuronal nicotinic receptors. Nicotinic transmission is totally unaffected after a 60-min exposure to 1.2 μ*M* α-bungarotoxin. After a subsequent exposure to 75 n*M* κ-bungarotoxin in the continued presence of α-bungarotoxin, the response is completely blocked. Recovery of the nicotinic response following exposure to κ-bungarotoxin requires a washout period of 5.3h. The experiment demonstrates that κ-bungarotoxin blocks nicotinic transmission by binding to a site that is not detected by high concentrations of α-bungarotoxin (modified from Chiappinelli, 1983).

such as the κ-neurotoxins, since most presynaptic toxins have an irreversible effect on neurotransmission.

2.5.3. β-Bungarotoxin

β-Bungarotoxin produces a complex blockade of nicotinic transmission in autonomic ganglia (Fig. 5). At low doses of β-bungarotoxin (25 n*M*), a complete block of transmission is observed after 10 min. Then, in the continuing presence of the toxin, the response partially or completely recovers. This apparent recovery is only transient, however. If the toxin is removed from the ganglion at this stage, a second blockade still occurs about 30 min later. The second blockade shows no recovery after many hours, and is assumed to be irreversible. Electrical transmission is also lost during the second β-bungarotoxin blockade (Fig. 5). The initial blockade is likely to be a depolarizing blockade produced by an increased release of acetylcholine from presynaptic nerve terminals. The second, irreversible blockade signals the destruction of the presynaptic terminal by the phospholipase A_2 activity of β-bungarotoxin. A similar sequence of events occurs when the neuromuscular junction is exposed to β-bungarotoxin (Abe et al., 1977).

2.5.4. Lophotoxin

Intracellular recording from autonomic ganglion neurons is very useful for examining the effects of neurotoxins. This is illustrated in Fig. 6, in which the effects of lophotoxin, a coral neurotoxin, are examined by intracellular recording in chick lumbar sympathetic ganglia. Recording from chick sympathetic ganglia is very similar to recording from the chick ciliary ganglion (Dryer and Chiappinelli, 1985b). For examining nicotinic transmission, 1.5 μ*M* atropine must be present throughout the experiment in order to block the action of ACh at muscarinic receptors (Chiappinelli and Dryer, 1984). The control response to carbachol is a depolarization associated with a decrease in input resistance (Fig. 6A). A similar effect is elicited by γ-aminobutyric acid (GABA), although a noncholinergic receptor is involved in this response (McEachern et al., 1985).

After exposure to lophotoxin, the carbachol response is almost completely blocked, whereas the response to GABA is unchanged (Fig. 6B). The cells remain healthy after lophotoxin exposure and continue to elicit overshooting action potentials in response to depolarizing current injected through the intracellular electrode

(Fig. 6B). Lophotoxin also has no effect on muscarinic receptors, tested by removing atropine from the perfusion solution (Fig. 6C).

Thus, the selectivity of lophotoxin is clearly pointed out by the intracellular records. Although nicotinic receptors are blocked by the toxin, muscarinic receptors and receptors for GABA are unaffected by lophotoxin. In addition, the toxin does not appear to have any nonspecific effects on the active or passive membrane properties of the neurons. This is the type of information required to ensure that the mechanism of action of a new toxin has been clarified and that the observed results are not the result of some nonselective toxic effect on neuronal membranes.

2.6. Use of Neurotoxins in Biochemical Studies of Cholinergic Neurons

The high-affinity nature of the binding observed with many neurotoxins makes them ideal for biochemical and anatomical studies. A case in point is the binding of α-bungarotoxin to chick ciliary ganglia. ^{125}I-α-Bungarotoxin is prepared as described in section 2.3. Binding can be examined either in whole ganglia (Chiappinelli, 1982, 1983) or in ganglia homogenized using ground-glass microhomogenizers (Chiappinelli and Giacobini, 1978). The buffer used in binding experiments is identical to the buffer used in the electrophysiological experiments, except that 1 mg/mL bovine serum albumin is added to reduce nonspecific binding of toxins to tubes and filters. Total binding in homogenates is measured by an ultrafiltration assay using Millipore EGWP cellulose acetate membrane filters (Chiappinelli and Giacobini, 1978). Nonspecific binding is taken as the amount of binding in the presence of a 100–1000-fold excess of nonradiolabeled toxin.

Binding sites identified with ^{125}I-α-bungarotoxin can be solubilized using 1% (vol/vol) Triton X-100. Because of the long dissociation constant ($t_{1/2}$ = 4 h), it is possible to examine the solubilized site while the bound ^{125}I-α-bungarotoxin serves as a radiolabeled marker. To extend this working time even longer, the solubilized binding site–^{125}I-α-bungarotoxin complexes are treated with 0.1% (vol/vol) glutaraldehyde for 20 min at 22°C. This treatment covalently cross-links many of the toxin molecules to the binding sites. Sucrose density gradient sedimentation can then be carried out on the solubilized binding site, to obtain an approximate measure of the molecular weight of the protein. This is done in a 5–20% (wt/vol) linear sucrose gradient [5mL total volume, 50

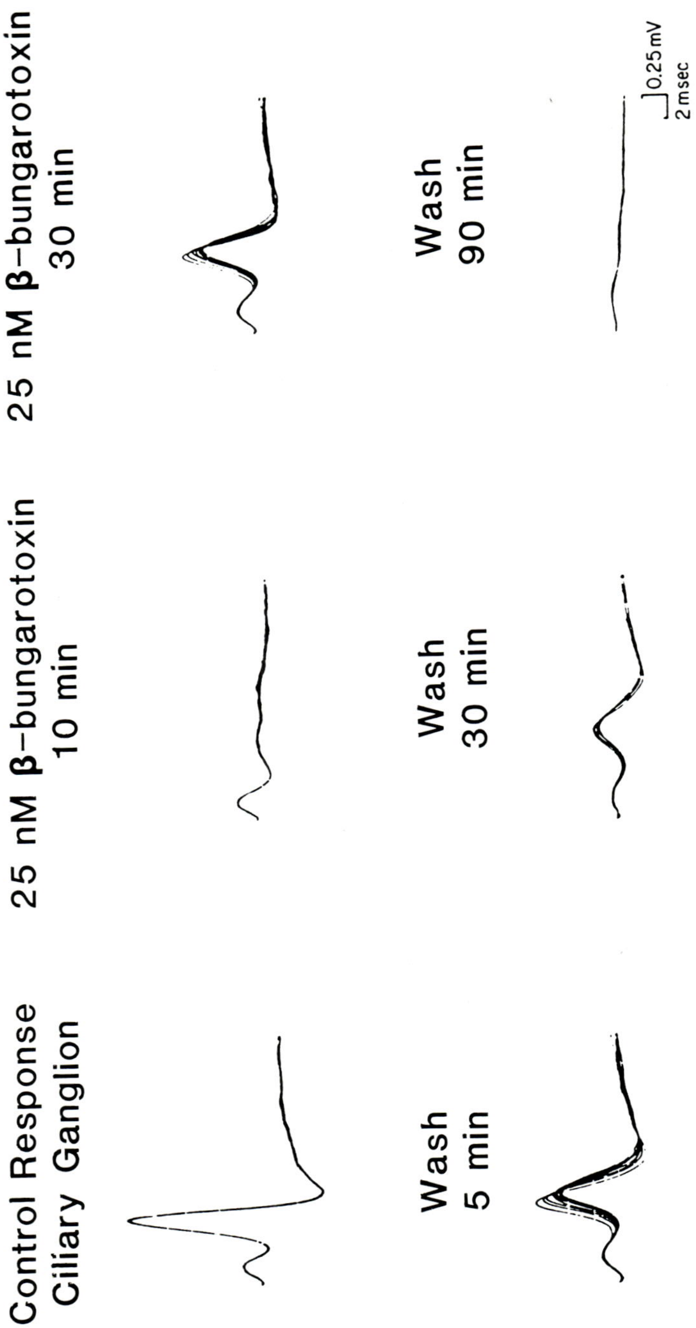
Control Response
Ciliary Ganglion
25 nM β–bungarotoxin
10 min
25 nM β–bungarotoxin
30 min
Wash
5 min
Wash
30 min
Wash
90 min
0.25 mV
2 msec

Fig. 5. Complex effect of β-bungarotoxin on transmission in the chick ciliary ganglion. Preganglionic nerve stimulation at 1 Hz results in two compound action potentials recorded extracellularly from the ciliary nerve. The smaller, shorter-latency response is caused by ciliary neurons electrically activated by gap junctions (Martin and Pilar, 1963), whereas the larger, longer-latency response is mediated by neuronal nicotinic receptors. After 10 min exposure to 25 n*M* β-bungarotoxin, the nicotinic response is blocked, but it recovers considerably after an additional 20 min of exposure to the toxin (upper right). The toxin is washed out of the solution bathing the ganglion after the 30 min exposure. During the washout period, the nicotinic response is blocked for a second time, and the electrical response also disappears. No recovery is seen from this second blockade of transmission by β-bungarotoxin (modified from Chiappinelli et al., 1981).

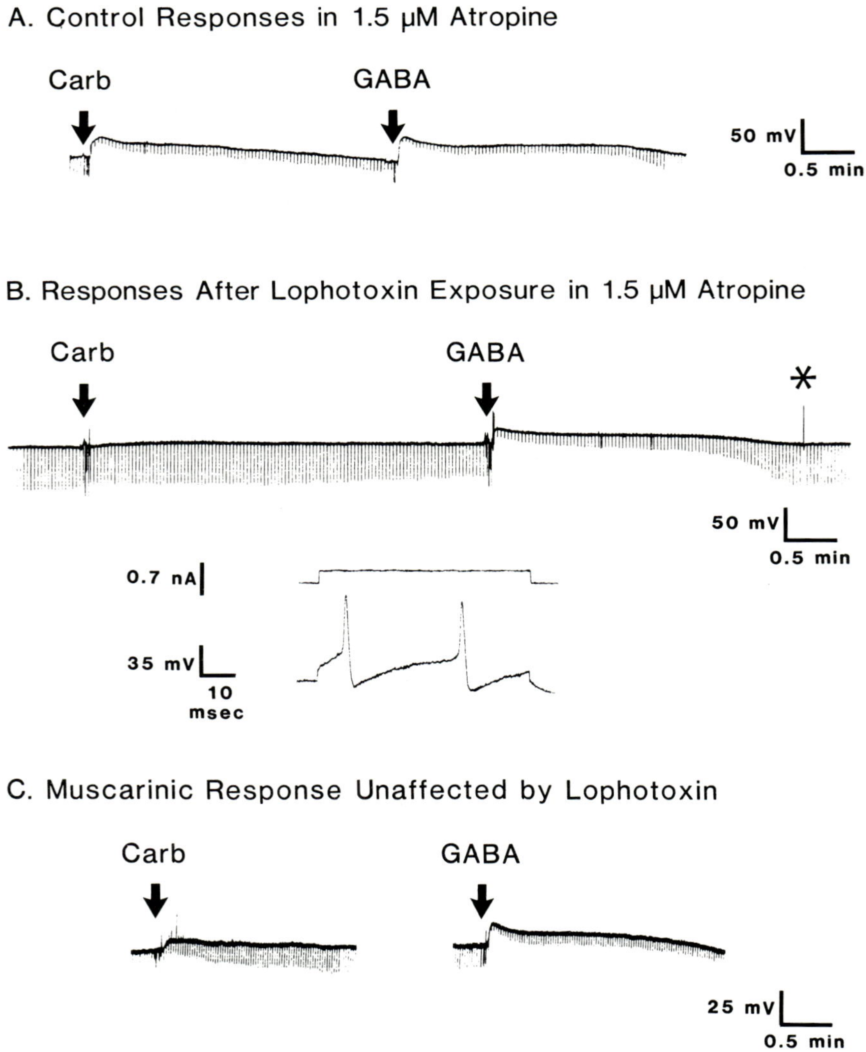

Fig. 6. Effects of lophotoxin examined by intracellular recording in chick lumbar sympathetic ganglia. (A) In the presence of 1.5 μ*M* atropine, pressure ejection of carbachol (Carb; 10 μL of 10^{-2} *M*) produces a depolarization in a sympathetic cell. This depolarization is associated with a decrease in input resistance, measured as the voltage response to constant pulses of hyperpolarizing current. γ-Aminobutyric acid (GABA; 10 μL of 10^{-2} *M*) also produces a depolarization with a decrease in input

m*M* potassium phosphate, pH 7.0, 1% (vol/vol) Triton X-100]. Centrifugation is carried out in a Beckman Ti55 rotor at 50,000 rpm for 5–10 h at 4°C. Fractions are collected at the end of the run by piercing the bottom of the centrifuge tube and allowing the tube to empty under the force of gravity. The ^{125}I-α-bungarotoxin binding site in chick ciliary ganglion has a sedimentation coefficient of 11S (Fig. 7). This is larger than the 9S peak observed for the monomeric nicotinic receptor from *Torpedo* electric organ (Fig. 7).

Neurotoxins such as ^{125}I-α-bungarotoxin can also be used in autoradiographic studies to determine the ultrastructural localization of sites identified by radiolabeled toxins. Procedures have been described for examining the localization of toxin binding in the vertebrate neuromuscular junction (Fertuck and Salpeter, 1976) and in autonomic ganglia (Loring et al., 1985).

3. Important Toxins Affecting Cholinergic Neurons

3.1. Sites Affected by Neurotoxins

Toxins interfere with cholinergic neurotransmission by a variety of mechanisms. Presynaptically acting toxins interfere with the

←

resistance. (B) The ganglia are subsequently exposed to 32 μ*M* lophotoxin for 20 min. Approximately 30 min after lophotoxin exposure, the response to carbachol (10 μL of 10^{-2}*M*) in another cell is almost completely blocked. Response to GABA (10 μL of 10^{-2} *M*) is similar to that seen before lophotoxin exposure. A depolarizing current pulse delivered after recovery from GABA (asterisk) is also shown on a greatly expanded time scale (lower record). The cell exhibits normal-appearing, overshooting action potentials in response to the depolarizing current pulse. (C) To determine whether lophotoxin has any effect on muscarinic responses in sympathetic ganglia, atropine was deleted from the physiological saline in another preparation. Following exposure to lophotoxin (32 μ*M* for 20 min), the response to carbachol (10 μL of 10^{-2}*M*) consists of a depolarization associated with an increase in input resistance. This response is muscarinic in nature (Dryer and Chiappinelli, 1985b). GABA (10 μL of 10^{-2}*M*) produces a depolarization with a decrease in input resistance similar to that seen before lophotoxin exposure. Resting membrane potential for the neurons shown is between –40 and –50 mV.

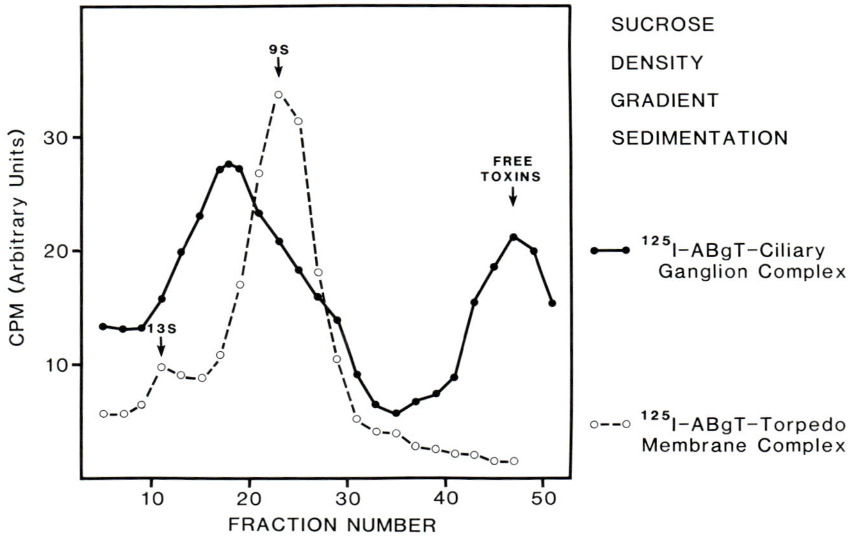

Fig. 7. Triton X-100-solubilized α-bungarotoxin binding sites. A 5–20% (wt/vol) sucrose gradient is used to determine the sedimentation coefficients of α-bungarotoxin binding sites in *Torpedo* electric organ membranes and ciliary ganglion homogenates. The *Torpedo* preparation exhibits a major radioactive peak at 9S, which represents the monomeric nicotinic receptor and a smaller peak at 13S caused by receptor dimers. The ganglion preparation exhibits a single, broad peak centered at 11S, as well as an additional peak containing free ^{125}I-α-bungarotoxin. The affinity of ^{125}I-α-bungarotoxin for the *Torpedo* receptor is so high and the off-rate is so long that no free ^{125}I-α-bungarotoxin appears in the *Torpedo* sample even though the gradient ran for 8 h. Although the ^{125}I-α-bungarotoxin binding site in ciliary ganglion resembles the *Torpedo* receptor in some respects, it does not appear to be the physiologically detected neuronal nicotinic receptor (*see* section 3.4.1.2.1).

synthesis or release of ACh from nerve terminals. Other toxins inhibit the degradative enzyme of ACh, acetylcholinesterase (EC 3.1.1.7). By far the largest group of toxins that interfere with transmission has selective effects on ACh receptors located on postsynaptic membranes. Figure 8 illustrates the sites affected by many of these toxins. The remainder of this review will examine the properties of some representative toxins affecting cholinergic neurons.

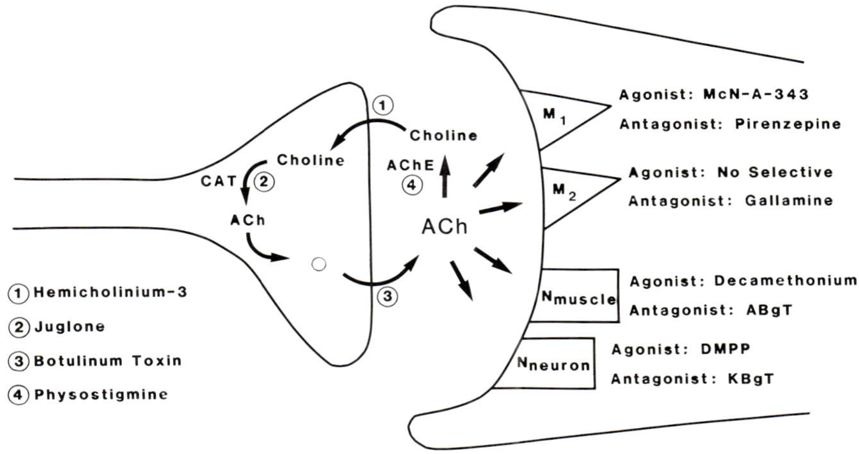

Fig. 8. Sites affected by neurotoxins. In this schematic diagram, a cholinergic nerve terminal (left) synapses on a composite postsynaptic element (right) containing four types of cholinergic receptors. Two types of muscarinic receptors (M_1 and M_2) and two types of nicotinic receptors (N_{muscle} and N_{neuron}) are shown, along with representative toxins that are selective for these receptor subtypes. The numbers refer to (1) high-affinity choline uptake; (2) inhibition of choline acetyltransferase (CAT) activity; (3) release of acetylcholine (ACh); and (4) inhibition of acetylcholinesterase (AChE) activity. Representative toxins interfering with these processes are indicated below the nerve terminal. The small open circle in the nerve terminal represents synaptic vesicles containing acetylcholine. Abbreviations: DMPP, 1,1-dimethyl-4-phenylpiperazinium; ABgT, α-bungarotoxin; KBgT, κ-bungarotoxin.

3.2. Toxins Affecting Presynaptic Nerve Terminals

3.2.1. Botulinum Toxin

A number of extremely toxic proteins are produced by various strains of the anaerobic bacteria, *Clostridium botulinum.* The most frequently studied of these pharmacologically similar toxins is botulinum toxin type A, which has a molecular weight of approximately 140,000 dalton and is composed of two subunits cross-linked by disulfide bonds (Tse et al., 1982).

Botulinum toxin is a potent blocker of cholinergic neurotransmission at the neuromuscular junction. In fact, the major symptom of botulism is muscle paralysis. Several pieces of evidence indicate that the blockade produced by botulinim toxin is of a presynaptic nature. The most direct evidence is that the muscle remains sensitive to applied cholinergic agonists following exposure to botulinum toxin. In addition, when recording from muscle cells, spontaneous miniature endplate potentials (mepps) continue to be observed, indicating that the muscle still responds to released ACh (Harris and Miledi, 1971; Spitzer, 1972). Botulinum toxin decreases the amount of ACh released from presynaptic terminals. Although the toxin may actually interfere to some extent with the synthesis of ACh, this decreased release of ACh is caused not by depletion of transmitter, but rather by a blockade of the release mechanism itself (Gundersen, 1980).

A requirement for the release of ACh from presynaptic terminals is a voltage-dependent Ca^{2+} influx triggered by depolarization. This Ca^{2+} influx is generally thought to be relatively normal after exposure to botulinum toxin (Cull-Candy et al., 1976; Gundersen et al., 1982), although Hirokawa and Heuser (1981) present ultrastructural evidence that Ca^{2+} influx is below normal during repetitive nerve stimulation in toxin-treated muscle. A larger than normal influx of Ca^{2+} is required for normal ACh release in toxin-treated muscle (Cull-Candy et al., 1976). Thus, drugs that enhance Ca^{2+} entry into nerve terminals, such as 4-aminopyridine, the calcium ionophore A-23187, and black widow spider venom, can partially reverse the effects of botulinum toxin (Lundh et al., 1976; Hirokawa and Heuser, 1981; MacKenzie et al., 1982).

Media that suppress transmitter release retard the blockade produced by botulinum toxin (Simpson, 1973). The blockade is therefore to some extent use-dependent, although the actual binding of the toxin to presynaptic sites is not (Simpson, 1974). There is thus a two-stage process observed: an initial stage in which botulinum toxin binds to specific sites on the nerve terminal membrane (Dolly et al., 1984), followed by a stage in which the toxin is internalized (Simpson, 1980) and an irreversible blockade of transmitter release occurs. The initial stage accounts for the long latency period (up to 30 min) of the blockade.

Botulinum toxin affects postganglionic cholinergic terminals at somewhat higher doses (5–10 n*M*) than those that are effective at the neuromuscular junction (Bigalke and Habermann, 1980). Postganglionic cholinergic terminals are susceptible regardless of whether they are parasympathetic or sympathetic in origin (Ambache, 1951). Preganglionic cholinergic terminals are also very sensitive to the action of botulinum toxin (Ambache, 1951; Kupfer,

1958). In contrast, botulinum toxin has little effect on postganglionic noradrenergic terminals (Ambache, 1951) or on nonadrenergic, noncholinergic transmission in the guinea pig ileum (MacKenzie et al., 1982). Thus botulinum toxin, at least at low doses, is relatively selective for cholinergic nerve terminals. This feature distinguishes botulinum toxin from the chemically similar tetanus toxin produced by *Clostridium tetani*. Tetanus toxin shows little selectivity for particular types of nerve terminals, and its blocking action on inhibitory terminals containing GABA and glycine results in the muscle spasms characteristic of tetanus (Mellanby and Green, 1981). Tetanus toxin blocks cholinergic terminals as well, but compared to botulinum toxin, it is five-fold less potent at postganglionic cholinergic terminals and 500-fold less potent at the motoneuron nerve terminal (Bigalke and Habermann, 1980).

Botulinum toxin can be used to chemically denervate skeletal muscle. The effects of acute botulinum toxin poisoning resemble those of acute denervation, whereas chronic exposure to the toxin mimics chronic denervation in that ACh receptors increase in number and distribution and the muscle becomes supersensitive to Ach (Simpson, 1977).

3.2.2. Black Widow Spider Venom

The venom of the black widow spider, *Latrodectus mactans*, contains a number of toxins with differing pharmacological actions (Frontali et al., 1976). A component variously referred to as α-latrotoxin (Tzeng and Siekevitz, 1978), B_5 fraction (Frontali et al., 1976), or fraction 8 (Ornberg et al., 1976) produces a massive release of ACh at the vertebrate neuromuscular junction. α-Latrotoxin has a molecular weight of approximately 130,000 dalton and is devoid of enzymatic activity. Upon binding to a presynaptic nerve terminal, α-latrotoxin behaves like a divalent cation ionophore, allowing Ca^{2+} and other divalent cations to flow into the nerve terminal (Finkelstein et al., 1976). The increased intracellular Ca^{2+} results in a virtually total depletion of synaptic vesicles from the presynaptic terminal and an eventual blockade of cholinergic transmission (Frontali et al., 1976). The vesicle-depleted terminals become swollen and the mitochondria within have an enlarged appearance and contain prominent electron-dense granules after fixation, presumably because of Ca^{2+} accumulation (Hirokawa and Heuser, 1981).

The effect of crude black widow spider venom at the neuromuscular junction is similar to that of the purified α-latrotoxin

(Clark et al., 1972). The venom also produces a massive release of ACh from cholinergic terminals in the rat superior cervical ganglion (Paggi and Rossi, 1971; Pumplin and McClure, 1977). α-Latrotoxin (Tzeng and Siekevitz, 1978) and crude venom (Frontali et al., 1973) release catecholamines from presynaptic terminals, indicating that spider venom is not selective for cholinergic terminals. α-Latrotoxin also enhances the release of dopamine from cultured PC12 cells, a clonal cell line derived from a rat pheochromocytoma (Meldolesi et al., 1983).

3.2.3. β-Bungarotoxin

β-Bungarotoxins are purified from the venom of the banded krait, *Bungarus multicinctus* (*see* Fig. 1A). The complete amino acid sequences of five different β-bungarotoxins have been reported (Kondo et al., 1978, 1982a,b). These toxins all consist of two dissimilar polypeptide chains, cross-linked by a disulfide bond. The A chain contains 120 amino acid residues and strongly resembles snake venom phospholipases A_2 (EC 3.1.1.4). The B chain contains 60 residues and does not closely resemble any other snake venom protein. The β-bungarotoxin used by most investigators is probably a mixture of several of these closely related β-bungarotoxins.

β-Bungarotoxin produces a complex blockade of transmission at the neuromuscular junction. Initially, there is either a facilitation or an inhibition of neuromuscular transmission and a four- to fivefold increase in mepp frequency. After 10–20 min, transmission and mepp frequency return to relatively normal levels. Finally, after 30–60 min, neuromuscular transmission is irreversibly blocked and the presynaptic terminals are heavily damaged or completely destroyed (Chang et al., 1973; Chang and Huang, 1974; Abe et al., 1977). The second, irreversible stage of blockade takes place even when unbound β-bungarotoxin is washed off the preparation at the end of the period of the initial effects (Chang and Huang, 1974).

β-Bungarotoxin appears to bind selectively to certain sites on presynaptic nerve terminals (Othman et al., 1982). This binding is accompanied by the initial effects of the toxin on mepp frequency and quantal content (Oberg and Kelly, 1976). Although this initial binding is not dependent on the enzymatic activity of the toxin, the later, irreversible action of β-bungarotoxin does not occur if the toxin's phospholipase A_2 activity is inhibited (Strong et al., 1976; Howard and Truog, 1977). Thus the B chain of the toxin is presumed to function as a selective probe for some presynaptic membrane component, and the A chain functions to nonselectively destroy the terminal by virtue of its potent phospholipase A_2 activity. Other snake venom phospholipases A_2, which lack this B chain, do not exhibit effects similar to β-bungarotoxin (Strong et al., 1976). The presynaptic sites identified by radiolabeled β-bungarotoxin are not recognized by α-latrotoxin or botulinum toxin (Othman et al., 1982), and β-bungarotoxin can partially reverse the reduction in mepps caused by botulinum toxin (Tse et al., 1986).

β-Bungarotoxin also affects cholinergic transmission in autonomic ganglia. The biphasic nature of the ganglionic blockade (*see* Fig. 5) is similar to the two-stage action of the toxin at the neuromuscular junction. β-Bungarotoxin causes an increase in the spontaneous release of ACh and a blockade of the stimulation-evoked release of ACh in preganglionic terminals (Kato et al., 1977). The toxin has little or no effect on adrenergic terminals or postganglionic cholinergic nerves (Kato et al., 1977).

Several studies have examined the possible use of β-bungarotoxin as a selective degenerative toxin. Following injection into the rat brain, β-bungarotoxin was found to be 10^6 times more potent than kainic acid in destroying neurons (Hanley and Emson, 1979). There appeared to be little selectivity for cholinergic neurons, however. In the chick retina, β-bungarotoxin destroyed both cholinergic and GABA-containing neurons at doses that had little effect on adrenergic cells (Rehm et al., 1982). Thus, although β-bungarotoxin may exhibit some selectivity in destroying neurons (Hirokawa, 1978), it cannot be used to selectively eliminate cholinergic neurons. The choline mustard analog AF64A has also been proposed as a degenerative toxin selective for cholinergic terminals (Mantione et al., 1981). AF64A appears to destroy cholinergic nerve terminals both in the central nervous system (Mantione et al., 1981) and in the autonomic nervous system (Kessler, 1985).

3.2.4. Other Toxins with Presynaptic Effects

Hemicholinium-3, a synthetic compound, is a potent inhibitor of ACh synthesis in nerve terminals. Hemicholinium-3 acts by interfering with the high-affinity uptake of choline (Birks and MacIntosh, 1961). Although ACh is depleted in hemicholinium-3-treated terminals, synaptic vesicles largely devoid of ACh content remain in the terminals and are released upon exposure to black widow spider venom (Ceccarelli and Hurlbut, 1975).

Several synthetic compounds interfere with the function of choline acetyltransferase (Smith et al., 1967; Rowell and Chiou, 1976). Juglone, an extract of walnut hulls, also inhibits choline acetyltransferase activity in vitro (Haubrich and Wang, 1976). Although some of these inhibitors are effective in reducing choline acetyltransferase activity in vivo, they do not always decrease brain ACh levels, indicating that, under some conditions, the enzyme is not rate-limiting in ACh synthesis (Carson et al., 1972; Haubrich and Wang, 1976; Budai et al., 1986). High-affinity choline uptake is at times the rate-limiting step in ACh synthesis, and this may explain why hemicholinium-3 rapidly exhausts ACh from stimulated terminals (Ceccarelli and Hurlbut, 1975).

A number of naturally occurring toxins interfere with the normal functioning of neuronal sodium channels (Catterall et al., 1982). Some of these, such as the scorpion venom peptide tityustoxin, can increase the release of ACh at presynaptic terminals (Warnick et al., 1976). The reason for this enhanced release of ACh is a persistent depolarization of the terminal secondary to the activation of sodium channels. These neurotoxins do not show any selectivity for ACh-containing terminals.

3.3. Acetylcholinesterase Inhibitors

Acetylcholinesterase, which is present in high concentrations at cholinergic synapses, rapidly hydrolyzes any free ACh in the synaptic cleft. Thus, the lifetime of ACh after it has been released from presynaptic terminals is extremely short, on the order of 200 μs (Colquhoun, 1979). If the activity of this degradative enzyme is inhibited, ACh levels in the cleft build up, resulting in an initial facilitation of cholinergic transmission. After continued enzyme inhibition, blockade of transmission can occur because of desensitization of postsynaptic receptors by excess transmitter. A number of agents are potent anticholinesterases. They affect all types of cholinergic synapses, except when physical barriers such as the blood–brain barrier prevent them from acting at certain sites. They are effective in essentially all vertebrate and invertebrate species, and a major use is as insecticides (Chadwick, 1963).

3.3.1. Physostigmine

Physostigmine, also known as eserine, is purified from the seed of *Physostigma venenosum,* a plant that grows in West Africa. Being a tertiary amine, physostigmine readily crosses the blood–

brain barrier (Karczmar, 1967). The actions of physostigmine are considered to be reversible, in contrast to the "irreversible" organophosphorus agents described below. A number of related compounds have been synthesized. The most important is neostigmine, which does not cross into the central nervous system. At high concentrations, neostigmine and other quaternary ammonium-containing agents exhibit an affinity for certain ACh receptors. For example, at the toad neuromuscular junction, perfusion with neostigmine at doses below 2 mg/mL (6.6 μM) increases the amplitude of miniature end-plate currents. At doses above 2 mg/mL, however, neostigmine decreases the amplitude of miniature end-plate currents (Gage and McBurney, 1975), probably as a result of direct inhibitory action on nicotinic receptors.

3.3.2. Organophosphorus Agents

A number of synthetic organophosphorus compounds are extremely potent and essentially irreversible (actually very slowly reversible) inhibitors of acetylcholinesterase (Taylor, 1985b). Among these are the "nerve gases" soman, sarin, and tabun, which are volatile oils. The insecticides malathion and parathion are members of this group, and accidental poisoning with anticholinesterases used as insecticides is perhaps the most commonly seen toxic effect of the drugs. Experimentally, the drugs are frequently used when the effects of applied ACh are being examined. Alternatively, one can eliminate the need for anticholinesterase exposure by using a cholinomimetic such as carbachol, which is a poor substrate for acetylcholinesterase.

Diisopropylfluorophosphate (DFP) is a commonly used organophosphorus anticholinesterase. DFP exposure followed by nerve stimulation results in a characteristic necrosis of muscle tissue. Leonard and Salpeter (1979) determined that this myopathy was caused not by DFP exposure itself, but by the increased agonist concentrations observed in the presence of DFP. This agonist-induced myopathy requires the presence of functional nicotinic receptors and may be caused by excessive Ca^{2+} influx (Leonard and Salpeter, 1979).

3.4. Toxins Affecting Acetylcholine Receptors

Following its release from presynaptic terminals, ACh interacts with selective receptors located on nerve and muscle cell membranes. These ACh receptors are not all alike. The two major categories of ACh receptors are defined pharmacologically by their interactions with a pair of naturally occurring toxins. Nicotinic receptors are sensitive to the actions of nicotine, but are unaffected by muscarine. Conversely, muscarinic receptors are readily identified by muscarine, but are not detected by nicotine. Subcategories of these receptor types also exist, defined by their pharmacological profiles as well as their locations.

Although some agonists, such as ACh and carbachol, bind with similar affinities to both nicotinic and muscarinic receptors, cholinergic antagonists are selective for certain types of receptors. This selectivity is an important property since it can provide valuable information concerning the differences between ACh receptors.

3.4.1. Nicotinic Receptors

Nicotinic ACh receptors are found in junctions between motoneuron terminals and vertebrate skeletal muscle (Fertuck and Salpeter, 1976). Electric organs of certain electric fish that are derived from skeletal muscle contain very large amounts of these muscle nicotinic receptors (Popot and Changeux, 1984). Neuronal nicotinic receptors are found in vertebrates in autonomic ganglia and in the central nervous system (Brown, 1979). Nicotinic receptors are also localized on some invertebrate neurons. Neuronal nicotinic receptors can be distinguished pharmacologically from the muscle nicotinic receptors, even though both classes of receptors share a number of properties.

In vertebrates, nicotinic receptors are involved in an excitatory response mediated by an increase in Na^+ and K^+ conductance. A cation-selective channel, which is an intimate part of the nicotinic receptor, opens when cholinergic agonists bind to specific sites on the receptor (Popot and Changeux, 1984). Cholinergic antagonists bind to either the ACh recognition sites on the receptor or the ion channel itself. The former antagonists are competitive with ACh, whereas the latter are noncompetitive.

The insect nicotinic receptor functions much like the vertebrate receptor (Sattelle et al., 1983). In *Aplysia* and other molluscs, there appear to be three separate nicotinic receptors mediating increases in the conductances of Na^+, K^+, and Cl^-, respectively (Kehoe, 1972a,b).

3.4.1.1. Toxins Affecting All Nicotinic Receptors.

3.4.1.1.1. Nicotine. Nicotine is isolated from the leaves of the tobacco plant, *Nicotiana tabacum*. The purified alkaloid is a volatile liquid that readily dissolves in water (Taylor, 1985c). Nicotine is a potent insecticide, causing death by excessive excitation of neuronal nicotinic receptors (Gepner et al., 1978). Abuse of tobacco by humans is largely related to the actions of nicotine on central and peripheral neuronal nicotinic receptors (Rosecrans, 1979).

The initial effects of nicotine on all types of nicotinic receptors are excitatory. The drug acts as a cholinomimetic, activating receptors and eliciting action potentials in postganglionic neurons and skeletal muscle. With continued exposure to nicotine, transmission is blocked, either by excessive receptor stimulation (depolarizing blockade) or by desensitization, in which nicotine remains bound to receptors, but the receptors are in an inactive conformation (Taylor, 1985c). Thus, nicotine is a complex agonist/antagonist at nicotinic receptors, a property that limits the usefulness of the drug in a number of experimental situations.

Radiolabeled nicotine has been used to characterize nicotinic receptors in both muscle and neuronal tissue. It is not a particularly good choice for this purpose in muscle, where the snake α-neurotoxins (*see* section 3.4.1.2.1) are better ligands. ^{3}H-Nicotine appears to be a good ligand for characterizing neuronal nicotinic receptors. Clarke et al. (1985) found that the localization of ^{3}H-nicotine binding sites in rat brain coincided almost exactly with that of ^{3}H-ACh binding sites measured in the presence of the muscarinic antagonist atropine. The central ^{3}H-nicotine binding site is very sensitive to blockade by nicotinic agonists, whereas nicotinic antagonists are generally less potent at this site (Marks and Collins, 1982).

3.4.1.1.2. *d*-Tubocurarine. *d*-Tubocurarine is an alkaloid, containing six carbon or carbon/nitrogen rings and a quaternary nitrogen, that is purified from the South American plant *Chondrodendron tomentosum* (Taylor, 1985d). It has long been used by the indigenous populations as an arrow poison for killing wild game. *d*-Tubocurarine is a potent antagonist of muscle nicotinic receptors, and blocks neuronal nicotinic receptors at slightly higher (5–10-fold) concentrations. The toxin is a nondepolarizing, competitive antagonist at most nicotinic receptors. It can also act as an agonist in embryonic muscle, however (Ziskind and Dennis, 1978; Morris et al., 1983) and has some noncompetitive blocking effects related to interaction with the ion channel associated with the nicotinic receptor (Katz and Miledi, 1978; Shaker et al., 1982).

Neubig and Cohen (1979) found that ^{3}H-*d*-tubocurarine bound with two very different affinities (K_{d1} = 33 nM; K_{d2} = 7.7 μM) to sites on *Torpedo* electric organ nicotinic receptors. In contrast, both of these sites were seen with a single affinity by ^{3}H-ACh or ^{3}H-α-neurotoxin, indicating that the two ACh binding sites contained on each muscle nicotinic receptor are very similar, but not identical. *d*-Tubocurarine blocks the function of the receptor under conditions in which only the high-affinity site is occupied by the drug, suggesting that both sites need to be occupied by agonist for a functional response to occur (Neubig and Cohen, 1979).

3.4.1.1.3. Lophotoxin. Lophotoxin is a diterpene lactone isolated from the gorgonian coral *Lophogorgia rigida* (Fenical et al., 1981). Lophotoxin produces an apparently irreversible blockade of nicotinic transmission at the vertebrate neuromuscular junction (Culver and Jacobs, 1981). This blockade is caused by the binding of the toxin to ACh recognition sites on the muscle nicotinic receptor (Culver et al., 1984). As with *d*-tubocurarine (Neubig and Cohen, 1979), lophotoxin binds with high affinity to only one of the ACh sites on the muscle receptor. This site is the one seen with low affinity by *d*-tubocurarine (Culver et al., 1984). Lophotoxin does not exhibit any presynaptic effects, nor does it bind to the ion channel associated with the nicotinic receptor (Atchison et al., 1984).

Lophotoxin blocks neuronal nicotinic receptors in frog sympathetic ganglia and in guinea pig and rabbit ileal strips (Langdon and Jacobs, 1985). In chick autonomic ganglia, the toxin blocks nicotinic receptors without affecting muscarinic or GABA receptors (Sorenson et al., 1987; *see* Fig. 6). The effects of low doses of lophotoxin (1 μM) in chick ganglia are partially reversible after 3–5 h of washing out the toxin, whereas no recovery is seen when higher doses of the toxin are used. As in muscle (Culver et al., 1984), the effects of lophotoxin in autonomic ganglia can be partially prevented by concurrent exposure of cells to *d*-tubocurarine, whereas noncompetitive ion channel blockers (local anesthetics) do not protect against lophc toxin (Sorenson et al., 1987). Lophotoxin blocks nicotinic receptors in muscle and nervous tissue at essentially identical doses (Sorenson et al., 1987).

3.4.1.1.4. Conus Toxins. The fish-hunting cone snails such as *Conus geographus* produce a variety of neurotoxins (Olivera et al., 1985). The α-conotoxins (conotoxins GI, GIA, GII, MI) completely abolish neuromuscular transmission at low doses (3–4 μM) by binding to muscle nicotinic receptors (Gray et al., 1981; McManus and Musick, 1985). This binding is readily reversible upon washing off the toxins. The amino acid sequences of the α-conotoxins (Gray et al., 1981; Olivera et al., 1985) exhibit considerable homology. The toxins are monomeric peptide chains containing between 13 and 15 amino acid residues and are internally cross-linked by two disulfide bonds. The effects of the α-conotoxins on neuronal nicotinic receptors remain to be elaborated.

A second variety of *Conus* toxins, the ω-conotoxins, should also be mentioned in connection with cholinergic neurons. These small peptides (containing 25–29 amino acids) are potent blockers of presynaptic voltage-sensitive Ca^{2+} channels (Olivera et al., 1985), and thus prevent the release of ACh from presynaptic terminals at the neuromuscular junction.

3.4.1.1.5. Histrionicotoxin. Histrionicotoxin is a spiropiperidine alkaloid found in the skin of *Dendrobates histrionicus,* the poison-dart frog of Central America. Histrionicotoxin and its saturated derivative, perhydrohistrionicotoxin, block nicotinic transmission at the neuromuscular junction at low doses (1 μ*M*). Unlike the toxins mentioned above, however, histrionicotoxin is a noncompetitive antagonist at nicotinic receptors (Albuquerque et al., 1973; Kato and Changeux, 1976; Eldefrawi et al., 1977). Histrionicotoxin appears to bind to the ion channel associated with the nicotinic receptor, thus preventing ion translocation. The binding site for the toxin resides on the purified receptor, but is an allosteric site distinct from the ACh recognition sites. Local anesthetics also bind to this ion channel site and in so doing block nicotinic transmission (Krodel et al., 1979). Perhydrohistrionicotoxin blocks nicotinic transmission in chick autonomic ganglia, but higher doses (32 μ *M*) of the toxin are required (Chiappinelli, unpublished observations). Binding of histrionicotoxin derivatives to muscarinic receptors has also been reported (Cremo and Schimerlik, 1983).

3.4.1.2. Toxins Selective for Muscle Nicotinic Receptors. 3.4.1.2.1. Snake Venom α-Neurotoxins. The recent history of research on the muscle nicotinic receptor is intimately tied to that of the α-neurotoxins, or postsynaptic neurotoxins, purified from the venom of elapid and hydrophid snakes. Soon after α-neurotoxins were first isolated, they were found to bind with extremely high affinity ($K_d = 10^{-12}$ to $10^{-9} M$) and specificity to the ACh recognition sites on muscle nicotinic receptors (Changeux et al., 1970). Occupation by α-neurotoxin of either one or both ACh sites on a muscle receptor blocks the functional response of the receptor to nicotinic agonists (Neubig and Cohen, 1980). The very slow off-rates observed (up to several days; Blanchard et al., 1979) make the toxins ideal probes for localization studies, electrophysiological investigations, and the purification of the muscle nicotinic receptor. It was the α-neurotoxins, combined with the specialized muscle of the electric fish electric organ, that have made the muscle nicotinic receptor the best characterized of any neurotransmitter receptor (reviewed in Stroud, 1983; Popot and Changeux, 1984; Raftery et al., 1984).

The amino acid sequences of over 70 α-neurotoxins from 25 species of snakes are now known (Mebs, 1985). Although there is considerable homology between all α-neurotoxins, two subgroups can be described on the basis of chain length and the number of intrachain disulfide bonds (Karlsson, 1979). The long α-neurotoxins consist of 66–74 amino acid residues in a single peptide chain cross-linked by five disulfide bonds. The short α-neurotoxins consist of 60–62 amino acid residues in a single chain cross-linked by four disulfide bonds. α-Bungarotoxin, a long α-neurotoxin purified from the venom of *Bungarus multicinctus* (Mebs et al., 1972; *see* Fig. 1A), is among the most potent and the most frequently used of the α-neurotoxins. The amino acid sequence of α-bungarotoxin is shown in Fig. 9.

Not all muscle nicotinic receptors are blocked by α-neurotoxins. In the snakes and advanced lizards, these toxins do not bind to skeletal muscle nicotinic receptors, or else bind with relatively low affinity (Burden et al., 1975). α-Neurotoxins also do not block nicotinic receptors in most vertebrate neuronal systems examined. Neuronal systems that are insensitive to α-neurotoxins include autonomic and central neurons from birds and mammals (reviewed in Schmidt et al., 1980; Morley and Kemp, 1981; Chiappinelli, 1986). An example is shown in Fig. 4, in which 1 μ *M* α-bungarotoxin has no effect on nicotinic transmission in a chick autonomic ganglion. Radiolabeled α-neurotoxins bind to pharmacologically nicotinic sites in these tissues, but immunological (Patrick and Stallcup, 1977a,b), pharmacological (Chiappinelli, 1983; Loring et al., 1984) and anatomical (Schwartz et al., 1982; Clarke et al., 1985; Loring et al., 1985) evidence indicates that these sites are not likely to be involved in physiologically detected nicotinic neuronal transmission. The sites do, however, exhibit a number of properties that indicate they are closely related to muscle nicotinic receptors, whatever their function is eventually determined to be (Conti-Tronconi et al., 1985; Kemp et al., 1985).

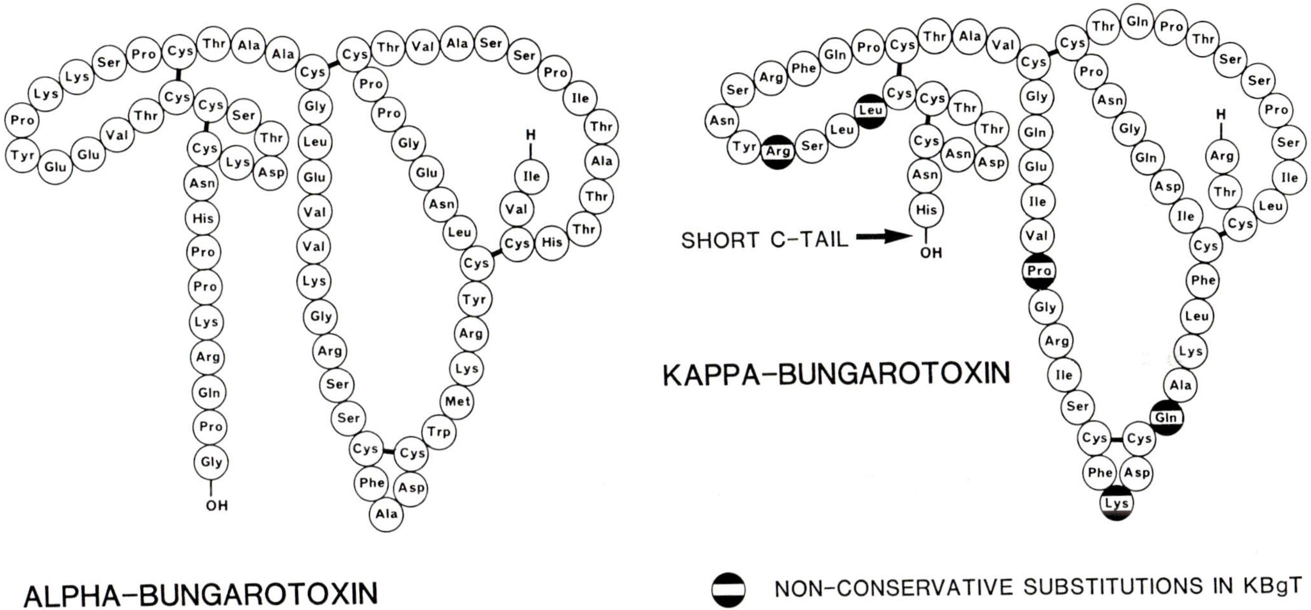

Fig. 9. Complete amino acid sequences of α-bungarotoxin and κ-bungarotoxin. These closely related neurotoxins share 47% homology, which includes all ten cysteinyl residues involved in internal disulfide bridges. Nonconservative substitutions in κ-bungarotoxin, which distinguish the toxin from all of the α-neurotoxins, are highlighted. The highlighted glutaminyl residue (Gln) is invariantly a tryptophanyl residue (Trp) in the α-neurotoxins, where it is considered essential for α-neurotoxin physiological activity. Sequence of α-bungarotoxin is from Mebs et al. (1971), and the sequence of κ-bungarotoxin is from Grant and Chiappinelli (1985).

Certain α-neurotoxins do block nicotinic transmission in frog sympathetic ganglia (Marshall, 1981) and central neurons of amphibians and teleosts (Miledi and Szczepaniak, 1975; Freeman et al., 1980). Invertebrate nicotinic receptors are also blocked by certain α-neurotoxins (Kehoe et al., 1976; Ono and Salvaterra, 1981; Sattelle et al., 1983; David and Sattelle, 1984).

3.4.1.3. Toxins Selective for Neuronal Nicotinic Receptors. A number of synthetic compounds can be used to distinguish neuronal nicotinic receptors from muscle receptors. Trimethaphan, tetraethylammonium, and hexamethonium are all more potent at blocking nicotinic receptors in autonomic ganglia than at blocking muscle nicotinic receptors (Taylor, 1985c). The recently discovered κ-neurotoxins share this selectivity for neuronal nicotinic receptors.

3.4.1.3.1. Snake Venom κ-Neurotoxins. The first κ-neurotoxin to be completely sequenced is κ-bungarotoxin (Grant and Chiappinelli, 1985; Fig. 9). This 66-amino-acid peptide is isolated from the venom of *Bungarus multicinctus* (Chiappinelli, 1983; *see* Fig. 1). κ-Bungarotoxin exhibits considerable homology (47%) with the long α-neurotoxins, such as α-bungarotoxin. Compared to all of the fully sequenced α-neurotoxins, however, κ-bungarotoxin has several nonconservatively substituted amino acid residues (Fig. 9). Most notable of these is the lack of an otherwise invariant tryptophanyl residue, which has been replaced by a glutaminyl residue in κ-bungarotoxin. In addition, the carboxy-terminal tail of κ-bungarotoxin is considerably reduced compared to that of the long α-neurotoxins. κ-Bungarotoxin exists in physiological solutions entirely as a dimer of identical subunits (Chiappinelli and Lee, 1985), whereas the α-neurotoxins are monomeric under similar conditions.

The physiological effects of κ-bungarotoxin are in many ways opposite those of α-bungarotoxin. Whereas 1 μ*M* α-bungarotoxin has no effect on neuronal nicotinic transmission in chick autonomic ganglia, 75 n*M* κ-bungarotoxin completely blocks neuronal transmission (Chiappinelli, 1983; *see* Fig. 4). Conversely, α-bungarotoxin is 200 times more potent than κ-bungarotoxin in binding to chick skeletal muscle nicotinic receptors (Chiappinelli et al., 1985; Wolf et al., 1988).

Considerable evidence indicates that κ-bungarotoxin blocks ganglionic transmission by binding to the acetylcholine recognition sites on neuronal nicotinic receptors. *d*-Tubocurarine (Chiappinelli, 1983) and trimethaphan (Chiappinelli, 1986) both protect against the effects of κ-bungarotoxin, whereas noncompetitive antagonists do not (Sorenson et al., 1987). Intracellular recording demonstrates that κ-bungarotoxin blocks nicotinic responses to carbachol, although having no effect on other active or passive properties of ganglion neurons (Dryer and Chiappinelli, 1983; Chiappinelli and Dryer, 1984).

Two other peptides isolated from *Bungarus multicinctus* venom share many biochemical and pharmacological properties with κ-bungarotoxin (Ravdin and Berg, 1979; Loring et al., 1984). A preliminary report (Loring and Zigmond, 1985) indicates that the sequence of one of these, termed toxin F, is likely to be identical to the complete amino acid sequence previously reported for κ-bungarotoxin (Grant and Chiappinelli, 1985). Although amino acid sequence data are currently unavailable for the other peptide, termed Bgt 3.1, this toxin shares the same isoelectric point (Loring et al., 1984) and apparent molecular weight (Ravdin and Berg, 1979) with κ-bungarotoxin and toxin F. Thus, both toxin F and Bgt 3.1 may eventually be shown to be identical to κ-bungarotoxin.

A κ-neurotoxin, which is not identical to κ-bungarotoxin, has recently been isolated from the venom of *Bungarus flaviceps*. κ-Flavitoxin has a pharmacological profile essentially identical to that of κ-bungarotoxin and blocks neuronal nicotinic receptors at doses as low as 50 n*M* (Chiappinelli et al., 1986). The complete amino acid sequence of κ-flavitoxin (Grant et al., 1988) exhibits 85% homology with κ-bungarotoxin, confirming that a new class of neuronal nicotinic receptor probes, the κ-neurotoxins, can be described on the basis of sequence homology and pharmacological properties.

Radiolabeled κ-bungarotoxin (Chiappinelli, 1983), toxin F (Loring et al., 1984), and κ-flavitoxin (Chiappinelli et al., 1986) all identify a nicotinic site in autonomic ganglia that is distinct from the ^{125}I-α-bungarotoxin binding site, and which appears to be associated with the neuronal nicotinic receptor. This presumptive nicotinic receptor is localized preferentially at synapses in the ganglion (Loring and Zigmond, 1985), whereas the ^{125}I-α-bungarotoxin site is not (Loring et al., 1985). Thus, it is possible that the κ-neurotoxins will be as useful in characterizing neuronal nicotinic receptors as the closely related α-neurotoxins have been in characterizing muscle nicotinic receptors.

3.4.1.3.2. Surugatoxin. Surugatoxin is an alkaloid isolated from the Japanese ivory mollusc, *Babylonia japonica* (Kosuge et al., 1972). Surugatoxin (0.1–2 μ*M*) reversibly blocks nicotinic transmission in mammalian sympathetic ganglia, although having little effect on muscle nicotinic receptors (Hayashi and Yamada, 1975; Brown et al., 1976). Surugatoxin antagonizes the depolarizing actions of nicotinic agonists, and has no apparent presynaptic effects (Brown et al., 1976). The toxin does not block muscarinic or GABA receptors (Brown et al., 1976). A related compound, neosurugatoxin, has similar selectivity for neuronal nicotinic receptors (Hayashi et al., 1984). Neosurugatoxin is a potent inhibitor (IC_{50} = 70 nM) of ^{3}H-nicotine binding in rat brain homogenates, while having no effect at concentrations up to 10 μ M on ^{3}H-3-quinuclidinyl benzilate (^{3}H-QNB) binding to muscarinic receptors (Hayashi et al., 1984).

Thus, surugatoxin and neosurugatoxin appear to be highly selective, potent antagonists at neuronal nicotinic receptors. The relatively long on and off rates of the toxins (blockade in 30 min at 1 μ*M*, recovery in 45 min; Brown et al., 1976) indicate that they should be useful probes in identifying and purifying nicotinic receptors in autonomic ganglia and the central nervous system. Radiolabeled derivatives of these toxins need to be prepared and tested in binding assays.

3.4.2. Muscarinic Receptors

Muscarinic receptors in vertebrates are found on central and ganglionic neurons and on cardiac and smooth muscle cells. Invertebrate neurons possess muscarinic receptors as well (Jones and Sumikawa, 1981). Classical naturally occurring toxins such as muscarine and atropine do not distinguish between these various muscarinic receptors, although several recently synthesized compounds do. Thus, one can refer to M_1 muscarinic receptors located on autonomic ganglion and central neurons and M_2 muscarinic receptors found in smooth muscle and on central neurons (Hammer and Giachetti, 1982; Vickroy et al., 1984). Cardiac muscle may contain a third type of muscarinic receptor (Birdsall and Hulme, 1983).

In contrast to the direct linkage between nicotinic receptors and their associated ion channels, muscarinic receptors appear to carry out their physiological functions through a variety of intracellular second messengers, including cyclic GMP and phosphatidylinositol (Birdsall and Hulme, 1983; Gil and Wolfe, 1985). Perhaps for this reason, the electrophysiological study of muscarinic responses can be complex. For example, muscarinic agonists can either increase or decrease K^+ conductance in bullfrog sympathetic ganglion neurons depending on the experimental conditions chosen (Horn and Dodd, 1981; Kuffler and Sejnowski, 1983). These conductance changes produce slow inhibitory or slow excitatory postsynaptic potentials, respectively. In the same ganglion, activation of presynaptic muscarinic receptors does not alter membrane potential, but rather inhibits the release of ACh from nerve terminals (Koketsu and Yamada, 1982).

3.4.2.1. Classical Muscarinic Agonists. Muscarine is a quaternary ammonium compound isolated from the mushroom *Amanita muscaria.* Its selective cholinomimetic action forms the basis for the definition of muscarinic receptors. Pilocarpine, a tertiary amine purified from South American *Pilocarpus* shrubs, is also a highly selective cholinomimetic that acts only on muscarinic receptors. Arecoline, found in the betel nut, or seed of *Areca catechu,* has been used as a euphoretic by natives of the East Indies for centuries (Taylor, 1985a). Although primarily a muscarinic agonist, arecoline is not as selective as muscarine or pilocarpine, since it exhibits some cholinomimetic effects at nicotinic receptors.

3.4.2.2. Classical Muscarinic Antagonists. A number of toxins that are potent antagonists at muscarinic receptors have been isolated from the deadly nightshade plant, *Atropa belladonna.* Atropine and scopolamine are the most important of these structurally related toxins. Atropine and scopolamine are competitive antagonists of acetylcholine at muscarinic receptors (Weiner, 1985). Like the classical agonists, they do not differentiate between various classes of muscarinic receptors. A quaternary ammonium derivative of scopolamine, *N*-methylscopolamine, is an excellent ligand for identifying muscarinic receptors (Dombrowski et al., 1983). Radiolabeled *N*-methylscopolamine recognizes all muscarinic receptors. Another valuable probe for muscarinic receptors is 3-quinuclidinyl benzilate (QNB). ^{3}H-QNB is a nonselective antagonist at muscarinic receptors (Snyder et al., 1975; Kobayashi et al., 1978).

3.4.2.3. Selective Muscarinic Agents. A number of synthetic compounds are useful in distinguishing subtypes of muscarinic receptors. For example, 4-(*m*-chlorophenylcarbamoyloxy)-2-butynyl-trimethylammonium chloride (McN-A-343) is a selective M_1 receptor agonist (Hammer and Giachetti, 1982). Pirenzepine is a selective M_1 antagonist (Hammer et al., 1980). ^{3}H-Pirenzepine is an excellent ligand for quantitating M_1 receptors (Vickroy et al., 1984). Drugs selective for M_2 receptors are also available, although their selectivity for M_2 receptors is not as good as that of McN-A-

343 and pirenzepine for M_1 receptors (Birdsall and Hulme, 1983). The classification of muscarinic receptor subtypes is currently provisional. It has been suggested that these subtypes do not represent different proteins, but rather that the observed data are caused by differences in the various intracellular effector systems to which the receptors are coupled (Ehlert et al., 1981; Gil and Wolfe, 1985).

4. Conclusions

A remarkable variety of bacterial, animal, and plant toxins interfere with cholinergic transmission. The selectivity of these toxins for particular enzymes or receptors subserving ACh function is the most important property of these valuable probes. Pharmacological studies utilizing the toxins have played a crucial role in helping to characterize the neurobiology of cholinergic neurotransmission. Radiolabeled derivatives of these toxins have been used in quantitative studies of nicotinic and muscarinic receptors. Recently isolated substances, such as the *Conus* toxins, lophotoxin, and the snake venom κ-neurotoxins, expand the usefulness of this collection of toxins. These recent discoveries also serve to inform the neuroscientist that the final chapter on naturally occurring neurotoxins will not be written for many years to come.

Acknowledgments

The editorial assistance of Maggie Klevorn and Kathleen Wolf is gratefully acknowledged. Research in the author's laboratory is supported by National Institutes of Health Grant NS 17574.

References

Abe T., Alema S., and Miledi R. (1977) Isolation and characterization of presynaptically acting neurotoxins from the venom of *Bungarus* snakes. *Eur. J. Biochem.* **80,** 1–12.

Albuquerque E. X., Barnard E. A., Chiu T. H., Lapa A. J., Dolly J. O., Jansson S.-E., Daly J., and Witkop B. (1973) Acetylcholine receptor and ion conductance modulator sites at the murine neuromuscular junction: Evidence from specific toxin reactions. *Proc. Natl. Acad. Sci. USA* **70,** 949–953.

Ambache N. (1951) A further survey of the action of *Clostridium botulinum* toxin upon different types of autonomic nerve fibre. *J. Physiol.* (Lond.) **113,** 1–17.

Atchison W. D., Narahashi T., and Vogel S. M. (1984) Endplate blocking actions of lophotoxin. *Br. J. Pharmacol.* **82,** 667–672.

Bigalke H. and Habermann E. (1980) Blockade by tetanus and botulinum A toxin of postganglionic cholinergic nerve endings in the myenteric plexus. *Naunyn Schmiedebergs Arch. Pharmacol.* **312,** 255–263.

Birdsall N. J. M. and Hulme E. C. (1983) Muscarinic receptor subclasses. *Trends Pharmacol. Sci.* **6,** 459–463.

Birks R. and MacIntosh F. C. (1961) Acetylcholine metabolism of a sympathetic ganglion. *Can. J. Biochem. Phys.* **39,** 787–827.

Blanchard S. G., Quast U., Reed K., Lee T., Schimerlik M. I., Vandlen R., Claudio T., Strader C. D., Moore H-P. H., and Raftery M. A. (1979) Interaction of [^{125}I]-α-bungarotoxin with acetylcholine receptor from *Torpedo californica. Biochemistry* **18,** 1875–1883.

Brown D. A. (1979) Neurotoxins and the Ganglionic (C6) Type of Nicotinic Receptor, in *Advances in Cytopharmacology* vol. 3 *Neurotoxins: Tools in Neurobiology* (Ceccarelli B. and Clementi F., eds.) Raven, New York.

Brown D. A., Garthwaite J., Hayashi E., and Yamada S. (1976) Action of surugatoxin on nicotinic receptors in the superior cervical ganglion of the rat. *Br. J. Pharmacol.* **58,** 157–159.

Budai D., Ricny J., Kasa P., and Tucek S. (1986) 4-(l-Naphthylvinyl) pyridine decreases brain acetylcholine *in vivo,* but does not alter the level of acetyl-CoA. *J. Neurochem.* **46,** 990–992.

Burden S. J., Hartzell H. C., and Yoshikami D. (1975) Acetylcholine receptors at neuromuscular synapses: Phylogenetic differences detected by snake α-neurotoxins. *Proc. Natl. Acad. Sci. USA* **72,** 3245–3249.

Carson V. G., Jenden D. J., and Cho A. K. (1972) The *in vivo* effects of some *in vitro* choline acetyltransferase inhibitors. *Proc. West. Pharmacol. Soc.* **15,** 127–131.

Catterall W. A., Hartshorne R. P., and Beneski D. A. (1982) Molecular properties of neurotoxin receptor sites associated with solium channels from mammalian brain. *Toxicon* **20,** 27–40.

Ceccarelli B. and Hurlbut W. P. (1975) The effects of prolonged repetitive stimulation in hemicholinium on the frog neuromuscular junction. *J. Physiol.* (Lond.) **247,** 163–188.

Chadwick L. E. (1963) Actions on Insects and Other Invertebrates, in *Cholinesterases and Anticholinesterase Agents* vol. 15 *Handbuch der Ex-*

perimentellen Pharmakologie (Koelle G. B., ed.) Springer-Verlag, Berlin.

Chang C. C. and Huang M. C. (1974) Comparison of the presynaptic actions of botulinum toxin and β-bungarotoxin on neuromuscular transmission. *Naunyn Schmiedebergs Arch. Pharmacol.* **282,** 129–142.

Chang C. C., Chen T. F., and Lee C. Y. (1973) Studies on the presynaptic effect of β-bungarotoxin on neuromuscular transmission. *J. Pharmacol. Exp. Ther.* **184,** 339–345.

Changeux J. P., Kasai M., and Lee C. Y. (1970) Use of a snake venom toxin to characterize the cholinergic receptor protein. *Proc. Natl. Acad. Sci. USA* **67,** 1241–1247.

Chiappinelli V. A. (1982) Blockade of ganglionic transmission during synaptogenesis decreases α-bungarotoxin binding in the chick ciliary ganglion and iris. *Dev. Brain Res.* **5,** 118–121.

Chiappinelli V. A. (1983) Kappa-bungarotoxin: A probe for the neuronal nicotinic receptor in the avian ciliary ganglion. *Brain Res.* **277,** 9–22.

Chiappinelli V. A. (1986) Actions of snake venom toxins on neuronal nicotinic receptors and other neuronal receptors. *Pharmacol. Ther.* **31,** 1–32.

Chiappinelli V. A. and Dryer S. E. (1984) Nicotinic transmission in sympathetic ganglia: Blockade by the snake venom neurotoxin kappa-bungarotoxin. *Neurosci. Lett.* **50,** 239–244.

Chiappinelli V. A. and Giacobini E. (1978) Time course of appearance of α-bungarotoxin binding sites during development of chick ciliary ganglion and iris. *Neurochem. Res.* **3,** 465–478.

Chiappinelli V. A. and Lee J. C. (1985) Kappa-bungarotoxin: Self-association of a neuronal nicotinic receptor probe. *J. Biol. Chem.* **260,** 6182–6186.

Chiappinelli V. A., Cohen J. B., and Zigmond R. E. (1981) The effects of α- and β-neurotoxins from the venoms of various snakes on transmission in autonomic ganglia. *Brain Res.* **211,** 107–126.

Chiappinelli V. A., Wolf K., and Ciarleglio A. (1985) Kappa-bungarotoxin: Binding of a neuronal nicotinic receptor probe to chick optic lobe and skeletal muscle. *Soc. Neurosci. Abs.* **11,** 92.

Chiappinelli V. A., Wolf K., DeBin J., and Holt I. L. (1986) Kappa-flavitoxin: Purification of a neuronal nicotinic receptor probe that is structurally related to kappa-bungarotoxin. *Fed. Proc.* **45,** 923.

Clark A. W., Hurlbut W. P., and Mauro A. (1972) Changes in the fine structure of the neuromuscular junction of the frog caused by black widow spider venom. *J. Cell. Biol.* **52,** 1–14.

Clarke P. B. S., Schwartz R. D., Paul S. M., Pert C. B., and Pert A. (1985) Nicotinic binding in rat brain: Autoradiographic comparison of

[^{3}H]acetylcholine, [^{3}H]nicotine, and [^{125}I]α-bungarotoxin. *J. Neurosci.* **5,** 1307–1315.

Colquhoun D. (1979) The Link Between Drug Binding and Response: Theories and Observations, in *The Receptors: A Comprehensive Treatise* vol. I (O'Brien R. D., ed.) Plenum, New York.

Conti-Tronconi B. M., Dunn S. M. J., Barnard E. A., Dolly J. O., Lai F. A., Ray N., and Raftery M. A. (1985) Brain and muscle nicotinic acetylcholine receptors are different but homologous proteins. *Proc. Natl. Acad. Sci. USA* **82,** 5208–5212.

Cremo C. and Shimerlik M. I. (1983) Histrionicotoxin and alkylguanidine interactions with the solubilized and membrane-bound muscarinic acetylcholine receptor from porcine atria. *Arch. Biochem. Biophys.* **224,** 506–514.

Crowcroft P. J., Holman M. E., and Szurszewski J. H. (1971) Excitatory input from the distal colon to the inferior mesenteric ganglion in the guinea-pig. *J. Physiol.* (Lond.) **219,** 443–461.

Crowcroft P. J. and Szurszewski J. H. (1971) A study of the inferior mesenteric and pelvic ganglia of guinea-pigs with intracellular electrodes. *J. Physiol.* (Lond.) **219,** 421–441.

Cull-Candy S. G., Lundh H., and Thesleff S. (1976) Effects of botulinum toxin on neuromuscular transmission in the rat. *J. Physiol.* (Lond.) **260,** 177–203.

Culver P. and Jacobs R. S. (1981) Lophotoxin: A neuromuscular acting toxin from the sea whip *(Lophogorgia rigida). Toxicon* **19,** 825–830.

Culver P., Fenical W., and Taylor P. (1984) Lophotoxin irreversibly inactivates the nicotinic acetylcholine receptor by preferential association at one of the two primary agonist sites. *J. Biol. Chem.* **259,** 3763–3770.

David J. A. and Sattelle D. B. (1984) Actions of cholinergic pharmacological agents on the cell body membrane of the fast coxal depressor motoneurone of the cockroach *(Periplaneta americana). J. Exp. Biol.* **108,** 119–136.

Dolly J. O., Black J., Williams R. S., and Melling J. (1984) Acceptors for botulinum neurotoxin reside on motor nerve terminals and mediate its internalization. *Nature* **307,** 457–460.

Dombrowski A. M., Jerkins A. A., and Kauffman F. C. (1983) Muscarinic receptor binding and oxidative enzyme activities in the adult rat superior cervical ganglion: Effects of 6-hydroxydopamine and nerve growth factor. *J. Neurosci.* **3,** 1963–1970.

Dryer S. E. and Chiappinelli V. A. (1983) Kappa-bungarotoxin: An intracellular study demonstrating blockade of neuronal nicotinic receptors by a snake neurotoxin. *Brain Res.* **289,** 317–321.

Dryer S. E. and Chiappinelli V. A. (1985a) Substance P depolarizes nerve terminals in an autonomic ganglion. *Brain Res.* **336,** 190–194.

Dryer S. E. and Chiappinelli V. A. (1985b) An intracellular study of synaptic transmission and dendritic morphology in sympathetic neurons of the chick embryo. *Dev. Brain Res.* **22,** 99–111.

Dryer S. E. and Chiappinelli V. A. (1985c) Properties of choroid and ciliary neurons in the avian ciliary ganglion and evidence for substance P as a neurotransmitter. *J. Neurosci.* **5,** 2654–2661.

Ehlert F. J., Roeske W. R., and Yamamura H. I. (1981) Muscarinic receptor: Regulation by guanine nucleotides, ions, and N-ethylmaleimide. *Fed. Proc.* **40,** 153–159.

Eldefrawi A. T., Eldefrawi M. E., Albuquerque E. X., Oliveira A. C., Mansour N., Adler M., Daly J. W., Brown G. B., Burgermeister W., and Witkop B. (1977) Perhydrohistrionicotoxin: A potential ligand for the ion conductance modulator of the acetylcholine receptor. *Proc. Natl. Acad. Sci. USA* **74,** 2172–2176.

Fairbanks G., Steck T. L., and Wallach D. F. (1971) Electrophoretic analysis of the major polypeptides of the human erythrocyte membrane. *Biochemistry* **10,** 2606–2617.

Fenical W., Okuda R. K., Bandurraga M. W., Culver P., and Jacobs R. S. (1981) Lophotoxin: A novel neuromuscular toxin from Pacific sea whips of the genus *Lophogorgia. Science* **212,** 1512–1514.

Fertuck H. C. and Salpeter M. M. (1976) Quantitation of junctional and extrajunctional acetylcholine receptors by electron microscope autoradiography after ^{125}I-α-bungarotoxin binding at mouse neuromuscular junctions. *J. Cell Biol.* **69,** 144–158.

Finkelstein A., Rubin L. L., and Tzeng M.-C. (1976) Black widow spider venom: Effect of purified toxin on lipid bilayer membranes. *Science* **193,** 1009–1011.

Freeman J. A., Schmidt J. T., and Oswald R. E. (1980) Effect of α-bungarotoxin on retinotectal synaptic transmission in the goldfish and the toad. *Neuroscience* **5,** 929–942.

Frontali N., Ceccarelli B., Gorio A., Mauro A., Siekevitz P., Tzeng M.-C., and Hurlbut W. P. (1976) Purification from black widow spider venom of a protein factor causing the depletion of synaptic vesicles at neuromuscular junctions. *J. Cell Biol.* **68,** 462–479.

Frontali N., Granata F., Traina M. E., and Bellino M. (1973) Catecholamine depleting effect of black widow spider venom on fibres innervating different guinea-pig tissues. *Experentia* **29,** 1525–1527.

Gage P. W. and McBurney R. N. (1975) Effects of membrane potential, temperature and neostigmine on the conductance change caused by a quantum of acetylcholine at the toad neuromuscular junction. *J. Physiol.* (Lond.) **244,** 385–407.

Gepner J. I., Hall L. M., and Sattelle D. B. (1978) Insect acetylcholine receptors as a site of insecticide action. *Nature* **276,** 188–190.

Gil D. W. and Wolfe B. B. (1985) Pirenzepine distinguishes between muscarinic receptor-mediated phosphoinositide breakdown and inhibition of adenylate cyclase. *J. Pharmacol. Exp. Ther.* **232,** 608–616.

Grant G. A. and Chiappinelli V. A. (1985) Kappa-bungarotoxin: Complete amino acid sequence of a neuronal nicotinic receptor probe. *Biochemistry* **24,** 1532–1537.

Grant G. A., Frazier M. W., and Chiappinelli V. A. (1988) The amino acid sequence of kappa-flavitoxin: establishment of a new family of snake venom neurotoxins. *Biochemistry* **27,** 3794–3798.

Gray W. R., Luque A., Olivera B. M., Barrett J., and Cruz L. J. (1981) Peptide toxins from *Conus geographus* venom. *J. Biol. Chem.* **256,** 4734–4740.

Gundersen C. B. (1980) The effects of botulinum toxin on the synthesis, storage and release of acetylcholine. *Prog. Neurobiol.* **14,** 99–119.

Gundersen C. B., Katz B., and Miledi R. (1982) The antagonism between botulinum toxin and calcium in motor nerve terminals. *Proc. Roy. Soc. Lond.* **B216,** 369–376.

Hammer R. and Giachetti A. (1982) Muscarinic receptor subtypes: M_1 and M_2 biochemical and functional characterization. *Life Sci.* **31,** 2991–2998.

Hammer R., Berrie C. P., Birdsall N. J. M., Burgen A. S. V., and Hulme E. C. (1980) Pirenzepine distinguishes between different subclasses of muscarinic receptors. *Nature* **283,** 90–92.

Hanley M. R. and Emson P. C. (1979) Neuronal degeneration induced by stereotaxic injection of β-bungarotoxin into rat brain. *Neurosci. Lett.* **11,** 143–148.

Hanley M. R., Eterovic V. A., Hawkes S. P., Hebert A. J., and Bennett E. L. (1977) Neurotoxins of *Bungarus multicinctus* venom. Purification and partial characterization. *Biochemistry* **16,** 5840–5849.

Harris, A. J. and Miledi R. (1971) The effect of type D botulinum toxin on frog neuromuscular junctions. *J. Physiol.* (Lond.) **217,** 497–515.

Haubrich D. R. and Wang P. F. L. (1976) Inhibition of acetylcholine synthesis by juglone and 4-(l-naphthylvinyl) pyridine. *Biochem. Pharmacol.* **25,** 669–672

Hayashi E. and Yamada S. (1975) Pharmacological studies on surugatoxin, the toxic principle from Japanese ivory mollusc *(Babylonia japonica). Br. J. Pharmacol.* **53,** 207–215.

Hayashi E., Isogai M., Kagawa Y., Takayanagi N., and Yamada S. (1984) Neosurugatoxin, a specific antagonist of nicotinic acetylcholine receptors. *J. Neurochem.* **42,** 1491–1494.

Hirokawa N. (1978) Characterization of various nervous tissues of the chick embryos through responses to chronic application and immunocytochemistry of β-bungarotoxin. *J. Comp. Neurol.* **180,** 449–466.

Hirokawa N. and Heuser J. E. (1981) Structural evidence that botulinum toxin blocks neuromuscular transmission by impairing the calcium influx that normally accompanies nerve depolarization. *J. Cell Biol.* **88,** 160–171.

Hoover D. B., Muth E. A., and Jacobowitz D. A. (1978) A mapping of the distribution of acetylcholine, choline acetyltransferase and acetylcholinesterase in discrete areas of rat brain. *Brain Res.* **153,** 295–306.

Horn J. P. and Dodd J. (1981) Monosynaptic muscarinic activation of K^+ conductance underlies the slow inhibitory postsynaptic potential in sympathetic ganglia. *Nature* **292,** 625–627.

Howard B. D. and Truog R. (1977) Relationship between the neurotoxicity and phospholipase A activity of β-bungarotoxin. *Biochemistry* **16,** 122–125.

Jones S. W. and Sumikawa K. (1981) Quinuclidinyl benzilate binding in house fly heads and rat brain. *J. Neurochem.* **36,** 454–459.

Karczmar A. G. (1967) Pharmacologic, Toxicologic, and Therapeutic Properties of Anticholinesterase Agents, in *Physiological Pharmacology* vol. 3 *The Nervous System* part C *Autonomic Nervous System Drugs* (Root W. S. and Hofmann F. G., eds.) Academic, New York.

Karlsson E. (1979) Chemistry of Protein Toxins in Snake Venoms, in *Snake Venoms, Handbook of Experimental Pharmacology* vol. 52 (Lee C. Y., ed.) Springer-Verlag, Berlin.

Kato G. and Changeux J.-P. (1976) Studies on the effect of histrionicotoxin on the monocellular electroplax from *Electrophorus electricus* and on the binding of ^{3}H-acetylcholine to membrane fragments from *Torpedo marmorata*. *Mol. Pharmacol.* **12,** 92–100.

Kato A. C., Pinto J. E. B., Glavinovic M., and Collier B. (1977) Action of a β-bungarotoxin on autonomic ganglia and adrenergic neurotransmission. *Can. J. Physiol. Pharmacol.* **55,** 574–584.

Katz B. and Miledi R. (1978) A re-examination of curare action at the motor endplate. *Proc. Roy. Soc. Lond.* **B203,** 119–133.

Kawa K. and Roper S. (1984) On the two subdivisions and intrinsic synaptic connexions in the submandibular ganglion of the rat. *J. Physiol.* (Lond.) **346,** 301–320.

Kehoe J. S. (1972a) Three acetylcholine receptors in *Aplysia* neurones. *J. Physiol.* (Lond.) **225,** 115–146.

Kehoe J. S. (1972b) The physiological role of three acetylcholine receptors in synaptic transmission in *Aplysia*. *J. Physiol.* (Lond.) **225,** 147–172.

Kehoe J., Sealock R., and Bon C. (1976) Effects of α-toxins from *Bungarus multicinctus* and *Bungarus caeruleus* on cholinergic responses in *Aplysia* neurons. *Brain Res.* **107,** 527–540.

Kemp G., Bentley L., McNamee M. G., and Morley B. J. (1985) Purification and characterization of the α-bungarotoxin binding protein from rat brain. *Brain Res.* **347,** 274–283.

Kessler J. A. (1985) Parasympathetic, sympathetic, and sensory interactions in the iris: Nerve growth factor regulates cholinergic ciliary ganglion innervation *in vivo. J. Neurosci.* **5,** 2719–2725.

Kobayashi R. M., Palkovits M., Hruska R. E., Rothschild R., and Yamamura H. I. (1978) Regional distribution of muscarinic cholinergic receptors in rat brain. *Brain Res.* **154,** 13–23.

Kochva E., Nakar O., and Ovadia M. (1983) Venom toxins: Plausible evolution from digestive enzymes. *Am. Zool.* **23,** 427–430.

Koketsu K. and Yamada M. (1982) Presynaptic muscarinic receptors inhibiting active acetylcholine release in the bullfrog sympathetic ganglion. *Br. J. Pharmacol.* **77,** 75–82.

Kondo K., Narita K., and Lee C.-Y. (1978) Amino acid sequences of the two polypeptide chains in β_1-bungarotoxin from the venom of *Bungarus multicinctus. J. Biochem.* **83,** 101–115.

Kondo K., Toda H., Narita K., and Lee C.-Y. (1982a) Amino acid sequence of β_2-bungarotoxin from *Bungarus multicinctus* venom. The amino acid substitutions in the B chains. *J. Biochem.* **91,** 1519–1530.

Kondo K., Toda H., Narita K., and Lee C.-Y. (1982b) Amino acid sequences of three β-bungarotoxins (β_3-, β_4-, and β_5-bungarotoxins) from *Bungarus multicinctus* venom. Amino acid substitutions in the A chains. *J. Biochem.* **91,** 1531–1548.

Kosuge T., Zenda H., and Ochiai A. (1972) Isolation and structure determination of a new marine toxin, surugatoxin, from the Japanese ivory shell, *Babylonia japonica. Tetrahedron Lett.* **25,** 2545–2548.

Krodel E. K., Beckman R. A., and Cohen J. B. (1979) Identification of a local anesthetic binding site in nicotinic postsynaptic membranes isolated from *Torpedo marmorata* electric tissue. *Mol. Pharmacol.* **15,** 294–312.

Kuffler S. W. and Sejnowski T. J. (1983) Peptidergic and muscarinic excitation at amphibian sympathetic synapses. *J. Physiol.* (Lond.) **341,** 257–278.

Kupfer C. (1958) Selective block of synaptic transmission in ciliary ganglion by type A botulinus toxin in rabbits. *Proc. Soc. Exp. Biol.* (New York) **99,** 474–476.

Langdon R. B. and Jacobs R. S. (1985) Irreversible autonomic actions by lophotoxin suggest utility as a probe for both C6 and C10 nicotinic receptors. *Brain Res.* **359,** 233–238.

Leonard J. P. and Salpeter M. M. (1979) Agonist-induced myopathy at the neuromuscular junction is mediated by calcium. *J. Cell. Biol.* **82,** 811–819.

Loring R. H. and Zigmond R. E. (1985) Amino acid sequence of a neurotoxin that blocks neuronal nicotinic receptors and the localization of its binding sites in chick ciliary ganglion. *Soc. Neurosci. Abs.* **11,** 92.

Loring R. H., Chiappinelli V. A., Zigmond R. E., and Cohen J. B. (1984) Characterization of a snake venom neurotoxin which blocks nicotinic transmission in the avian ciliary ganglion. *Neuroscience* **11,** 989–999.

Loring R. H., Dahm L. M. and Zigmond R. E. (1985) Localization of alpha-bungarotoxin binding sites in the ciliary ganglion of the embryonic chick: An autoradiographic study at the light and electron microscopic level. *Neuroscience* **14,** 645–660.

Lundh H., Cull-Candy S. G., Leander S., and Thesleff S. (1976) Restoration of transmitter release in botulinum-poisoned skeletal muscle. *Brain Res.* **110,** 194–198.

MacKenzie I., Burnstock G., and Dolly J. O. (1982) The effects of purified botulinum neurotoxin type A on cholinergic, adrenergic and nonadrenergic, atropine-resistant autonomic neuromuscular transmission. *Neuroscience* **7,** 997–1006.

Mantione C. R., Fisher A., and Hanin I. (1981) The AF64A-treated mouse: Possible model for central cholinergic hypofunction. *Science* **213,** 579–580.

Marks M. J. and Collins A. C. (1982) Characterization of nicotinic binding in mouse brain and comparison with the binding of α-bungarotoxin and quinuclidinyl benzilate. *Mol. Pharmacol.* **22,** 554–564.

Marshall L. M. (1981) Synaptic localization of α-bungarotoxin binding which blocks nicotinic transmission at frog sympathetic neurons. *Proc. Natl. Acad. Sci. USA* **78,** 1948–1952.

Martin A. R. and Pilar G. (1963) Dual mode of synaptic transmission in the avian ciliary ganglion. *J. Physiol.* (Lond.) **168,** 443–463.

McEachern A. E., Margiotta J. F., and Berg D. K. (1985) Gamma-aminobutyric acid receptors on chick ciliary ganglion neurons *in vivo* and in cell culture. *J. Neurosci.* **5,** 2690–2695.

McManus O. B. and Musick J. R. (1985) Postsynaptic block of frog neuromuscular transmission by conotoxin GI. *J. Neurosci.* **5,** 110–116.

Mebs D. (1985) *List of Biologically Active Components from Snake Venoms* pp. 1–141, Zentrum der Rechtsmedizin, University of Frankfurt, Frankfurt.

Mebs D., Narita K., Iwanaga S., Samejima Y., and Lee C. Y. (1971) Amino acid sequence of α-bungarotoxin from the venom of *Bungarus multicinctus. Biochem. Biophys. Res. Commun.* **44,** 711–716.

Mebs D., Narita K., Iwanaga S., Samejima Y., and Lee C.-Y. (1972) Purification, properties and amino acid sequence of α-bungarotoxin from the venom of *Bungarus multicinctus. Hoppe-Seyler's Z. Physiol. Chem.* **353,** 243–262.

Meldolesi J., Madeddu L., Torda M., Gatti G., and Niutta E. (1983) The effect of α-latrotoxin on the neurosecretory PC12 cell line: Studies on toxin binding and stimulation of transmitter release. *Neuroscience* **10,** 997–1009.

Mellanby J. and Green J. (1981) How does tetanus toxin act? *Neuroscience* **6,** 281–300.

Miledi R. and Szczepaniak A. C. (1975) Effect of *Dendroaspis* neurotoxins on synaptic transmission in the spinal cord of the frog. *Proc. Roy. Soc. Lond.* **B190,** 267–274.

Morley B. J. and Kemp G. E. (1981) Characterization of a putative nicotinic acetylcholine receptor in mammalian brain. *Brain Res. Rev.* **3,** 81–104.

Morris C. E., Wong B. S., Jackson M. B., and Lecar H. (1983) Single-channel currents activated by curare in cultured embryonic rat muscle. *J. Neurosci.* **3,** 2525–2531.

Neubig R. R. and Cohen J. B. (1979) Equilibrium binding of ^{3}H-tubocurarine and ^{3}H-acetylcholine by *Torpedo* postsynaptic membranes: Stoichiometry and ligand interactions. *Biochemistry* **18,** 5464–5475.

Neubig R. R. and Cohen J. B. (1980) Permeability control by cholinergic receptors in *Torpedo* postsynaptic membranes: Agonist dose–response relations measured at second and millisecond times. *Biochemistry* **19,** 2770–2779.

Oberg S. G. and Kelly R. B. (1976) The mechanism of β-bungarotoxin action I. Modification of transmitter release at the neuromuscular junction. *J. Neurobiol.* **7,** 129–141.

Olivera B. M., Gray W. R., Zeikus R., McIntosh J. M., Varga J., Rivier J., de Santos V., and Cruz L. J. (1985) Peptide neurotoxins from fish-hunting cone snails. *Science* **230,** 1338–1343.

Ono J. K. and Salvaterra P. M. (1981) Snake α-toxin effects on cholinergic and noncholinergic responses of *Aplysia californica. J. Neurosci.* **1,** 259–270.

Ornberg R. L., Smyth Jr., T., and Benton A. W. (1976) Isolation of a neurotoxin with a presynaptic action from the venom of the black widow spider (*Latrodectus mactans,* Fabr.) *Toxicon* **14,** 329–333.

Othman I. B., Spokes J. W., and Dolly J. O. (1982) Preparation of neurotoxic ^{3}H-β-bungarotoxin: Demonstration of saturable binding to brain synapses and its inhibition by toxin I. *Eur. J. Biochem.* **128,** 267–276.

Paggi P. and Rossi A. (1971) Effect of *Latrodectus mactans tredecimguttatus* venom on sympathetic ganglion isolated *in vitro*. *Toxicon* **9,** 265–269.

Patrick J. and Stallcup W. B. (1977a) α-Bungarotoxin binding and cholinergic receptor function on a rat sympathetic nerve line. *J. Biol. Chem.* **252,** 8629–8633.

Patrick J. and Stallcup W. B. (1977b) Immunological distinction between acetylcholine receptor and the α-bungarotoxin-binding component on sympathetic neurons. *Proc. Natl. Acad. Sci. USA* **74,** 4689–4692.

Perri V., Sacchi O., and Casella C. (1970) Electrical properties and synaptic connections of the sympathetic neurons in the rat and guinea-pig superior cervical ganglion. *Pflugers Arch.* **314,** 40–54.

Popot J.-L. and Changeux J.-P. (1984) Nicotinic receptor of acetylcholine: structure of an oligomeric integral membrane protein. *Physiol. Rev.* **64,** 1162–1239.

Pumplin D. W. and McClure W. O. (1977) The release of acetylcholine elicited by extracts of black widow spider glands: Studies using rat superior cervical ganglia and inhibitors of electrically stimulated release. *J. Pharmacol. Exp. Ther.* **201,** 312–319.

Quik M. and Lamarca M. V. (1982) Blockade of transmission in rat sympathetic ganglia by a toxin which co-purifies with α-bungarotoxin. *Brain Res.* **238,** 385–399.

Raftery M. A., Conti-Tronconi B. M., Dunn S. M. J., Crawford R. D., and Middlemas D. (1984) The nicotinic acetylcholine receptor: Its structure, multiple binding sites and cation transport properties. *Fund. Appl. Toxicol.* **4,** S34–S51.

Ravdin P. M. and Berg D. K. (1979) Inhibition of neuronal acetylcholine sensitivity by α-toxins from *Bungarus multicinctus* venom. *Proc. Natl. Acad. Sci. USA* **76,** 2072–2076.

Rehm H., Schafer T., and Betz H. (1982) β-Bungarotoxin-induced cell-death of neurons in chick retina. *Brain Res.* **250,** 309–319.

Rosecrans J. A. (1979) Nicotine as a Discriminative Stimulus to Behavior: Its Characterization and Relevance to Smoking Behavior, in *Cigarette Smoking as a Dependence Process* (Krasnegor N. A., ed.) National Institute on Drug Abuse, Department of Health, Education and Welfare Publication No. (ADM) 79-800, US Govt. Printing Office, Washington, DC.

Ross M. E., Park D. H., Teitelman G., Pickel V. M., Reis D. J., and Joh T. H. (1983) Immunohistochemical localization of choline acetyltransferase using a monoclonal antibody: A radioautographic method. *Neuroscience* **10,** 907–922.

Rowell P. P. and Chiou C. Y. (1976) Pharmacological studies of *N,-N*-dimethylaminoethyl chloroacetate and *N, N*-dimethylaminoethyl acrylate as inhibitors of choline acetyltransferase in isolated skeletal and smooth muscle preparations. *Pharmacology* **14,** 339–350.

Saiani L., Kageyama H., Conti-Tronconi B. M., and Guidotti A. (1984) Purification and characterization of a bungarotoxin polypeptide which blocks nicotinic receptor function in primary culture of adrenal chromaffin cells. *Mol. Pharmacol.* **25,** 327–334.

Sattelle D. B., Harrow I. D., Hue B., Pelhate M., Gepner J. I., and Hall L. M. (1983) Alpha-bungarotoxin blocks excitatory synaptic transmission between cercal sensory neurones and giant interneurone 2 of the cockroach, *Periplaneta americana. J. Exp. Biol.* **107,** 473–489.

Schmidt J., Hunt S., and Polz-Tejera G. (1980) Nicotinic Receptors of the Central and Autonomic Nervous System, in *Neurotransmitters, Receptors and Drug Action* (Essman W. B., ed.) Spectrum, Jamaica, New York.

Schwartz R. D., McGee Jr., R., and Kellar K. J. (1982) Nicotine cholinergic receptors labeled by [^{3}H]-acetylcholine in rat brain. *Mol. Pharmacol.* **22,** 56–62.

Shaker N., Eldefrawi A. T., Aguayo L. G., Warnick J. E., Albuquerque E. X., and Eldefrawi M. E. (1982) Interactions of *d*-tubocurarine with the nicotinic acetylcholine receptor/channel molecule. *J. Pharmacol. Exp. Ther.* **220,** 172–177.

Simpson L. L. (1973) The interaction between divalent cations and botulinum toxin type A in the paralysis of the rat phrenic nerve-hemidiaphragm preparation. *Neuropharmacol.* **12,** 165–176.

Simpson L. L. (1974) Studies on the binding of botulinum toxin type A to the rat phrenic nerve-hemidiaphragm preparation. *Neuropharmacol.* **13,** 683–691.

Simpson L. L. (1977) The effects of acute and chronic botulinum toxin treatment on receptor number, receptor distribution and tissue sensitivity in rat diaphragm. *J. Pharmacol. Exp. Ther.* **200,** 343–351.

Simpson L. L. (1980) Kinetic studies on the interaction between botulinum toxin type A and the cholinergic neuromuscular junction. *J. Pharmacol. Exp. Ther.* **212,** 16–21.

Smith J. C., Cavallito C. J., and Foldes F. F. (1967) Choline acetyltransferase inhibitors: A group of styryl-pyridine analogs. *Biochem. Pharmacol.* **16,** 2438–2441.

Snyder S. H., Chang K. J., Kuhar M. J., and Yamamura H. I. (1975) Biochemical identification of the mammalian muscarinic cholinergic receptor. *Fed. Proc.* **34,** 1915–1921.

Sorenson E. M., Culver P., and Chiappinelli V. A. (1987) Lophotoxin: Selective blockade of nicotinic transmission in autonomic ganglia by a coral neurotoxin. *Neuroscience* **20,** 875–884.

Spitzer N. (1972) Miniature endplate potentials at mammalian neuromuscular junctions poisoned by botulinum toxin. *Nature New Biol.* **237,** 26–27.

Strong P. N., Goerke J., Oberg S. G., and Kelly R. B. (1976) β-Bungarotoxin, a presynaptic toxin with enzymatic activity. *Proc. Natl. Acad. Sci. USA* **73,** 178–182.

Stroud R. M. (1983) Acetylcholine receptor structure. *Neurosci. Comm.* **1,** 124–133.

Strydom D. J. (1979) The Evolution of Toxins Found in Snake Venoms, in *Snake Venoms, Handbook of Experimental Pharmacology* vol. 52 (Lee C. Y., ed.) Springer-Verlag, Berlin.

Taylor P. (1985a) Cholinergic Agonists, in *The Pharmacological Basis of Therapeutics* (Gillman A. G., Goodman L. S., Rall T. W., and Murad F., eds.) Macmillan, New York.

Taylor P. (1985b) Anticholinesterase Agents, in *The Pharmacological Basis of Therapeutics* (Gilman A. G., Goodman L. S., Rall T. W., and Murad F., eds.) Macmillan, New York.

Taylor P. (1985c) Ganglionic Stimulating and Blocking Agents, in *The Pharmacological Basis of Therapeutics* (Gillman A. G., Goodman L. S., Rall T. W., and Murad F., eds.) Macmillan, New York.

Taylor P. (1985d) Neuromuscular Blocking Agents, in *The Pharmacological Basis of Therapeutics* (Gillman A. G., Goodman L. S., Rall T. W., and Murad F., eds.) Macmillan, New York.

Toldi J., Joo F., Adam G., Feher O., and Wolff J. R. (1983) Inhibition of synaptic transmission in the rat superior cervical ganglion by intracarotid infusion of bungarotoxin. *Brain Res.* **262,** 323–327.

Tse C. K., Dolly J. O., Hambleton P., Wray D., and Melling J. (1982) Preparation and characterization of homogenous neurotoxin type A from *Clostridium botulinum. Eur. J. Biochem.* **122,** 493–500.

Tse C. K., Wray D., Melling J., and Dolly J. O. (1986) Actions of β-bungarotoxin on spontaneous release of transmitter at muscle endplates treated with botulinum toxin. *Toxicon* **24,** 123–130.

Tzeng M.-C. and Siekevitz P. (1978) The effect of the purified major protein factor (α-latrotoxin) of black widow spider venom on the release of acetylcholine and norepinephrine from mouse cerebral cortex slices. *Brain Res.* **139,** 190–196.

Vickroy T. W., Watson M., Yamamura H. I., and Roeske W. R. (1984). Agonist binding to multiple muscarinic receptors. *Fed. Proc.* **43,** 2785–2790.

Warnick J. E., Albuquerque E. X., and Diniz C. R. (1976) Electrophysiological observations on the action of the purified scorpion venom, tityustoxin, on nerve and skeletal muscle of the rat. *J. Pharmacol. Exp. Ther.* **198,** 155–167.

Weber K. and Osborn M. (1969) The reliability of molecular weight determinations by dodecyl sulfate-polyacrylamide gel electrophoresis. *J. Biol. Chem.* **244,** 4406–4412.

Weiner N. (1985) Atropine, Scopolamine, and Related Antimuscarinic drugs, in *The Pharmacological Basis of Therapeutics* (Gilman A. G., Goodman L. S., Rall T. W., and Murad F., eds.) Macmillan, New York.

Wolf K. M., Ciarleglio A., and Chiappinelli V. A. (1988) Kappa-bungarotoxin: binding of a neuronal nicotinic receptor antagonist to chick optic lobe and skeletal muscle. *Brain Res.* **439,** 249–258.

Zaimis E. (1976) *Neuromuscular Junction,* Springer-Verlag, Berlin.

Ziskind L. and Dennis M. J. (1978) Depolarizing effect of curare on embryonic rate muscles. *Nature* **276,** 622–623.

GABA Antagonists

Their Use and Mechanisms of Action

M. Farrant and R. A. Webster

1. Introduction

As with any neurotransmitter, there are two basic ways to decrease the synaptic action of γ-aminobutyric acid (GABA) in order to establish its functional role at particular synapses in the nervous system. They are to reduce its availability and release through a presynaptic action or to reduce its effects with appropriate postsynaptic antagonists. The former might be achieved by specifically destroying GABA-containing neurons, inhibiting its synthesis, or interfering in some way with its actual release. All of these approaches, however, suffer from the disadvantage of a very slow onset and an effect that, even if reversible, is likely to be prolonged and not easily quantified. There are no specific neurotoxins for GABA neurons similar to 6-hydroxydopamine for catecholamine systems, and although antibodies raised to the GABA-synthesizing enzyme, glutamic acid decarboxylase (GAD), can be used to destroy (lyse) synaptosomes containing GABA in vitro (Docherty et al., 1983), the method is not applicable in vivo. Intracerebral injection of tetanus toxin has been shown to reduce GABA-mediated inhibition by reducing release of the transmitter (Collingridge and Davies, 1982), but this toxin has a number of actions, including an effect on glycine release, and it is obviously not specific for GABA. Many inhibitors of GAD have been described (*see* Woodbury, 1980); unfortunately, many of them not only reduce GABA levels, but also raise the concentration of its precursor glutamate. Their use also suffers from the disadvantage that the rate of decline of GABA depends on the rate of its utilization, which will vary from neuron to neuron, and we do not know to what extent GABA levels must be reduced before its release and function are significantly impaired.

From the brief discussion above, it is clear that any reduction of GABA function through a presynaptic action is unlikely to be sufficiently specific or quantitatively controllable at present to be of much value in elucidating the synaptic function of GABA. This is more likely to be achieved by drugs that block the postsynaptic action of GABA. A wide range of structurally dissimilar compounds has been shown to do this. In the following review we describe these compounds and discuss their likely mechanisms of action, selectivity, and utility as experimental "tools."

Depending on the level at which antagonism occurs, such agents may also be used to characterize receptors or transduction mechanisms involved in a GABA-mediated response. Indeed, the use of antagonists has led to the recognition in vertebrates of two distinct GABA receptor systems, classified as $GABA_A$ and $GABA_B$, with markedly different characteristics (*see* section 3.1). Although $GABA_B$ antagonists are generally referred to specifically as such, the term $GABA_A$ antagonist is rarely used and drugs acting at this receptor have simply been described as "GABA antagonists." This is in part a consequence of history, numerous "GABA antagonists" having been described prior to the recognition of this receptor division. Although $GABA_B$ receptors have been described only in vertebrates, receptors sharing certain pharmacological features with vertebrate $GABA_A$ receptors have been identified in a range of phyla. Throughout this review, however, emphasis has been placed on the pharmacology of vertebrate GABA receptors; information derived from the study of invertebrate preparations has been included primarily for reasons of historical completeness.

2. $GABA_A$ Antagonists

2.1. Picrotoxin and Bicuculline: Historical Background

$GABA_A$ antagonists may be grouped into two main classes: (1) those acting upon the GABA receptor itself and (2) those affecting, either directly or via some associated regulatory site, GABA-activated chloride (Cl^-) channels. Before discussing in detail the members of these two groups, it is worth considering from an historical viewpoint the evidence in favor of such a distinction and in particular the pharmacology of two drugs, bicuculline and picrotoxin, which have come to be regarded as "classical" GABA antagonists and prime examples of these two classes of compound.

Picrotoxin is a plant toxin originally isolated from the seeds of *Anamirta cocculus* and subsequently shown to be present in several other members of the menispermaceae family (*see* Porter, 1967). Early investigations using vertebrates highlighted the ability of picrotoxin to produce marked CNS exicitation. In the search for the basis of this potent convulsant activity, extensive studies were made of picrotoxin's action on anatomically simpler systems, notably in crustacea, in which it was shown to block several types of inhibitory synapse (Robbins and Van der Kloot, 1958; Grundfest et al., 1959). The fact that GABA inhibited contractions of crayfish muscle and produced membrane permeability changes resembling those resulting from inhibitory nerve stimulation (Biostel and Fatt, 1958) suggested a link between this amino acid and the action of picrotoxin, which was confirmed when the toxin was shown to antagonize the actions of exogenous GABA (Van der Kloot and Robbins, 1959). In fact, Elliot and Florey (1956) had earlier described a blocking effect of picrotoxin on the inhibition produced by Factor I at the crayfish stretch receptor. Later studies on insects (Usherwood and Grundfest, 1965) and crustacea (Takeuchi and Takeuchi, 1969) extended these observations by showing that picrotoxin blocked inhibitory postsynaptic potentials as well as the effects of exogenous GABA without affecting excitatory postsynaptic potentials or the response of the muscle to exogenous glutamate.

In the mammalian CNS, blockade of a GABA response by picrotoxin was first described by Galindo (1969) in a study of the actions of GABA on cuneate neurons of the cat. Subsequent investigations yielded evidence for antagonism of GABA by picrotoxin at various sites in the CNS, as well as complementary evidence indicating blockade of neurally evoked inhibition (for review, *see* Krnjevic, 1974; Kelly and Beart, 1975; Nistri and Constanti, 1979).

Bicuculline, a phthalide isoquinoline alkaloid isolated from corydalis species, is another plant toxin that possesses potent convulsant properties (Welch and Henderson, 1934). Its introduction as a GABA antagonist stemmed from the attempts of Curtis and colleagues in the late 1960s to find a specific antagonist for the depressant action of GABA on mammalian CNS neurons. Their interest eventually turned to several convulsant alkaloids in the hope of identifying a GABA antagonist analogous to the glycine antagonist, and alkaloid, strychnine. As a result, bicuculline was "rediscovered" and shown to block the actions of iontophoretically applied GABA on central neurons and to reduce strychnine-resistant inhibition (Curtis et al., 1970). These authors and numerous other groups subsequently confirmed the ability of bicuculline to antagonize GABA at numerous sites in the mammalian CNS and in several invertebrate muscle preparations (for reviews, *see* Krnjevic, 1974; Curtis and Johnston, 1974; Kelly and Beart, 1975; De Feudis, 1977; Usherwood, 1978; Nistri and Constanti, 1979).

Although electrophysiological experiments clearly demonstrated that both picrotoxin and bicuculline were able to antagonize the actions of GABA, the nature of their antagonism was seen to differ. Quantitative studies, initially performed on crustacean muscle, indicated that bicuculline behaved as a noncompetitive antagonist (Takeuchi and Onodera, 1972; Shank et al., 1974; Constanti, 1978), whereas, depending on the species used, picrotoxin was variously reported to act competitively (Grundfest et al., 1959; Shank et al., 1974), noncompetitively (Takeuchi and Takeuchi, 1969; Earl and Large, 1974), or in a "mixed" competitive plus noncompetitive fashion (Constanti, 1978; Adams et al., 1981).

In the vertebrate nervous system, a more consistent view of the actions of these two antagonists has emerged, and when they have been compared directly, picrotoxin and bicuculline have generally been seen to act as noncompetitive and competitive antagonists, respectively (Homma and Rovainen, 1978; Simmonds, 1978, 1980, 1982; Nicoll and Wojtowicz, 1980; Lebeda et al., 1982; Akaike and Oomura, 1984; Kaneko and Tachibana, 1986). Some evidence of mixed-type antagonism by picrotoxin is present in the GABA dose-response curves shown by Nicoll and Wojtowicz (1980) and Lebeda et al. (1982).

Classical receptor theory predicts that the mechanism of action of an antagonist should be reflected in the nature of its antagonism. Competitive antagonists bind to the same receptor as the agonist, but possess zero or limited efficacy, whereas noncompetitive antagonists interact with receptors distinct from those of the agonist (or a receptor subunit different from that occupied by the agonist), and so bring about a change in the apparent efficacy or affinity of the agonist. Although information derived solely from dose–response curves does not allow an unequivocal designation of the type of antagonism produced by an antagonist, the results outlined above would suggest that, in vertebrates at least, bicuculline acts at the GABA receptor, whereas picrotoxin acts at a different site (*see* Simmonds, 1983; Kemp et al., 1986). Support for this view has come from ligand binding studies showing that bicuculline, but not picrotoxin, is able to displace ^{3}H-GABA from vertebrate neuronal membranes (Zukin et al., 1974; Enna et al., 1977; Olsen et al., 1978; Ticku et al., 1978). In contrast, GABA blinding to invertebrate neuronal membranes is not affected by bicuculline (Mann and Enna, 1980), whereas GABA binding to crustacean muscle is only weakly and noncompetitively affected (Olsen et al., 1976). These findings are consistent with the results of electrophysiological studies showing that neuronal responses to GABA in locust ganglia and some of those in *Aplysia* are unaffected by bicuculline (Lees et al., 1987; Yarowsky and Carpenter, 1978), whereas the response of crustacean muscle is noncompetitively blocked (see Smart and Constanti, 1986). The differences in the action of bicuculline at invertebrate neuronal and muscle GABA receptors and at vertebrate GABA receptors most likely reflect basic differences in the receptor complex present at each site (*see* Simmonds, 1983; Lees et al., 1987).

At what site does picrotoxin act? An answer to this question was first proposed by Takeuchi and Takeuchi (1969). Earlier experiments had demonstrated that at both pre- and postsynaptic sites, GABA selectively increased membrane permeability to Cl^- ions (Boistel and Fatt, 1958; Dudel and Kuffler, 1961; Takeuchi and Takeuchi, 1966). The observation by Takeuchi and Takeuchi (1969) that on crayfish muscle the effectiveness of picrotoxin as a GABA antagonist (noncompetitive) was dependent upon the concentration of Cl^- ions in the bathing medium was interpreted as evidence for a direct action of the toxin on the GABA-activated Cl^- ionophore. Although the noncompetitive action of picrotoxin in vertebrates may be readily explained by such an effect, the aforementioned variation in the type of antagonism exhibited by the toxin in invertebrate systems would appear inconsistent with it having a single mode of action. Constanti (1978) suggested that the deviation from "true" noncompetitive antagonism on lobster muscle could be caused by a biphasic influence of GABA on the affinity of picrotoxin for its ionophore binding site. The mechanism of this antagonism was viewed as being analogous to "mixed" enzyme inhibition (*see* Smart and Constanti, 1981) and the kinetic model developed to encompass these ideas provided a good fit for the experimental data. It also offered an explanation (in terms of different antagonist equilibrium dissociation constants) for the differences seen in the action of picrotoxin at different invertebrate synapses, while retaining the idea of the antagonist only having a single site of action—namely the Cl^- channel.

2.2. Receptor Antagonists

2.2.1. Bicuculline and Related Phthalideisoquinolines

Several compounds closely related to bicuculline produce GABA antagonism. Early studies with bicuculline highlighted the problem of its low aqueous solubility (*see* section 2.5). Fortunately, quaternary salts of bicuculline are considerably more water-soluble than the parent compound and retain GABA antagonist properties. Thus, bicuculline methiodide (BMI; also known as *N*-methylbicuculline) was found to be a potent convulsant when given centrally (Pong and Graham, 1972), and bicuculline methochloride (BMC) was shown to be very effective when used iontophoretically to antagonize the actions of GABA in the cat spinal cord (Johnston et al., 1972).

The effects of bicuculline and its methohalide derivatives, whether examined electrophysiologically (Collins and Hill, 1974) or by receptor binding techniques (Enna et al., 1977, Mohler and Okada, 1977), are highly stereospecific. Isomers having the configuration 1S,9R around the C_1–C_9 bond [i.e., based on (+)-bicuculline] are active, whereas those with a 1R,9S configuration [(–)-bicuculline] are inactive. Similarly, of a number of related alkaloids tested as GABA antagonists, only those with the same absolute stereochemistry as (+)-bicuculline have been found to be effective (*see* Fig. 1). These are (+)-bicucine methyl ester, (+)-corlumine, and (+)-narcotine (Johnston et al., 1972; Johnston, 1976; Bowery et al., 1979).

As outlined above, there is good reason to suppose that the GABA antagonism produced by bicuculline in vertebrates follows a direct action upon the GABA receptor (*see* also Nowak et al., 1982). Unfortunately the selectivity of bicuculline and its analogs for this site is not absolute, since some of their observed effects are not consistent with GABA antagonism. Bicuculline, bicuculline methohalides, and bicucine methyl ester competitively inhibit acetylcholine esterase (AChE) (Svenneby and Roberts, 1973; Olsen et al., 1976), but, since the related alkaloids bicucine, (+)-adlumine, and (–)-adlumine are equally effective enzyme inhibitors (Olsen et al., 1976; Johnston, 1976), yet inactive as GABA antagonists (Johnston, 1976), the two processes would appear to be unrelated. Bicuculline and bicuculline methochloride also appear to excite central neurons (Johnston et al., 1972; Dray, 1975b; Hill et al., 1976) independently of any reduction in background inhibition; (–)-bicuculline methochloride, which is not a GABA an-

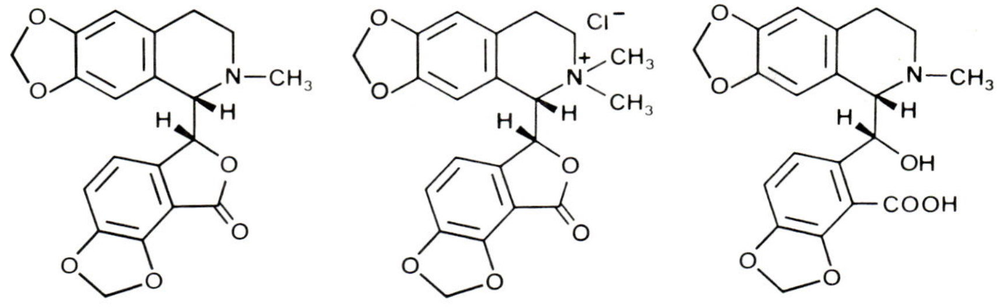

(+)-Bicuculline (+)-Bicuculline Methochloride (+)-Bicucine

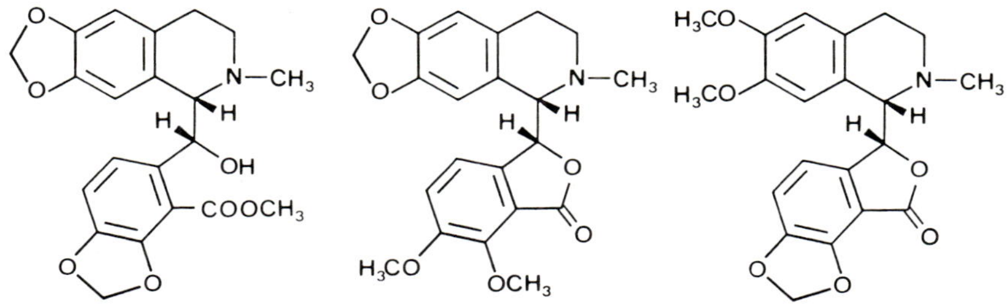

(+)-Bicucine Methyl Ester (+)-Narcotine (+)-Corlumine

Fig. 1. Bicuculline and related alkaloids.

tagonist, is also an excitant (Collins and Hill, 1974). This excitation is unlikely to result from AChE inhibition because although bicuculline can augment the effects of ACh, noncholinoceptive neurons may also be excited (Johnston, 1976). Direct membrane effects could play some part in such excitation; bicuculline has been shown to produce membrane depolarization and decrease membrane conductance in the lobster giant axon (Freeman, 1973), lobster muscle (Shank et al., 1974), crayfish muscle (Takeuchi and Onodera, 1972), and cultured mouse spinal cord neurons (Heyer et al., 1982). In the latter study, this membrane depolarization was suggested to result from a blockade of potassium conductance and to possibly play a role in the production of paroxysmal activity by bicuculine.

2.2.2. Securinine and Dihydrosecurinine

Securinine is one of a number of convulsant alkaloids present in the shrub *Securinega suffructicosa*. In a recent study (Beutler et al.,

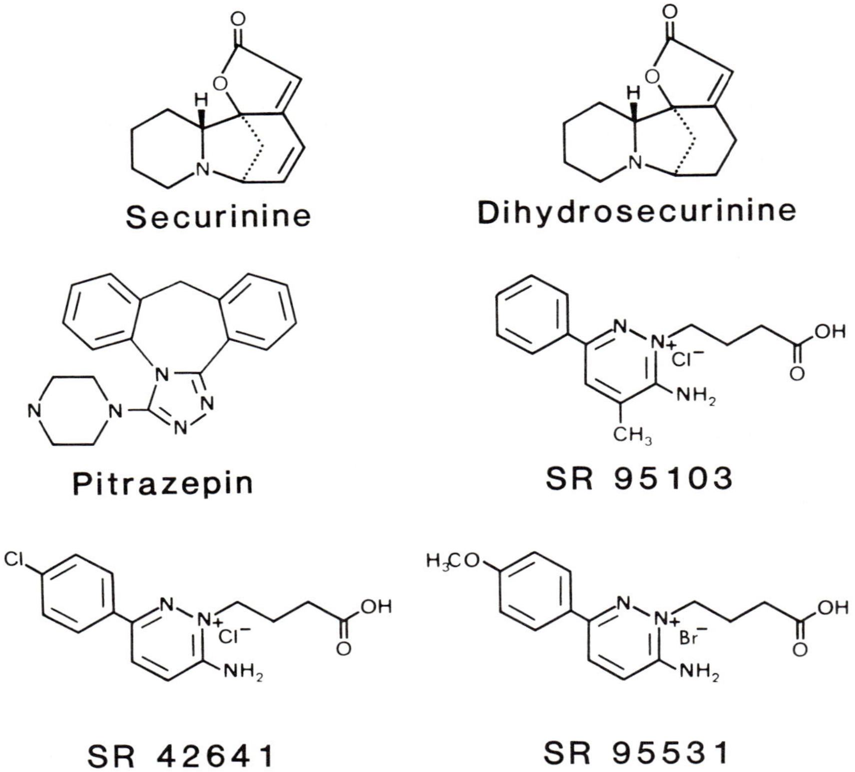

Fig. 2. Recently introduced $GABA_A$ receptor antagonists.

1985), four such alkaloids were examined as potential GABA or glycine antagonists. Of these, securinine and its analog dihydrosecurinine (*see* Fig. 2) were shown to inhibit ^{3}H-GABA binding and GABA-stimulated ^{3}H-diazepam binding to rat brain membranes. Both compounds also reversibly antagonized the depressant effects of GABA, but not glycine, on neurons in the cat spinal cord.

Despite the lack of quantitative evidence regarding the nature of the antagonism produced by these two agents, their ability to displace ^{3}H-GABA and the similarity of their action to that of bicuculline methochloride when applied iontophoretically would suggest that they are true receptor antagonists. In this context it is interesting to note (as pointed out by Beutler et al., 1985) that the fused indolizide ring structure of securinine, like bicuculline, has both tertiary amine and lactone groups.

Securinine and its dihydro analog appear selective for GABA receptors since neither affect the binding of benzodiazepine, cholinergic muscarinic, or β-adrenergic receptor ligands, nor the neuronal depressant action of glycine (Beutler et al., 1985). The convulsant potencies of both compounds are somewhat higher (relative to bicuculline), however, than would be expected from their ability to displace ^{3}H-GABA, suggesting an additional action at some other site. Although this discrepancy could also be explained by the ability of these compounds to more easily cross the blood–brain barrier (*see* section 2.5), it should be noted that allosecurinine and virosecurinine (isomers of securinine inactive at GABA sites) are also convulsants (Beutler et al., 1985).

2.2.3. Pitrazepin

Pitrazepin, 3-(piperazinyl-1)-9H-dibenz(c,f)triazolo(4,5-a)azepin (*see* Fig. 2), was demonstrated to be an antagonist at GABA receptors by Gahwiler et al. (1984). Quantitative analysis indicates that pitrazepin acts competitively and is approximately three times more potent (as an antagonist of isoguvacine) than bicuculline methochloride (Kemp et al., 1985, 1986). Consistent with these electrophysiological observations, pitrazepin potently displaces ^{3}H-muscimol (Gahwiler et al., 1984) and ^{3}H-bicuculline methochloride (Kemp et al., 1985) from rat brain membranes and antagonizes, apparently competitively, muscimol-induced enhancement of ^{3}H-diazepam binding and muscimol-induced reduction of ^{35}S-TBPS binding (a ligand for the picrotoxin site at the Cl^{-} ionophore; *see* section 2.4.2) (Braestrup and Nielsen, 1985). Pitrazepin, however, is not wholly selective for GABA receptors. It interacts, albeit weakly, with benzodiazepine receptors (Gahwiler et al., 1984; Braestrup and Nielsen, 1985), and at fairly low concentrations competitively inhibits the binding of ^{3}H-strychnine (Braestrup and Nielsen, 1985). Furthermore, pitrazepin reduces equally the inhibitory effects of GABA and glycine on the firing rate of cat spinal interneurons (Curtis and Gynther, 1986).

2.2.4. Arylamino-pyridazine Derivatives of GABA

Members of this recently introduced group of compounds are reported to interact with the GABA recognition site in a bicuculline-like manner (Chambon et al., 1985). The prototypic compound SR 95103 (2-(carboxy-3'-propyl)-3amino-4-methyl-6-phenylpyridazinium chloride) (*see* Fig. 2) displaces ^{3}H-GABA from neuronal membranes, blocks the GABA elicited enhancement of ^{3}H-diazepam binding, and competitively antagonizes the electrophysiological action of GABA on rat dorsal root ganglion cells in vitro (Chambon et al., 1985; Desarmenien et al., 1986). Substituted derivatives of SR 95103, such as SR 95531 (p-CH_3O) and SR 42641 (p-Cl), which lack the methyl group in the 4-position of the pyridazine ring, combine similar activity with even greater potency (Biziere et al., 1985; Desarmenien et al., 1987; Mienville and Vicini, 1987).

All three compounds appear highly selective for the $GABA_A$ receptor. In particular they do not affect the binding of ^{3}H-baclofen, ^{3}H-diazepam, ^{3}H-strychnine, or ^{3}H-glutamate and have no effect on Na^+-dependent uptake of GABA by synaptosomes, GABA transaminase, or glutamate decarboxylase activity (Chambon et al., 1985; Heaulme et al., 1986a). Recent iontophoretic experiments, however, have shown that SR 95103 antagonizes both GABA- and glycine-induced responses at several sites in the CNS. SR 95531 and SR 42641 are considerably more selective, blocking the depressant effect of glycine only rarely or at high ejecting currents (Gynther and Curtis, 1986; Michaud et al., 1986).

2.2.5. R 5135

The steroid derivative R 5135 (3a-hydroxy-16-imino-5β-17-aza-androstan-11-one) was proposed as a GABA antagonist on the basis of evidence from binding studies (Hunt and Clements-Jewery, 1981). R 5135 (*see* Fig. 3) potently displaces ^{3}H-GABA (Hunt and Clements-Jewery, 1981), antagonizes GABA-induced enhancement of benzodiazepine binding (Hunt and Clements-Jewery, 1981), and blocks the (presumed allosteric) displacement of ^{35}S-TBPS by GABA, muscimol, and other $GABA_A$ agonists (Squires et al., 1983). Although these activities point to a bicuculline-like effect on the GABA receptor, this is not the only site of action of R 5135. It also interacts with glycine and benzodiazepine receptors as judged by its ability to displace ^{3}H-strychnine and ^{3}H-diazepam (Hunt and Clements-Jewery, 1981).

GABA antagonism by R 5135 has not been examined in detail electrophysiologically. Results from iontophoretic experiments in the cat spinal cord reveal it to be nonselective (also affecting glycine responses), but shed no light on its mechanism of action (Curtis and Malik, 1985). It is perhaps relevant, however, to note that the glucocorticoids—prednisolone and hydrocortisone—have been reported to noncompetitively antagonize GABA-induced de-

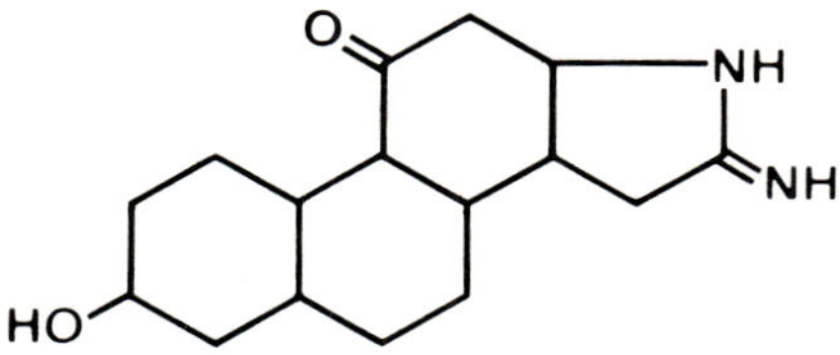

R 5135

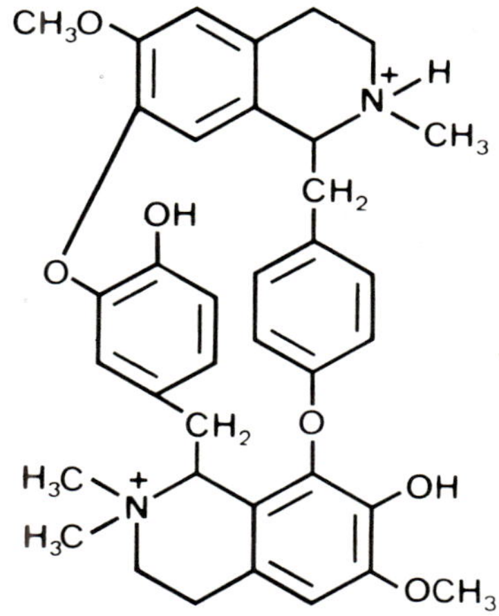

d-Tubocurarine

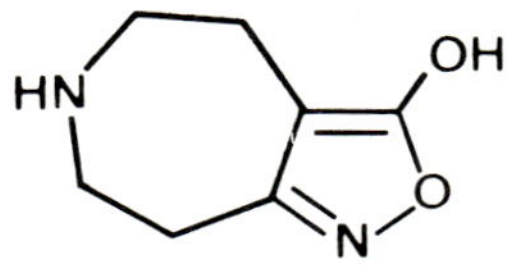

iso-THAZ

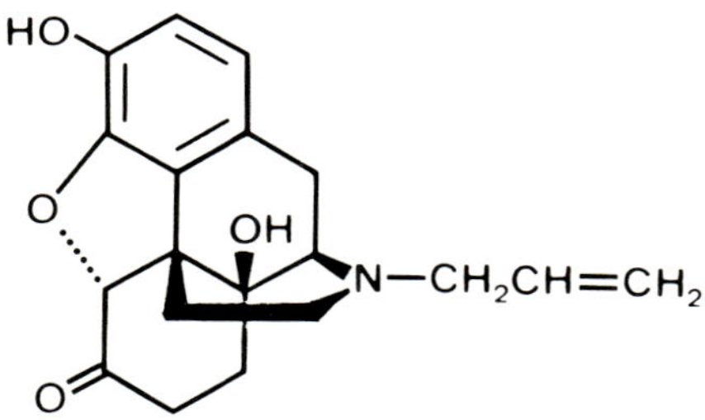

Naloxone

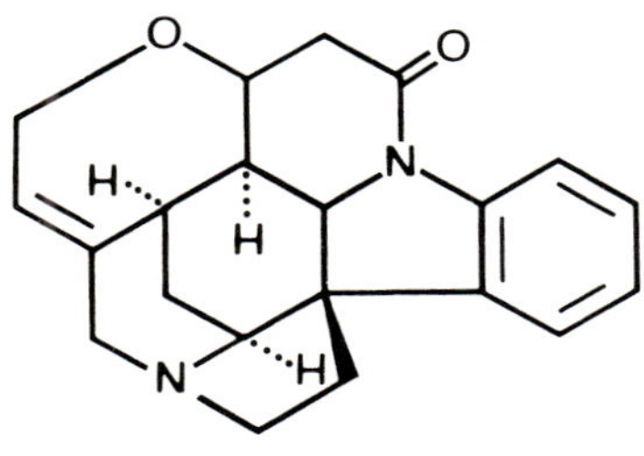

Strychnine

Fig. 3. Various nonspecific $GABA_A$ receptor antagonists.

polarization of bullfrog primary afferent neurons (Ariyoshi and Akasu, 1986).

2.2.6. Other Receptor Antagonists

Several other compounds (see Fig. 3) have been shown to antagonize GABA through an action at its receptor, but in view of their nonselective nature, they will be mentioned here only briefly.

Strychnine (*see* Nistri and Constanti, 1979; Dichter, 1980; Simmonds, 1982; Scholfield, 1982), *d*-tubocurarine (Nicoll, 1975; Hill et al., 1976; Lebeda et al., 1982; Schofield, 1982), and naloxone (Dingledine et al., 1978; Gruol et al., 1980; Simonds et al., 1983), in high concentrations, reduce the effects of exogenous GABA on a number of in vitro preparations and block GABA-mediated inhibition at several sites in the mammalian CNS. All three antagonists appear to act competitively. Thus strychnine, when tested on the rat cuneate nucleus slice in vitro (Simmonds, 1978, 1982) or isolated olfactory cortex (Scholfield, 1982), causes a parallel displacement of muscimol and GABA dose–response curves. Furthermore, the results of antagonist combination experiments carried out on the former preparation suggest that strychnine acts at the same site as bicuculline (Simmonds, 1982). Qualitatively similar effects are produced by *d*-tubocurarine on the rat hippocampal slice (Lebeda et al., 1982) and by naloxone on cultured mouse spinal cord neurons (Gruol et al., 1980). The competitive nature of the antagonism produced by these agents suggests an action at the receptor. This idea is supported by evidence from binding studies showing the ability of the compounds, at high concentrations, to displace ^{3}H-GABA or ^{3}H-BMI from rat brain membranes (Zukin et al., 1974; Mohler and Okada, 1977; Olsen et al., 1978; Dingledine et al., 1978).

2.3. Molecular Action of Receptor Antagonists

Both neurochemical and electrophysiological data support the idea of a competitive interaction between bicuculline and GABA at the vertebrate $GABA_A$ receptor. Curtis et al. (1970) first noted structural similarities between these two molecules and considered the cationic nitrogen and lactone component of bicuculline to compete with GABA for the same site. Since then a number of authors have used this relationship with bicuculline to examine possible "active conformations" of the flexible GABA molecule (*see* Kelly and Beart, 1975; Johnston, 1976; Andrews and Johnston, 1979). Fortunately, the competitive phenomena described in section 2.1 need not be explained exclusively by a single-site agonist–antagonist interaction, and indeed there is considerable evidence to suggest that such an explanation is unlikely.

Studies of ^{3}H-GABA binding in vitro have revealed the existence of two GABA sites—one high and one low affinity (for references, *see* Olsen and Snowman, 1983). ^{3}H-BMI or ^{3}H-BMC, on the other hand, bind only to a single site (Mohler and Okada, 1978; Olsen and Snowman, 1983). The binding of agonists and antagonists can also be distinguished by their response to "chaotropic" anions such as bromide, iodide, perchlorate, and thiocyanate.

These have little effect on agonist binding other than to decrease it, but enhance bicuculline binding or the potency of bicuculline to displace ^{3}H-GABA (Enna and Snyder, 1977; Mohler and Okada, 1978). Furthermore, the potencies of agonists in displacing ^{3}H-GABA are greater than their potencies in displacing ^{3}H-BMC, whereas antagonists (bicuculline, BMI, strychnine, R 5135) are more potent against ^{3}H-BMC binding than ^{3}H-GABA binding (Mohler and Okada, 1978; Olsen and Snowman, 1983). Similar results have been reported for ^{3}H-SR 95531 (Heaulme et al., 1986b).

The diazotizing reagent *p*-diazobenzenesulfonic acid (DSA) selectively eliminates low-affinity GABA binding sites (Maksay and Ticku, 1984a), whereas the thiocyanate anion eliminates the high-affinity sites (Browner et al., 1981). Thiocyanate treatment decreases the ^{3}H-GABA-displacing potency of agonists and raises that of BMC, *d*-tubocurarine, and R 5135, whereas DSA treatment has the opposite effects (Maksay and Ticku, 1984b). These observations led the latter authors to suggest that GABA antagonists bind preferentially to the low-affinity GABA site; the same suggestion was made by Olsen and Snowman (1983) from the results of their kinetic studies.

Based on the likely action of chaotropic anions, and with reference to the general receptor theory of Ariens et al. (1979), Maksay and Ticku (1984b) propose that the low-affinity GABA sites are associated with hydrophobic "accessory sites" that bind the hydrophobic ring systems of antagonists. Since these molecules lack the functional anion group considered necessary for agonist action, they do not interact with the proposed cationic region of the binding site. These results may be interpreted in accord with the hypothesis of Mohler and Okada (1978) that the GABA receptor exists in two interconvertible forms having inverse affinities for agonists and antagonists. Such a dual-receptor model, in which the sites for agonists and antagonists, although located on the same receptor protein, are different, is still capable of predicting the competitive phenomena observed (Ariens et al., 1979). This idea has been taken one step further by Olsen et al. (1984), who suggest that the high- and low-affinity sites represent *functional* states of the receptor–ionophore complex, the former associated with an open Cl^- channel and the latter with a closed channel. The recent development of new antagonists (securinine, pitrazepin, SR 95531) will allow this model to be tested further (e.g., Heaulme et al., 1986a,b) and enable more extensive studies of the structure-activity relationships of GABA receptor antagonists to be carried out (*see* Wermuth and Biziere, 1986).*

2.4. Ionophore Antagonists

2.4.1. Picrotoxinin and Related Sesquiterpenes

Picrotoxin is an equimolar mixture of two compounds differing only in the hydration of an isopropenyl group; these are picrotin and picrotoxinin (*see* Fig. 4). The latter is the active constituent, being 50 times more potent as a convulsant in mice than picrotin (Jarboe et al., 1968). Consistent with this observation, picrotoxinin, but not picrotin, is able to induce bursting activity in housefly flight motor neurons, block inhibitory synaptic transmission in the cockroach metathoracic coxal muscle (Ticku et al., 1978), and antagonize the action of muscimol in an in vitro preparation of the rat cuneate nucleus (Simmonds, 1982). The hydrogenated derivative of picrotoxinin, α-dihydropicrotoxinin (α-DHP), displays a pharmacological profile similar to its parent compound in several systems, but is slightly less potent (Jarboe et al., 1968; Ticku et al., 1978). Its labeled form, ^{3}H-DHP, was synthesized and found to bind reversibly and in a saturable fashion to a single site on crayfish muscle (Olsen et al., 1978) and rat brain membranes (Ticku et al., 1978). This has become known as the "picrotoxin site" (*see* Olsen, 1981; Ticku and Olsen, 1984). Three plant toxins structurally related to picrotoxinin and having similar convulsant activity (*see* Porter, 1967)—anisatin (shikimin), tutin, and coriamyrtin (*see* Fig. 4)—possess picrotoxin-like GABA antagonistic properties (Curtis et al., 1973; Nistri et al., 1974; Olsen et al., 1980; Kudo et al., 1981, 1984). Of this group, only anisatin has been tested quantitatively. In the frog spinal cord (Kudo et al., 1981), it was seen to produce a mixed form of noncompetitive antagonism, like picrotoxin, whereas bicuculline behaved as a classical, albeit weak, competitive antagonist.

In view of the close structural relationship between these compounds and picrotoxinin and their similarity of action, it would seem likely that they too act at the level of the Cl^- ionophore.

*A recent report on pyridazinyl-GABA derivatives suggests that thiocyanate sensitivity is related to the absence of a functional anionic group from the ligand and not its agonist/antagonist nature (Wermuth et al., 1987).

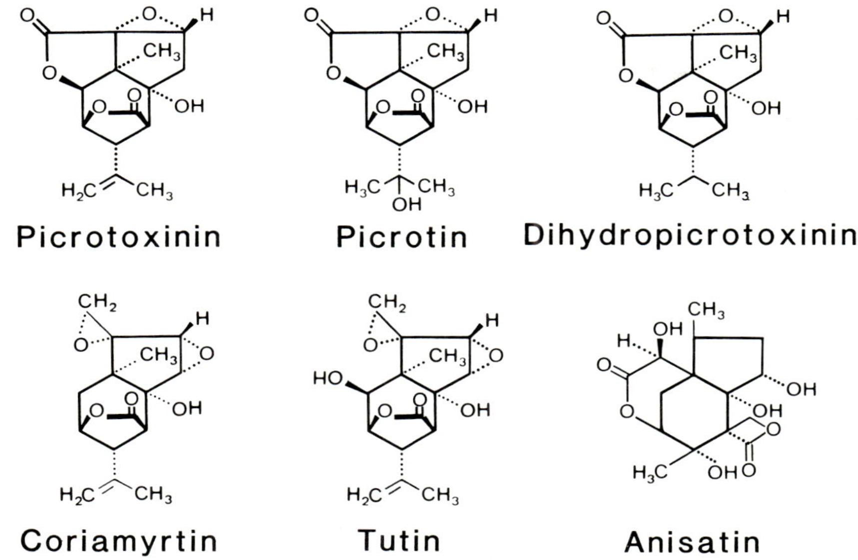

Fig. 4. Picrotoxinin and related compounds.

2.4.2. Cage Convulsants

Several forms of bi- or tricyclic ("cage") compounds (*see* Fig. 5) are highly toxic to mammals and produce convulsions and death in a manner similar to picrotoxin (*see* Bellet and Casida, 1973; Casida et al., 1976).

On the basis of a theoretical modeling of the GABA receptor, one of these compounds, tetramethylenedisulfotetramine (2,6-dithia-1,3,5,7-tetrazatricyclo [3.3.1.1.^{3,7}]decane-2,2,6,6-tetraoxide; TETS) was suggested to act as a GABA antagonist by Smythies (1974). A similar proposal was made by Curtis and Johnston, based on their observations in the cat spinal cord, as well as by Kerr (*see* Curtis and Johnston, 1974). TETS was subsequently shown to reversibly antagonize the actions of GABA on the rat superior cervical ganglion (Bowery et al., 1975) and crab neuromuscular junction (Large, 1975) in a picrotoxin-like "mixed" noncompetitive fashion. The selectivity of TETS, however, must be viewed as poor, since systemic administration in the rat reduces the neuronal depressant effects of both GABA and glycine (Collins et al., 1975; Dray, 1975a).

A particularly interesting series of cage convulsants is that formed by two types of 2,6,7-trioxabicyclo[2.2.2]octanes: the

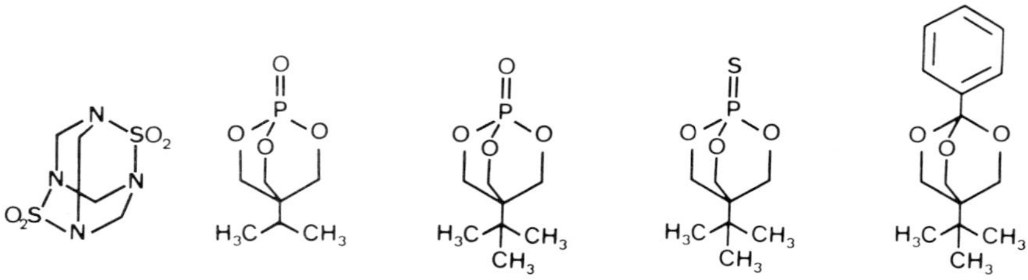

Fig. 5. Cage convulsants (see text for full names).

bicyclophosphorous esters and bicycloorthocarboxylates (Bellet and Casida, 1973; Casida et al., 1976, 1985). Several members of the former group, 4-alkyl-substituted derivatives of 1-phospha-2,6,7-trioxabicyclo[2.2.2.]octane-1-oxide (PTBO), have been reported to antagonize the actions of GABA (Bowery et al., 1976a,b, 1977). Comparative studies in the rat superior cervical ganglion and isolated frog spinal cord showed the isopropyl (IPTBO) and *t*-butyl (*t*-BPTBO or TBPO) derivatives to be the most potent, the latter being some four times more potent than bicuculline methochloride (Bowery et al., 1977). The nature of the antagonism produced by IPTBO was comparable to that produced by TETS. Unfortunately, this similarity extends to the ability of IPTBO to reduce the neuronal depressant action of glycine as well as GABA (Bowery et al., 1976a). Recently, TBPS (a bicyclophosphorous ester developed as a radioligand; *see below*) was shown to cause "mixed" inhibition of GABA-evoked currents in *Xenopus* oocytes induced to express GABA receptors by injection of chick brain mRNA (Van Renterghem et al., 1987).

Direct evidence for GABA antagonism produced by bicycloorthocarboxylates is lacking, although one member of this group, 4-*t*-butyl-1-(bromophenyl)-bicycloorthocarboxylate, has been shown to antagonize the stimulation-induced relaxation of cockroach coxal depressor muscle in common with picrotoxinin (Casida et al., 1985). More extensively studied, however, has been the potential of both bicycloorthocarboxylates and bicyclophosphates to act as radiolabeled probes for the GABA-activated Cl^- ionophore.

Initial attempts to label the "picrotoxin" binding site involved the use of ^{3}H-α-dihydropicrotoxinin (DHP) (*see* section 2.4.1). Although effective, a major problem with this ligand is its marked nonspecific binding (*see* Ticku and Maksay, 1983; Ticku and Olsen, 1984). In view of the high potency with which certain cage convulsants were seen to displace ^{3}H-DHP (Ticku and Olsen, 1979), new radioligands were sought from among this class of compound. Thus, *t*-^{35}S-butylbicyclophosphorothionate (TBPS or TBPT) was produced and shown to bind to the same site as ^{3}H-DHP, but with the advantages of higher affinity and lower levels of nonspecific binding (Squires et al., 1983). ^{35}S-TBPS has since been used successfully to examine the structure–activity relationships of ligands for the Cl^{-} ionophore binding site, as well as the interaction between this site and various components of the GABA–benzodiazepine receptor complex (Lawrence and Casida, 1983, 1984; Ramanjaneyulu and Ticku, 1984a,b; Squires et al., 1984; Seifert and Casida, 1985a,b; Trifiletti et al., 1985; Gallo et al., 1985; Maksay and Ticku, 1985; Maksay and Simonyi, 1985).

A third radioligand, *t*-^{3}H-butylbicycloorthobenzoate (4-*t*-butyl-1-phenyl-bicycloorthocarboxylate; TBOB), has been introduced that displays similar, but not identical, binding characteristics to ^{35}S-TBPS (Lawrence et al., 1985), and it seems that interesting advances are likely to be made in future studies with compounds of the same class. Addition of polar substituents to the phenyl ring of TBOB results in compounds with extremely high potencies in displacing ^{35}S-TBPS. Thus, 4-*t*-butyl-1-(4-cyanophenyl)-bicycloorthocarboxylate is approximately 40 times as potent as picrotoxinin in this assay. In addition the 4-*t*-butyl-1-(4-azidophenyl) analog has been developed as a potential photoaffinity label (Casida et al., 1985).

2.4.3. Caprolactam Derivatives

Caprolactam (hexahydro-2H-azepin-2-one), a simple seven-membered ring lactam, and certain of its derivatives (*see* Fig. 6) have convulsant activity when administered systemically (for references, *see* Kerr et al., 1976). Antagonism of GABA-mediated inhibition was suggested to underly these effects following the demonstration that caprolactams (in common with picrotoxin) block field potentials evoked in the substantia nigra by striatal stimulation (Kerr et al., 1976). Three compounds were shown to be effective: thiocaprolactam, 4,6,6-trimethyl-Δ3-caprolactam, and 4,6,6-trimethylcaprolactam. The latter derivative was also found to attenuate GABA-induced depolarizations in the isolated toad spinal cord (Kerr et al., 1976).

Subsequently, a range of methyl-substituted caprolactams were reported to produce picrotoxin-like partial inhibition of ^{3}H-muscimol binding and to potently inhibit ^{35}S-TBPS binding (Sker-

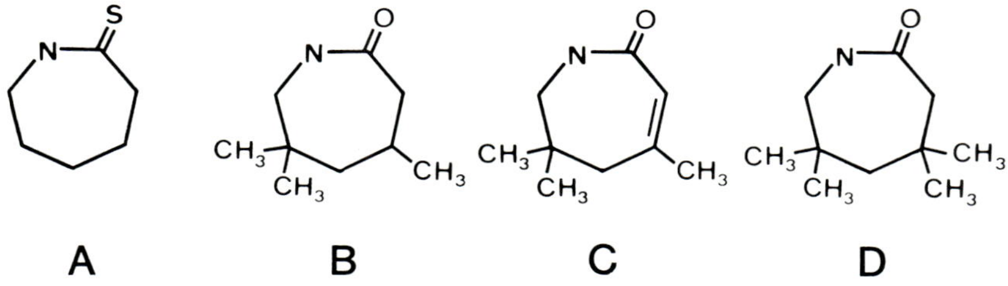

Fig. 6. Caprolactam derivatives: Thiocaprolactam (A), 4,6,6-trimethyl-caprolactam (B), 4,6,6-trimethyl-Δ3-caprolactam (C), and 4,4,6,6-tetramethyl-caprolactam (D).

ritt et al., 1985). The most potent displacer of ^{35}S-TBPS, 4,4,6,6-tetramethylcaprolactam, was also shown to reduce the response of cultured mouse spinal cord neurons to iontophoretically applied GABA. Not all caprolactam derivatives, however, are GABA antagonists. Those with alkyl substituents on carbons 4 and 6 of the ring are convulsants; other derivatives display depressant or muscle relaxant properties and are only weak displacers of ^{35}S-TBPS. One of these, 3,3-diallyl-6,6-dimethylhexahydro-2H-azepin-2,4-dione, actually enhances the response of mouse spinal cord neurons to GABA (Skerritt et al., 1985).

Although the GABA antagonism produced by caprolactams has not been subject to quantitative analysis, an action at the picrotoxin site would seem most likely. The relative potency of 4,4,6,6-tetramethylcaprolactam and picrotoxin in displacing ^{35}S-TBPS is paralleled by their potencies in antagonizing the action of GABA. Such effects of caprolactams appear fairly specific, since neither GABA- nor glutamate-uptake processes are affected. The reported inhibition of diazepam binding at high doses (Skerritt et al., 1985) may be due to an allosteric interaction.

2.4.4. Ro 5-3663 and Ro 5-4864

Ro 5-3663 (1,3-dihydro-5-methyl-2H-1,4-benzodiazepine-2-one) (*see* Fig. 7) is a convulsant with a potency comparable to that of picrotoxin (Schlosser and Franco, 1979). The results of a number of studies indicate that the most likely cause of Ro 5-3663-induced convulsions is a reduction in GABA-mediated inhibition. Thus, this 1,4-benzodiazepine has been found to block strychnine-resistant inhibition in the cat spinal cord, to antagonize the effects of GABA on both the cat superior cervical ganglion (Schlosser and Franco, 1979) and cultured rat hypothalamic cells (Geller et al.,

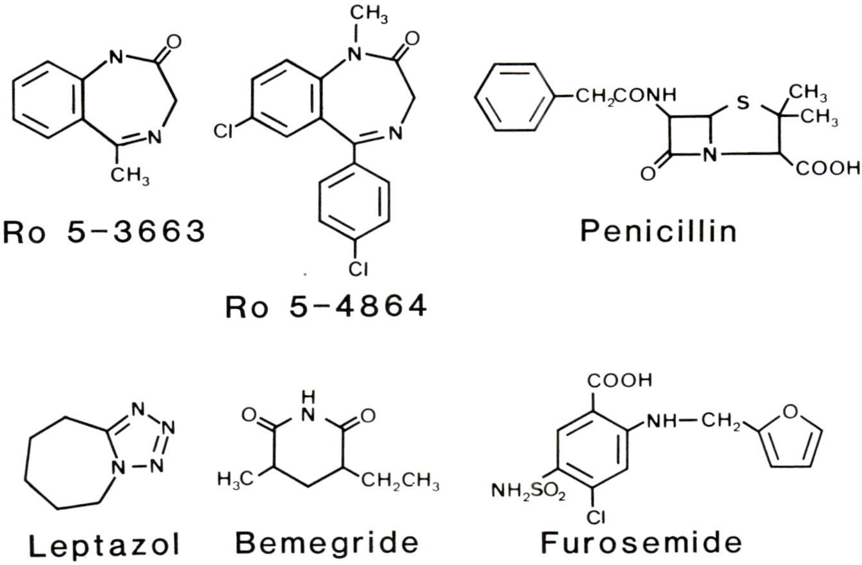

Fig. 7. Various "ionophore antagonists."

1980), and to attenuate muscimol-induced depolarization of the rat cuneate nucleus slice in vitro (Harrison and Simmonds, 1983).

Quantitative examination of the GABA antagonism produced by Ro 5-3663, alone and in combination with bicuculline or picrotoxin, points to it acting not at the GABA receptor, but at the Cl^- ionophore (Harrison and Simmonds, 1983). Consistent with this idea, Ro 5-3663 displaces both ^{3}H-DHP (Leeb-Lundberg et al., 1981) and ^{35}S-TBPS (Squires et al., 1983; Ramanjaneyulu and Ticku, 1984b) from rat brain membranes, but has only slight effects on the binding of ^{3}H-muscimol (Supavilai et al., 1982) and ^{3}H-flunitrazepam (Speth et al., 1979).

Ro 5-3663 is more potent than picrotoxinin at displacing ^{3}H-DHP (*see* Ticku and Olsen, 1984), but is only a weak antagonist of muscimol (considerably less potent than picrotoxin) in the cuneate slice preparation (Harrison and Simmonds, 1983). This fact, together with the observation of a degree of mutual antagonism between Ro 5-3663 and picrotoxin when used in combination against muscimol, led Harrison and Simmonds (1983) to suggest that Ro 5-3663, although having high affinity for the picrotoxin site, "is less effective than picrotoxin in influencing ionophore function," i.e., it behaves as a partial agonist. Some support for this theory may be seen in the actions of Ro 5-3663 exhibited in binding studies. Under appropriate ionic conditions and at physiological temperatures, antagonists such as picrotoxin, anisatin, and the bicyclophosphate esters partially inhibit (by an allosteric mechanism) ^{3}H-muscimol binding and GABA-induced enhancement of ^{3}H-benzodiazepine binding (Fujimoto and Okabayashi, 1980; Karobath et al., 1981; Supavilai et al., 1982; Matsuomo and Fukuda, 1983; Quast and Brenner, 1983). By comparison however, Ro 5-3663 has only weak effects on both of these processes (Supavilai et al., 1982; O'Brien and Spirt, 1980). As noted by Harrison and Simmonds (1983), the discrepancy between the ^{3}H-DHP-displacing and GABA-antagonistic abilities of Ro 5-3663 could also be accounted for by the apparent heterogeneity of DHP sites (*see* Squires, 1984), in that certain of these sites may be unrelated to the Cl^- ionophore. Interestingly, relative to picrotoxin, the potency of Ro 5-3663 in displacing ^{35}S-TBPS bears a much greater similarity to its potency as a muscimol antagonist than does its ability to displace ^{3}H-DHP.

What little evidence there is concerning the selectivity of Ro 5-3663 points to it having a specific action against GABA. It has no effect on strychnine-sensitive inhibition in the cat spinal cord (Schlosser and Franco, 1979) or on the depressant action of glycine on cultured rat hypothalamic neurons (Geller et al., 1980). In view of the poor correlation between the GABA-antagonistic (Harrison and Simmonds, 1983) and convulsant (Schlosser and Franco, 1979) potencies of Ro 5-3663, however, some further action cannot be ruled out.

Ro 5-4864 (*p*-Cl derivative of diazepam) (*see* Fig. 7), another atypical 1,4-benzodiazepine, may also have an action at the GABA-activated Cl^- ionophore. It produces behavioral effects similar to those elicited by picrotoxin (Pellow and File, 1984) and causes convulsions in a number of experimental animals (Weissman et al., 1983, 1984; Massotti and Lucantoni, 1986). Antagonism of GABA would seem to underly these actions of Ro 5-4864 since it reduces evoked inhibition in the cat spinal cord (Pieri et al., 1983; Polc and Schaffner, 1983), increases the firing rate of cells in the substantia nigra of the rat (Weissman et al., 1984), attenuates the effects of muscimol on the rat cuneate nucleus slice preparation (Simmonds, 1984), and reduces the response of cultured mouse spinal cord neurons to iontophoretically applied GABA (Skerritt et al., 1984).

Although Ro 5-4864, like Ro 5-3663, is virtually inactive at displacing benzodiazepines from their classical "central-type" binding sites, it has a high affinity for the "peripheral-type" or "nonneuronal" benzodiazepine receptor (*see* Squires, 1984; Haefely et al., 1985). Such sites are, however, unlikely to mediate the GABA antagonism produced by Ro 5-4864 (*see* Pellow and File, 1984; File and Pellow, 1985; Weissman et al., 1984). In fact GABA or muscimol antagonism in vitro is not evident below concentrations of Ro 5-4864, which would fully saturate the high-affinity "peripheral-type" sites (Skerritt et al., 1984; Simmonds, 1984).

An action at the Cl^- ionophore was suggested by Ticku and Ramanjaneyulu (1984) on the basis of the ability of Ro 5-4864 to competitively displace ^{35}S -TBPS from rat brain membranes. A similar conclusion has been reached by Weissman et al. (1985), who observed a marked correlation between the potencies of a series of Ro 5-4864 derivatives to inhibit ^{35}S-TBPS binding and their convulsant potencies. Although there are certain differences between the pharmacological profiles of Ro 5-4864 and recognized ligands for the "picrotoxin-site" (*see* Massotti and Lucantoni, 1986), this proposal is appealing. The IC_{50} for ^{35}S-TBPS displacement by Ro 5-4864 is in good agreement with the concentrations required to antagonize muscimol (Simmonds, 1984), and preliminary evidence has been presented that the hyposensitivity of one strain of mice (BALB/cBY) to the convulsant effects of Ro 5-4864 is associated with a similar hyposensitivity to picrotoxinin and leptazol (Seale et al., 1985), which seems to involve qualitative changes in the functioning of the Cl^- ionophore (Schwartz et al., 1985b).

2.4.5. Penicillin

Penicillin (sodium benzylpenicillin; PCN) is widely used to induce focal or generalized seizure activity in experimental models of epilepsy. It is now clear that this epileptogenic effect is caused, at least in part, by the ability of PCN to reduce GABA-mediated inhibition. This mechanism of action was first suggested by Davidoff (1972a), who showed that PCN blocked presynaptic inhibition in the isolated toad spinal cord and antagonized the depolarizing action of GABA on dorsal root terminals. Evidence in support of this proposal has since come from studies on invertebrate preparations (Wilson and Escueta, 1974; Hochner et al., 1976,) and mammalian neurons, both in vivo (Curtis et al., 1972; Davidoff, 1972b; Clarke and Hill, 1972; Hill et al., 1973; Duijn et al., 1973) and in vitro (Macdonald and Barker, 1977, 1978; Pickles and Simmonds, 1980; Dingledine and Gjerstad, 1980; Barker et al., 1983; Mathers, 1985).

The nature of the GABA antagonism produced by PCN has been the subject of much debate. In the report of their early iontophoretic study Curtis et al. (1972) noted structural similarities between PCN and bicuculline that led them to suggest an action of PCN at the GABA receptor. A similar conclusion was reached by Macdonald and Barker (1977) on the basis of the selective and apparently competitive antagonism of GABA by PCN on mouse spinal cord neurons in culture. However, in the detailed investigation carried out by Hochner et al. (1976) on the effects of PCN on crustacean muscle, a dual form of antagonism was observed—weak competitive inhibition in combination with powerful noncompetitive inhibition. The former effect was suggested to involve an action at the receptor, whereas the latter was ascribed to block of the GABA-activated Cl^- channel. One alternative explanation of these results would be a "mixed" noncompetitive antagonism as proposed for picrotoxin by Constanti (1978) (*see* section 2.1). Such a "single-site" interpretation would be in accord with the findings of Wilson and Escueta (1974) and Pellmar and Wilson (1977b) in *Aplysia* that GABA responses not involving changes in Cl^- conductance are unaffected by PCN, whereas Cl^--mediated responses to GABA and other transmitters are selectively blocked. An action of PCN on the Cl^- ionophore was also suggested by Pickles and Simmonds (1980) from their observations on the isolated rat cuneate nucleus in which clear differences were seen between the action of PCN and other GABA antagonists, including picrotoxin and bicuculline (Simmonds, 1980, 1982). Such differences were confirmed in studies by Barker et al. (1983), who applied fluctuation analysis to the effect of antagonists on GABA-induced responses in cultured mouse spinal cord neurons and found that PCN produced complex changes in the kinetics of channel activation, whereas picrotoxin and bicuculline affected neither the conductance nor the mean open-time of GABA-activated channels. Mathers, using an outside-out patch clamp technique to record from the same cells, has reported that PCN causes a significant reduction in the average open-time of GABA-induced membrane channels (Mathers, 1985; Chow and Mathers, 1986); this is consistent with an ability of PCN to block the GABA-activated Cl^- ionophore (*see also* Smart and Constanti, 1986; Akaike et al., 1987; Yakushiji et al., 1987).

Unfortunately, the action of PCN is not limited to this particular channel. In invertebrates, nonsynaptic Cl^- channels and Cl^- channels associated with other neurotransmitter receptors are also susceptible to its blocking action (for discussion, *see* Pellmar and Wilson, 1977b), although, in contrast to what one might expect of a truly nonspecific Cl^- channel blocker, PCN does not antagonize the actions of glycine on cultured mouse spinal cord neurons (Macdonald and Barker, 1977, 1978), even if it does so in the cuneate nucleus of the cat (Hill et al., 1976). PCN has also been shown to have a direct membrane depolarizing action related to blockade of K^+ conductance (Heyer et al., 1982).

Although there are no structure–activity studies of penicillin-derivatives with regard to GABA antagonism, information is available concerning the epileptogenic activity of such compounds. Based on studies using the in vitro hippocampal slice, Esplin et al. (1985) have identified an absolute requirement for the thiazolidine ring, β-lactam ring, and side chain substitution on C_6 (*see* Fig. 7). Results from earlier investigation conform to this theory. Thus, of a number of β-lactam antibiotics tested as cortical convulsants by Gutnick et al. (1976), only phenoxymethylpenicillin displayed a potency comparable to that of PCN. Similarly, both the convulsant (Gutnick and Prince, 1971) and GABA antagonistic (Davidoff, 1972a) properties of PCN are lost following cleavage of the β-lactam ring by penicillinase.

2.4.6. Pentamethylenetetrazole and Bemegride

Pentamethylenetetrazole (PTZ, pentalenetetrazole, leptazol), like penicillin, is widely used to induce seizures in experimental animals and is routinely employed in the screening of potential anticonvulsant drugs. PTZ has a variety of actions (*see* Woodbury, 1980), but there is considerable evidence to suggest that it may cause convulsions by reducing GABA-mediated inhibition through an action at the GABA-receptor-linked Cl^- ionophore.

In the isolated frog spinal cord, PTZ blocks inhibition at primary afferents (Nicoll and Padjen, 1976), although high concentrations are required. With similar potency, PTZ also blocks postsynaptic inhibition and the effects of applied GABA on guinea pig olfactory cortex slices (Scholfield, 1982) and attenuates the effects of GABA and muscimol on rat cuneate nucleus slices (Simmonds, 1978, 1982). The need for high concentrations may explain the lack of GABA antagonism by PTZ when applied by iontophoresis in vivo (Hill et al., 1976). Both iontophoretically and bath-applied PTZ, however, is effective in reducing the response of cultured mouse spinal cord neurons to GABA (Macdonald and Barker, 1977; Barker et al., 1983).

Quantitative studies by Simmonds (1978, 1982) indicate that the antagonism of GABA by PTZ is noncompetitive and is caused by an interaction of PTZ with the picrotoxin site of the GABA–receptor complex. This finding is in agreement with the results of Pellmar and Wilson (1977a), who described a selective Cl^- channel-blocking action of PTZ on neurons of *Aplysia*, and is also supported by more recent ligand binding studies. Thus, PTZ competitively inhibits the binding of ^{3}H-α-DHP (Ticku et al., 1978; Ticku and Olsen, 1979) and ^{35}S-TBPS (Squires et al., 1983, 1984; Ramanjaneyulu and Ticku, 1984b) to rat brain membranes. This ability is shared by a number of tetrazole derivatives, the most potent of these, namely, 8-*t*-butyl-PTZ, 6,6'-dichloromethylenetetrazole, 6,6'dichloro-PTZ, 8-sec-butyl-PTZ, and 1-cyano-5-methyltetrazole, being approximately 15–150 times more potent than PTZ (Squires et al., 1984; Ramanjaneyulu and Ticku, 1984b). Although there is no information available concerning GABA antagonism by these compounds, all five cause convulsions in mice, and the convulsant potencies of twelve such tetrazoles, including PTZ, are highly correlated with their potencies in displacing ^{35}S-TBPS (Squires et al., 1984). PTZ and its analogs also competitively inhibit ^{3}H-diazepam binding (Muller et al., 1978; Rehavi et al., 1982; Ramanjaneyulu and Ticku, 1984a), but at higher concentrations than those required for ^{35}S-TBPS displacement. It is not clear whether this action represents true competition for the benzodiazepine site or an allosteric effect resulting from an action at the picrotoxin site.

The convulsant bemegride displays a similar pharmacological profile to PTZ both in electrophysiological (Simmonds, 1978; Scholfield, 1982) and ligand binding studies (Squires et al., 1983; Ramanjaneyulu and Ticku, 1984b) and would appear to act at the same site (*see* Ticku and Maksay, 1983).

2.4.7. Other Ionophore Antagonists

A number of drugs known to inhibit active Cl^- transport across cell membranes (for references, *see* Gallagher et al., 1983) have been found to attenuate the responses of several in vitro preparations to GABA by a mechanism seemingly independent of any effect on the transmembrane Cl^- gradient. This type of action has been described for furosemide (*see* Fig. 7) on isolated frog spinal cord (Nicoll, 1978), rat cuneate nucleus slice (Simmonds, 1982), and cat dorsal root ganglion (DRG) cells (Gallagher et al., 1983); for SITS

(4-acetamido-4'-isothiocyano-2,2'-disulfonic stilbene) on cat (Gallagher et al., 1983) and bullfrog DRG cells (Akaike and Oomura, 1984); for bumetanide on cat DRG cells (Gallagher et al., 1983); and for phlordzin on cat DRG cells (Stevens et al., 1985a). Also, furosemide and piretanide block GABA-induced contraction of the guinea pig ileum (Krantis and Kerr, 1981), while DIDS (4,4'-di-isothiocyano-2,2'-disulfonic stilbene) and DNDS (4,4'-dinitro-2,2'disulfonic stilbene) have been shown to block barbiturate-stimulated $^{36}Cl^-$ efflux from rat brain synaptoneurosomes (Schwartz et al., 1985a).

Furosemide (Krantis and Kerr, 1981; Simmonds, 1982), piretanide (Krantis and Kerr, 1981), and phloridzin (Stevens et al., 1985a) depress muscimol and GABA dose–response curves without producing any parallel shift; i.e., they act in an apparently noncompetitive fashion. The action of furosemide on the cuneate nucleus preparation resembles that of PCN (Pickles and Simmonds, 1980), and it has been suggested that Cl^- pump inhibitors block GABA-activated Cl^- conductance in a similar manner (Stevens et al., 1985a). Unfortunately there is no information from binding studies regarding the site of action of these compounds and a dissociation between effects on Cl^- transport and Cl^- conductance is not evident for all pump inhibitors since piretanide reduces GABA responses and alters the GABA reversal potential of frog DRG cells (Wojtowicz and Nicoll, 1982).

Certain chlorinated hydrocarbon insecticides and α-cyano-3-phenoxybenzyl pyrethroids have been recognized as having an action at the GABA-activated Cl^- ionophore. Compounds of both classes are highly insecticidal and produce toxic symptoms in mammals similar to those brought about by picrotoxin and bicyclophosphorous esters (Lawrence and Casida, 1983, 1984). Three types of chlorinated hydrocarbon insecticides; hexachlorocyclohexane (e.g. lindane—the γ-isomer), polychlorobornane components of toxaphene (e.g., 2,2,5-*endo*,6*exo*, 8,9,9,10-octachlorobornane), and chlorinated cyclodienes (e.g., heptachlor epoxide, 12-ketoendrin) potently, competitively, and stereospecifically inhibit the binding to rat brain membranes of ^{3}H-DHP (Matsumura and Tanaka, 1984) and ^{35}S-TBPS (Lawrence and Casida, 1984). Functional studies have shown various chlorinated hydrocarbon insecticides to block GABA-induced $^{36}Cl^-$ uptake into cockroach coxal muscles (Matsumura and Tanaka, 1984), and mouse and rat brain vesicles (Bloomquist and Soderlund, 1985; Eldefrawi and Eldefrawi, 1987). In addition, lindane has been demonstrated to block GABA-induced depolarization of rat superior cervical ganglion (Matsumura and Tanaka, 1984) and GABA-activated Cl^- currents in voltage-clamped dorsal root ganglion neurones (Ogata et al., 1987).

Like the chlorinated hydrocarbon insecticides, α-cyanophenoxybenzyl pyrethroids (e.g., cypermethrin, deltamethrin, fenvalerate) inhibit ^3H -DHP (Leeb-Lundberg et al., 1981) and ^{35}S -TBPS (Squires et al., 1983; Lawrence and Casida, 1983) binding with potencies related to their toxicity (Lawrence and Casida, 1983). Although only limited studies have been carried out, they too have been found to block GABA responses in certain preparations. Thus, they inhibit the conductance increase induced in crayfish claw opener muscle by bath-applied GABA (Gammon and Casida, 1983) and block $^{36}Cl^-$ uptake into rat brain vessicles (Eldefrawi and Eldefrawi, 1987). There are also differences between the action of the cyano-pyrethroids and the polychlorocycloalkane insecticides. Inhibition of ^{35}S -TBPS binding by the former is noncompetitive or "mixed" (Lawrence and Casida, 1983), reaches a maximum at 60–80%, and is partially reversed by the GABA receptor antagonist R 5135 (*see* section 2.2.5) (Squires et al., 1983). This latter effect may result from the presence of residual GABA in the membrane preparation since GABA has been shown to facilitate cypermethrin displacement of ^{35}S-TBPS from EDTA/water-dialyzed membranes (Seifert and Casida, 1985b). In contrast, GABA reduces the displacement of ^{35}S-TBPS by unlabeled TBPS and has no effect on displacement induced by photoheptachlor epoxide (Seifert and Casida, 1985b). These differences have led to the suggestion that cyano-pyrethroids interact with a part of the GABA receptor complex that is distinct from, but closely related to, the TBPS domain but do so only during conformational changes induced by GABA (*see* Seifert and Casida, 1985a,b). Deltamethrin, unlike lindane, does not block GABA responses of cultured rat dorsal root ganglion neurones (Ogata et al., 1987).

Recently, tremorgenic toxins of fungal origin have been shown to inhibit ^{35}S-TBPS binding to, and $^{36}Cl^-$ uptake by, rat brain microsacs with micromolar potencies (Gant et al., 1987). Complexities in the inhibition of TBPS binding by one of these toxins, paspalinine, has led to the suggestion that their site of action at the $GABA_A$ receptor complex is distinct from that of TBPS (Gant et al., 1987).

2.5. Molecular Action of Ionophore Antagonists

Although the apparently noncompetitive (*see* section 2.1) and Cl^- ion-dependent (Takeuchi and Takeuchi, 1969) antagonism of GABA by picrotoxin is considered to result from an interaction with the GABA-activated Cl^- ionophore, the exact nature of this process remains unresolved. Simplistically, two distinct mechanisms may be envisaged. One involves a "plugging" action in which the ionophore is physically occluded by the presence of picrotoxin. The other requires that blockade is achieved indirectly, through a conformational change in the ionophore brought about by the antagonist binding at some site distinct from the channel lumen.

The first of these mechanisms was favored by Klunk et al. (1983), who used data from an earlier structure–activity study of picrotoxin-related convulsants by Jarboe et al. (1968), together with their own results with γ-butyrolactones, postulated to act like picrotoxin, to produce a model of the picrotoxin binding site. Based on a computer-assisted construction of electron density volume maps of the various molecules and their association with an idealized Cl^- channel, it was proposed that alkyl substituents in the β-position of the γ-butyrolactone ring of active picrotoxin-like compounds directly block the passage of Cl^- ions through the Cl^- channel.

This model was questioned by Ticku and Maksay (1983) since increasing the size of the hydrophobic alkyl group on bicyclophosphate esters should not lead to the observed increase in their affinity for the picrotoxin site if this is a hydrophilic Cl^- channel. They viewed β-substituents of the convulsants as interacting instead with an allosteric hydrophobic regulatory site governing the state of the channel. In their model, the picrotoxin binding site accepts the semiplanar 5–6-membered ring of its ligands with polar heteroatoms, ester, or amide bonds fixing the ring to the receptor via hydrogen bonding or charge transfer reactions, while the ability of each compound to affect the ionophore is determined by the size and position of its ring substituents. These ideas would also appear to hold true for the potent bicycloorthocarboxylate convulsants (*see* section 2.4.2) and for tetrazole derivatives (*see* section 2.4.6). However, chlorinated hydrocarbon insecticides such as heptachlor epoxide and lindane (*see* section 2.4.7), although having carbon skeletons that would allow them to interact with the proposed receptor (*see* Matsumura and Tanaka, 1984), lack the alkyl substituent considered necessary for activity and, if orientated analogously to picrotoxinin, would overlap the hypothesized regulatory site with only one or two Cl substituents.

Despite this discrepancy, the allosteric model put forward by Ticku and Maksay is appealing. In particular, it is consistent with the explanation offered by Constanti (1978) for the observed "saturation" of GABA antagonism at high picrotoxin concentrations; namely, that block of the ionophore by picrotoxin is preceded by the formation of "an intermediate bound-unblocked state." The kinetic model that gave rise to this suggestion was developed following an analysis of the action of picrotoxin in blocking the response of crustacean muscle to GABA. Recently this action has been reexamined in detail by Smart and Constanti (1986), who conclude that picrotoxin, by binding at a hydrophobic site unconnected with the channel lumen, causes the ionophore to close or to be stabilized in a closed state. Their suggestion that picrotoxin can reduce the rate constant for channel activation (transition from agonist-bound-closed to open) is supported by data from fluctuation analysis of GABA-induced membrane noise in cultured mouse spinal cord neurons (Barker et al., 1983), rat hippocampal neurons (Segal and Barker, 1984) and dissociated rat sympathetic ganglion cells (Cull-Candy et al., 1988) showing that picrotoxin depresses membrane responses to GABA with little effect on the mean open-state conductance or lifetime of individual ion channels. Furthermore, patch clamp recordings of single GABA-activated channel currents in bullfrog dorsal root ganglion cells exposed to picrotoxin (Akaike and Oomura, 1984; Akaike et al., 1985) show none of the "chatter" that, by analogy with nicotinic systems (Neher and Steinbach, 1978; Ogden et al., 1981), should be seen if picrotoxin transiently plugged the open channel. Instead, picrotoxin appears to attenuate the GABA response solely by decreasing the probability or rate of channel opening (*see also* Adams et al., 1981; McBurney, 1984).

Other indirect evidence argues against a simple "plugging" mechanism. As pointed out by Neher and Steinbach (1978), stabilization of a transmitter-operated ion channel in an open (but blocked) state should result in the associated receptor having a higher affinity for agonists. In fact picrotoxin, anisatin, and IPTBO have been shown to *reduce,* in a temperature- and anion-dependent fashion, the affinity of GABA and muscimol for the GABA receptor (Fujimoto and Okabayashi, 1980; Supavilai et al., 1982; Matsuomo and Fukuda, 1982; Quast and Brenner, 1983). That the picrotoxin binding site is unlikely to lie within the channel lumen has also been inferred from studies on neuromuscular preparations from two species of spiny lobster (Albert et al., 1986) in which GABA-induced conductance increases in the foregut and opener muscles are insensitive to picrotoxin, yet in other respects closely resemble the picrotoxin-sensitive responses seen at other crustacean neuromuscular junctions. Thus, changes in the picrotoxin site do not seem to affect the functional properties of the Cl^- ionophore.

Further information regarding the action of picrotoxin has come from the work of Akaike and colleagues on bullfrog dorsal root ganglion cells. Internal perfusion of picrotoxin diminished the Cl^- current induced by external perfusion of GABA, and this antagonism was found to be voltage-dependent with greater block occurring at more depolarized membrane potentials (Akaike and Oomura, 1984; Akaike et al., 1985). These findings are intriguing, if somewhat difficult to interpret. The voltage-dependent block seen with internal application suggests a "channel-plugging" process, but since picrotoxinin is an uncharged molecule, its movement into the channel lumen under the direct control of the transmembrane electric field is unlikely. More plausible is an allosteric mechanism (as indicated above), in which either, (1) the picrotoxin-induced transition of the ionophore from an antagonist-bound-unblocked state to a blocked state is in some way voltage-dependent or (2) conformational changes in the GABA-receptor–ionophore complex caused by alterations in the membrane potential lead to steric exposure of the picrotoxin binding site. The former is not favoured since it would imply a voltage dependence opposite to that of the normally gated ionophore (Cull-Candy and Miledi, 1981; Segal and Barker, 1984). With regard to the second alternative, it is interesting to note that GABA and benzodiazepines, presumably by causing a conformational change in the receptor complex, alter the rate (association and dissociation) of TBPS binding to rat brain membranes (Seifert and Casida, 1985a; Maksay and Simonyi, 1985). Consistent with this finding, TBPS block of GABA responses evoked from mRNA-injected *Xenopus* oocytes exhibits a marked use dependency of both onset and recovery (Van Renterghem et al., 1987). A conclusion similar to that of Smart and Constanti (1986) (concerning picrotoxin) was reached with regard to the mechanism of action of TBPS; namely, that the preferential binding of the antagonist to a closed-liganded form of the receptor complex leads to its stabilization in this state. The action of TBPS (an uncharged molecule) was also found to be independent of the direction of Cl^- flow. Interestingly, a recent report from the laboratory of Akaike now reports a similar voltage-independent blocking effect of picrotoxin on bullfrog dorsal root ganglion cells (Yakushiji et al., 1987 cf. Akaike and Oomura, 1984; Akaike et al., 1985). Furthermore, in rat sympathetic ganglion cells, GABA-induced whole-cell currents and the properties of GABA-activated single channels (derived from noise analysis) are not affected by intracellularly applied picrotoxin (Cull-Candy et al., 1988). These discrepancies remain to be resolved.

Finally, based on measurements of the dose dependency of GABA-induced membrane noise, Yasui et al., (1985) have postulated the existence of three types of $GABA_A$–receptor-channel complex (designated α_1, α_2, and β) with different functional properties, only two of which (α_1 and α_2) are blocked by picrotoxin. A preferential effect of lindane on the initial peak component of GABA-induced currents recorded from rat dorsal root ganglion cells has also been interpreted as possible evidence for the existence of more than one type of $GABA_A$ receptor (Ogata et al., 1987). Similar variations in TBPS block during desensitizing GABA responses have been observed in *Xenopus* oocytes (Van Renterghem et al., 1987). These authors, however, prefer to consider two macroscopic conductance states: a high one prior to desensitization and a low one produced during desensitization, the former being TBPS sensitive.

Although such detailed information concerning the mode of action of other ionophore antagonists is not available, in view of the general similarity of their macroscopic actions to those of picrotoxin and TBPS, it is likely that they exert a similar effect on the Cl^- channel. Exceptions include penicillin and certain Cl^- pump inhibitors whose properties are somewhat different from those of other ionophore antagonists (*see* sections 2.4.5 and 2.4.7). In fact, penicillin may act by binding to a cationic site within the channel lumen, thus blocking the passage of the Cl^- ion (Smart and Constanti, 1986). Recently described kinetics, voltage dependence, and wash-out characteristics of GABA antagonism by penicillin in bullfrog dorsal root ganglion neurons (Akaike et al., 1987; Yakushiji et al., 1987) support such a proposal.

2.6. Practical Aspects: Choice of Antagonist

For the purposes of assessing the *physiological role* of GABA in a particular system, any agent that blocks the postsynaptic effects of

GABA, regardless of its mode of action, will suffice. This is obviously not true for studies aimed at characterizing the receptor or transduction processes involved in a given response when the choice of "antagonist" is critically dependent on its mechanism of action. In both cases, however, specificity is of prime importance. Finally, factors such as solubility, stability, and pharmacokinetic properties need to be considered.

Methohalide derivatives of bicuculline remain the most widely used GABA antagonists for a number of reasons. Unlike bicuculline they are freely water soluble, dissociating to the *N*-methylbicuculline cation. At physiological pH, bicuculline is only weakly protonated and thus less suitable than BMI or BMC for iontophoretic application. The quaternary derivatives, however, have the disadvantage that they do not easily cross the blood–brain barrier, and for systemic administration bicuculline itself must be employed. The preparation and use of bicuculline solutions require care since at physiological pH hydrolysis can occur, leading to the formation of inactive bicucine (*see* Fig. 1). Fortunately solutions of BMI and BMC are considerably more stable (Olsen et al., 1975).

The recently developed receptor antagonists securinine, SR 95103, and pitrazepin have differing properties that are likely to make them valuable alternatives to bicuculline and its derivatives. Securinine and SR 95103 are soluble in aqueous media and stable over a wide pH range (Chambon et al., 1985; Beutler et al., 1985). Both are suitable for iontophoretic application, although only the former crosses the blood–brain barrier easily. Pitrazepin, which is dissolved in lactic acid prior to dilution, has a prolonged action (bath application) (Gahwiler et al., 1984) that is in contrast to the fast dissociation kinetics reported for SR 95103 (Chambon et al., 1985).

Of the ionophore antagonists, picrotoxin and related compounds, caprolactam derivatives, and bicyclophosphorous esters have been used most extensively. All three types of antagonist are relatively water-soluble and cross the blood–brain barrier, but since they lack any charge, systemic or bath application, rather than iontophoresis, is usually employed. Picrotoxin may be administered by electroosmosis with moderate success. TETS is practically insoluble in water and, although it can be dissolved in acetone (prior to dilution), its low solubility even under these conditions makes it unsuitable for iontophoretic use (Dray, 1975a). Bicycloorthocarboxylates are also poorly soluble, being used with ethanol or dimethylsulfoxide vehicles (Casida et al., 1985 Van Renterghem et al., 1987).

The selectivity of each GABA antagonist has been referred to above (sections 2.2.1–2.2.6 and 2.4.1–2.4.7), but one aspect of this deserves further attention; that is the ability of antagonists of discriminate between GABA- and glycine-like amino acids. The glycine antagonist strychnine, at high concentrations, is able to block GABA responses (*see* section 2.2.6). Conversely, a number of "GABA antagonists," including R 5135, pitrazepin, IPTBO, TETS, *d*-tubocurarine, and even bicuculline (*see* above and Hill et al., 1976), have been reported to block glycine responses or affect the binding of glycine. Several other compounds appear to block inhibitory amino acid responses indiscriminately: isoTHAZ (a bicyclic analog of isomuscimol), 3PYOL (a bicyclic analog of homomuscimol), and AMBD (6-amino-methyl-3-methyl-4H-1,2,4-benzothiadiazine-1,1-dioxide hydrochloride) (Krogsgaard-Larsen et al., 1982). Also, bicuculline stereospecifically inhibits the binding of ^{3}H-strychnine (Goldinger and Muller, 1980).

Two possible explanations for these effects are, (1) that bicuculline and other antagonists bind nonspecifically to both GABA and glycine receptors or (2) that the two receptor systems are linked in some mechanistic way. A common ionophore has been suggested on the basis of electrophysiological results (Barker and McBurney, 1979; Hamill et al., 1983) but this does not now seem likely; rather, separate GABA receptor- and glycine receptor-linked ionophores appear to possess extraordinarily similar ion permeability and conductance properties (Bormann et al., 1987). The first explanation, on the other hand, would be consistent with the general theory of low specificity of receptor antagonists (Ariens et al., 1979). Indeed, given the considerable homology between the primary structures of $GABA_A$ receptor subunits (α and β) and that of the strychnine-binding subunit of the glycine receptor (*see* Barnard et al., 1987), this degree of pharmacological overlap does not seem wholly surprising. Whatever the explanation is for the lack of specificity seen with certain "GABA antagonists", it is clear that care should be taken when interpreting the results of studies aimed at discerning the nature of the transmitter at a given site. The use of the newer more selective receptor antagonists, securinine and SR 95531, may be indicated in such circumstances. As with all antagonists, however, strict attention should always be given to the doses employed.

3. $GABA_B$ Antagonists

3.1. Historical Background

The ability of GABA to depolarize ganglion cell bodies and the axons of autonomic nerves (Bowery and Brown, 1974) led Bowery and Hudson to suggest that if GABA receptors were present at the nerve terminals of these cells, and capable of mediating a similar effect, their activation would reduce the evoked release of neurotransmitter. To test this hypothesis, they examined the effect of GABA on the release of ^{3}H-noradrenaline from isolated atria and vas deferens of the rat. GABA was indeed found to diminish the stimulated release of ^{3}H-noradrenaline, but this action was not antagonized by bicuculline or a number of other "GABA antagonists," including picrotoxin, isopropylbicyclophosphate, leptazol, penicillin, bemegride, tubocurarine, and furosemide (Bowery and Hudson, 1979; Bowery et al., 1981). The suggested presence of a novel GABA receptor was supported by the fact that the action of GABA was not mimicked by a range of potent agonists at bicuculline-sensitive sites. Thus isoguvacine, piperidine-4-sulfonic acid, 3-aminopropane sulfonic acid, muscimol, and THIP were inactive. In contrast, baclofen, the *p*-chlorophenyl derivative of GABA (*see* Fig. 8) was found to be active at this site, the (–)-isomer being more effective than the (+)-isomer and approximately equipotent with GABA. The marked pharmacological differences between the bicuculline-sensitive and bicuculline-resistant GABA receptors led Hill and Bowery (1981) to classify them as $GABA_A$ and $GABA_B$ sites, respecitvely.

$GABA_B$ sites have now been demonstrated in a variety of peripheral tissues. They are present on neuronal inputs to heart, intestine, bladder, vas deferens, anococcygeus, and vascular muscle (Bowery et al., 1981; Kapilta et al., 1982; Giotti et al., 1983a; Ong and Kerr, 1983; Luzzi et al., 1985; Allan and Dickenson, 1986; Maggi et al., 1984, 1985; Muhyaddin et al., 1982; Anwar and Mason, 1982), where their action is to reduce neurotransmitter release and thus the postsynaptic response.

$GABA_B$ receptors are also present in the mammalian CNS. Their characterization depended heavily on studies with the now recognized $GABA_B$ receptor ligand, baclofen. This compound, when administered systemically or in low iontophoretic "doses," specifically reduces synaptic excitation at several sites in the CNS

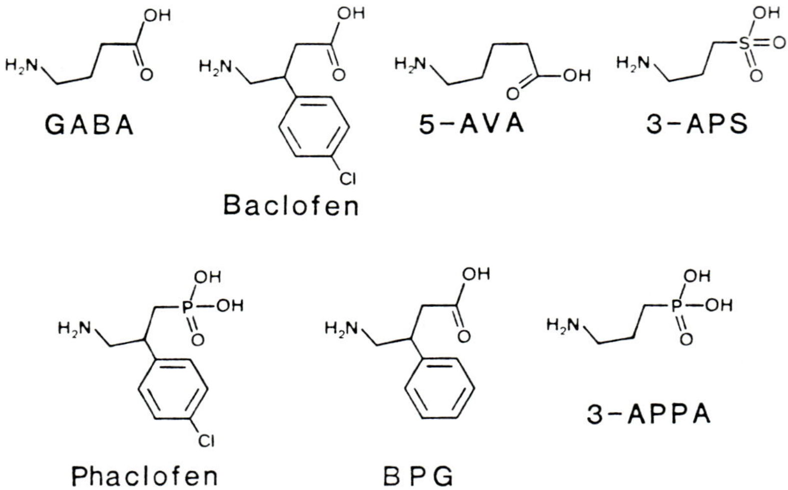

Fig. 8. GABA, baclofen, and proposed $GABA_B$ antagonists.

(Fox et al., 1978; Ono et al., 1979; Davies, 1981). This effect was suggested to result from a suppression of transmitter release, an idea supported by studies demonstrating the ability of baclofen to reduce the electrically evoked release of glutamate and aspartate from slices of rat cortex and cat spinal cord (Potashner, 1979; Johnston et al., 1980). At the same time, using larger iontophoretic ejecting currents, baclofen was shown to possess a nonspecific depressant action on neuronal firing (Curtis et al., 1974; Davies and Watkins, 1974; Olpe et al., 1978; Fox et al., 1978), which was bicuculline-insensitive (Curtis et al., 1974; Davies and Watkins, 1974). More recently, the presence of $GABA_B$ sites in the mammalian brain has been demonstrated from studies of transmitter release (Bowery et al., 1980) as well as by ligand binding (Hill and Bowery, 1981; Bowery et al., 1983) and by autoradiographic (Bowery et al., 1983; Bowery et al., 1984) and electrophysiological (Newberry and Nicoll, 1985; Inoue et al., 1985a, b, c; Gallagher et al., 1984) methods. These $GABA_B$ sites offer a likely explanation for both the pre- and postsynaptic actions of baclofen already described.

The need for specific $GABA_B$ receptor antagonists is widely acknowledged. Such compounds would enable further characterization of this receptor system and investigation of the physiological role of these sites. Unfortunately no truly satisfactory $GABA_B$ antagonist has been described. Five compounds, however, have received attention as possible antagonists and may provide clues to the future development of this class of pharmacological agent. These are 5-aminovaleric acid, 3-aminopropanesulfonic acid, 3-aminopropylphosphonic acid, β-phenyl-GABA and β-(*p*-chlorophenyl)-3-aminopropylphosphonic acid (phaclofen) (*see* Fig. 8).*

3.2. Receptor Antagonists

3.2.1. 5-Aminovaleric Acid

This compound, a GABA homolog, was proposed as a $GABA_B$ antagonist by Muhyaddin et al. (1982). Contraction of the field-stimulated anococcygeus muscle was shown by these workers to be reduced by GABA and (–)-baclofen in a manner consistent with $GABA_B$ receptor activation. 5-Aminovaleric acid (5-AVA) blocked, albeit weakly, the response to GABA and baclofen in a competitive fashion. Such competitive antagonism by 5-AVA of $GABA_B$ receptor-mediated responses has also been recorded in the guinea pig ileum, where 5-AVA antagonizes the inhibition of transmurally stimulated contractions produced by both GABA (Luzzi et al., 1985) and ethylenediamine (Kerr and Ong, 1984). Similarly, 5-AVA has been shown to block the picrotoxin- and bicuculline-insensitive GABA- and baclofen-induced relaxation of guinea pig distal colon both in vitro and in vivo (Giotti et al., 1985). In the CNS 5-AVA also acts as a $GABA_B$ antagonist. Population spikes recorded in the CAl area of the hippocampus following Schaffer collateral stimulation are reduced by (–)-baclofen in the presence of bicuculline and this inhibitory action of baclofen is antagonized by 5-AVA (Nakahiro et al., 1985).

Unfortunately, antagonism at $GABA_B$ receptors is not the only action of 5-AVA. Bowery and Brown (1974) found 5-AVA to act as an agonist at $GABA_A$ receptors in sympathetic ganglia. $GABA_A$-mediated effects of 5-AVA have also been observed in the unstimulated guinea pig ileum (Luzzi et al., 1985) and in the hippocampal slice (Nakahiro et al., 1985). Additional evidence from binding studies supports the contention that 5-AVA interacts with $GABA_A$ receptors, for although 5-AVA has an IC_{50} of 50–100 μM for displacing ^3H-GABA specifically bound to $GABA_B$ sites (Bowery et al., 1983; Nakahiro et al., 1985), it has an approximately 10-fold

*Z-5-aminopent-2-enoic acid, a conformationally restricted analog of 5-AVA, has also been shown to act as a $GABA_B$ antagonist (Dickenson et al., 1988).

higher affinity for $GABA_A$ sites (Olsen et al., 1979; Nakahiro et al., 1985). Thus, although 5-AVA appears to act selectively at GABA receptors, it has agonist and antagonist effects at $GABA_A$ and $GABA_B$ sites, respectively, and, as pointed out by Nakahiro et al. (1985), "can not be used to define $GABA_B$ sites selectively unless $GABA_A$ sites are blocked by other substances." There is also some suggestion from the study of Muhyaddin et al. (1982), referred to above, that 5-AVA may possess slight $GABA_B$ agonist activity.

3.2.2. 3-Aminopropane Sulfonic Acid

3-Aminopropane sulfonic acid (3-APS; homotaurine), the sulfonic analog of GABA, was first reported to possess $GABA_B$ receptor antagonist properties by Giotti et al. (1983b), who showed that the action of GABA and baclofen in reducing field-stimulated contractions of the guinea pig ileum was competitively antagonized by 3-APS. Maggi et al. (1985) have observed a similar action of 3-APS in detrusor strips from newborn rats. In contrast, a slight $GABA_B$ agonist action of 3-APS was noted by Giotti et al. (1983a). 3-APS is also a potent agonist at bicuculline-sensitive $GABA_A$ receptors, and binding studies have shown it to have a much greater affinity for these sites than $GABA_B$ sites (Bowery et al., 1983).

3.2.3. Functional or Pharmacological Antagonism?

It is interesting to note that both 5-AVA and 3-APS have a higher affinity for $GABA_A$ than $GABA_B$ receptors and opposite actions at these two sites, i.e., agonist at $GABA_A$ and antagonist at $GABA_B$ receptors. In fact, a possible interaction between these two systems has been proposed by Allan and Dickenson (1986) to explain their findings that in the field-stimulated guinea pig ileum the $GABA_B$ antagonism shown by 5-AVA, 3-APS, and isoguvacine, is not only reduced by the $GABA_A$ antagonist bicuculline, but produced by these agents with potencies that parallel their potencies in displacing ^3H -GABA from $GABA_A$ binding sites (Bowery et al., 1983). Although this may suggest an involvement of $GABA_A$ receptors in $GABA_B$ antagonist action, this would not be surprising and might apply only to tissues like the guinea pig ileum, in which $GABA_A$ and $GABA_B$ agonists have essentially opposite effects and are thought to act by increasing and decreasing, respectively, the release of acetylcholine from cholinergic postganglionic neurons (see Giotti et al., 1983a).

Certainly, in the rat anococcygeus muscle preparation, in which there is no evidence for the existence of $GABA_A$ receptors, 5-AVA still antagonizes the action of baclofen (Muhyaddin et al., 1982), and even in the guinea pig ileum preparation used by Allan and Dickenson it has been shown by other workers that the antagonism by 5-AVA (Luzzi et al., 1985) and 3-APS (Giotti et al., 1983b) of the depression of field-stimulated twitch contractions by GABA ($GABA_B$ receptors) is not blocked by doses of picrotoxin capable of antagonizing the $GABA_A$-mediated contractions produced by these compounds on their own. Similarly, in the hippocampal slice the antagonism of baclofen-induced population spike suppression by 5-AVA was achieved in the presence of bicuculline (Nakahiro et al., 1985). In this preparation, in the absence of bicuculline, 5-AVA itself suppresses the population spike, making any form of functional antagonism highly unlikely. Clearly, further studies are required before these discrepancies can be resolved. Until then the use of 3-APS and 5-AVA as "$GABA_B$-antagonists" should be viewed with caution.

3.2.4. Phaclofen, β-phenyl-GABA and 3-Aminopropylphosphonic Acid

β-(p-chlorophenyl)-3-aminopropylphosphonic acid (phaclofen), the phosphonic analog of baclofen, has recently been shown to block the GABA- and baclofen-induced depression of twitch contractions of the field-stimulated guinea-pig ileum without affecting $GABA_A$ receptor-mediated contractions caused by GABA (Kerr et al., 1987). In the cat spinal cord, this compound prevents the action of baclofen in reducing monosynaptic excitation of interneurons but not its postsynaptic depressant effects (Kerr et al., 1987). β-phenyl-GABA (BPG) and 3-aminopropylphosphonic acid (3-APPA) exhibit phaclofen-like, baclofen/$GABA_B$ antagonist actions in the gut but produce weak baclofen-like effects in the cat spinal cord which, at presumed presynaptic sites, are phaclofen-sensitive (Ong et al., 1987; Kerr et al., 1987). Additionally, BPG is able to depress ileal twitch contractions itself in a bicuculline-, 5-AVA- and phaclofen-insensitive manner and this effect is not prevented by prior desensitization to GABA or baclofen (Ong et al., 1987).

These somewhat complex results suggest a possible heterogeneity of baclofen-sensitive receptors and indicate the need for a degree of caution when ascribing the effects of baclofen to

$GABA_B$ receptor activation. Further studies with these agents should enable a better understanding of baclofen and $GABA_B$ receptors while providing the basis for the development of more selective and potent compounds.

3.3. $GABA_B$ Receptor Mechanisms: Noncompetitive Antagonists?

No Cl-dependency has been reported for any of the effects associated with $GABA_B$ receptor activation. In addition, such effects are not blocked by "GABA antagonists" thought to act at the level of the Cl^- channel. These observations indicate that $GABA_A$ and $GABA_B$ receptors are linked to different effector mechanisms.

In cell bodies of primary afferent fibers (Dunlap and Fischbach, 1981; Dunlap, 1981; Desarmenien et al., 1984) and neurons of the myenteric plexus (Cherubini and North, 1984), GABA and baclofen reduce the duration of the action potential, which might suggest that these agonists diminish voltage-sensitive calcium (Ca^{2+}) conductance. This shortening of the Ca^{2+} component of the action potential has been recognized as a possible mechanism for the decrease in transmitter release observed at peripheral and central synapses following $GABA_B$ receptor activation. In addition, Newberry and Nicoll (1984, 1985) have described a baclofen- and GABA-induced potassium (K^+) conductance in hippocampal pyramidal cells that may directly regulate intrinsic neuronal excitability. Similar results have been reported by Inoue et al. (1985a, b, c) and Gahwiler and Brown (1985) using the same tissue. This postsynaptic K^+-mediated hyperpolarizing action of baclofen has also been observed in cells of the substantia nigra zona compacta (Pinnock, 1984), dorsolateral septal nucleus (Gallagher et al., 1984; Stevens et al., 1985b), and cortex (Howe et al., 1987).

It is possible that the same ionic mechanism underlies both the pre- and postsynaptic actions of $GABA_B$ agonists (Gahwiler and Brown, 1985; Inoue et al., 1985b), so that the effect of baclofen on transmitter release may result from an increased K^+ conductance leading to a reduction in Ca^{2+} influx rather than a direct action on Ca^{2+} conductance itself. Experiments on chick, mouse, and cat dorsal root ganglion neurons argue against this however: using whole-cell voltage clamp techniques, GABA and baclofen have been shown to directly interfere with a calcium conductance in these cells (Deisz and Lux, 1985; Cottrell and Green, 1986; Robertson and Taylor, 1986). The fact that baclofen has no effect on calcium currents recorded from cultured hippocampal cells (Gahwiler and Brown, 1985;) and none on the calcium component of action potentials recorded from rat neocortical neurons (Howe et al., 1987) suggests that, if $GABA_B$ receptors are linked to calcium channels in the CNS, then this may occur only at presynaptic terminals and not at the soma as found for cells in the peripheral nervous system. Nevertheless, it appears that $GABA_B$ receptors can be linked to at least two separate ionic mechanisms—one involving the closure of voltage activated Ca^{2+} channels and the other involving the activation of a K^+ conductance. Interestingly, GTP-binding proteins have been implicated in both responses (Holz et al., 1986; Andrade et al., 1986; Dolphin and Scott, 1987).

3.3.1. 4-Aminopyridine and Tetraethylammonium

The effects of drugs thought to affect K^+ conductance have been tested on the postsynaptic action of baclofen on neurons of the hippocampus (Inoue et al., 1985a) and dorsolateral septal nucleus (Stevens et al., 1985b) in vitro. In both preparations TEA caused membrane depolarization and the appearance of spontaneous action potentials, but only in the latter did it affect the response to baclofen. Lower concentrations of 4-AP blocked the action of baclofen in both tissues. 4-AP did not cause a change in the membrane potential, but an increase in the occurrence of spontaneous action potentials was observed. The resistance of the baclofen response to cadmium and manganese (Newberry and Nicoll, 1985; Stevens et al., 1985b) or omission of Ca^{2+} from the medium (Inoue et al., 1985a) indicates that the action of baclofen is not mediated by a Ca^{2+}-dependent K^+ conductance (but, *see* Blaxter et al., 1986). Nor is it likely to involve the "delayed rectifier" or the "I_A" current (*see* Stevens et al., 1985b; Inoue et al., 1985a for discussion). Instead, these studies suggest that the postsynaptic effect of baclofen is mediated via a novel voltage-dependent K^+ channel that is sensitive to 4-AP. This drug may prove a useful tool for further studies of postsynaptic $GABA_B$ responses. It should be noted, however, that in the hippocampal slice the hyperpolarizing action of bath-applied GABA, unlike that of baclofen, is not blocked by 4-AP (Inoue et al., 1985a). This is unexpected since the hyperpolarizing action of GABA in the presence of bicuculline and picrotoxin shows the same reversal potential as that of the hyperpolarization induced by baclofen (Inoue et al., 1985c). One possible explanation for this discrepancy is that a somatic Cl^--mediated event (Newberry and Nicoll, 1985) may have contributed to the GABA-induced hyperpolarization observed in the study of Inoue et al. (1985a).

Two further problems remain. One is the ability of 4-AP to block the presynaptic action of baclofen (Inoue et al., 1985b). 4-AP is known to increase transmitter release at central and peripheral synapses. This phenomenon may be caused by increased Ca^{2+} influx at the nerve terminal during action potentials prolonged by block of voltage-sensitive K^+ channels or a direct facilitatory action of 4-AP on Ca^{2+} entry (Rogawski and Barker, 1983). Clearly either of these mechanisms could result in a functional antagonism of the depressant effect of baclofen on synaptic transmission. The other concerns the ability of high concentrations of bicuculline to block the postsynaptic effects of baclofen (Inoue et al., 1985b; Newberry and Nicoll; 1985). It is not clear whether this represents a genuine pharmacological difference between pre- and postsynaptic $GABA_B$ receptors (section 3.2.4) or, as pointed out by Newberry and Nicoll (1985), reflects the ability of bicuculline to block K^+ channels at these concentrations (Heyer et al., 1982).

References

Adams P. R., Constanti A., and Banks F. W. (1981) Voltage clamp analysis of inhibitory synaptic action in crayfish stretch receptor neurons. *Fed. Proc.* **40,** 2637–2641.

Akaike N. and Oomura Y. (1984) GABA-activated chloride channels in internally perfused frog dorsal root ganglion cells. *Biochem. Res.* **5** (suppl.), 115–132.

Akaike N., Hattori K., Oomura Y., and Carpenter D. O. (1985) Bicuculline and picrotoxin block γ-aminobutyric acid-gated Cl^- conductance by different mechanisms. *Experientia* **41,** 70–71.

Akaike N., Yakushiji T., Tokutomi N., and Carpenter D. O. (1987) Multiple mechanisms of antagonism of γ-aminobutyric acid (GABA) responses. *Cell. Mol. Neurobiol.* **7,** 97–103.

Albert J., Lingle C. J., Marder E., and O'Neil M. B. (1986) A GABA-activated chloride-conductance not blocked by picrotoxin on spiny lobster neuromuscular preparations. *Br. J. Pharmacol.* **87,** 771–779.

Allan R. D. and Dickenson H. W. (1986) Evidence that antagonism by δ-aminovaleric acid of $GABA_B$ receptors in the guinea-pig ileum may be due to an interaction between $GABA_A$ and $GABA_B$ receptors. *Eur. J. Pharmacol.* **120,** 119–122.

Andrade R., Malenka R. C., and Nicoll R. A. (1986) A G protein couples serotonin and $GABA_B$ receptors to the same channels in hippocampus. *Science* **234,** 1261–1265.

Andrews P. R. and Johnston G. A. R. (1979) Commentary: GABA agonists and antagonists. *Biochem. Pharmacol.* **28,** 2697–2702.

Anwar N. and Mason D. F. J. (1982) Two actions of γ-aminobutyric acid on the responses of the isolated basilar artery from the rabbit. *Br. J. Pharmacol.* **75,** 177–181.

Ariens E. J., Beld A. J., Miranda J. F. R., and Simonis A. M. (1979) The Pharmacon-Receptor-Effector Concept: A Basis for Understanding the Transmission of Information in Biological Systems, in *The Receptors* vol. 1 (O'Brien R. D., ed.) Plenum, New York.

Ariyoshi M. and Akasu T. (1986) Glucocorticoid modulates the sensitivity of the $GABA_A$ receptor on primary afferent neurons of bullfrogs. *Brain Res.* **367,** 332–336.

Barker J. L. and McBurney R. N. (1979) GABA and glycine may share the same conductance channel on cultured mammalian neurones. *Nature* **277,** 234–236.

Barker J. L., McBurney R. N., and Mathers D. A. (1983) Convulsant-induced depression of amino acid responses in cultured mouse spinal neurones studied under voltage clamp. *Br. J. Pharmacol.* **80,** 619–629.

Barnard E. A., Darlison M. G., and Seeburg P. (1987) Molecular biology of the $GABA_A$ receptor: the receptor/channel superfamily. *Trends NeuroSci.* **10,** 502–509.

Bellet E. M. and Casida J. E. (1973) Bicyclic phosphorus esters: High toxicity without cholinesterase inhibition. *Science* **118,** 1135–1136.

Beutler J. A., Karbon E. W., Brubaker A. N., Malik R., Curtis D. R., and Enna S. J. (1985) Securinine alkaloids: A new class of GABA receptor antagonist. *Brain Res.* **330,** 135–140.

Biziere K., Heaulme M., Chambon J. C., Molimard J. C., Bourguignon J. J., and Wermuth C. C. (1985) Aminopyridazines: Two structural analogs of SR 95103 are potent GABA-A antagonists. *Pharmacologist* **27,** 520.

Blaxter T. J., Carlen P. L., Davies M. F., and Kujtan P. W. (1986) γ-Aminobutyric acid hyperpolarizes rat hippocampal pyramidal cells through a calcium-dependent potassium conductance. *J. Physiol.* **373,** 181–194.

Bloomquist J. R. and Soderlund D. M. (1985) Neurotoxic insecticides inhibit GABA-dependent chloride uptake by mouse brain vessicles. *Biochem. Biophys. Res. Commun.* **133,** 37–43.

Boistel J. and Fatt P. (1958) Membrane permeability change during inhibitory transmitter action in crustacean muscle. *J Physiol.* **144,** 176–191.

Bormann J., Hamill O. P., and Sakmann B. (1987) Mechanism of ion permeation through channels gated by glycine and γ-aminobutyric acid in mouse cultured spinal neurones. *J. Physiol.* **385,** 243–286.

Bowery N. G. and Brown D. A. (1974) Depolarizing actions of γ-aminobutyric acid and related compounds on rat superior cervical ganglia in vitro. *Br. J. Pharmacol.* **50,** 205–218.

Bowery N. G. and Hudson A. L. (1979) γ-Aminobutyric acid reduces the evoked release of ^{3}H-noradrenaline from sympathetic nerve terminals. *Br. J. Pharmacol.* **66,** 108P.

Bowery N. G., Brown D. A., and Collins J. F. (1975) Tetramethylenedisulphotetramine: An inhibitor of γ-aminobutyric acid-induced depolarization of the isolated superior cervical ganglion of the rat. *Br. J. Pharmacol.* **53,** 422–424.

Bowery N. G., Collins J. F., Cryer G., Inch T. D., and McLaughlin N. J. (1979) The GABA Receptor: Stereospecificity and Structure–Activity Studies, in *GABA—Biochemistry and CNS Functions* (Mandel P., DeFeudis F. V., and Mook J., eds.) Plenum, New York.

Bowery N. G., Collins J. F., and Hill R. G. (1976a) Bicyclic phosphorous esters that are potent convulsants and GABA antagonist. *Nature* **261,** 601–603.

Bowery N. G., Collins J. F., Hill R. G., and Pearson S. (1976b) GABA antagonism as a possible basis for the convulsant action of a series of bicyclic phosphorous esters. *Br. J. Pharmacol.* **57,** 435–436.

Bowery N. G., Collins J. F., Hill R. G., and Pearson S. (1977) *t*-Butyl bicyclophosphate: A convulsant and GABA antagonist more potent than bicuculline. *Br. J. Pharmacol.* **60,** 275–276.

Bowery N. G., Doble A., Hill D. R., Hudson A. L., Shaw J. S., Turnbull M. J., and Warrington R. (1981) Bicuculline-insensitive GABA receptors on peripheral autonomic nerve terminals. *Eur. J. Pharmacol.* **71,** 53–70.

Bowery N. G., Hill D. R., and Hudson A. L. (1983) Characteristics of $GABA_B$ receptor binding sites on rat whole brain synaptic membranes. *Br. J. Pharmacol.* **78,** 191–206.

Bowery N. G., Hill D. R., Hudson A. L., Doble A., Middlemiss D. N., Shaw J., and Turnbull M. (1980) (–)Baclofen decreases neurotransmitter release in the mammalian CNS by an action by a novel GABA receptor. *Nature* **283,** 92–94.

Bowery N. G., Price G. W., Hudson A. L., Hill D. R., Wilkin G. P., and Turnbull M. J. (1984) GABA receptor multiplicity: Visualization of different receptor types in the mammalian CNS. *Neuropharmacology* **23,** 219–231.

Braestrup C. and Nielsen M. (1985) Interaction of pitrazepin with the GABA/benzodiazepine receptor complex and with glycine receptors. *Eur. J. Pharmacol.* **118,** 115–121.

Browner M., Feranky J. W., and Enna S. (1981) Biochemical identification of pharmacologically and functionally distinct GABA receptors in rat brain. *J. Neurosci.* **1,** 514–518.

Casida J. E., Eto M., Moscioni A. D., Engel J. L., Milbrath D. S., and Verkade J. G. (1976) Structure-toxicity relationships of 2,6,7-trioxabicyclo [2.2.2] octanes and related compounds. *Toxicol. Appl. Pharmacol.* **36,** 261–279.

Casida J. E., Palmer C. J., and Cole L. M. (1985) Bicycloorthocarboxylate convulsants: Potent $GABA_A$ receptor antagonists. *Mol. Pharmacol.* **28,** 246–253.

Chambon J. P., Feltz P., Heaulme M., Pestle S., Schlichter R., Biziere K., and Wermuth C. G. (1985) An arylaminopyridazine derivative of γ-aminobutyric acid (GABA) is a selective and competitive antagonist at the $GABA_A$ receptor site. *Proc. Natl. Acad. Sci. USA* **82,** 1832–1836.

Cherubini E. and North R. A. (1984) Actions of γ-aminobutyric acid on neurones of guinea-pig myenteric plexus. *Br. J. Pharmacol.* **82,** 93–100.

Chow P. and Mathers D. (1986) Convulsant doses of penicillin shorten the lifetime of GABA-induced channels in cultured central neurones. *Br. J. Pharmacol.* **88,** 541–547.

Clarke G. and Hill R. G. (1972) Effects of a focal penicillin lesion on responses of rabbit cortical neurones to putative neurotransmitters. *Br. J. Pharmacol.* **44,** 435–441.

Collingridge G. L. and Davies J. (1982) Reversible effects of tetanus toxin on striatal-evoked responses and $[^3H]$-γ-aminobutyric acid release in the rat substantia nigra. *Br. J. Pharmacol.* **76,** 403–411.

Collins J. F. and Hill R. G. (1974) (+) and (–)Bicuculline methochloride as optical isomers of a GABA antagonist. *Nature* **249,** 845–847.

Collins J. F., Hill R. G., and Roberts F. (1975) A study of tetramethylenedisulphotetramine (TETS) and related compounds as antagonists of presynaptic inhibition and microiontophoretically applied γ-aminobutyric acid (GABA) and glycine in the rat cuneate nucleus. *Br. J. Pharmacol.* **54,** 239–240P.

Constanti A. (1978) The "mixed" effect of picrotoxin on the GABA dose/conductance relation recorded from lobster muscle. *Neuropharmacology* **17,** 159–167.

Cottrell G. A. and Green K. A. (1986) Action of baclofen on voltage-dependent currents in mouse dorsal root ganglion neurones. *J. Physiol.* **372,** 49P.

Cull-Candy S. G. and Miledi R. (1981) Junctional and extrajunctional membrane channels activated by GABA in locust muscle fibres. *Proc. Roy. Soc. Lond.* **B211,** 527–535.

Cull-Candy S. G., Mathie, A., and Newland C. F. (1988) Influence of picrotoxin on ion channels activated by γ-aminobutyric acid in dissociated sympathetic neurones of the rat. *J. Physiol.* In press.

Curtis D. R. and Johnston G. A. R. (1974) Amino acid transmitters in the mammalian central nervous system. *Ergeb. Physiol.* **69,** 97–188.

Curtis D. R. and Malik R. (1985) Glycine antagonism by RU 5135. *Eur. J. Pharmacol.* **110,** 383–384.

Curtis D. R., Davies J., Game C. J. A., Johnston G. A. R., and McCulloch R. M. (1973) Central actions of shikimin and tutin. *Brain Res.* **63,** 419–423.

Curtis D. R., Duggan A. W., Felix D., and Johnston G. A. R. (1970) GABA bicuculline and central inhibition. *Nature* **226,** 1222–1224.

Curtis D. R., Game C. J. A., Johnston G. A. R., and McCulloch R. M. (1974) Central effects of β-(*p*-chlorophenyl)-γ-aminobutyric acid. *Brain Res.* **70,** 493–499.

Curtis D. R., Game C. J. A., Johnston G. A. R., McCulloch R. M., and MacLachlan R. M. (1972) Convulsive action of penicillin. *Brain Res.* **43,** 242–245.

Curtis D. R and Gynther B. D. (1986) Pitrazepin: a central glycine and GABA antagonist. *Eur. J. Pharmacol.* **131,** 311–313.

Davidoff R. A. (1972a) Penicillin and presynaptic inhibition in the amphibian spinal cord. *Brain Res.* **36,** 218–222.

Davidoff R. A. (1972b) Penicillin and inhibition in the cat spinal cord. *Brain Res.* **45,** 638–642.

Davies J. (1981) Selective depression of synaptic excitation in cat spinal neurones by baclofen: An ionotophoretic study. *Br. J. Pharmacol.* **72,** 373–384.

Davies J. and Watkins J. C. (1974) The action of β-phenyl-GABA derivatives on neurones of the cat cerebral cortex. *Brain Res.* **70,** 501–505.

De Feudis F. V. (1977) GABA-Receptors in the vertebrate nervous system. *Prog. Neurobiol.* **9,** 123–145.

Deisz R. A. and Lux H. D. (1985) γ-Aminobutyric acid-induced depression of calcium currents of chick sensory neurons. *Neurosci. Lett.* **56,** 205–210.

Desarmenien M., Desaulles E., Feltz P., and Hamann M. (1987) Electrophysiological study of SR 42641, a novel aminopyridazine derivative of GABA: antagonist properties and receptor selectivity of $GABA_A$ versus $GABA_B$ responses. *Br. J. Pharmacol.* **90,** 287–298.

Desarmenien M., Feltz P., Occhipinti G., Santangelo F., and Schlichter R. (1984) Coexistence of $GABA_A$ and $GABA_B$ receptors of Aδ and C primary afferents. *Br. J. Pharmacol.* **81,** 327–333.

Dichter M. A. (1980) Physiological identification of GABA as the inhibitory transmitter for mammalian cortical neurons in culture. *Brain Res.* **190,** 111–121.

Dickenson H. W., Allan R. D., Ong J., and Johnston G. A. R., (1988) $GABA_B$ receptor antagonist and $GABA_A$ receptor agonist properties of a δ-aminovaleric acid derivative, *Z*-5-aminopent-2-enoic acid. *Neurosci. Lett.* **86,** 351–355.

Dingledine R. and Gjerstad L. (1980) Reduced inhibition during epileptiform activity in the in vitro hippocampal slice. *J. Physiol.* **305,** 297–313.

Dingledine R., Iversen L. L., and Breuker E. (1978) Naloxone as a GABA antagonist: Evidence from iontophoretic, receptor binding and convulsant studies. *Eur. J. Pharmacol.* **47,** 19–27.

Docherty M., Bradford H. F., Anderton B., and Wu J-Y. (1983) Specific lysis of GABAergic synaptosomes by an antiserum to glutamate decarboxylase. *FEBS Lett.* **152,** 57–61.

Dolphin A. C. and Scott R. H. (1987) Calcium channel currents and their inhibition by (-)baclofen in rat sensory neurones: modulation by guanine nucleotides. *J. Physiol.* **386,** 1–17.

Dray A. (1975a) Tetramethylenedisulphotetramine and amino acid inhibition in the rat brain. *Neuropharmacology* **14,** 703–705.

Dray A. (1975b) Comparison of bicuculline methochloride with bicuculline and picrotoxin as antagonists of amino acid and monoamine depression on neurones in the rat brain stem. *Neuropharmacology* **14,** 887–891.

Dudel J. and Kuffler S. W. (1961) Presynaptic inhibition at the crayfish neuromuscular junction. *J. Physiol.* **155,** 543–562.

Duijn H., Schwartzkroin P. A., and Prince D. A. (1973) Action of penicillin on inhibitory processes in the cats cortex. *Brain Res.* **53,** 470–476.

Dunlap K. (1981) Two types of γ-aminobutyric acid receptor on embryonic sensory neurones. *Br. J. Pharmacol.* **74,** 579–585.

Dunlap K. and Fischbach G. D. (1981) Neurotransmitters decrease the calcium conductance activated by depolarization of embryonic chick sensory neurones. *J. Physiol.* **317,** 519–535.

Earl J. and Large W. A. (1974) Electrophysiological investigation of GABA-mediated inhibition at the hermit crab neuromuscular junction. *J. Physiol.* **236,** 113–127.

Eldefrawi A. T. and Eldefrawi M. E. (1987) Receptors for γ-aminobutyric acid and voltage-dependent chloride channels as targets for drugs and toxicants. *FASEB J.* **1,** 262–271.

Elliot K. A. C. and Florey E. (1956) Factor I—Inhibitory factor from brain. Assay. Conditions in brain. Stimulating and antagonizing substances. *J. Neurochem.* **1,** 181–191.

Enna S. J. and Snyder S. H. (1977) Influences of ions, enzyme and detergents on γ-aminobutyric acid-receptor binding in synaptic membranes of rat brain. *Mol. Pharmacol.* **13,** 442–453.

Enna S. J., Collins J. F., and Snyder S. H. (1977) Stereospecificity and structure–activity requirements of GABA receptor binding in rat brain. *Brain Res.* **124,** 185–190.

Esplin B., Theoret Y., Seward E., and Capek R. (1985) Epileptogenic action of penicillin derivatives: Structure–activity relationship. *Neuropharmacology* **24,** 571–575.

File S. E. and Pellow S. (1985) The anxiogenic action of Ro 5-4864 in the social interaction test: Effects of chlordiazepoxide, Ro 15-1788 and CGS 8216. *Naunyn Schmiedebergs Arch. Pharmacol.* **328,** 225–228.

Fox S., Krnjevic J., Morris M. E., Puil E., and Werman R. (1978) Action of baclofen on mammalian synaptic transmission. *Neuroscience* **3,** 495–515.

Freeman A. R. (1973) Electrophysiological analysis of the actions of strychnine, bicuculline and picrotoxin on the axonal membrane. *J. Neurobiol.* **4,** 567–582.

Fujimoto M. and Okabayashi T. (1980) Effect of picrotoxin on benzodiazepine receptors and GABA receptors with reference to the effect of Cl^- ion. *Life Sci.* **28.** 895–901.

Gahwiler B. H. and Brown D. A. (1985) $GABA_B$-receptor-activated K^+ current in voltage-clamped CA_3 pyramidal cells in hippocampal cultures. *Proc. Natl. Acad. Sci. USA* **82,** 1558–1562.

Gahwiler B. H., Maurer R., and Wüthrich H. J. (1984) Pitrazepin, a novel $GABA_A$ antagonist. *Neurosci. Lett.* **45,** 311–316.

Galindo A. (1969) GABA-picrotoxin interaction in the mammalian central nervous system. *Brain Res.* **14,** 763–767.

Gallagher J. P., Nakamura J., and Shinnick-Gallagher P. (1983) The effects of temperature, pH and Cl-pump inhibitors on GABA responses recorded from cat dorsal root ganglia. *Brain Res.* **267,** 249–259.

Gallagher J. P., Stevens D. R., and Shinnick-Gallagher P. (1984) Actions of GABA and baclofen on neurons of the dorsolateral septal nucleus (DLSN) in vitro. *Neuropharmacology* **23,** 825–826.

Gallo V., Wise B. C., Vaccarino F., and Guidotti A. (1985) γ-Aminobutyric acid- and benzodiazepine-induced modulation of $[^{35}S]$-*t*-butylbicyclophosphorothionate binding to cerebellar granule cells. *J. Neurosci.* **5,** 2432–2438.

Gammon D. and Casida J. E. (1983) Pyrethroids of the most potent class antagonize GABA action at the crayfish neuromuscular junction. *Neurosci. Lett.* **40,** 163–168.

Gant D. B., Cole R. J., Valdes J. J., Eldefrawi M. E., and Eldefrawi A. T. (1987) Action of tremorgenic mycotoxins on $GABA_A$ receptor. *Life Sci.* **41,** 2207–2214.

Geller H. M., Hoffer B. J., and Taylor D. A. (1980) Electrophysiological actions of benzodiazepines. *Fed. Proc.* **39,** 3016–3023.

Giotti A., Luzzi S., Spagnesi S., and Zilletti L. (1983a) $GABA_A$ and $GABA_B$ receptor-mediated effects in guinea pig ileum. *Br. J. Pharmacol.* **78,** 469–478.

Giotti A., Luzzi F., Spagnesi S., and Zilletti L (1983b) Homotaurine: A $GABA_B$ antagonist in guinea-pig ileum. *Br. J. Pharmacol.* **79,** 855–862.

Giotti A., Luzzi S., Maggi C. A., Spagnesi S., and Zilletti L. (1985) Modulatory activity of $GABA_B$ receptors on cholinergic tone in guinea-pig distal colon. *Br. J. Pharmacol.* **84,** 883–895.

Goldinger A. and Muller W. E. (1980) Stereospecific interaction of bicuculline with specific [^{3}H]-strychnine binding to rat spinal cord synaptosomal membranes. *Neurosci. Lett.* **16,** 91–95.

Grundfest H., Reuben J. P., and Rickles W. H. (1959) The electrophysiology and pharmacology of lobster neuromuscular synapses. *J. Gen. Physiol.* **42,** 1301–1323.

Gruol D. L., Barker J. L., and Smith T. G. (1980) Naloxone antagonism of GABA-evoked membrane polarizations in cultured mouse spinal cord neurons. *Brain Res.* **198,** 323–332.

Gutnick M. J. and Prince D. A. (1971) Penicillinase and the convulsant action of penicillin. *Neurobiology* **21,** 759–764.

Gutnick M. J., VanDiujn H., and Citri N. (1976) Relative convulsant potencies of structural analogues of penicillin. *Brain Res.* **114,** 139–143.

Gynther B. D. and Curtis D. R. (1986) Pyridazinyl-GABA derivatives as GABA and glycine antagonists in the spinal cord of the cat. *Neurosci. Lett.* **68,** 211–215.

Haefely W., Kyburz E., Gerecke M., and Mohler H. (1985) Recent advances in the molecular pharmacology of benzodiazepine receptors and in the structure–activity relationships of their agonists and antagonists. *Adv. Drug Res.* **14,** 166–322.

Hammill O. P., Bormann J., and Sakmann B. (1983) Activation of multiple-conductance state chloride channels in spinal neurones by glycine and GABA. *Nature* **305,** 805–808.

Harrison N. L. and Simmonds M. A. (1983) The picrotoxin-like action of a convulsant benzodiazepine, Ro 5-3663. *Eur J. Pharmacol.* **87,** 155–158.

Heaulme M., Chambon J. P., Leyris R., Molimard J-C., Wermuth C. G., and Biziere K. (1986a) Biochemical characterization of the interaction of three pyridazinyl-GABA derivatives with the GABA-A receptor site. *Brain Res.* **384,** 224–231.

Heaulme M., Chambon J. P., Leyris R., Wermuth C. G., and Biziere K. (1986b) Specific binding of a phenyl-pyridazinium derivative endowed with $GABA_A$ receptor antagonist activity to rat brain. *Neuropharmacology* **25,** 1279–1283.

Heyer E. J., Nowak L. M., and Macdonald R. L. (1982) Membrane depolarization and prolongation of calcium-dependent action potentials of mouse neurons in cell culture by two convulsants: Bicuculline and penicillin. *Brain Res.* **232,** 41–56.

Hill D. R. and Bowery N. G. (1981) ^{3}H-Baclofen and ^{3}H-GABA bind to bicuculline-insensitive $GABA_B$ sites in rat brain. *Nature* **290,** 149–152.

Hill R. G., Simmonds M. A., and Straughan D. W. (1973) A comparative study of some convulsant substances as γ-aminobutyric acid antagonists in the feline cerebral cortex. *Br. J. Pharmacol.* **49,** 37–51.

Hill R. G., Simmonds M. A., and Straughan D. W. (1976) Antagonism of γ-aminobutyric acid and glycine by convulsants in the cuneate nucleus of cat. *Br. J. Pharmacol.* **56,** 9–21.

Hochner B., Spira M. E., and Werman R. (1976) Penicillin decreases chloride conductance in crustacean muscle: A model for the epileptic neuron. *Brain Res.* **107,** 85–103.

Holz G. G., Rane S. G., and Dunlap K. (1986) GTP-binding proteins mediate transmitter inhibition of voltage-dependent calcium channels. *Nature* **319,** 670–672.

Homma S. and Rovainen C. M. (1978) Conductance increases produced by glycine and γ-aminobutyric acid in lamprey interneurones. *J. Physiol.* **279,** 231–252.

Howe J. R., Sutor B., and Zieglgansberger W. (1987) Baclofen reduces post-synaptic potentials of rat cortical neurones by an action other than its hyperpolarizing action. *J. Physiol.* **384,** 539–569.

Hunt P. and Clements-Jewery S. (1981) A steroid derivative, R 5135, antagonizes the GABA/benzodiazepine receptor interaction. *Neuropharmacology* **20,** 357–361.

Inoue M., Matsuo T., and Ogata N. (1985a) Baclofen activates voltage-dependent and 4-aminopyridine sensitive K^+ conductance in guinea-pig hippocampal pyramidal cells maintained in vitro *Br. J. Pharmacol.* **84,** 833–842.

Inoue N. M., Matsuo T., and Ogata N. (1985b) Characterization of pre- and postsynaptic actions of (–)-baclofen in the guinea pig hippocampus in vitro. *Br. J. Pharmacol.* **84,** 843–852.

Inoue M., Matsuo T., and Ogata N. (1985c) Possible involvement of K^+-conductance in the action of γ-aminobutyric acid in the guinea-pig hippocampus. *Br. J. Pharmacol.* **86,** 515–524.

Jarboe C. H., Porter L. A., and Buckler R. T. (1968) Structural aspects of picrotoxinin action. *J. Med. Chem.* **11,** 729–731.

Johnston G. A. R. (1976) Physiologic Pharmacology of GABA and Its Antagonists in the Vertebrate Nervous System, in *GABA in Nervous System Function* (Roberts E., Chase T. N., and Tower D. B., eds.) Raven, New York.

Johnston G. A. R., Beart P. M., Curtis D. R., Game C. J. A., McCulloch R. M., and Maclachlan R. M. (1972) Bicuculine methochloride as a GABA antagonist. *Nature New Biol.* **240,** 219–220.

Johnston G. A. R., Hailstone M. M., and Freeman C. G. (1980) Baclofen: Stereo-selective inhibition of excitant amino acid release. *J. Pharm. Pharmacol.* **32,** 230–231.

Kaneko A. and Tachibana M. (1986) Effects of γ-aminobutyric acid on isolated cone photoreceptors of the turtle retina. *J. Physiol.* **373,** 443–461.

Kaplita P. V., Waters D. H., and Triggle D. J. (1982) γ-Aminobutyric acid action in guinea-pig ileal myenteric plexus. *Eur. J. Pharmacol.* **79,** 43–51.

Karobath M., Drexler G., and Supavilai P. (1981) Modulation by picrotoxin and IPTBO of ^{3}H-flunitrazepam binding to the GABA/benzodiazepine receptor complex of rat cerebellum. *Life Sci.* **28,** 307–313.

Kelly J. S. and Beart P. M. (1975) Amino Acid Receptors in CNS. II. GABA in Supraspinal Regions, in *Handbook of Psychopharmacology* vol. 4 *Amino Acid Neurotransmitters* (Iversen L. L., Iversen S. D., and Snyder S. H., eds.) Plenum, New York.

Kemp J. A., Marshall G. R., Wong E. H. F., and Woodruff G. N. (1985) Pharmacological studies on pitrazepin, a $GABA_A$ receptor antagonist. *Br. J. Pharmacol.* **85,** 237P.

Kemp J. A., Marshall G. R., and Woodruff G. N. (1986) Quantitative evaluation of the potencies of GABA-receptor agonists and antagonists using the rat hippocampal slice preparation. *Br. J. Pharmacol.* **87,** 677–684.

Kerr D. I. B. and Ong J. (1984) Evidence that ethylenediamine acts in the isolated ileum on the guinea-pig by releasing endogenous GABA. *Br. J. Pharmacol.* **83,** 169–177.

Kerr D. I. B., Dennis B. J., Breuker E. L. M., Prager R. H., Ward A. D., and Duong T. (1976) Antagonism of GABA-mediated inhibition in the central nervous system. *Brain. Res.* **110,** 413–416.

Kerr D. I. B., Ong J., Prager R. H., Gynther B. D., and Curtis D. R. (1987) Phaclofen: a peripheral and central baclofen antagonist. *Brain Res.* **405,** 150–154.

Klunk W. E., Kalman B. L., Ferrendelli J. A., and Covey D. F. (1983) Computer-assisted modeling of the picrotoxinin and γ-butyrolactone receptor site. *Mol. Pharmacol.* **23,** 511–518.

Krantis A. and Kerr D. I. B. (1981) GABA induced excitatory responses in the guinea-pig small intestine are antagonized by bicuculline, picrotoxin and chloride ion blockers. *Naunyn Schmiedebergs Arch. Pharmacol.* **317,** 257–261.

Krnjevic K. (1974) Chemical nature of synaptic transmission in vertebrates. *Physiol. Rev.* **54,** 418–540.

Krogsgaard-Larsen P., Hjeds H., Curtis D. R., Leah J. D., and Peet M. J. (1982) Glycine antagonists structurally related to muscimol THIP, or isoguvacine. *J. Neurochem.* **39,** 1319–1324.

Kudo Y., Niwa H., Tanaka A., and Yamada K. (1984) Actions of picrotoxinin and related compounds on the frog spinal cord: The role of a hydroxyl-group at the 6 position in antagonizing the actions of amino acids and presynaptic inhibition. *Br. J. Pharmacol.* **81,** 373–380.

Kudo Y., Oka J-I., and Yamada K. (1981) Anisatin a potent GABA antagonist, isolated from *Illicium anisatum. Neurosci. Lett.* **25,** 83–88.

Large W. A. (1975) Effect of tetramethylenedisulphotetramine on the membrane conductance increase produced by γ-amino-butyric acid at the crab neuromuscular junction. *Br. J. Pharmacol.* **53,** 598–599.

Lawrence L. J. and Casida J. E. (1983) Stereospecific action of pyrethroid insecticides on the γ-aminobutyric acid receptor-ionophore complex. *Science* **221,** 1399–1401.

Lawrence L. J. and Casida J. E. (1984) Interactions of lindane, toxaphene and cyclodienes with brain-specific *t*-butylbicycloorthophosphorothionate receptor. *Life Sci.* **35,** 171–178.

Lawrence L. L., Palmer C. J., Gee K. W., Wang X., Yamamura H. I., and Casida J. E. (1985) *t*-[^{3}H] Butylbicycloorthobenzoate: New radioligand probe for the γ-aminobutyric acid-regulated chloride ionophore. *J. Neurochem.* **45,** 798–804.

Lebeda F. J., Hablitz J. J., and Johnston D. (1982) Antagonism of GABA-mediated responses by *d*-tubocurarine in hippocampal neurons. *J. Neurophysiol.* **48,** 622–632.

Leeb-Lundberg F., Napias C., and Olsen R. W. (1981) Dihydropicrotoxinin binding sites in mammalian brain: Interaction with convulsant and depressant benzodiazepines. *Brain Res.* **216,** 339–408.

Lees G., Beadle D. J., Neuman R., and Benson J. A. (1987) Responses to GABA by isolated insect neuronal somata: pharmacology and modulation by a benzodiazepine and a barbiturate. *Brain Res.* **401,** 267–278.

Luzzi S., Maggi C. A., Spagnesi S., Santicioli P., and Zilletti L. (1985) 5-Aminovaleric acid interactions with $GABA_A$ and $GABA_B$ receptors in guinea-pig ileum. *J. Auton. Pharmacol.* **5,** 65–69.

Macdonald R. L. and Barker J. L. (1977) Pentylenetetrazol and penicillin are selective antagonists of GABA-mediated postsynaptic inhibition in cultured mammalian neurons. *Nature* **267,** 720–721.

Macdonald R. L. and Barker J. L. (1978) Specific antagonism of GABA-mediated postsynaptic inhibition in cultured mammalian spinal cord neurons: A common mode of convulsant action. *Neurology* **28,** 325–330.

Maggi C. A., Santicioli P., and Meli A. (1984) $GABA_B$ receptor mediated inhibition of field stimulation induced contractions of detrusor muscle strips from newborn rats. *J. Auton. Pharmacol.* **4,** 45–51.

Maggi C. A., Santicioli P., and Meli A. (1985) $GABA_A$ and $GABA_B$ receptors in detrusor strips from guinea-pig bladder dome. *J. Auton. Pharmacol.* **5,** 55–64.

Maksay G. and Simonyi M. (1985) Benzodiazepine anticonvulsants accelerate and β-carboline convulsants decelerate the kinetics of $[^{35}S]$-TBPS binding at the chloride ionophore. *Eur. J. Pharmacol.* **117,** 275–278.

Maksay G. and Ticku M. K. (1984a) Characterization of γ-aminobutyric acid–benzodiazepine receptor complexes by protection against inactivation by group-specific reagents. *J. Neurochem.* **42,** 1715–1727.

Maksay G. and Ticku M. K. (1984b) Diazotization and thiocyanate differentiate agonists from antagonists for the high- and low-affinity receptors of γ-aminobutyric acid. *J. Neurochem.* **43,** 261–268.

Maksay G. and Ticku M. K. (1985) Dissociation of $[^{35}S]$*t*-Butylbicyclophosphorothionate binding differentiates convulsants and depressant drugs that modulate GABAergic transmission. *J. Neurochem.* **44,** 480–486.

Mann E. and Enna S. J. (1980) Phylogenetic distribution of bicuculline-sensitive γ-aminobutyric acid (GABA) receptor binding. *Brain Res.* **184,** 367–373.

Massotti M. and Lucantoni D. (1986) The peripheral benzodiazepine receptor ligand Ro 5-4864 induces supraspinal convulsions in rabbits. Reversal by the central benzodiazepine antagonist Ro 15-1788. *Psychopharmacology* **88,** 336–340.

Mathers D. A. (1985) Penicillin shortens the mean open time of GABA induced membrane channels in cultured spinal neurons. *Soc. Neurosci. Abstr.* **11,** 7.

Matsumura F. and Tanaka K. (1984) Molecular Basis of Neuroexcitatory Actions of Cyclodiene-Type Insecticides, in *Cellular and Molecular Neurotoxicology* (Narahashi T., ed.) Raven, New York.

Matsuomo K. and Fukuda H. (1983) Anisatin modulation of GABA- and pentobarbital-induced enhancement of diazepam binding in rat brain. *Neurosci. Lett.* **32,** 175–179.

McBurney R. N. (1984) Membrane Actions of GABA in Cultured Central Neurones, in *Actions and Interactions of GABA and Benzodiazepines* (Bowery N. G., ed.) Raven, New York.

Michaud J. C., Mienville J. M., Chambon J. P., and Biziere K. (1986) Interactions between three pyridazinyl-GABA derivatives and central GABA and glycine receptors in the rat, an in vivo microiontophoretic study. *Neuropharmacology* **25,** 1197–1203.

Mienville J-M. and Vicini S. (1987) A pyridazinyl derivative of gamma-aminobutyric acid (GABA), SR 95531, is a potent antagonist of Cl^- channel opening regulated by GABA receptors. *Neuropharmacology* **26,** 779–783.

Mohler H. and Okada T. (1977) GABA receptor binding with ^{3}H-bicuculline-methiodide in rat CNS. *Nature* **267,** 65–67.

Mohler H. and Okada T. (1978) Properties of γ-aminobutyric acid receptor binding with (+)[^{3}H]bicuculline methiodide in rat cerebellum. *Mol. Pharmacol.* **14,** 256–265.

Muhyaddin M., Roberts P. J., and Woodruff G. N. (1982) Presynaptic γ-aminobutyric receptors in the rat anococcygeus muscle and their antagonism by 5-aminovaleric acid. *Br. J. Pharmacol.* **77,** 163–168.

Muller W. E., Schalfer M., and Wollert U. (1978) Benzodiazepine receptor binding: The interaction of some nonbenzodiazepine drugs with specific [^{3}H]diazepam binding to rat brain synaptosomal membranes. *Naunyn Schmiedebergs Arch. Pharmacol.* **305,** 23–28.

Nakahiro M., Saito K., Yamada I., and Yoshida H. (1985) Antagonistic effect of δ-aminovaleric acid on bicuculline-insensitive γ-aminobutyric $acid_B$ ($GABA_B$) sites in the rats brain. *Neurosci. Lett.* **57,** 263–266.

Neher E. and Steinbach J. H. (1978) Local anaesthetics transiently block currents through single acetylcholine-receptor channels. *J. Physiol.* **277,** 153–176.

Newberry N. R. and Nicoll R. A. (1984) Direct hyperpolarizing action of baclofen on hippocampal pyramidal cells. *Nature* **308,** 450–452.

Newberry N. R. and Nicoll R. A. (1985) Comparison of the action of baclofen with γ-aminobutyric acid on rat hippocampal pyramidal cells in vitro. *J. Physiol.* **360,** 161–185.

Nicoll R. A. (1975) The action of acetylcholine antagonists on amino acid response in the frog spinal cord in vitro. *Br. J. Pharmacol.* **55,** 449–458.

Nicoll R. A. (1978) The blockade of GABA mediated responses in the frog spinal cord by ammonium ions and furosemide. *J. Physiol.* **283,** 121–132.

Nicoll R. A. and Padjen A. (1976) Pentylenetetrazol: An antagonist of GABA at primary afferents of the isolated frog spinal cord. *Neuropharmacology.* **15,** 69–71.

Nicoll R. A. and Wojtowicz J. M. (1980) The effects of pentobarbital and related compounds on frog motoneurons. *Brain Res.* **191,** 225–237.

Nistri A. and Constanti A. (1979) Pharmacological characterization of different types of GABA and glutamate receptors in vertebrates and invertebrates. Prog. *Neurobiol.* **13,** 117–235.

Nistri A., Constanti A., and Quilliam J. P. (1974) Central inhibition, GABA and tutin. *Lancet* **i,** 996–997.

Nowak L. M., Young A. B., and Macdonald R. L. (1982) GABA and bicuculline actions on mouse spinal cord and cortical neurons in cell culture. *Brain Res.* **224,** 155–164.

O'Brien R. A. and Spirt N. (1980) The inhibition of GABA stimulated benzodiazepine binding by a convulsant benzodiazepine. *Life Sci.* **26,** 1441–1448.

Ogata N., Vogel S. M., and Narahashi T. (1987) Lindane but not deltamethrin blocks a component of GABA-activated chloride channels. *Soc. Neurosci. Abstr.* **13,** 65.

Ogden D. C., Siegelbaum S. A., and Colquhoun D. (1981) Block of acetylcholine-activated ion channels by an uncharged local anaesthetic. *Nature* **289,** 596–598.

Olpe H. R., Demieville H., Baltzer V., Bencze W. L., Koella W. P., Wolf P., and Haas H. L. (1978) The biological activity of *d*- and *l*-baclofen (lioresal). *Eur. J. Pharmacol.* **52,** 133–136.

Olsen R. W. (1981) The GABA postsynaptic membrane receptor-ionophore complex. *Mol. Cell. Biochem.* **39,** 261–279.

Olsen R. W. and Snowman A. M. (1983) [^{3}H]Bicuculline methochloride binding to low-affinity γ-aminobutyric acid receptor sites. *J. Neurochem.* **41,** 1653–1663.

Olsen R. W., Ban M., and Miller T. (1976) Studies on the neuropharmacological activity of bicuculline and related compounds. *Brain Res.* **102,** 283–299.

Olsen R. W., Ban M., Miller T., and Johnston G. A. R. (1975) Chemical instability of the GABA antagonist bicuculline under physiological conditions. *Brain Res.* **98,** 383–387.

Olsen R. W., Leeb-Lundberg F., and Nopias C. (1980) Picrotoxin and convulsant binding sites in mammalian brain. *Brain Res. Bull.* **5** (suppl. 2), 217–227.

Olsen R. W., Ticku M. K., and Miller T. (1978) Dihydropicrotoxinin binding to crayfish muscle sites possibly related to γ-aminobutyric acid receptor-ionophores. *Mol. Pharmacol.* **14,** 381–390.

Olsen R. W., Ticku M. J., Greenlee D., and Van Ness P. (1979) GABA Receptor and Ionophore Binding Sites: Interaction with Various Drugs, in *GABA-Neurotransmitters. Pharmacochemical Biochemical and Pharmacological Aspects* (Krogsgaard-Larsen P., Scheel-Kruger J., and Kofod H., eds.) Academic, New York.

Olsen R. W., Wong E. H. F., Stauber G. B., and King R. G. (1984) Biochemical pharmacology of the γ-aminobutyric acid receptor/ionophore protein. *Fed. Proc.* **43,** 2773–2778.

Ong J. and Kerr D. I. B. (1983) $GABA_A$ and $GABA_B$-receptor-mediated modification of intestinal motility. *Eur. J. Pharmacol.* **86,** 9–17.

Ong J., Kerr D. I. B., and Johnston G. A. R. (1987) Differing actions of β-phenyl-GABA and baclofen in the guinea pig isolated ileum. *Neurosci. Lett.* **77,** 109–112.

Ono H., Fukuda H., and Kudo Y. (1979) Mechanism of depressant action of baclofen on the spinal reflex in the rat. *Neuropharmacology* **18,** 647–653.

Pellmar T. C. and Wilson W. A. (1977a) Synaptic mechanism of pentylenetetrazole: Selectivity for chloride conductance. *Science* **197,** 912–914.

Pellmar T. C. and Wilson W. A. (1977b) Penicillin effects on iontophoretic responses in *Aplysia californica. Brain Res.* **136,** 89–101.

Pellow S. and File S. E. (1984) Behavioural actions of Ro 5-4864: A peripheral type benzodiazepine? *Life Sci.* **35,** 229–240.

Pickles H. G. and Simmonds M. A. (1980) Antagonism by penicillin of γ-aminobutyric acid depolarizations at presynaptic sites in rat olfactory cortex and cuneate nucleus in vitro. *Neuropharmacology* **19,** 35–38.

Pieri L., Polc P., Bonetti E. P., Burkard W., Cumin R., and Haefely W. (1983) Some pharmacological effects of Ro 5-4864, a specific ligand of the peripheral type benzodiazepine binding sites. *Naunyn Schmiedebergs Arch. Pharmacol.* **322,** R95.

Pinnock R. D. (1984) Hyperpolarizing action of baclofen on neurons in the rat substantia nigra slice. *Brain Res.* **322,** 337–340.

Polc P. and Schaffner R. (1983) Electrophysiological effects of the peripheral-type benzodiazepine receptor ligand Ro 5-4864 in cat spinal cord and rat hippocampus. *Neurosci. Lett.* **14** (suppl.), S288.

Pong S. F. and Graham L. T. (1972) *N*-methylbicuculline, a convulsant more potent than bicuculline. *Brain Res.* **42,** 486–490.

Porter L. A. (1967) Picrotoxinin and related substances. *Chem. Rev.* **67,** 441–464.

Potashner S. J. (1979) Baclofen: Effects on amino acid release and metabolism in slices of guinea pig cerebral cortex. *J. Neurochem.* **32,** 103–109.

Quast U. and Brenner O. (1983) Modulation of [^{3}H]muscimol binding in rat cerebellar and cerebral cortical membranes by picrotoxin, pentobarbitone and etomidate. *J. Neurochem.* **41,** 418–425.

Ramanjaneyulu R. and Ticku M. K. (1984a) Interactions of pentamethyleneletrazole and tetrazole analogues with the picrotoxinin site of the benzodiazepine-GABA receptor-ionophore complex. *Eur. J. Pharmacol.* **98,** 337–345.

Ramanjaneyulu R. and Ticku M. K. (1984b) Binding characteristics and interactions of depressant drugs with [^{35}S]*t*-butylbicyclophosphorothionate, a ligand that binds to the picrotoxinin site. *J. Neurochem.* **42,** 221–229.

Rehavi M. P., Skolnick P., and Paul S. M. (1982) Effects of tetrazole derivatives on [^{3}H]diazepam binding in vitro: Correlation with convulsant potency. *Eur. J. Pharmacol.* **23,** 326–336.

Robbins J. and Van der Kloot W. G. (1958) The effect of picrotoxin on peripheral inhibition in crayfish. *J. Physiol.* **143,** 541–552.

Robertson B. and Taylor W. R. (1986) Effects of $GABA_B$ receptor activation on calcium and potassium current in cat dorsal root ganglion (DRG) neurones in vitro. *Neurosci. Lett.* **23** (suppl.), S74.

Rogawski M. A. and Barker J. L. (1983) Effects of 4-aminopyridine on calcium action potentials and calcium currents under voltage clamp in spinal neurons. *Brain Res.* **280,** 180–185.

Schlosser W. and Franco S. (1979) Reduction of γ-aminobutyric acid (GABA)-mediated transmission by a convulsant benzodiazepine. *J. Pharmacol. Exp. Ther.* **211,** 290–295.

Scholfield C. N. (1982) Antagonism of γ-aminobutyric acid and muscimol by picrotoxin, bicuculline, strychnine, bemegride, leptazol, *d*-tubocurarine and theophylline in the isolated olfactory cortex. *Naunyn Schmiedebergs Arch. Pharmacol.* **318,** 274–280.

Schwartz R. D., Jackson J. A., Weigert D., Skolnick P., and Paul S. M. (1985a) Characterization of barbiturate-stimulate chloride efflux from rat brain synaptoneurosomes. *J. Neurosci.* **5,** 2963–2970.

Schwartz R. D., Seale T. W., Skolnick P., and Paul S. M. (1985b) Picrotoxin reversal of barbiturate-stimulated chloride efflux from synaptoneurosomes parallels differences in in vivo convulsant sensitivity between two inbred strains of mice. *Soc. Neurosci. Abstr.* **11,** 275.

Seale T. W., Bolger G. T., and Skolnick P. (1985) Inherent differences in convulsant sensitivity to Ro 5-4864, picrotoxin and pentylenetetrazole among inbred stains of mice. *Soc. Neurosci. Abstr.* **11,** 275.

Segal M. and Barker J. L. (1984) Rat hippocampal neurons in culture: Voltage clamp analysis of inhibitory synaptic connections. *J. Neurophysiol.* **52,** 469–487.

Seifert J. and Casida J. E. (1985a) Solubilization and detergent effects on interactions of some drugs and insecticides with the *t*-butylbicyclophosphorothionate binding site within the γ-aminobutyric acid receptor–ionophore complex. *J. Neurochem.* **44,** 110–116.

Seifert J. and Casida J. E. (1985b) Regulation of [^{35}S]-*t*-Butylbicyclophosphorothionate binding sites in rat brain by GABA, pyrethroid and barbiturate. *Eur. J. Pharmacol.* **115,** 191–198.

Shank R. P., Pong S. F., Freeman A. R., and Graham L. T. (1974) Bicuculline and picrotoxin as antagonists of γ-aminobutyrate and neuromuscular inhibition in lobster. *Brain Res.* **72,** 71–78.

Simmonds M. A. (1978) Presynaptic actions of γ-aminobutyric acid and some antagonists in a slice preparation of cuneate nucleus. *Br. J. Pharmacol.* **63,** 495–502.

Simmonds M. A. (1980) Evidence that bicuculline and picrotoxin act at separate sites to antagonize γ-aminobutyric acid in rat cuneate nucleus. *Neuropharmacology* **19,** 39–45.

Simmonds M. A. (1982) Classification of some GABA antagonists with regard to site of action and potency in slices of rat cuneate nucleus. *Eur. J. Pharmacol.* **80,** 347–358.

Simmonds M. A. (1983) Multiple GABA receptors and associated regulatory sites. *Trends NeuroSci.* **6,** 279–281.

Simmonds M. A. (1984) Interactions of the benzodiazepine Ro 5-4864 with the $GABA_A$ receptor complex. *Br. J. Pharmacol.* **82,** 198P.

Simonds W. F., Booth A. M., Thor K. B., Ostrowski N. L., Nagel J. R., and de Groat W. C. (1983) Parasympathetic ganglia: Naloxone antagonizes inhibition by leucine-enkephalin and GABA. *Brain Res.* **271,** 365–370.

Skerritt J. H., Johnston G. A. R., ChenChow S., Macdonald R. L., Prager R. H., and Ward A. D. (1985) Differential modulation of γ-aminobutyric acid receptors by caprolactam derivatives with central nervous system depressant or convulsant activity. *Brain Res.* **331,** 225–233.

Skerritt J. H., Werz M. A., McLean M. J., and Macdonald R. L. (1984) Diazepam and its anomalous *p*-chloro-derivative Ro 5-4864: Comparative effects on mouse neurons in cell culture. *Brain Res.* **310,** 99–105.

Smart T. G. and Constanti A. (1981) A re-examination of the GABA-inhibitory action of bicuculline on lobster muscle. *Eur. J. Pharmacol.* **70,** 25–33.

Smart T. G. and Constanti A. (1986) Studies on the mechanism of action of picrotoxinin and other convulsants at the crustacean muscle GABA receptor. *Proc. Roy. Soc. Lond.* **B277,** 191–216.

Smythies J. R. (1974) Relationships between the chemical structure and biological activity of convulsants. *Ann. Rev. Pharmacol.* **14,** 9–22.

Speth R. C., Wastek G. J., and Yamamura H. I. (1979) Benzodiazepine receptors: Temperature dependence of [^{3}H]flunitrazepam binding. *Life Sci.* **24,** 351–358.

Squires R. F. (1984) Benzodiazepine Receptors, in *Handbook of Neurochemistry* 2nd Ed., vol. 6 *Receptors in the Nervous System* (Lajtha A., ed.) Plenum, New York.

Squires R. F., Casida J. E., Richardson M., and Saederup E. (1983) [^{35}S]-*t*-Butylbicycloorthophosphorothionate binds with high affinity to brain-specific sites coupled to γ-aminobutyric acid A and ion recognition sites. *Mol. Pharmacol.* **23,** 326–336.

Squires R. F., Saederup E., Crawley J. N., Skolnick P., and Paul S. M. (1984) Convulsant potencies of tetrazoles are highly correlated with actions on GABA/benzodiazepine/picrotoxin receptor complexes in brain. *Life Sci.* **35,** 1439–1444.

Stevens D. R., Gallagher J. P., and Shinnick-Gallagher P. (1985a) Studies on depression of γ-aminobutyric acid potentials by phoridzin in cat dorsal root ganglion cells in vitro. *Neurosci. Lett.* **60,** 115–119.

Stevens D. R., Gallagher J. P., and Shinnick-Gallagher P. (1985b) Further studies on the action of baclofen on neurons of the dorsolateral septal nucleus of the rat, in vitro. *Brain Res.* **358,** 360–363.

Supavilai P., Mannonen A., Collins J. F., and Karobath M. (1982) Anion-dependent modulation of [^{3}H]muscimol binding and of GABA-stimulated [^{3}H]flunitrazepam binding by picrotoxin and related CNS convulsants. *Eur. J. Pharmacol.* **81,** 687–691.

Svenneby G. and Roberts E. (1973) Bicuculline and *N*-methylbicuculline-competitive inhibitors of brain acetylcholinesterase. *J. Neurochem.* **21,** 1025–1026.

Takeuchi A. and Onodera K. (1972) Effect of bicuculline on the GABA receptor of the crayfish neuromuscular junction. *Nature New Biol.* **236,** 55–56.

Takeuchi A. and Takeuchi N. (1966) On the permeability of the presynaptic terminal of the crayfish neuromuscular junction during synaptic inhibition and the action of γ-aminobutyric acid. *J. Physiol.* **183,** 433–449.

Takeuchi A. and Takeuchi N. (1969) A study of the action of picrotoxin on the inhibitory neuromuscular function of the crayfish. *J. Physiol.* **205,** 377–392.

Ticku M. K. and Maksay G. (1983) Convulsant/depressant site of action at the allosteric benzodiazepine-GABA receptor–ionophore complex. *Life Sci.* **33,** 2363–2375.

Ticku M. K. and Olsen R. W. (1979) Cage convulsants inhibit picrotoxinin binding. *Neuropharmacology* **18,** 315–318.

Ticku M. K. and Olsen R. W. (1984) Picrotoxinin Binding Sites in Brain, in *Brain Receptor Methodologies* part B *Amino Acids. Peptides. Psychoactive*

Drugs (Marangs P. J., Campbell I. C., and Cohen R. M. eds.) Academic, London.

Ticku M. K. and Ramanjaneyulu R. (1984) Ro 5-4864 inhibits the binding of ^{35}S-*t*-butylbicyclophosphorothionate to rat brain membranes. *Life Sci.* **34,** 631–638.

Ticku M. K., Ban M., and Olsen R. W. (1978) Binding of [^{3}H]α-dihydropicrotoxinin, a γ-aminobutyric acid synaptic antagonist to rat brain membranes. *Mol. Pharmacol.* **14,** 391–402.

Trifiletti R. R., Snowman A. M., and Snyder S. H. (1985) Barbiturate recognition site on the GABA/benzodiazepine receptor complex is distinct from the picrotoxinin/TBPS recognition site. *Eur. J. Pharmacol.* **106,** 441–447.

Usherwood P. N. R. (1978) Amino acids as neurotransmitters. *Adv. Comp. Physiol. Biochem.* **7,** 227–309.

Usherwood P. N. R. and Grundfest H. (1965) Peripheral inhibition in skeletal muscle of insects. *J. Neurophysiol.* **28,** 497–518.

Van der Kloot W. G. and Robbins J. (1959) The effects of γ-aminobutyric acid and picrotoxin on the junctional potential and the contraction of crayfish muscle. *Experientia* **15,** 35–36.

Van Renterghem C., Bible G., Moss S., Smart T. G., Constanti A., Brown D. A., and Barnard E. A. (1987) GABA receptors induced in *Xenopus* oocytes by chick brain mRNA: evaluation of TBPS as a use-dependent channel blocker. *Mol. Brain Res.* **2,** 21–31.

Weissman B. A., Cott J., Hommer D., Paul S., and Skolnick P. (1984) Electrophysiological and pharmacological actions of the convulsant benzodiazepine Ro 5-4864. *Eur. J. Pharmacol.* **97,** 257–263.

Weissman B. A., Cott J., Jackson J. A., Bolger G. T., Weber K. H., Horst W. D., Paul S. M., and Skolnick P. (1985) "Peripheral-type" binding sites for benzodiazepines in brain: Relationship to the convulsant actions of Ro 5-4864. *J. Neurochem.* **44,** 1494–1499.

Weissman B. A., Cott J., Paul S. M., and Skolnick P. (1983) Ro 5-4864: A potent benzodiazepine convulsant. *Eur. J. Pharmacol.* **90,** 149–150.

Welch A. D. and Henderson V. E. (1934) A comparative study of hydrastine, bicuculline and adlumine. *J. Pharmacol. Exp. Ther.* **51,** 482–491.

Wermuth C. G. and Biziere K. (1986) Pyridazinyl-GABA derivatives: A new class of synthetic $GABA_A$ antagonists. *Trends Neurosci.* **7, 421–424**

Wermuth C. G., Chambon J.-P., Heavlone M., Melikian A., Schlewer G., Leyris R., and Biziere K. (1987) The sensitivity of γ-amino-butyric acid antagonists to thiocyanate is related to the absence of a functional amionic group in their structure. *Eur. J. Pharmacol.* **144,** 375–378.

Wilson W. and Escueta A. V. (1974) Common synaptic effects of pentylenetetrazol and penicillin. *Brain Res.* **72,** 168–171.

Wojtowicz J. M. and Nicoll R. A. (1982) Selective action of piretanide on primary afferent GABA responses in the frog spinal cord. *Brain Res.* **236,** 173–181.

Woodbury D. M. (1980) Convulsant drugs: Mechanisms of action. *Adv. Neurol.* **27,** 249–303.

Yakushiji T., Tokutomi N., Akaike N., and Carpenter D. O. (1987) Antagonists of GABA responses, studied using internally perfused frog dorsal root ganglion neurons. *Neuroscience* **22,** 1123–1133.

Yarowsky P. J. and Carpenter, D. O. (1978) Receptors for gammaaminobutyric acid (GABA) on *Aplysia* neurons. *Brain Res.* **144,** 75–94.

Yasui S., Ishizuka S., and Akaike N. (1985) GABA activates different types of chloride conducting receptor–ionophore complexes in a dose dependent manner. *Brain Res.* **344,** 176–180.

Zukin S. R., Young A. B., and Snyder S. H. (1974) γ-Aminobutyric acid binding to receptor sites in rat central nervous system. *Proc. Natl. Acad. Sci. USA* **71,** 4802–4807.

Specificity of Drugs as Tools in Psychopharmacology

Drugs That Interfere with Monoamine Storage

S. Garattini and T. Mennini

1. Introduction

Most studies concerning the mechanism of action of psychotropic drugs are based on pharmacological interactions. A drug known to block, for instance, the dopaminegic system is utilized to establish whether another drug exerts its pharmacological effects via that monoaminergic system. *d*-Amphetamine is believed to act on the dopaminergic system because some, but not all, pharmacological effects are inhibited by antidopaminergic agents.

This methodology, which has permitted considerable progress in understanding the mechanism of action of drugs, must, however, be used with criticism. The selectivity of drugs on exerting a given biochemical effect must always be established under the precise experimental conditions employed since it is very difficult to extrapolate across doses, routes of administrations, times after dosing, animal species, tissues, and so on. There is evidence that the selectivity of drugs is considerably affected by the experimental condition. Furthermore it must always be considered that drug interactions may take place not only at the "target," but also at other sites and through other mechanisms. An example is the "metabolic" interaction: A drug may change the effect of another drug because it affects its disposition as regards organ distribution, protein binding, and/or formation of active metabolites. It is in fact very difficult to discuss synergism or antagonism between two drugs without knowing their concentrations in tissue(s), where the effect occurs.

This review will give some examples, taken mostly from work done in this laboratory, to illustrate some of the above points, with the aim of cautioning against simplified and sometimes simplicistic conclusions from pharmacological experiments.

2. *d*-Amphetamine Inhibition of Monoamine Uptake

d-Amphetamine is a drug tool used to study catecholaminergic effects elicited by release of noradrenaline (N) and dopamine (D). In addition, this drug inhibits catecholamine uptake and, in vitro, is a specific inhibitor of N uptake (Garattini et al., 1978). Its IC_{50} is 80 n*M* for inhibition of N uptake by rat brain synaptosomes, whereas a concentration of 500 n*M* is required to inhibit D uptake by 50%, and 12 μ*M* is the concentration giving half maximal inhibition of serotonin [5-hydroxytryptamine (5-HT)] uptake (Garattini et al., 1978). The dose–effect curves are reported in Fig. 1.

Different doses of amphetamine result in brain levels in the range of those inhibiting monoamine uptake in vitro (Fig. 1). Brain amphetamine concentrations 2 h after ip injection of 0.7 mg/kg (Cho et al., 1973) are enough, however, to give considerable inhibition of N uptake without changing 5-HT uptake, inhibiting D uptake only by about 30%. At higher dose (1.8 mg/kg) D uptake is affected to a considerably greater degree (Jori et al., 1978), and at a dose of 15 mg/kg (Belvedere et al., 1973), the brain levels can inhibit the uptake of all three amines. This may be important in explaining the different effects obtained with amphetamine in rats: 1.25 mg/kg can reduce food intake, an effect related to the activation of noradrenergic system (Quattrone et al., 1977); 10 mg/kg induces stereotyped behavior, typical of dopaminergic stimulation (Quattrone et al., 1977). An increase in the turnover of 5-HT in the brain has also been reported (Reid, 1970) after high doses of amphetamine. The fact that different doses of *d*-amphetamine interact with different monaminergic systems is also reflected in the biochemical and behavioral effects induced by repeated treatment with this drug.

The effects of long-term treatment with different doses of *d*-amphetamine indicate that monoamine pathways are differently affected: two daily injections of 1.25 mg/kg of *d*-amphetamine for 28 d did not alter the number of binding sites for 5-HT and catecholamines, and no tolerance arose to the anorectic activity (Bendotti et al., 1982). Similar treatment, but with 10 mg/kg, significantly reduced the number of β-noradrenergic receptors in hippocampus and D receptors in striatum and nucleus accumbens (Bendotti et al. 1982). With this treatment, the anorectic effect of 1.25 mg/kg *d*-amphetamine was not modified, but there was markedly less motor hyperactivity induced by this drug, and the stereotyped licking and biting induced by apomorphine was completely block-

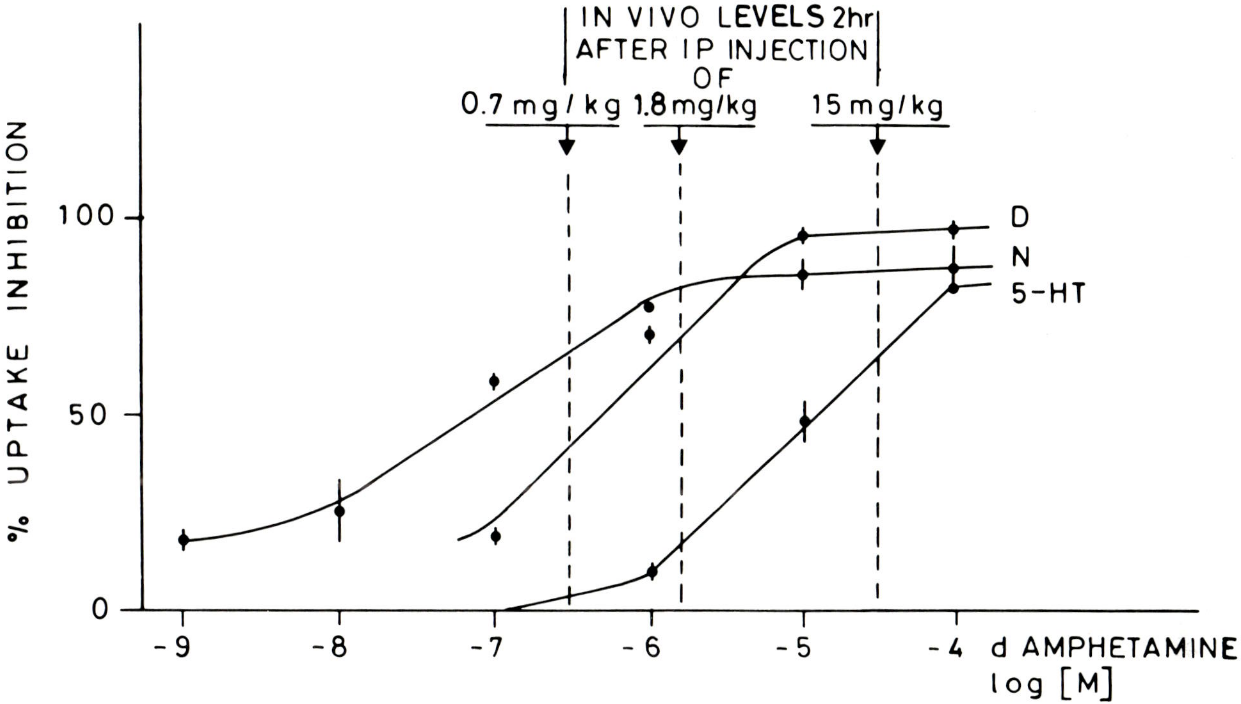

Fig. 1. Log dose–response curves for *d*-amphetamine inhibition of monoamine uptake. The dotted lines represent the concentration of *d*-amphetamine reached in the brain 2 h after ip injection of 0.7, 1.8, and 15 mg/kg of *d*-amphetamine sulfate in rats. The concentrations are calculated from the levels of the drug per gram of brain tissue, assuming a density of 1. Each point represents the mean ± SE of four replications (from Garattini et al., 1978).

ed (Bendotti et al., 1982). These effects cannot be easily extrapolated across animal species or even across different strains of the same species since it is well established, for instance, that certain strains of mice are resistant to some effects of *d*-amphetamine (Weaver and Kerley, 1962; Caccia et al., 1973; Dolfini et al., 1969a,b).

3. *d*-Fenfluramine as a Selective Indirect 5-HT Agonist

d-Fenfluramine (*d*F) is considered a tool drug to study serotonergic effects. This drug has been characterized in vitro as an indirect 5-HT agonist, and in fact inhibits 5-HT uptake and induces 5-HT release by rat brain synaptosomes (Garattini et al., 1979; Mennini et al., 1981; Borroni et al., 1983).

As for *d*-amphetamine, this selectivity of action depends on the dose (Invernizzi et al., 1986). Four hours after an oral dose of 2.5 mg/kg to rats, *d*F reduces 5-HT and raised 5-hydroxyindole-3-acetic acid (5-HIAA) concentrations without affecting other monoaminergic systems (Table 1). If a dose of 20 mg/kg is administered, however, N concentrations are also lowered and 3-methoxy-4-hydroxyphenylethylene glycol sulfate (MOPEG SO_4) is raised in the brain of treated rats (Table 1).

Repeated doses of 2.5 mg/kg of *d*F result in a decrease of 5-HT1 and 5-HT2 receptors in rat cortex without affecting receptors for antidepressants, BDZs, and β-noradrenoceptors (Table 2).

4. Amineptine Interaction with Dopaminergic System

Amineptine is a new antidepressant agent (Rosier, 1979; Samanin et al., 1977; Poignant, 1979; Borsini et al., 1981) that differs from other antidepressants in its selectivity on the dopaminergic system (Garattini et al., 1984). It has been reported to inhibit D uptake in vitro (Ceci et al., 1986) and in vivo (Garattini et al., 1984), without affecting N neurons (Garattini et al., 1984). An ip injection of 40 mg/kg to rats resulted in a persistent, clear-cut decrease of 3,4-dihydroxyphenylacetic acid (DOPAC) levels in the nucleus accumbens (Fig. 2a), as revealed by in vivo voltammetry recordings (De Simoni et al., 1986), whereas in the striatum only a slight and transient decrease of DOPAC levels could be documented after in vivo treatment (De Simoni et al., 1986) (Fig. 2b)

Table 1
Monoamine and Metabolite Levels in Striatum and Rest of Brain 4 h After Oral *d*-Fenfluramine[a]

	Dose, mg/kg	Striatum, ng/g[b]				Rest of Brain, ng/g			
		3MT	HVA	DOPAC	D	5-HT	5-HIAA	$MOPEG\text{-}SO_4$	N
Control		15 ± 1	390 ± 11	493 ± 11	8049 ± 374	442 ± 18	181 ± 13	104 ± 6	433 ± 14
d-Fenfluramine	2.5	16 ± 2	412 ± 31	491 ± 51	7621 ± 359	366 ± 8[c]	134 ± 7[c]	80 ± 5	437 ± 30
d-Fenfluramine	5.0	15 ± 2	380 ± 15	456 ± 24	7149 ± 346	300 ± 13[c]	128 ± 9[c]	86 ± 5	431 ± 14
d-Fenfluramine	10.0	16 ± 3	384 ± 23	481 ± 18	7548 ± 446	242 ± 16[c]	126 ± 10[c]	90 ± 9	408 ± 10
d-Fenfluramine	20.0	18 ± 2	401 ± 34	446 ± 32	7348 ± 226	236 ± 21[c]	122 ± 6[c]	130 ± 5[c]	361 ± 24[c]

[a] ±SE.
[b] Each value is the mean of seven determinations; from Invernizzi et al. (1986).
[c] $p < 0.01$ vs control (Dunnett's test).

Table 2

Effect of Long-Term *d*-Fenfluramine Treatment on Rat Brain Receptors[a]

Treatment	pmol bound/g tissue				
	$[^3H]$-5-HT	$[^3H]$-Ketanserin	$[^3H]$-Dihydro-alprenolol	$[^3H]$-Fluni-trazepam	$[^3H]$-Imipramine
Vehicle	16.2 ± 0.14	14.7 ± 2.30	9.8 ± 1.72	85.8 ± 7.3	19.4 ± 22
d-Fenfluramine 2.5 mg/kg × 2 28 d	13.1 ± 0.83[b]	6.2 ± 1.62[b]	8.8 ± 1.84	90.9 ± 4.5	19.0 ± 56

[a]Animals were killed 5 d after last treatment; all assays were done in cortex. Data are means ±SD of four animals per group and represent B_{max} values, with the exception of $[^3H]$-imipramine binding, which was determined using 12 n*M* ligand (from Mennini, unpublished results).

[b]$p < 0.01$ vs. Vehicle.

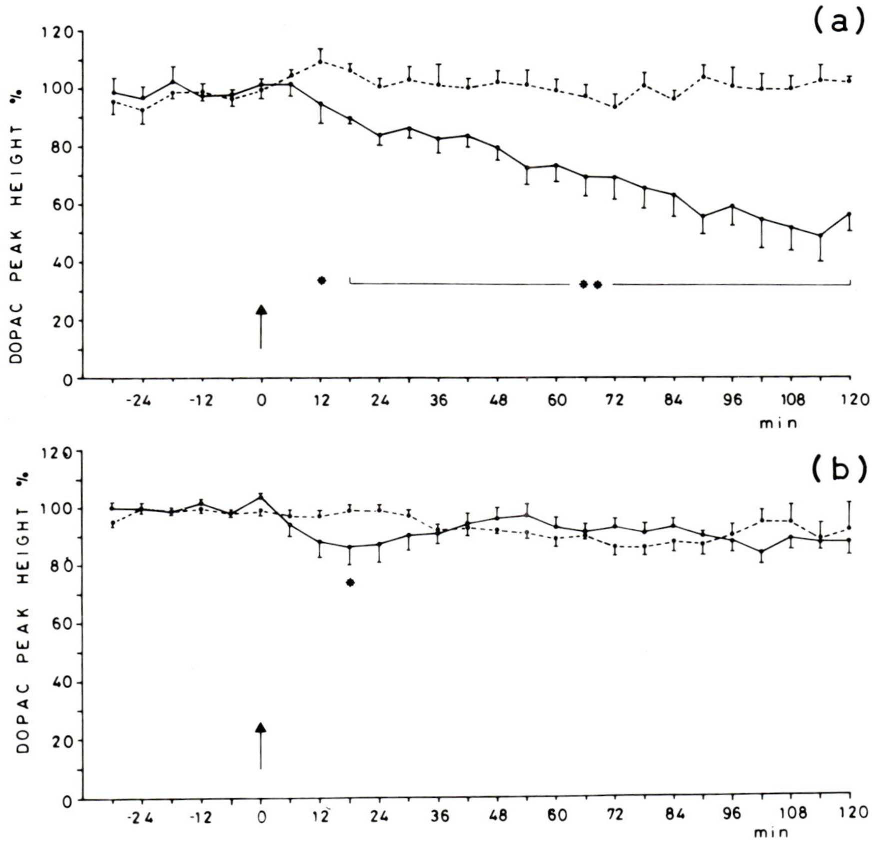

Fig. 2. DOPAC levels in the nucleus accumbens (a) and in the caudate (b) of unanesthetized, freely moving rats: effect of amineptine, 40 mg/kg ip (solid line, $n = 6$) and saline (broken line, $n = 5$). The arrow indicates the time of injection. DOPAC was recorded every 2 min, but the levels are shown here every 6 min. Friedman test: ns. $^*p < 0.05$; ** $p < 0.01$ vs saline, Mann-Whitney test (from De Simoni et al., 1986).

Thus for amineptine, as discussed later for buspirone, it appears that the interaction with a particular neuronal system in the brain occurs specifically in a selected brain region responsible for the drug's pharmacological activity. It would therefore be inappropriate to use amineptine as a tool drug to obtain striatal dopaminergic activation at doses between 20 and 40 mg/kg because it is selective for the limbic area and practically inactive on the striatum.

5. Reserpine-Induced Storage Impairment in Release Experiments

Reserpine is known to interfere with the storage of monamines in nerve terminals. It has been successfully used as a tool to determine whether a drug acts on the amine storage pool in vivo and in vitro. An example of the in vitro interaction is monoamine release in synaptosomes from rats pretreated with reserpine in order to impair the storage vesicles. Such experiments have clearly indicated that *d*-amphetamine releases D from a reserpine-insensitive pool (Farnebo, 1971), and the in vitro results are confirmed by in vivo findings that the behavioral effects of amphetamine are still present, and even potentiated, in rats pretreated with reserpine (Van Rossum et al. 1962; Svensson, 1970).

Similar results have been obtained for the serotonergic system, in which reserpine pretreatment results in different interactions with drugs that release 5-HT (Mennini et al., 1981). Thus it has been found that dF and its metabolite *d*-norfenfluramine (dNF) release 5-HT from different pools (Mennini et al. 1981). The releasing activity of *d*F is virtually eliminated by reserpine, indicating that in normal synaptosomes *d*F releases 5-HT from a reserpine-sensitive pool. In contrast, the effect of *d*NF was strongly enhanced in synaptosomes from reserpine-treated animals. One possible explanation is that in reserpinized tissue labeled 5-HT accumulated is exclusively bound to extravesicular sites for which *d*NF has high affinity. This interpretation seems that most likely, since in vivo too, pretreatment of rats with reserpine strongly potentiated the decrease of food intake induced by dNF (Borsini et al., 1982).

The anorectic effect of *d*F was not modified by reserpine treatment, and these data seem to be in contrast with the in vitro results. However, dF is rapidly metabolized to *d*NF (Caccia et al., 1979, 1981), so inhibition of the parent compound's effect and potentiation of the metabolite's effect could occur after reserpine treatment, thus making it difficult to assess the interaction with *d*F.

6. Mechanism of Anxiolysis by Buspirone

Buspirone is a new, non-BDZ anticonflict agent (Weissman et al., 1984), with reported antianxiety activity (Seidel et al., 1985).

Table 3
Effect of Buspirone and 1PP on Different Binding Sites in the Rat Brain[a]

Receptor	[^{3}H]-Ligand	Buspirone	1PP
5-HT1	[^{3}H]-5-HT	1.2×10^{-6}	1.9×10^{-6}
5-HT1A	[^{3}H]-PAT	7.0×19^{-7}	7.5×10^{-6}
5-HT1B	[^{125}I]-CYP	6.5×10^{-5}	4.5×10^{-6}
5-HT2	[^{3}H]-SPI	0.7×10^{-6}	$>10^{-5}$
	[^{3}H]-KET	1.3×10^{-6}	$>10^{-4}$
α_1-N	[^{3}H]-WB4101	7.7×10^{-7}	7.7×10^{-6}
α_2-N	[^{3}H]-CLO	1.1×10^{-5}	2.1×10^{-8}
	[^{3}H]-YOH	3.2×10^{-6}	2.0×10^{-7}
β-N	[^{3}H]-DHA	1.6×10^{-5}	1.3×10^{-6}
D_1	[^{3}H]-DA	4.9×10^{-4}	4.7×10^{-4}
D_2	[^{3}H]-SPI	3.0×10^{-7}	$>10^{-4}$
D_{ag}	[^{3}H]-ADTN	4.0×10^{-6}	$>10^{-5}$

[a]K_i (*M*) were calculated from IC_{50} values. Standard deviations were always lower than 15%. IC_{50} values were calculated by nonlinear fitting of curves for inhibition of specific ^{3}H-ligand binding, using 6–8 drug concentrations. The following conditions were used for receptor binding assays: [^{3}H]-5-HT 2 n*M*, cortex 100 vol, 10 μ*M* 5-HT for nonspecific binding; [^{3}H]-PAT 2 n*M*, cortex 100 vol, 10 μ*M* 5-HT; [^{125}I]-iodocyanopindolol 0.15 n*M* in presence of 30 μ*M* (–)-isoprenaline, cortex 500 vol, 10 μ*M* 5-HT; [^{3}H]-spiperone 0.7 n*M*, cortex 100 vol, 1 μ*M* ketanserin; [^{3}H]-ketanserin 0.6 n*M*, cortex 200 vol, 1 μ*M* methysergide; [^{3}H]-WB4101 0.22 n*M*, cortex 100 vol, 100 μ*M* (–)-N; [^{3}H]-clonidine 0.4 n*M* cortex 100 vol, 100 μ*M* (–)-N; [^{3}H]-Yohimbine 2.2 n*M*, cortex 200 vol, 100 μ*M* (–)-N; [^{3}H]-dihydroalprenolol 1 n*M*, cortex 100 vol, 1 μ*M* (±)-propranolol; [^{3}H]-dopamine 1 n*M*, striatum 400 vol, 10 μ*M* (+)-butaclamol; [^{3}H]-spiperone 0.22 n*M*, striatum 400 vol, 1 μ*M* (+)-butaclamol; [^{3}H]-ADTN 8 n*M*, striatum 120 vol, 1 μ*M* (+)-butaclamol. Final incubation volume was 1 mL. All radiochemicals were purchased from N.E.N. Incubations were carried out for 15 min at 37°C (5-HT1, 5-HT1A, 5-HT2, D_2 and D_{ag}), for 20 min at 25°C (α_1-N, α_2-N, and β-N), or for 30 min at 25°C (D_1) or for 90 min at 37°C ($5HT_1B$). (Mennini, unpublished results).

Unlike benzodiazepines (BDZs), it does not interact with the GABA–Cl–BDZ receptor complex (Riblet et al., 1980; Garattini et al., 1982). This fact prompted several investigators to characterize the possible "target" of buspirone activity using in vitro receptor binding techniques (Glaser and Traber, 1983; Wood et al., 1983; Mennini et al., 1987). However, the in vivo relevance of these data, summarized in Table 3, can be questioned considering the particular pharmacokinetics of the drug. In fact, after iv injection of 10

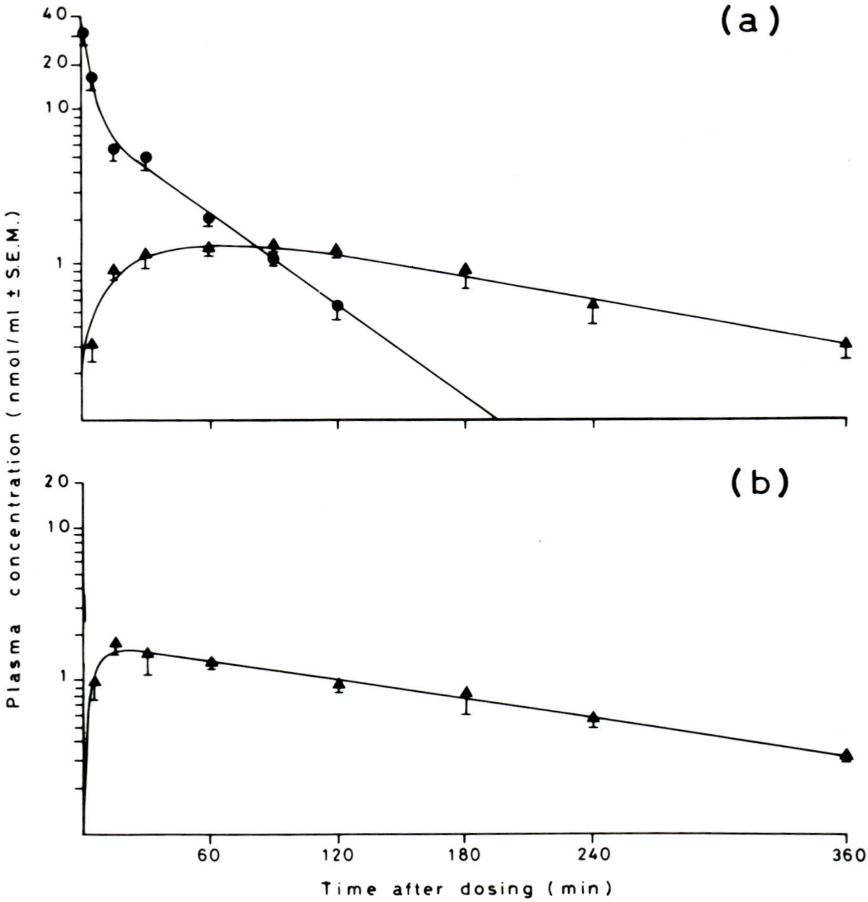

Fig. 3. Plasma concentration–time curves of buspirone and 1-(2-pyrimidinyl)-piperazine after iv (a) and po (b) administration of buspirone to rats. Buspirone, (●—●); 1PP, (▲—▲). Dose of buspirone hydrochloride was 10 mg/kg (from Caccia et al., 1983).

mg/kg of buspirone to rats, the drug is rapidly cleared from blood with a $t_{1/2}$ of 30 min (Fig. 3a) (Caccia et al., 1982, 1983). After the same dose given orally (Fig. 3b), the drug is not detectable in blood or brain within the limits of sensitivity of the method (Caccia et al., 1983). The principal metabolite formed is the 1-pyrimidylpiperazine (1PP).

Thus, in order to extrapolate the possible mechanism of action from in vitro data, it is important to consider the effect of the metabolite on receptors and to compare these data with the brain concentrations of parent compound and metabolite after in vivo administration. Since buspirone is active in the rat conflict test at doses as low as 1 mg/kg orally (Riblet et al., 1982), brain concentrations of the parent compound and metabolite are probably sufficient to act at 5-HT1A and α_2-noradrenergic receptors, respectively.

The pharmacological activity may be the result of the complex interaction between parent drug and metabolite. In fact although drug interactions with particular subtypes of 5-HT receptors have been described as producing anxiolytic activity (Glaser and Traber, 1985), an activation of 5-HT receptors *per se* does not seem to be responsible for this pharmacological activity. In fact 5-HT antagonists are reportedly active in the rat conflict test (Stein et al., 1973; Geller et al., 1974; Robichaud and Sledge, 1969), whereas agonists result in suppression of this behhavior (Stein et al., 1973), and BDZ anxiolytics do not affect 5-HT metabolism (Rastogi et al., 1977). The same applies to the activity of 1PP as an α_2-noradrenoceptor antagonist (Sanghera and German, 1983) since there is a decrease, not an increase, in the output of the noradrenergic locus ceruleus with anxiolytic agents like diazepam (Sanghera and German, 1983). It is therefore possible that the combined activity of the two compounds results in the final pharmacological effect.

Another important point raised by studies on the mechanism of action of buspirone is the matter of regional differences. The affinity of buspirone for 5-HT receptors markedly differs if one considers just three brain regions (Mennini et al., 1986a), being highest in the hippocampus, intermediate in the cortex, and lowest in the striatum (Table 4).

This may be caused by the regional heterogeneity of 5-HT receptor subtypes or their environment to which buspirone may bind allosterically. Considering the in vivo concentrations of buspirone after oral administration, the drug seems to act on 5-HT receptors in hippocampus, but not in striatum. This is confirmed in vivo, since a dose-dependent decrease of 5-HIAA can be obtained in hippocampus (Mennini et al. 1986a), but not in striatum (Cimino et al., 1983), of buspirone-treated rats.

Table 4
Inhibition of $[^3H]$-5-HT Binding in Three Rat Brain Regions[a]

	Hippocampus		Cortex		Striatum	
Drug	K_i,*M*	Slope	K_i,*M*	Slope	K_i,*M*	Slope
Buspirone	$3.4 \pm 0.7 \times 10^{-8}$	0.6 ± 0.1^b	$1.0 \pm 0.3 \times 10^{-6}$	0.5 ± 0.1^b	$2.0 \pm 1.1 \times 10^{-5}$	0.5 ± 0.1^b
1PP	$1.1 \pm 0.1 \times 10^{-6}$	0.9 ± 0.1	$1.9 \pm 0.7 \times 10^{-6}$	0.9 ± 0.6	$1.1 \pm 0.2 \times 10^{-6}$	0.8 ± 0.1

[a] $[^3H]$-5-HT concentration was 2 n*M*; nonspecific binding was determined in the presence of 10 μ*M* unlabeled 5-HT; 8–10 drug concentrations were used for each curve. The values were obtained as parameters (±SD) analyzing the displacing curves with the logistic function

[b] $p < 0.01$ different from 5-HT slope in the same area (Student's *t*-test) (Mennini, unpublished results).

7. Neurotoxicity Induced by Monosodium Glutamate (MSG)

Glutamate is utilized as a tool to produce selective neurotoxicity in the nucleus arcuatus of the hypothalamus (NAH). The sensitivity of animals to the neurotoxic effects of MSG at high doses differs considerably, however, depending on its route of administration (Garattini et al., 1979). For instance, in 10-d-old mice, an ip dose of 0.4 g MSG/kg caused hypothalamic lesions in 100% of animals (Takasaki, 1978); the corresponding sc dose is 0.7 g/kg (Takasaki, 1978), and between 1 (Olney and Ho, 1970) and 1.5 g/kg (Takasaki, 1978) is required when MSG is given by gavage. In newborn rats, too, 0.5 g/kg given sc caused lesions of the NAH, whereas 2 g/kg by gavage is neurotoxic in only 60% of the animals (Heywood and Worden, 1979). It is of interest to recall that in adult rats and mice the LD_{50} of MSG rises in the following order in relation to the route of administration: iv $<$ ip $<$ sc $<$ po (Moriyuki and Ichimura, 1978; Pinto-Scognamiglio et al., 1972). MSG given with the diet has never been found to induce lesions at any dose tested (Heywood and Worden, 1979).

Several reports now show that there is a clear-cut difference in the toxicity of MSG when given parenterally or by gavage, on the one hand, and by dietary administration on the other. The neurotoxic, endocrine, and behavioral effects of MSG when given at high doses by gavage or parenterally to sensitive species are completely lacking when the compound is given *ad libitum* with the food (Garattini et al., 1979). The explanation is relatively simple and lies in the kinetics of MSG (Garattini, 1979). In a typical experiment (Airoldi et al., 1979), MSG was given by gavage (dissolved in water, 10%) or in the food of mice trained to eat their meals in 30 min. Plasma levels attained a peak of 2.08 μmol/mL (12 times the basal levels) when MSG was given by gavage, and 0.27 μmol/mL (1.6 times basal level) when mixed with the food. The elevation of plasma glutamic acid (GA) in relation to time (area under curve, AUC) was 63 times higher when MSG was given at the same dose by gavage than when given with food (Bizzi et al., 1977). Therefore the lack of toxicity found in various experimental conditions when MSG is given with the diet is probably fully explained by the fact that the levels of plasma GA, the agent probably responsible for neurotoxicity (Garattini, 1979), are considerably lower than those known to produce brain lesions when MSG is given by other routes of administration. This seems to be related to the ample capacity of the intestinal mucosa and liver to metabolize MSG when it is present in large amounts, at a rate determined by food intake, along with other amino acid constituents (Garattini, 1979).

Carbohydrates are another major food component buffering the rise in plasma GA after MSG consumption, just as they buffer the effects of other amino acids (Hardcastle et al., 1968). Consumption of 400 mg sucrose/kg halves the peak plasma GA levels occurring after consumption of 100 mg MSG/kg (Stegink et al., 1972). Glucose, or sugars that can be transformed into glucose, may provide pyruvate for faster transamination of GA to alanine in the intestinal mucosa (Fang et al., 1977).

8. Potency of Sulpiride in Inhibiting Amphetamine-Induced Hyperactivity

The benzamide group of drugs has been classified with the neuroleptic agents, since agents from this series inhibit DA function, particularly outside the blood–brain barrier (Fang et al., 1977; Horowsky and Graf, 1976; Jenner et al., 1978; Prieto et al. 1977; Valenzuela, 1976). The central activity spectrum of the benzamides differs markedly, however, when they are administered by a peripheral route (Jenner et al., 1978), and this may reflect different abilities to penetrate cerebral tissue.

Table 5 reports the findings of a study (Costall et al., 1978) designed to compare the antagonistic effects of some neuroleptics when administered peripherally or intracerebrally against amphetamine-induced hyperactivity. The hyperactivity induced by 1.5 mg/kg ip amphetamine was inhibited by fluphenazine injected either systemically or into the nucleus accumbens, the selective area responsible for all antiamphetamine effect of neuroleptics (Costall and Naylor, 1976).

When administered peripherally, sulpiride (1–40 mg/kg) failed to reduce amphetamine hyperactivity, although 1.25–5 μg administered into the nucleus accumbens caused dose-related antagonism of the amphetamine response, with the larger dose causing virtually complete inhibition.

Similar observations were reported for tiapride, whereas metoclopramide, sultopride, and clebopride were active against amphetamine both by the ip and intraaccumbens routes. Thus the

Table 5
Effect on Amphetamine Hyperactivity of Fluphenazine and Some Benzamide Derivatives Injectied Peripherally or into the Nucleus Accumbens[a]

Drug	Peripherally		Nucleus accumbens	
	Dose, mg/kg	Inhibition amphetamine hyperactivity, %	Dose, μg	Inhibition amphetamine hyperactivity, %
Fluphena-zine	0.01	0		
	0.04	40	0.2	2
	0.1	90	0.4	17
	0.4	100	0.8	100
Metoclo-pramide	0.4	4		
	1.0	30	20.0	7
	4.0	60	40.0	20
	10.0	94	80.0	50
Sultopride	0.4	0		
	1.0	21	0.625	22
	4.0	40	1.25	60
	10.0	100	2.5	94
Clebo-pride	0.04	0		
	0.1	14	2.5	6
	0.4	43	5.0	41
	1.0	88	10.0	86
Tiapride	1.0	0		
	4.0	0	0.31	5
	10.0	0	0.625	63
	40.0	18	1.25	96
Sulpiride	1.0	0		
	4.0	0	1.25	8
	10.0	0	2.5	74
	40.0	0	5.0	97

[a]Amphetamine and the potential antagonists were administered together, amphetamine (1.5 mg/kg) by the ip route and the antagonists ip (mg/kg) or directly into the nucleus accumbens (μg). The effects of fluphenazine, sultopride, and clebopride were significant by both routes ($p < 0.05$–0.001). Tiapride and sulpiride significantly ($p < 0.001$) antagonized amphetamine hyperactivity only when administered intraaccumbens. Metoclopramide was more active by the peripheral ($p < 0.05$–0.001) than by the intraaccumbens route ($p < 0.05$ only). SEM values for metoclopramide were large (+20%), whereas for other agents they were in the range of 9–17%. There were 6–8 rats in each group. Intraaccumbens solvent failed to antagonize amphetamine hyperactivity (from Costall et al., 1978).

low efficacy of sulpiride or tiapride on peripheral administration may partly be related to their low ability to penetrate the blood–brain barrier (Frigerio and Pantarotto, 1977). Comparison of drug efficacy is therefore completely different for the two routes of administration: sulpiride and tiapride are only weakly active given intraperitoneally, but when given intracerebrally they are as potent as fluphenazine in inhibiting amphetamine hyperactivity.

9. Amphetamine and Tricyclic Antidepressants

Potentiation by tricyclic antidepressant drugs (TADs) of amphetamine-induced locomotor hyperactivity or hyperthermia has been suggested as a test to differentiate this class of psychotropic drugs (Morpurgo and Theobald, 1965) from neuroleptics, which are characterized by the fact that they block several actions of amphetamine (Bradley and Hance, 1957; Janssen et al., 1967). The hypothesis to explain the interaction between TAD and amphetamine was based on the fact that TADs, by blocking N uptake at nerve terminals (Carlsson et al., 1966, 1969), would increase the concentrations of N released by amphetamine at the receptors. This attractive hypothesis is no longer tenable, at least in a simple way, because TADs interact with amphetamine by raising its concentration in the brain (Valzelli et al., 1967). A series of studies has been conducted in this respect on the interaction between desipramine and amphetamine (Garattini et al., 1976). It was shown that the prolongation of amphetamine hyperthermia was temporally related to an increased brain level of amphetamine in rats (Valzelli et al., 1967). This increase is related to the fact that desipramine blocks amphetamine hydroxylation (Dingell and Bass, 1969; Consolo et al., 1967) by liver microsomal enzymes (Axelrod, 1954), thus reducing its major metabolic pathway in rats. Since hyperthermia is directly related to brain amphetamine concentrations (Fig. 4), it appears that when equal concentrations of amphetamine in brain are considered, animals pretreated with desipramine are actually less responsive than controls (Garattini, 1969). That the potentiating effect of desipramine is almost entirely related to an inhibition of amphetamine hydroxylation is indirectly confirmed by the fact that desipramine does not potentiate amphetamine and does not raise brain amphetamine concentrations in mice (Dolfini et al., 1969c), species in which amphetamine is only poorly hydroxylated (Dring et al., 1966).

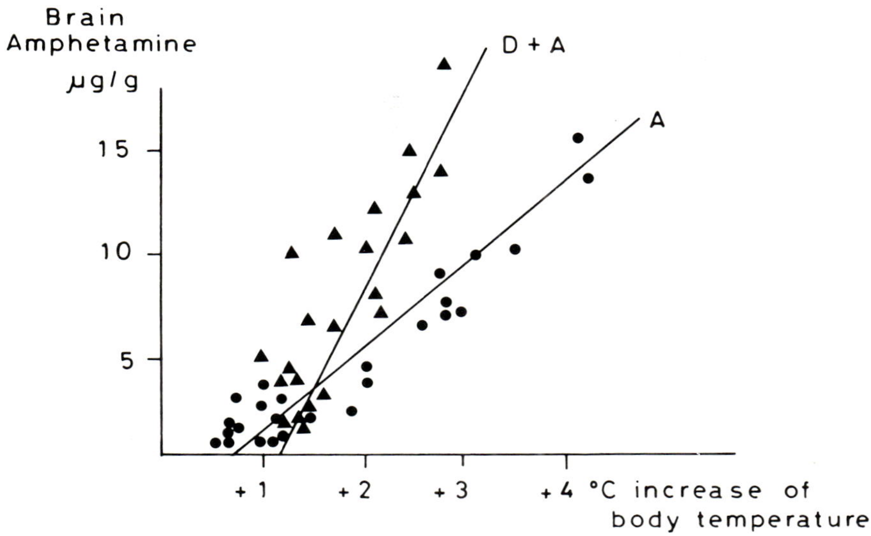

Fig. 4. Correlation between hyperthermia and levels of brain amphetamine in rats treated with amphetamine alone (A) of with a combination of desipramine and amphetamine (D + A) (from Garattini et al., 1976).

The interaction between TAD and amphetamine in rats can be minimized when amphetamine is given orally, because TADs block gastrointestinal motility and reduce the oral absorption of several drugs in relation to their anticholinergic activity (Consolo and Garattini, 1969). At the same time, therefore, there is less amphetamine being absorbed, but also less drug being metabolized in the presence of TAD pretreatment. A recent paper, however, reports potentiation of amphetamine's effects after chronic TAD treatment (Spyraki and Fibiger, 1981). In these conditions, no differences were found in brain concentrations of amphetamine, so the effect of TAD may not be simply related to metabolic interaction (Spyraki and Fibiger, 1981). Probably a supersensitivity of postsynaptic D receptors is responsible for this potentiation of amphetamine effects (Spyraki and Fibiger, 1981).

10. Drugs as Ligands in Receptor Binding Assays

The advent of receptor binding assays has heralded a new approach in psychopharmacology: the labeling of receptors for

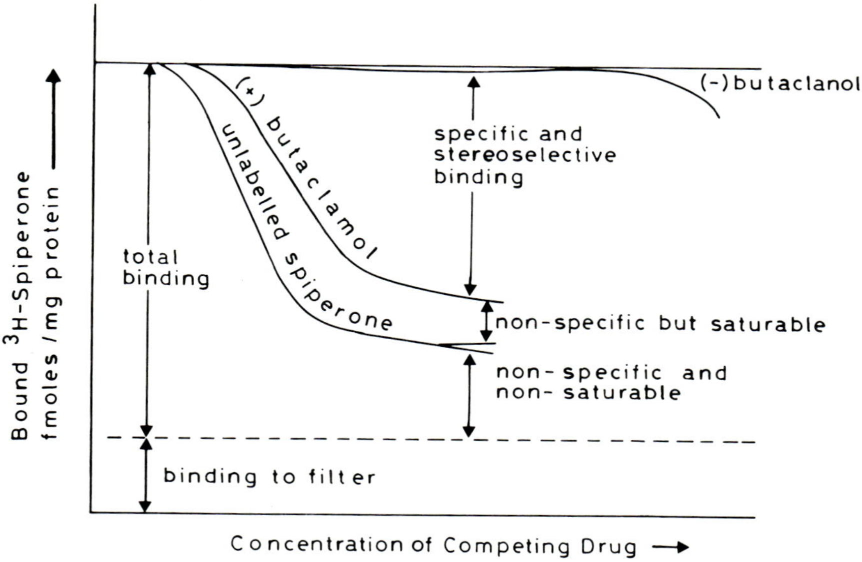

Fig. 5. Evaluation of binding of [^{3}H]-spiperone to dopamine receptors in the striatum (modified from Seeman, 1981).

neurotransmitters with radiochemical compounds at high specific activity (Snyder, 1984). With a few exceptions, however, such as [^{3}H]5-HT, the molecules available for receptor binding are principally drugs believed to act selectively at a neurotransmitter receptor.

Examples are antagonists for α_1, α_2, and β-adrenoceptors, and the use of neuroleptics for D_2 receptors. In such cases, care must be taken to establish that the binding assay does represent the receptor for the neurotransmitter and not a drug or a chemical binding site unrelated to the neurotransmitter receptor. Moreover, since the specific binding of a ligand is indirectly determined by evaluation of nonspecific binding measured in the presence of a high concentration of an unlabeled compound, the choice of this compound becomes decisive in defining the binding under study.

Figure 5 presents an example of evaluation of the binding of [^{3}H]spiperone, a spirodecanone derivative, to D receptors in the striatum (Seeman, 1981). If unlabeled spiperone is used, specific binding amounts to more than 90% of total binding. Only a portion (80%) of this, however, is shared by another neuroleptic of a different chemical class, e.g., (+)-butaclamol. This indicates that in

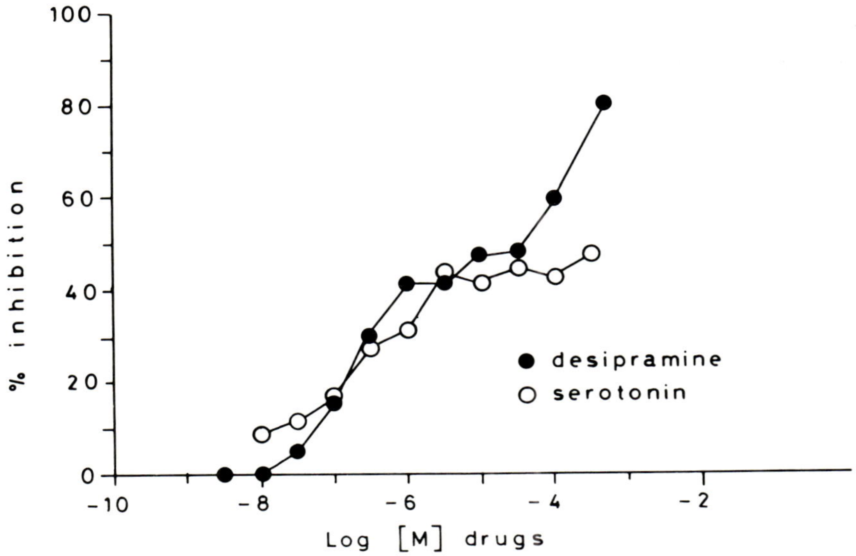

Fig. 6. In vitro inhibition of [^{3}H]-imipramine binding (Mennini, unpublished results).

the first case, the spirodecanone binding sites are also evaluated. Moreover spiperone and certain other neuroleptics interact with high affinity at both D and 5-HT receptors (Leysen et al., 1978), so stereospecific binding as defined by the difference between (+)- and (–)-butaclamol is likely to include both D and 5-HT receptor components (Quik et al., 1978). Thus the use of isomers of butaclamol to define "specific" binding of [^{3}H]-spiperone exclusively to D receptors would clearly yield misleading results (Quik et al., 1978). Thus more selective drugs (Quik et al., 1978; Leysen et al., 1981; List and Seeman, 1981) must be used to define [^{3}H]-spiperone binding to D or 5-HT receptors; these include ADTN and sulpiride or ketanserine, respectively.

Another example is given by [^{3}H]-imipramine, used to label the 5-HT presynaptic uptake carrier (Langer et al., 1980). Figure 6 shows that 5-HT inhibits [^{3}H]-imipramine binding only by 50%, whereas desmethylimipramine (DMI) completely inhibits it. This means that only a portion of total imipramine binding is related to the 5-HT uptake site, the other being a drug-binding site not located on 5-HT nerve terminals. In fact, if [^{3}H]-imipramine binding is determined in the presence of an excess of DMI, lesioning of serotonergic neurons results in only about a 40% decrease of specific imipramine binding (Mennini et al., 1986b; Severson et al., 1986). If an excess of unlabeled 5-HT is used in place of DMI, an approximately 60% decrease of imipramine binding is obtained after the serotonergic lesion (Severson et al., 1986).

Since [^{3}H]-imipramine binding has been studied in different physiological and pathological conditions, such as depression, chronic treatment with antidepressants, and aging, the choice of the drug used to define "specific" imipramine binding becomes particularly crucial in correlations with the 5-HT uptake.

For instance, a decrease of the maximal velocity of the 5-HT uptake system was reported in 24-mo-old rats (Brunello et al., 1985), together with an increase of [^{3}H]-imipramine binding sites (Brunello et al., 1985). The effect of aging on total [^{3}H]-imipramine binding was also reported by others and related to an increase in the myelin content of aged brain (Severson, 1986).

When only the portion of [^{3}H]-imipramine binding related to 5-HT uptake is considered, however, an age-related decrease in the density of the physiologically significant component of [^{3}H]-imipramine binding has been documented (Severson, 1986), in agreement with an age-related decrease in 5-HT uptake (Brunello et al., 1985).

11. Concluding Remarks

The examples reported, though only briefly discussed, should give some indication of the complexity of the problems encountered when a drug is utilized as a tool to achieve specific modifications of a particular neurotransmitter that, in turn, may help to clarify the mechanism underlying the action of another drug given concomitantly or in succession, in vitro or in vivo.

The question of the dose or concentration of the drug utilized as a tool is always crucial. The examples given for *d*-amphetamine and *d*-fenfluramine indicate that their selectivity of effects on a single monoaminergic system and 5-HT, respectively, is limited to a narrow range of doses that must be validated in the system (animal species, strain, route of administration, time, brain area, and so on) actually employed for the experiment.

The effect of the drug tool should not be generalized to the whole brain because in some cases—but not always—the effect is dependent on the discrete brain area considered. Amineptine and buspirone are examples of this selectivity for rat nucleus accumbens D and for hippocampal 5-HT, respectively.

In other cases, when biologically active metabolites are present in the brain, the effect of the parent drug may be modulated by the metabolite's effect. *d*-Fenfluramine releases 5-HT only from the reserpine-sensitive pool, but its metabolite *d*-norfenfluramine may release 5-HT, even at lower concentrations, from the reserpine resistant pool. Buspirone affects serotoninergic receptors (mostly 5-HT1A), but its metabolite 1PP has high affinity for α_2-adrenergic receptors, an effect lacking for the parent compounds. Under these conditions any in vitro study aimed at understanding the mechanism of action of a drug may be misleading if proper consideration is not given to the presence of the metabolite(s).

The route of administration can also sometimes determine the potency of an effect; examples are the potency of GA in producing selective brain lesions when given orally by gavage or mixed with food and the different antagonism between sulpiride and *d*-amphetamine. The matter may be further complicated by the presence of metabolites. In the case of buspirone, iv injection results in the simultaneous presence in the brain of the parent compound and its metabolite 1PP, whereas oral administration results only in the presence of the metabolite. The consequences of these two situations on the interaction of buspirone with brain monoamine receptors can be quite different.

The possibility that a tool drug may interact with the clearance or metabolism of other drugs must also be considered to avoid incorrect conclusions, such as in the case of the interaction of tricyclic antidepressant agents and *d*-amphetamine in rats.

Finally, the use of drugs as ligands for rceptor sites must also be interpreted with great caution as shown by the cases of spiperone and imipramine.

In summary, the real problem is that there is no such thing as a tool drug: there are only drugs that become proper tools when they are fully characterized under the conditions of each specific experiment.

References

Airoldi L., Bizzi A., Salmona M., and Garattini S. (1979) Attempts To Establish the Safety Margin for Neurotoxicity of Monosodium Glutamate, in *Glutamic Acid: Advances in Biochemistry and Physiology* (Filer L. J., Jr., Garattini S., Kare M. R., Reynolds W. A., and Wurtman R. J., eds.) Raven, New York, pp. 321–331.

Axelrod J. (1954) Studies on sympathomimetic amines. II. The biotransformation and physiological disposition of d-amphetamine, d-p-hydroxyamphetamine and d-methamphetamine. *J. Pharmacol. Exp. Ther.* **110,** 315–326.

Belvedere G., Caccia S., Frigerio A., and Jori A. (1973) A specific gas chromatographic method for the detection of p-hydroxyamphetamine and p-hydroxynorephedrine in brain tissue. *J. Chromatogr.* **84,** 355–360.

Bendotti C., Borsini F., Cotecchia S., De Blasi A., Mennini T., and Samanin R. (1982) d-Amphetamine induced anorexia and motor behavior after chronic treatment in rats: Relationship with changes in the number of catecholamine receptor sites in the brain. *Arch. Int. Pharmacodyn. Ther.* **260,** 36–49.

Bizzi A., Veneroni E., Salmona M., and Garattini S. (1977) Kinetics of monosodium glutamate in relation to its neurotoxicity. *Toxicol. Lett.* **1,** 123–130.

Borroni E., Ceci A., Garattini S., and Mennini T. (1983) Differences between d-fenfluramine and d-norfenfluramine in serotonin presynaptic mechanisms. *J. Neurochem.* **40,** 891–893.

Borsini F., Bendotti C., Aleotti A., Samanin R., and Garattini S. (1982) d-Fenfluramine and d-norfenfluramine reduce food intake by acting on different serotonin mechanisms in the rat brain. *Pharmacol. Res. Commun.* **14,** 671–678.

Borsini F., Bendotti C., Velkov V., Rech R., and Samanin R. (1981) Immobility test: Effects of 5-hydroxytryptaminergic drugs and role of catecholamines in the activity of some antidepressants. *J. Pharm. Pharmacol.* **33,** 33–37.

Bradley P. B. and Hance A. J. (1957) The effect of chlorpromazine and methopromazine on the electrical activity of the brain in the cat. *Electroencephalogr. Clin. Neurophysiol.* **9,** 191–215.

Brunello N., Riva M., Volterra A., and Racagni G. (1985) Age-related changes in 5HT uptake and [^{3}H]imipramine binding sites in rat cerebral cortex. *Eur. J. Pharmacol.* **110,** 393–394.

Caccia S., Ballabio M., and De Ponte P. (1979) Pharmacokinetics of fenfluramine enantiomers in man. *Eur. J. Drug Metab. Pharmacokinet.* **4,** 129–132.

Caccia S., Cecchetti G., Garattini S., and Jori A. (1973) Interaction of (+)-amphetamine with cerebral dopaminergic neurones, in two strains of mice, that show different temperature responses to this drug. *Br. J. Pharmacol.* **49,** 400–406.

Caccia S., Dagnino G., Garattini S., Guiso G., Madonna R., and Zanini M. G. (1981) Kinetics of fenfluramine isomers in the rat. *Eur. J. Drug Metab. Pharmacokinet.* **6,** 297–301.

Caccia S., Garattini S., Mancinelli A., and Muglia M. (1982) Identification and quantitation of 1-(2-pyrimidinyl)piperazine, an active metabolite of the anxiolytic agent buspirone, in rat plasma and brain. *J. Chromatogr.* **252,** 310–314.

Caccia S., Muglia M., Mancinelli A., and Garattini S. (1983) Disposition and metabolism of buspirone and its metabolite 1-(2-pyrimidinyl)-piperazine in the rat. *Xenobiotica* **13,** 147–153.

Carlsson A., Corrodi H., Fuxe K., and Hokfelt T. (1969) Effects of some antidepressant drugs on the depletion of intraneuronal brain catecholamine stores caused by 4,α-dimethyl-meta-tyramine. *Eur. J. Pharmacol.* **5,** 367–373.

Carlsson A., Fuxe K., Hamberger B., and Lindqvist M. (1966) Biochemical and histochemical studies on the effects of imipramine-like drugs and (+)-amphetamine on central and peripheral catecholamine neurons. *Acta Physiol. Scand.* **67,** 481–497.

Ceci A., Garattini S., Gobbi M., and Mennini T. (1986) Effect of long term amineptine treatment on pre- and postsynaptic mechanisms in rat brain. *Br. J. Pharmacol.* **88,** 269–275.

Cho A. K., Hodshon B. J., Lindeke B., and Miwa G. T. (1973) Application of quantitative GC-mass spectrometry to study of pharmacokinetics of amphetamine and phentermine. *J. Pharm. Sci.* **62,** 1491–1494.

Cimino M., Ponzio F., Achilli G., Vantini G., Perego C., Algeri S., and Garattini S. (1983) Dopaminergic effects of buspirone, a novel anxiolytic agent. *Biochem. Pharmacol.* **32,** 1069–1074.

Consolo S. and Garattini S. (1969) Effect of desipramine on intestinal absorption of phenylbutazone and other drugs. *Eur. J. Pharmacol.* **6,** 322–326.

Consolo S., Dolfini E., Garattini S., and Valzelli L. (1967) Desipramine and amphetamine metabolism. *J. Pharm. Pharmacol.* **19,** 253–256.

Costall B. and Naylor R. J. (1976) A comparison of the abilities of typical neuroleptic agents and of thioridazine, clozapine, sulpiride and metoclopramide to antagonise the hyperactivity induced by dopamine applied intracerebrally to areas of the extrapyramidal and mesolimbic systems. *Eur. J. Pharmacol.* **40,** 9–19.

Costall B., Fortune D. H., and Naylor R. J. (1978) Differential activities of some benzamide derivatives on peripheral and intracerebral administration. *J. Pharm. Pharmacol.* **30,** 796–797.

De Simoni M. G., Dal Toso G., Algeri S., and Ponzio F (1986) Differences in the effect of the antidepressant amineptine on striatal and limbic DOPAC measured by HPLC-ECD and in vivo voltammetry. *Eur. J. Pharmacol.* 123, 433–439.

Dingell J. V. and Bass A. D. (1969) Inhibition of the hepatic metabolism of amphetamine by despriamine. *Biochem. Pharmacol.* **18,** 1535–1538.

Dolfini E., Garattini S., and Valzelli L. (1969a) Different sensitivity to amphetamine of three strains of mice. *Eur. J. Pharmacol.* **7,** 220–223.

Dolfini E., Garattini S., and Valzelli L. (1969b) Activity of (+)-amphetamine at different environmental temperatures in three strains of mice. *J. Pharm. Pharmacol.* **21,** 871–872.

Dolfini E., Tansella M., Valzelli L., and Garattini S. (1969c) Further studies on the interaction between desipramine and amphetamine. *Eur. J. Pharmacol.* **5,** 185–190.

Dring L. G., Smith R. L., and Williams R. T. (1966) The fate of amphetamine in man and other mammals. *J. Pharm. Pharmacol.* **18,** 402–404.

Fang V. S., Zimo D. A., and Byyny R. (1977) Pituitary response to metoclopramide in the rat. *J. Endocrinol.* **74,** 155–156.

Farnebo L-O. (1971) Effect of d-amphetamine on spontaneous and stimulation-induced release of catecholamines. *Acta Physiol. Scand.* suppl. **371,** 45–52.

Frigerio A. and Pantarotto C. (1977) Quantitative determination of sulpyrid in biological samples from rats by gas-liquid chromatography and chemical ionization-mass fragmentography. *J. Chromatogr.* **130,** 361–367.

Garattini S. (1969) Importance of a Knowledge of Drug Metabolism for the Assessment of Drug Interactions, in *Importance of Fundamental Principles in Drug Evaluation* (Tedeschi D. H. and Tedeschi R. E., eds.) Raven, New York, pp. 129–139.

Garattini S. (1979) Evaluation of the Neurotoxic Effects of Glutamic Acid. In *Nutrition and the Brain* vol. 4 (Wurtman R. J. and Wurtman J. J., eds.) Raven, New York, pp. 79–124.

Garattini S., Borroni E., Mennini T., and Samanin R. (1978) Differences and Similarities Among Anorectic Agents, in *Central Mechanisms of Anorectic Drugs* (Garattini S. and Samanin R., eds.) Raven, New York, pp. 127–143.

Garattini S., Caccia S., and Mennini T. (1982) Notes on buspirone's mechanisms of action. *J. Clin. Psychiat.* **43** (12 sec. 2), 19–22.

Garattini S., Caccia S., Mennini T., Samanin R., Consolo S., and Ladinsky H. (1979) Biochemical pharmacology of the anorectic drug fenfluramine: A review. *Curr. Med. Res. Opin.* **6** (suppl. 1), 15–27.

Garattini S., Jori A., and Samanin R. (1976) Interactions of various drugs with amphetamine. *Ann. NY Acad. Sci.* **281,** 409–425.

Garattini S., Mennini T., and Ponzio F. (1984) Neurochemical studies on amineptine. Presented at Dopamina e Depressione, Meeting Roma, October 11, 1984.

Geller I., Hartmann R. J., Croy D. J., and Haber B. (1974) Attenuation of conflict behavior with cinanserin, a serotonin antagonist: Reversal of the effect with 5-hydroxytryptophan and α-methyltryptamine. *Res. Commun. Chem. Pathol. Pharmacol.* **7,** 165–174.

Glaser T. and Traber J. (1983) Buspirone: action on serotonin receptors in calf hippocampus. *Eur. J. Pharmacol.* **88,** 137, 138.

Glaser T. and Traber J. (1985) Binding of the putative anxiolytic TVX Q 7821 to hippocampal 5-hydroxytryptamine (5-HT) recognition sites. *Naunyn Schmiedebergs Arch. Pharmacol.* **329,** 211–215.

Hardcastle P. T., Newey H., and Smyth D. H. (1968) Factors influencing the effect of glucose on intestinal transport of amino acids. *J. Physiol.* **196,** 33P–34P.

Heywood R. and Worden A. N. (1979) Glutamate Toxicity in Laboratory Animals, in Glutamic Acid Advances in Biochemistry and Physiology (Filer L. J., Jr., Garattini S., Kare M. R., Reynolds W. A., and Wurtman R. J. eds.) Raven, New York, pp. 203–215.

Horowski R. and Graf K.-J. (1976) Influence of dopaminergic agonists and antagonists on serum prolactin concentrations in the rat. *Neuroendocrinology* **22,** 273–286.

Invernizzi R., Berettera C., Garattini S., and Samanin R. (1986) d- And l-isomers of fenfluramine differ markedly in their interaction with brain serotonin and catecholamines in the rat. *Eur. J. Pharmacol.* **120,** 9–15.

Janssen P. A. J., Niemegeers C. J. E., Schellekens K. H. L., and Lenaerts F. M. (1967) Is it possible to predict the clinical effects of neuroleptic drugs (major tranquillizers) from animal data? *Arzneimittelforsch.* **17,** 841–854.

Jenner P., Elliott P. N. C., Clow A., Reavill C., and Marsden C. D. (1978) A comparison of in vitro and in vivo dopamine receptor antagonism produced by substituted benzamide drugs. *J. Pharm. Pharmacol.* **30,** 46–48.

Jori A., Caccia S., and De Ponte P. (1978) Differences in the availability of d- and l-enantiomers after administration of racemic amphetamine to rats. *Xenobiotica* **8,** 589–595.

Langer S. Z., Moret C., Raisman R., Dubocovich M. C., and Briley M. (1980) High-affinity [^{3}H]imipramine binding in rat hypothalamus. Association with uptake of serotonin but not of norepinephrine. *Science* **210,** 1133–1135.

Leysen J. E., Awouters F., Kennis L., Laduron P. M., Vandenberk J., and Janssen P. A. J. (1981) Receptor binding profile of R 41 468, a novel antagonist at 5-HT_2 receptors. *Life Sci.* **28,** 1015–1022.

Leysen J. E., Niemegeers C. J. E., Tollenaere J. P., and Laduron P. M. (1978) Serotonergic component of neuroleptic receptors. *Nature* **272,** 168–171.

List S. J. and Seeman P. (1981) Resolution of dopamine and serotonin receptor components of [^{3}H]spiperone binding to rat brain regions. *Proc. Natl. Acad. Sci. USA* **78,** 2620–2624.

Mennini T., Borroni E., Samanin R., and Garattini S. (1981) Evidence of the existence of two different intraneuronal pools from which pharmacological agents can release serotonin. *Neurochem. Int.* **3,** 289–294.

Mennini T., Gobbi M., and Garattini S. (1987) Effect of buspirone, MJ13805 and 1PP on different binding sites in rat brain. *Psychopharmacology,* in press.

Mennini T., Gobbi M., Ponzio F., and Garattini S. (1986a) Neurochemical effects of buspirone in rat hippocampus: Evidence for selective activation of 5HT neurons. *Arch. Int. Pharmacodyn. Ther.* **279,** 40–49.

Mennini T., Gobbi M., and Romandini S. (1986b) Localization of $GABA_A$ and $GABA_B$ receptor subtypes on serotonergic neurons. *Brain Res.* **371,** 372–375.

Moriyuki H. and Ichimura, M. (1978) Acute toxicity of monosodium L-glutamate in mice and rats. *Pharmacometrics* **15,** 433–436.

Morpurgo C. and Theobald W. (1965) Influence of imipramine-like compounds and chlorpromazine on the reserpine-hypothermia in mice and the amphetamine-hyperthermia in rats. *Med. Pharmacol. Exp.* **12,** 226–232.

Olney J. W. and Ho O-L. (1970) Brain damage in infant mice following oral intake of glutamate, aspartate or cysteine. *Nature* **227,** 609–611.

Pinto-Scognamiglio W., Amorico L., and Gatti G. L. (1972) Esperienze di tossicità e di tolleranza al monosodioglutammato con un saggio di condizionamento di salvaguardia. *Farmaco* (Prat.) **27,** 19–27.

Poignant J-C. (1979) Pharmacologie du survector *Psychol. Méd.* **11A,** 29–48.

Prieto J., Moragues J., Spickett R. G., Vega A., Colombo M., Salazar W., and Roberts D. J. (1977) Synthesis and pharmacological properties of a series of antidopaminergic piperidyl benzamides *J. Pharm. Pharmacol.* **29,** 147–152.

Quattrone A., Bendotti C., Recchia M., and Samanin R. (1977) Various effects of d-amphetamine in rats with selective lesions of brain noradrenaline-containing neurons or treated with penfluridol. *Commun. Psychopharmacol.* **1,** 525–531.

Quik M., Iversen L. L., Larder A., and Mackay A. V. P. (1978) Use of ADTN to define specific ^{3}H-spiperone binding to receptors in brain. *Nature* **274,** 513–514.

Rastogi R. B., Agarwal R. A., Lapierre Y. D., and Singhal R. L. (1977) Effects of acute diazepam and clobazam on spontaneous locomotor activity and central amine metabolism in rats. *Eur. J. Pharmacol.* **43,** 91–98.

Reid W. D. (1970) Turnover rate of brain 5-hydroxytryptamine increased by d-amphetamine. *Br. J. Pharmacol.* **40,** 483–491.

Riblet L. A., Allen L. E., Hyslop D. K., Taylor D. P., and Wilderman R. C. (1980) Pharmacologic activity of buspirone, a novel non-benzodiazepine antianxiety agent. *Fed. Proc.* **39,** 752.

Riblet L. A., Taylor D. P., Eison M. S., and Stanton H. C. (1982) Pharmacology and neurochemistry of buspirone. *J. Clin. Psychiatry* **43** (12 sec. 2), 11–16.

Robichaud R. C. and Sledge K. L. (1969) The effects of p-chlorophenylalanine on experimentally induced conflict in the rat. *Life Sci.* **8, pt. I,** 965–969.

Rosier Y. (1979) L'emploi de l'amineptine dans le traitement de la dépression. *Psychol. Méd.* **11,** 211–223.

Samanin R., Jori A., Bernasconi S., Morpurgo E., and Garattini S. (1977) Biochemical and pharmacological studies on amineptine (S 1694) and (+)-amphetamine in the rat. *J. Pharm. Pharmacol.* **29,** 555–558.

Sanghera M. K. and German D. C. (1983) The effects of benzodiazepine and non-benzodiazepine anxiolytics on locus coeruleus unit activity. *J. Neural Transm.* **57,** 267–279.

Seeman P. (1981) Brain dopamine receptors. *Pharmacol. Rev.* **32,** 229–313.

Seidel W. F., Cohen S. A., Bliwise N. G., and Dement W. C. (1985) Buspirone: An anxiolytic without sedative effect. *Psychopharmacology* **87,** 371–373.

Severson J. A. (1986) ^{3}H-imipramine binding in aged mouse brain: Regulation by ions and serotonin. *Neurobiol. Aging* **7,** 83–87.

Severson J. A., Woodward J. J., and Wilcox R. E. (1986) Subdivision of mouse brain [^{3}H]imipramine binding based on ion dependence and serotonin sensitivity. *J. Neurochem.* **46,** 1743–1754.

Snyder S. H. (1984) Drug and neurotransmitter receptors in the brain. *Science* **224,** 22–31.

Spyraki C. and Fibiger H. C. (1981) Behavioural evidence for supersensitivity of postsynaptic dopamine receptors in the mesolimbic system after chronic administration of desipramine. *Eur. J. Pharmacol.* **74,** 195–206.

Stegink L. D., Filer L. J., Jr., and Baker G. L. (1972) Monosodium glutamate: Effect on plasma and breast milk amino acid levels in lactating women. *Proc. Soc. Exp. Biol. Med.* **140,** 836–841.

Stein L., Wise C. D., and Berger B. D. (1973) Antianxiety Action of Benzodiazepines: Decrease in Activity of Serotonin Neurons in the

Punishment System, in *The Benzodiazepines* (Garattini S., Mussini E., and Randall L. O., eds.) Raven, New York, pp. 299–326.

Svensson T. H. (1970) The effect of inhibition of catecholamine synthesis on dexamphetamine induced central stimulation. *Eur. J. Pharmacol.* **12,** 161–166.

Takasaki Y. (1978) Studies on brain lesion by administration of monosodium L-glutamate to mice. I. Brain lesions in infant mice caused by administration of monosodium L-glutamate. *Toxicology* **9,** 293–305.

Valenzuela J. E. (1976) Dopamine as a possible neurotransmitter in gastric relaxation. *Gastroenterology* **71,** 1019–1022.

Valzelli L., Consolo S., and Morpurgo C. (1967) Influence of Imipramine-Like Drugs on the Metabolism of Amphetamine, in *Antidepressant Drugs* (Garattini S. and Dukes M. N. G., eds.) Excerpta Medica, Amsterdam, pp. 61–69.

Van Rossum J. M., Van der Schoot J. B., and Hurkmanns J. A. Th. (1962) Mechanism of action of cocaine and amphetamine in the brain. *Experientia* **18,** 229–231.

Weaver L. C. and Kerley T. L. (1962) Strain difference in response of mice to d-amphetamine. *J. Pharmacol. Exp. Ther.* **135,** 240–244.

Weissman B. A., Barrett J. E., Brady L. S., Witkin J. M., Mendelson W. B., Paul S. M., and Skolnick P. (1984) Behavioral and neurochemical studies on the anticonflict actions of buspirone. *Drug Dev. Res.* **4,** 83–93.

Wood P. L., Nair N. P. V., Lal S., and Etienne P. (1983) Buspirone: A potential atypical neuroleptic. *Life Sci.* **33,** 269–273.

Determination of Monoamine Turnover by Blockade of Their Synthesis or Metabolism

Maria Hadjiconstantinou and Norton H. Neff

1. Introduction

The neurotransmitter amines, dopamine (DA), norepinephrine (NE), epinephrine (EPI), and serotonin (5-hydroxytryptamine, 5-HT) are continuously released and destroyed at various rates, yet their content in neuronal tissue remains relatively constant. Thus, it is not possible to evaluate neuronal function solely from the content of the transmitter in a tissue. A better estimate of neuron function can be obtained from the rate of formation of the amines. The hypothesis is that there is increased amine formation and turnover when neurons are actively firing. Presently, there are two general techniques for estimating the rate of amine formation (*see* reviews by Costa and Neff, 1970: Neff, 1972; Costa et al., 1975, and Korf 1985). There are the steady-state methods where an isotope-labeled amine (radioactive or stable isotope) or a precursor is administered, and following uptake into nerve endings, the decline of the specific radioactivity of the transmitter is followed with time. The rate of loss of radioactive amine under steady-state conditions is proportional to its turnover. The higher the neuronal firing rate, the faster the rate of decline of the specific radioactivity of the transmitter amine.

Another, and perhaps simpler, procedure for estimating the rate of formation of amine transmitter is to block one of the enzymatic events required for their formation or metabolism with specific pharmacologic agents, and measure either the accumulation of a substrate or the decline of the product of the enzymatic event. For example, following the administration of a monoamine oxidase inhibitor, 5-HT increases with time, whereas 5-hydroxyindoleacetic acid (5-HIAA) decreases with time (Neff and Tozer, 1968). The rate of conversion of 5-HT to 5-HIAA can be calculated from the slopes of the curves. A faster accumulation of an intermediate reactant or decline of a product indicates a more

Table 1[a]
Approximate Turnover Rates of the Biogenic Amines in Selected Regions of Brain

Neurotransmitter	Range, nmol/g tissue/h
	Striatum
DA	10–30
5-HT	1–2
	Hypothalamic nuclei
NE	1–10
EPI	0.025–0.050
DA	1–3

[a]This table lists ranges of values that the authors have determined over the years for rats using the perturbation techniques. A more complete list of brain regions together with a comparison with results of the steady-state procedure can be found in the article by Korf (1985).

rapid rate of formation of the amine. This article will focus on the perturbation or nonsteady-state procedures for estimating transmitter turnover. The reader is referred to a recent article by Korf (1985) in this series for a comparison of the turnover rates found for different regions of brain using both steady-state and nonsteady-state methods. In Table 1, we present a general range of turnover values for rat striatum and hypothalamus that we have determined in our laboratory over the years using different strains of animals.

The perturbation procedures evolved about three decades ago, when it was realized that the content of the amines in tissues changed after administering drugs. In 1958, Udenfriend and Weissbach estimated 5-HT formation from its rise in tissue after administering a monoamine oxidase inhibitor. There are now potent drugs available to block all of the major enzymatic events required for the formation, catabolism, and transport of acidic metabolites out of the brain for the catecholamines and 5-HT.

The term turnover refers to renewal of a substance (Zilversmith et al., 1943). Turnover rate implies renewal during steady-state conditions, i.e, influx and efflux of the substance are equal, and the tissue content of the substance does not change during the measurement procedure. With the nonsteady-state procedures, homeostasis is disrupted, and tissue substrate content changes. It is assumed that neuronal events continue normally after blockade of the enzymatic events, even though intermediates increase and products decrease. This may not be true for many of the biogenic amines, as it has been shown that they do not accumulate in all tissues after inhibition of monoamine oxidase, most likely because of local and neuronal feedback systems that come into play when transmitter content changes (Carlsson and Lindqvist 1963; Carlsson et al., 1976).

Kinetic models of physiological systems are difficult to formulate because of the following (Zilversmith et al., 1943):

1. The systems are not fully understood
2. The sites for measurement are restricted
3. The times when measurements can be made are restricted.
4. The data contain fluctuations
5. The number of studies that can be performed on a single subject is limited and
6. The results obtained using different procedures may be related in a complicated manner.

Thus, the best approach is to choose the simplest model, test the model, and modify it as needed to fit the data and existing hypotheses. Moreover, deviations from a simple model are easy to recognize.

2. Blockade of the Rate-Limiting Initial Enzymatic Step for the Synthesis of the Biogenic Amines with Drugs

The procedures we will describe are based on an open single-compartmemt model. The rate-limiting event for catecholamine synthesis is the hydroxylation of tyrosine to form 3,4-dihydroxyphenylalanine (DOPA) (Levitt et al., 1965). For 5-HT, the hydroxylation of tryptophan to form 5-hydroxytryptophan (5-HTP) is apparently rate-limiting (Lovenberg et al., 1967), although the enzyme does not appear to be saturated with tryptophan. To perform these studies, drugs that block tyrosine hydroxylase (alpha-methyltyrosine methyl ester, 200 mg/kg) (Brodie et al., 1966) or tryptophan hydroxylase (*p*-chlorophenylalanine, 300 mg/kg) (Koe and Weissman, 1966) are administered to the animals. The decline of the amines is then followed with time (Fig.

1). The tissue contents of the hydroxylated amino acids are low and barely detectable by most analytical procedures. Therefore, they are not considered to contribute significantly to the pool of amines after blocking the hydroxylating enzymes. The validity of this procedure is based on the following assumptions: (1) The drugs completely and rapidly inactivate the hydroxylating enzymes; and (2) The drugs do not release the biogenic amines or change their catabolic pathways. These assumptions are not easy to evaluate under normal situations and are even more difficult to evaluate after administering experimental drugs. The same general assumptions are made when drugs are used to block any pathway to estimate turnover rates.

Biogenic amines decline exponentially with time (Fig. 1) following the administration of a blocking drug. Traditionally, the data have been plotted on semilogarithmic graph paper (Brodie et al., 1966). The data imply that the rate of decline of the biogenic amine is proportional to the biogenic amine content remaining in the brain sample. The Eq. (1) that describes Fig. 1 is

$$\log (A)_t = \log (A)_0 - Kt \tag{1}$$

where $(A)_0$ is the tissue content of the amine before administering the blocking drug, which should be the extrapolated zero time value for the amine, $(A)_t$ is the level of the amine at any time t after the administration of the blocking drug, and K is 0.434 times the rate constant for decline, k. The turnover rate is calculated by multiplying k by the steady-state content of the amine. The steady-state tissue content of the amine is usually based on wet tissue weight or mg protein. There could, therefore, be considerable differences between rate calculations from laboratory to laboratory depending on the precision and accuracy of dissecting brain regions (*see* Table 1). For example, the inclusion of more or less tissue that does not contain the biogenic amines could make a large difference in the calculated rates. The value of k can be determined statistically or by drawing the best-fit line through the data points and estimating the half-life for decline ($t_{1/2}$) and using the relationship $k = 0.693/t_{1/2}$. Another estimate of turnover is the reciprocal of k, the turnover time (T_t). The T_t is the time required to synthesize the quantity of substance present in the tissue. The quantity of tissue assayed does not enter into this calculation.

A typical turnover study for the catecholamines is performed by administering alpha-methyltyrosine (or the methyl ester), 200 mg/kg ip, to a group of 20 rats, after which groups of five animals

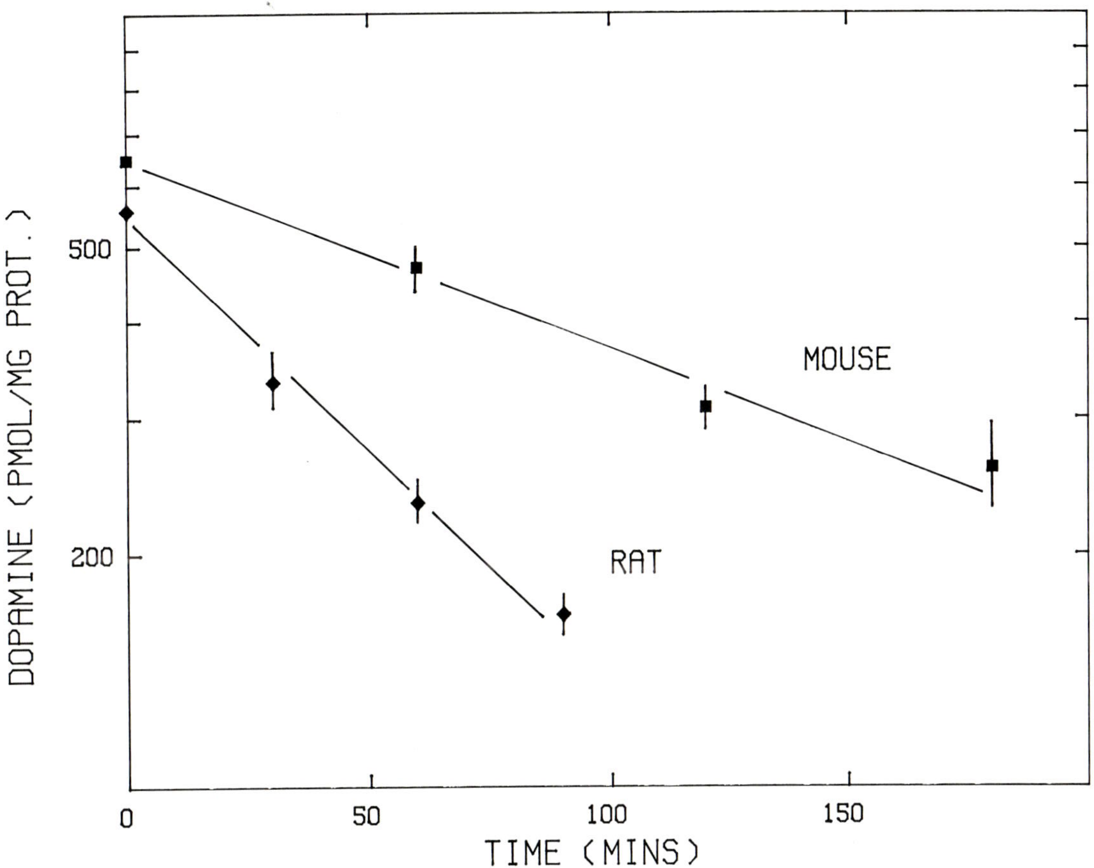

Fig. 1. Decline of dopamine from the rat or mouse striatum after administering alpha-methyltyrosine (200 mg/Kg, ip) at time zero. The. calculated rate of turnover is about 3.5 pmol/mg prot./min for the mouse and about 8 pmol/mg prot./min for the rat.

each are euthanized at selected times over several hours after the injection (Fig. 1). The times for euthanasia depends on the anticipated rate of decline of the amine.

For 5-HT turnover in the brain, either *p*-chlorophenylalanine or alpha-propyldopacetamide administration results in the decline of 5-HT (Koe and Weismann, 1966; Corrodi and Fuxe, 1968). The rates calculated by this procedure are lower than by other procedures. The reason for the discrepancy might be that the enzyme is not immediately and completely inhibited by the drugs.

The availability of amino acid substrates, tyrosine and tryptophan, can influence the rate of formation of the biogenic amines (Tagliamonte et al., 1971; Korf et al., 1972; Wurtman et al., 1981). Under normal conditions, tryptophan hydroxylase is not saturated, and injection of tryptophan results in elevated brain 5-HT content. Tryptophan is bound to plasma proteins, and drugs that disrupt the binding can alter brain 5-HT content independently of the turnover rate (Tagliamonte et al., 1971). Wurtman and collaborators (1981) have reported that tyrosine may become rate-limiting when neurons are rapidly firing; thus, the administration of tyrosine can enhance catecholamine turnover in actively firing neurons. Molecular oxygen is also a substrate for the hydroxylating enzymes, and any procedure or drug that interferes with the availability of oxygen could alter neurotransmitter turnover rates (Diaz et al., 1968; Carlsson et al., 1977).

There are a number of other factors that must be considered when designing turnover studies as well. Environmental temperature influences turnover rates. There is an increase of NE turnover in the brain of cold-exposed animals that is probably related to increased neuronal firing rates (Reid et al., 1968). During periods of prolonged anesthesia, body temperature drops and catecholamine turnover decreases (Eakins et al., 1968). If the body temperature is maintained, so is turnover. Some drugs might affect turnover because they change body temperature. Stress of all types alters amine turnover (Thierry et al., 1968a,b; Corrodi et al., 1968). It is necessary to have a uniform colony of animals before starting a study. Neurotransmitter turnover rates vary in Sprague-Dawley rats obtained from different suppliers, and there is a difference in the steady-state biogenic amine content (Brodie et al., 1966). In addition, the content of amine oscillates during the day (Manshardt and Wurtman, 1968) and throughout the year (Valzelli and Garattini, 1968). Furthermore, housing animals singly or in groups affects behavior and the rate of formation of the biogenic amines (Vazelli and Garattini, 1968).

As mentioned previously, the decline of amine appears to be exponential after blockade of hydroxylation. Thus, some investigators do not construct a complete decline curve, but pick a single time point after blocking hydroxylation to euthanize animals for analysis. The amine content is then compared with an appropriate control group. This design can serve as a rapid screening procedure for a series of potential psychoactive drugs. There is the caution, however, that a drug could change the decline characteristics and result in misinterpretation of results from a single time point.

Another procedure for estimating NE and EPI turnover that is not commonly used, consists of administering the DA beta-hydroxylase inhibitor FLA-63 (bis[4-methyl-1-homopiperazinylthiocarbonyl] disulfide) and following the decline of the amines (Svensson and Waldeck, 1969). EPI turnover has been estimated from its fall after administering the phenylethanolamine-*N*-methyltransferase inhibitor 2,3-dichloro-alpha-methylbenzylamine (Fuller et al., 1981).

3. Blockade of Aromatic L-Amino Acid Decarboxylase

The second enzyme in the sequence leading to the production of the amine transmitters is aromatic L-amino acid decarboxylase. Blockade of this enzyme with drugs [NSD-1015 (3-hydroxybenzylhydrazine) or Ro4-4602 (benserazide)] results in a rise of the hydroxylated amino acid intermediates 5-HTP and DOPA (Carlsson et al., 1972). The rate of rise is directly related to the rate of formation of the amines. This is a rather popular procedure of estimating turnover.

4. Blockade of Monoamine Oxidase

Inhibition of monoamine oxidase with pargyline results in a rise of 5-HT, DA, and 3-methoxytyramine (3-MT), and a fall of 5-HIAA, 3,4-dihydroxyphenlacetic acid (DOPAC), and homovanillic acid (HVA). The rise of the amines, primarily 5-HT, has been used as a measure of their rates of formation (Neff and

Tozer, 1968). It must be remembered, however, that many nerve endings have inhibitory autoreceptors that could potentially suppress the release and synthesis of the amines as their concentration rises (Kehr et al., 1975; Bunney et al, 1973). The methylated metabolite of DA, 3-MT, rises in the brain after inhibition of monoamine oxidase (DiGiulio et al., 1978), but some investigators doubt that it is a major metabolite under normal circumstances.

5-HIAA (Neff and Tozer, 1968) and DOPAC (Spano and Neff, 1971) content decline exponentially with time after blocking monoamine oxidase. The data can be fitted to Eq. (1) and the rate constant for decline, *k*, determined as already discussed. The rate of loss of the acid metabolites can be determined by multiplying *k* by the steady-state tissue content of the acid metabolite. As the systems are assumed to be in equilibrium, the rate of loss of the acids should be equal to their rate of formation. There are reports that the decline of 5-HIAA is not exponential in all situations, and that the actual calculated values for decline may not therefore be meaningful (Morot-Gaudry et al., 1974).

The efflux of the acid metabolites from the brain is probably dependent on diffusion to a site where transport of acids takes place. The choroid plexus transports materials from the cerebrospinal fluid to the blood (Sampath and Neff 1974; 1976). There are probably sites within brain tissue where transport of acidic biogenic metabolites takes place as well. Generally, the content of neurotransmitter remains relatively constant with time, but the content of the acid metabolites does not. Brain acid metabolite content appears to vary directly with the turnover of the transmitter (Neff et al., 1967). Thus, some investigators use acid metabolite content as a measure of transmitter turnover. When this procedure is used, it is important to determine if a test compound under study directly blocks acid transport from the brain.

5. Blockade of the Transport System that Rids the Brain of Acidic Neurotransmitter Metabolites

Probenecid blocks the transport of many endogenous acidic compounds from the brain (Neff et al., 1967) and cerebrospinal fluid (Sampath and Neff 1974; 1976). When transport is blocked, the acidic compounds increase, and the increase is directly related to the turnover of the respective neurotransmitter (Neff et al., 1967). This procedure has been used to estimate the turnover of

5-HT (5-HIAA) (Neff et al., 1967), DA (DOPAC and HVA) (Spano and Neff, 1971), and NE [3-methoxy-4-hydroxyphenylethyleneglycol (MOPEG) and 3-4,dihydroxyphenylethyleneglycol (DOPEG) and their sulfate conjugate] (Meek and Neff, 1973) in regions of the brain from experimental animals and in spinal fluid from humans (Olsson and Roos, 1968).

6. Conclusions

What is the best method to measure neurotransmitter turnover? In our opinion, the best method is the simplest method that gives the most information with the least expense and effort, and with minimal use of animals. The perturbation methods, in most cases, meet these criteria. By using the procedures described in sections 3–5 above, both 5-HT and catecholamine turnover can be studied in a single sample. Tissue can be assayed for both transmitter amines and metabolites in the same samples by HPLC with an electrochemical detector. The steady-state methods require that a radioactive or stable isotope be administered to animals, and that expensive equipment or complicated extraction procedures be used to assay the tissue. In some studies, however, the steady-state procedures are the methods of choice depending on the information needed to test a hypothesis.

References

Brodie, B. B., Costa E., Dlabac A., Neff, N. H., and Smookler H. H. (1966) Application of steady-state kinetics to the estimation of synthesis rate and turnover time of tissue catecholamines. *J. Pharmacol. Exp. Ther.* **154,** 493–498.

Bunney G.S., Walters H. R., Roth R. H., and Aghajanian G. K. (1973) Dopaminergic neurons: Effect of antipsychotic drugs and amphetamine on single cell activity. *J. Pharmacol. Exp. Ther.* **185,** 560–571.

Carlsson A. and Lindquist M. (1963) Effect of chlorpromazine or Haloperidol on formation of 3-methoxytyramine and normetanephrine in mouse brain *Acta. Pharmacol. Toxicol.* **20,** 140–144.

Carlsson A., Davis J. N., Kehr W, Lindqvist M., and Atack C. V. A. (1972) Simultaneuous measurement of tyrosine and tryptophan hydroxylation activities in brain *in vivo* using an inhibitor of the aromatic amino acid decarboxylase. *Nauyn-Schmiedeberg's Arch. Pharmacol.* **275,** 153–168.

Carlsson A., Holmin T., Lindqvist M., and Siesjo B. K. (1977) Effect of hypercapnia and hypocapnia on tryptophan and tyrosine hydroxylation in rat brain. *Acta Physiol. Scand.* **99,** 503–509.

Carlsson A., Kehr W., and Lindqvist M. (1976) The role of intraneuronal amine levels in the feedback control of dopamine, noradrenaline and 5-hydroxytryptamine synthesis in rat brain. *J. Neural. Transm.* **39,** 1–19.

Corrodi H., Fuxe K., and Hokfelt, T. (1968) The effect of immobilization stress on the activity of central monoamine neurons. *Life Sci.* **7,** 107–111.

Costa E., Koslow S. H., and LeFevre H. (1975) Mass fragmentgraphy: A tool for studying transmitter function at the synaptic level, in *Handbook of Psychopharmacology* (Iversen L. L., Iversen S. D. and Snyder S. H., eds). Vol. 1, pp. 1–24. Plenum, New York.

Costa E. and Neff N. H. (1970) Estimation of turnover rates to study the metabolic regulation of the steady-state level of neuronal monoamines, in *Handbook of Neurochemistry,* (Lajtha, A., ed.) Vol. 4, pp. 45–90. Plenum, New York.

Diaz P. M., Ngai S. H., and Costa E. (1968) Effect of oxygen on brain serotonin metabolism in rats. *Am. J.Physiol.* **214,** 591–595.

DiGiulio A. M., Gropetti A., Cattabeni F., Galli C. L., Maggi A., Algeri S., and Ponzio F. (1978) Significance of dopamine metabolites in the evolution of drugs acting on dopaminergic neurons. *Eur. J. Pharmacol.* **52,** 201–207.

Eakins K. E., Costa E., Katy R. L., and Reyes C. L. (1968) Effect of pentobarbital anesthesia on the turnover of ^{3}H-noradrenoradrenaline in peripheral tissues of cats. *Life Sci.,* **7,** 71–76.

Fuller R. W., Perry K. W., Hemrick S. K., and Molloy, B. B. (1981) Lowering of brain epinephrine by inhibition of norepinephrine *N*-methyltransferase in rats, in catecholamines, in *Basic and Clinical Frontiers* (Usdin E., Kopin I. J., and Barchas J., eds.) pp. 186–188. Pergamon, New York.

Kehr W., Carlsson A., and Lindqvist N. (1975) Biochemical aspects of dopamine agonists, in *Advances in Neurology* (Colne T., Chase N., and Barbeau A, eds.) Vol. 9, pp. 185–195. Raven, New York.

Koe B. K. and Weissman A. (1966) *P*-chlorophenylalanine: A specific depletor of brain serotonin. *J. Pharmacol. Exp. Ther.* **154,** 499–516.

Korf J. (1985) Turnover rate assessments of cerebral neurotransmitter amines and acetylcholine, in *Neuromethods 2. Amines and Their Metabolites* (Boulton A. A., Baker G. B., and Baker J. M., eds.) pp. 407–456. Humana, Clifton, New Jersey.

Korf J., van Praag H. M., and Sebens J. B. (1972) Serum tryptophan decreased tryptophan increased and brain serotonin synthesis unchanged after probenecid loading. *Brain Res.* **42,** 239–242.

Levitt M., Spector S., Sjoerdsma A, and Undenfriend S. (1965) Elucidation of the rate-limiting step in norepinephrine biosynthesis in the perfused guinea pig heart. *J. Pharmacol. Exp. Ther.* **148,** 1–6.

Lovenberg W., Jequier E., and Sjoerdsma A. (1967) Tryptophan hydroxylation: Measurements in pineal gland, brain stem and carcinoid tumor. *Science* **155,** 217–219.

Manshardt J. and Wurtman R. J. (1968) Daily rhythm in the noradrenaline content of rat hypothalamus. *Nature* **217,** 574–517.

Meek J. L. and Neff N. H. (1973) The rate of formation of 3-methoxy,4-hydroxyphenylethyleneglycol sulfate in brain as an estimate of the rate of formation of norepinephrine. *J. Pharmacol. Exp. Ther.* **184,** 570–575.

Morot-Gaudry Y., Hamon M., Bourgoin S., Ley J. P., and Glowinski, J. (1974) Estimation of the rate of 5-HT synthesis in the mouse brain by various methods. *Naunyn-Schmiedeberg's Arch. Pharmacol.* **282,** 223–238.

Neff N. H. (1972) Methods for estimating the *in vivo* rate of formation of norepinephrine dopamine and serotonin in nerve tissue, in *The Thyroid and Biogenic Amines* (Rall T. and Kopin I. J., eds.) pp. 604–640. North-Holland, New York.

Neff N. H. and Tozer T. N. (1968) In vivo measurement of brain serotonin turnover. *Adv. Pharmacol.* **619,** 97–109.

Neff N. H., Tozer T. N., and Brodie B. B. (1967) Application of steady-state kinetics to studies of the transfer of 5-hydroxyindoleacetic acid from brain to plasma. *J. Pharmacol. Exp. Ther.* **158,** 214–218.

Olsson R. and Roos B. E. (1968) Concentrations and 5-hydroxyindoleacetic acid and homovanillic acid in the cerebrospinal fluid after treatment with probenecid in patients with Parkinson's disease. *Nature* **219,** 502–506.

Reid W. D., Volicer L., Smookler H., Beaver M. A., and Brodie B. B. (1968) Brain amines and temperature regulation. *Pharmacology* **1,** 329–234.

Sampath S. S. and Neff N. H. (1974) The elimination of 5-hydroxyindoleacetic acid from cerebrospinal fluid: characteristics of the acid transport system of the chloroid plexus. J. Pharmacol. Exp. Ther. **181,** 410–414.

Sampath S. S. and Neff N. H. (1976) Transport of 3-methoxy-4-hydroxymandelic acid by isolated rat choroid plexus. *Neurochem. Res.* **1,** 47–52.

Spano P. F and Neff N. H. (1971) A procedure for the simultaneous determination of dopamine, 3-methoxy-4-hydroxyphenylacetic acid and 3,4-dihydroxyphenylacetic acid in brain. *Anal. Biochem.* **42,** 113–118.

Svensson T. H. and Waldeck B. (1969) On the significance of central noradrenaline for motor activity: Experiments with a new dopamine β-hydroxylase inhibitor. *Eur. J. Pharmacol.* **7,** 278–282.

Tagliamonte A., Tagliamonte P., Perez-Cruet J., and Gessa G. L. (1971) Increase of brain tryptophan caused by drugs which stimulate serotonin synthesis. *Nature New Biol.* **229,** 125–126.

Thierry A. M., Fekete M., and Glowinski J. (1968a) Effects of stress on the metabolism of noradrenaline, dopamine and serotonin (5-HT) in, *The Central Nervous System of the Rat II: Modification of Serotonin Metabolism. Eur. J. Pharmacol.* **4,** 384–389.

Thierry A. M., Javoy F, Glowinski J., and Kety S. S. (1968b) Effects of stress on the metbolism of norepinephrine, dopamine and serotonin, in the central nervous system of the rat I: Modification of norepinephrine turnover. *J. Pharmacol. Exp. Ther.* **163,** 163–171.

Undenfriend S. and Weissbach H. (1958) Turnover of 5-hydroxytryptamine (serotonin) in tissues. *Proc. Soc. Exp. Biol. Med.* **967,** 748–751.

Valzelli L. and Garattini S. (1968) Behavioral changes and 5-hydroxytryptamine turnover in animals. *Adv. Pharmacol.* **6B,** 249–260.

Wurtman R. J., Hefti F., and Melamed E. (1981) Precursor control of neurotransmitter synthesis. *Pharmacol. Rev.* **32,** 315–335.

Zilversmith D. B., Enterman C., and Fishler M. C. (1943) On the calculation of "turnover time" and "turnover rate" from experiments involving the use of labeling agents. *J. Gen. Physiol.* **26,** 325–331.

Dopamine Receptors and Neuroleptic Drugs

Pierre M. Laduron

1. Introduction

What we know today about dopamine (DA) receptors stems directly from the discovery of potent and selective neuroleptic drugs. These compounds are not only essential tools in neuropharmacology, but are also guides that neurobiologists follow to identify, isolate, and even dismantle the molecular structure of brain neuroreceptors. For many people, it is not clear why potent neurotransmitter antagonists are an absolute requirement for these studies when the endogenous neurotransmitters are available. In fact pharmacologists have known for a long time that "classical" neurotransmitters [acetylcholine (ACh), dopamine (DA), noradrenaline (NA), 5-hydroxytryptamine (5-HT), gamma-aminobutyric acid (GABA), and so on] elicit a physiological response at micromolar concentrations (Table 1). Since binding studies are not feasible at such high concentrations because of high nonspecific binding, one needs substances that are endowed with high-affinity properties. Potent neuroleptics fulfil these conditions: they possess high affinity at nanomolar concentrations. In fact, the problem is not new; α-bungarotoxin was an absolute requirement for looking at the molecular structure of the nicotinic receptor (Changeux, 1981).

All the receptors are not identical. For instance, hormone or neuropeptide receptors are different from receptors for "classical" neurotransmitters. Table 1 shows some differences in terms of affinity for endogenous ligands and ligands used in binding studies. Hormones and neuropeptides display high-affinity properties and may, therefore, be used as such in binding studies. Classical neurotransmitters possess too low an affinity to become appropriate ligands in binding studies. These receptors require drugs or toxins for their identification through in vitro binding assays. Such a requirement is still more imperative for ions that have only millimolar affinity for the channels (Table 1). Here, again, only

Table 1
Receptor Affinity Toward Endogenous Ligands

	Receptor affinity	Binding ligand	Tissue concentration
Hormone	High, n*M*	Hormone	Low, ng/g
Neuropeptide	High, n*M*	Neuropeptide	Low, ng/g
"Classical" neurotransmitter (ACh, DA, 5-HT, NA, . . .)	Low, μ*M*	Drug (toxin)	High, μg/g
Ion	Very low, m*M*	Toxin (drug)	Very high, mg/g

toxins or drugs (calcium channel blockers) can be used as ligands to identify and isolate the receptor protein of ion channels.

Therefore, in terms of binding affinity, drugs and toxins behave like hormones or neuropeptides. If "classical" neurotransmitters are known to bind weakly to their receptor sites, one may assume that the release of higher concentrations of transmitter in the synaptic cleft will counterbalance the low-affinity properties. This low affinity, which represents a major drawback in binding studies, is, however, the most efficient way to elicit physiological responses in term of seconds or milliseconds. High-affinity properties are more compatible with long-lasting effects. Consequently, the grade of affinity of receptor sites is related to the duration of the physiological response.

Why have neuroleptic drugs played a prominent role in the identification, isolation, and even purification of DA receptors? Simple because they bind with high affinity to DA receptors. Consequently these drugs are keystones in neuropharmacology and neurobiology. It is quite difficult to imagine the enormous difficulties that would be raised in exploring the dopaminergic systems without neuroleptic drugs.

The aim of the present paper is not to provide a survey of neuroleptic and DA receptors; for this the reader is referred to numerous reviews (Seeman, 1980; Creese et al., 1983; Stoof and Kebabian, 1984; Leysen, 1984). I would like to emphasize the main steps that led to the establishment of the DA receptor concept and to show the prominent role played by neuroleptic drugs in the identification, isolation, and purification of DA receptors. The field of DA receptors is certainly one that has generated numerous controversies, however; we will see that the mistakes began when researchers abandoned the use of neuroleptics for studying DA receptors.

2. The Concept of Dopamine Receptor

2.1. Historical Background

As already emphasized, neuroleptic drugs have played a prominent role in defining the DA receptor concept; they were necessary at each step in earlier pharmacological experiments, as well as in the recent binding studies. Of course, the DA receptor concept also takes advantage of certain favorable circumstances, such as the disposition in the brain of DA, which is mainly restricted to a few brain areas, principally the nigrostriatal and mesolimbic systems. This has certainly made the problem easier to solve than when studying serotonin (5-hydroxytryptamine, 5-HT) receptors. The fact that DA and its receptor are present in high amounts in the striatum has greatly facilitated all the biochemical studies devoted to this neurotransmitter.

The concept of the DA receptor appeared progressively and is essentially derived from five different approaches: (1) the anatomy of dopaminergic neurons in the brain, (2) the discovery of neuroleptic drugs, (3) the pharmacology of DA agonists and antagonists, (4) the increased DA turnover after administration of neuroleptics, and finally, (5) the biochemical identification of DA receptors.

2.1.1. Dopaminergic Neurons in Brain

In 1957, DA, which was first considered to be a precursor of NA, was discovered in the brain (Montagu, 1957; Weil-Malherbe and Bone, 1957). Two years later, the basal ganglia were recognized as the main brain areas for DA (Bertler and Rosengren, 1959). Because of its high concentration in the striatum DA was suspected to play an important role as neurotransmitter. This was definitely proved when lower concentrations of DA were found in the brains of patients with Parkinson's disease (Ehringer and Hornykiewicz, 1960) (Fig. 1). This finding led to the subsequent introduction of L-Dopa in the treatment of the Parkinson's disease. Interestingly,

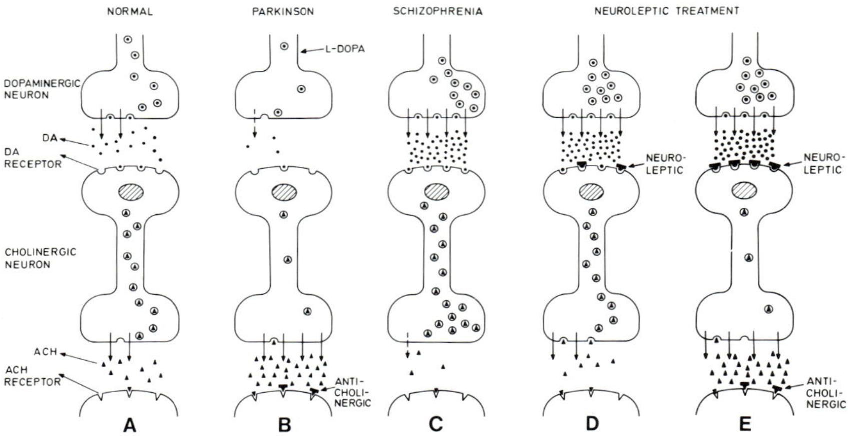

Fig. 1. Scheme illustrating dopaminergic neurotransmission in normal and pathological conditions. Dopamine receptors are presumably located on cholinergic neurons where DA elicits a negative effect on acetylcholine release (Stadler et al., 1973; Guyenet et al., 1975). As a consequence of this, a lack of DA in Parkinson's disease, leads to an increase of acetylcholine release. Neuroleptic drugs are known to block DA receptors and to produce extrapyramidal side effects (or parkinsonism), especially when the drug is given in excess or not adapted to the patient. The idea that schizophrenia could be caused by an hyperstimulation of dopaminergic systems is not yet supported by direct experimental evidence.

when psychotic patients receive neuroleptic drugs, extrapyramidal side effects (Parkinsonism) may occur, especially when the drug is given in excess or when the dose is not adapted to the patient (Fig. 1).

Our knowledge of the central DA neurons markedly progressed when the Falck-Hillarp method was applied to the visualization of DA at the cellular level. The pioneering work identified cell bodies containing DA in the substantia nigra (Dahlstrom and Fuxe, 1964); they were found to project to the striatum and the mesolimbic forebrain (Bjorklund and Lindvall, 1984). The subsequent identification of DA neurons in the cortex represented an important step in the identification of these neurons in the brain (Thierry et al., 1973) because it is quite difficult to imagine that schizophrenia is purely a "striatal" disease; upper brain areas, namely the cortex, should be involved in the etiology of this psychotic disorder. Nevertheless, because of its high DA content, the caudate nucleus remains the brain area "par excellence" in which to study the dopaminergic system.

2.1.2. Discovery of Neuroleptic Drugs

Chlorpromazine, a phenothiazine derivative related to the antihistamine drug promethazine, was first used in anesthesia before becoming the first neuroleptic drug. In 1952, Delay et al. reported the first clinical observations on the antipsychotic effects of chlorpromazine. In 1957, Paul Janssen discovered haloperidol and the butyrophenone series (Janssen, 1970). Clinically, haloperidol was rapidly recognized as a more selective antipsychotic drug; the reason for this was simply that it is a more selective DA antagonist. As a consequence, haloperidol became the drug of choice for pharmacologists to use to block central DA receptors. Later on, it was even used as the ligand to identify biochemically DA receptors. It is not an exaggeration to state that haloperidol has been a keystone leading to the DA receptor concept.

2.1.3. Pharmacology of Dopamine Agonists and Antagonists

After the introduction of chlorpromazine and haloperidol in the treatment of psychotic disorders, numerous compounds belonging to different chemical classes were selected pharmacologically as potential antipsychotic drugs. Although neuroleptic drugs as such may elicit different types of behavioral effects, often called "neuroleptic" effects (catalepsy, hypomotility, ptosis, and so on), two compounds, amphetamine and apomorphine, played a prominent role in assessing the antidopaminergic properties of antipsychotic drugs.

First, the stereotypies induced by amphetamine were found to be blocked by neuroleptic drugs (Janssen et al., 1965). A large body of evidence indicated that the stereotyped behavior induced by amphetamine is mediated through DA release in the brain (Randrup and Munkvad, 1961). In addition to this, it was believed that the psychotic syndrome elicited by amphetamine in humans might have some similarities with that observed in schizophrenic patients. In contrast, the abnormal behavioral effects induced by apomorphine were found to be the consequence of a direct interaction of apomorphine on DA receptors (Ernst and Smelik, 1966). In fact, apomorphine can mimic the effects of DA. The same is true in the dog, in which neuroleptic drugs are known to antagonize apomorphine-induced emesis (Niemegeers and Janssen, 1979).

This pharmacological test seems to be one of the most sensitive for neuroleptic drugs.

According to the DA hypothesis of schizophrenia, the psychotic syndrome should be caused by an overstimulation of the dopaminergic system. Until now there has been no direct evidence supporting this view; only two things are certain—neuroleptic drugs can greatly alleviate psychotic symptoms in schizophrenic patients (Fig. 1), and these drugs are definitively DA antagonists, as suggested 20 yr ago (Van Rossum, 1966).

2.1.4. Increased Dopamine Turnover with Neuroleptics

More direct evidence for an interaction of neuroleptics with the dopaminergic system was provided by the observation that DA metabolites, (DOPAC) and homovanillic acid (HVA), increase in both the basal ganglia (Carlsson and Lindqvist, 1963) and mesolimbic system (Anden, 1972) after administration of neuroleptics. This enhancement of DA turnover, which was confirmed for many drugs in different ways, was interpreted as a consequence of DA receptor blockade. Interestingly, increased HVA was also observed in the cortex of animals treated with neuroleptics, but in this brain area there was no tolerance to HVA increase, as was normally observed in the striatum (Scatton et al., 1976; Laduron et al., 1977). After chronic treatment, a tolerance phenomenon develops to the effects of neuroleptics so that the HVA increase in the striatum is less marked than after acute treatment; this is compatible with the clinical observation that certain side effects tend to decrease during chronic administration of neuroleptics (Ayd, 1974). In contrast, the lack of the tolerance phenomenon to HVA increase in the frontal cortex could explain the absence of tolerance to the antipsychotic effects of neuroleptics if one accepts the idea that the antischizophrenic activity of these drugs occurs in the cortex.

2.1.5. Biochemical Identification of Dopamine Receptors

In 1972, a first attempt to identify DA receptors biochemically was reported when the adenylate cyclase sensitive to DA was found to represent a possible target for antipsychotic drugs (Kebabian et al., 1972). Indeed, this enzyme occurs specifically in brain dopaminergic areas, and only the production of cyclic nucleotide that is stimulated by DA is antagonized by certain neuroleptics (Clement-Cormier et al., 1974). Thereafter this site was classified as DA D_1 subtype (Kebabian and Calne, 1979).

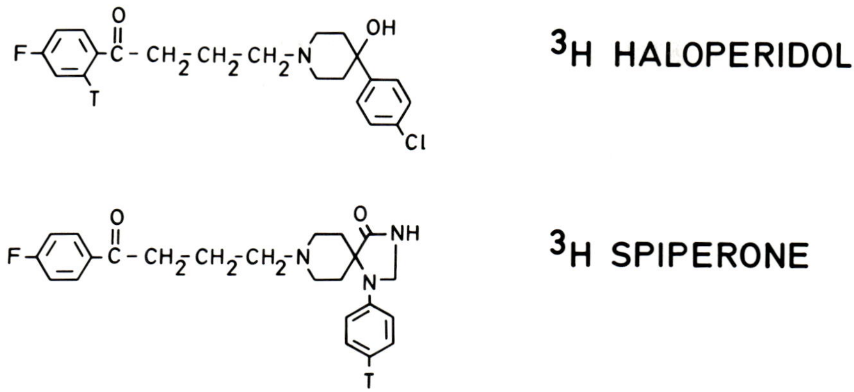

Fig. 2. Chemical structure of the main ligands of dopamine receptors.

In 1975, the first in vitro binding assay for DA receptors was reported independently by two different groups (Seeman et al., 1975; Creese et al., 1975). ^{3}H-Haloperidol (Fig. 2) was used as ligand at a nanomolar concentration, and the specific binding on striatal membranes was defined as the difference in the radioactivity bound in the presence of (+)-butaclamol (a potent neuroleptic) and (–)-butaclamol (its inactive enantiomer). This site was called the DA D_2 receptor (Kebabian and Calne, 1979). In 1978, a new ligand, ^{3}H-spiperone (Fig. 2), also belonging to the butyrophenone series, was proposed as the ligand of choice for DA receptors (Laduron and Leysen, 1977; Leysen et al., 1978a), as well as for serotonin S_2 receptors (Leysen et al., 1978b). As we will discuss later, this ligand possesses considerable advantages.

From the foregoing, it is obvious that the concept of a DA receptor was born progressively through various different approaches—pharmacological, anatomical, and biochemical. A common feature of all these studies, however, is the fact that potent and selective neuroleptic drugs were always an absolute requirement.

2.2. Definition of the Dopamine Receptor

The word receptor has been used with many different meanings by so many scientists that it is impossible to discuss all the biological processes in which receptors seem to be involved. In many cases, the term is misused or its meaning largely deviates

Table 2
Functional Response of Dopamine Agonists

Emesis
Stereotypy
Hypermotility
Turning behavior
Regulation of prolactin secretion
Neurogenic vasodilation
Antiparkinsonism
Psychosis

from the original definition provided in the early twentieth century by John Newport Langley (1905). Studying the effects of nicotine and curare on muscle contraction, Langley came to the conclusion that both must compete at the same "receptive substance"; the receptor field was born. According to Langley's definition, a receptor is a site of competition for agonist and antagonist; the agonist produces a stimulus that is transmitted within the cell and leads to a physiological response. As we have already emphasized many times (Laduron, 1981, 1982, 1984a), the physiological response is essential when defining a receptor; a binding site is not necessarily equivalent to a receptor site. In the absence of a physiological response, one has to prefer the terms acceptor site, binding site, or recognition site. This membrane protein, although specific for ligands, does not transmit a signal from the exterior to the interior of the cell. Sometimes we must look back at the pioneering work in order to clearly explain the present state of art; Langley's definition may be adopted for the DA receptor. Consequently the DA receptor is a site of competition for DA and neuroleptic drugs; the physiological responses elicited by DA and DA agonists are blocked by neuroleptic drugs or DA antagonists.

Table 2 shows the different effects of DA and other dopaminomimetic agents; emesis and stereotypy are of great importance from a pharmacological point of view, whereas prolactin secretion, antiparkinson effects, and psychosis (side effects of L-Dopa therapy) are the most relevant clinically. In all the biochemical studies, it is necessary to correlate the observed in vitro effects with functional responses in vivo.

For some people, the foregoing definition of the DA receptor or of any other receptor may appear much too conservative or too pharmacological. Historically the receptor concept originated from pharmacological studies on the muscular contraction (Langley, 1905). Moreover such a restrictive attitude is, I think, the best way to be on one's guard against the appeal of the multiple-site concept.

3. In Vitro Characterization of Dopamine Receptors

3.1. Dopamine-Sensitive Adenylate Cyclase (Dopamine D_1 Site)

A DA-stimulated adenylate cyclase was first identified in mammalian superior cervical ganglion (Kebabian and Greengard, 1971), retina (Brown and Markman, 1972), and the rat striatum (Kebabian et al., 1972; Clement-Cormier et al., 1974). This site was believed to be the DA receptor because it was stimulated by DA and its agonists, and the effect was antagonized by some DA antagonists like chlorpromazine, haloperidol, (+)-butaclamol, and other phenothiazine and thioxanthene derivatives (Kebabian et al., 1972; Clement-Cormier et al., 1974; Iversen et al., 1976). There were some inconsistencies concerning the drug potency observed in vitro and in vivo for compounds like haloperidol and pimozide, however. Thereafter it became obvious that compounds belonging to the diphenylpiperidine series (pimozide or penfluridol) (Laduron, 1976), domperidone (Laduron and Leysen, 1979; Laduron, 1980), sulpiride (Trabucchi et al., 1975), and other benzamides (Jenner and Marsden, 1979) were weakly active or inactive on the cyclase. Table 3 shows clearly that numerous potent DA antagonists are poorly active or inactive on the DA D_1 site, although they compete, even at a nanomolar concentration with ^{3}H-haloperidol binding; there was no correlation between the inhibitory effect of neuroleptic drugs on the cyclase and their pharmacological activity or their clinical potency (Laduron, 1980, 1982; Seeman, 1980).

One of the major sources of confusion surrounding the DA receptor originated from the idea that the increase of the cyclic AMP production observed in vitro with DA represents a physiological response to DA (Kebabian and Calne, 1979). In fact, the stimulation of cyclic AMP by DA is a biochemical effect for which one needs a find a physiological response in vivo. In this regard, more than 10 yr after its discovery, the DA-stimulated adenylate cyclase is still in search of a function. The secretion of parathor-

Table 3
Comparison of IC_{50} Values (*M*) on Adenylate Cyclase (D_1) and Haloperidol Binding (D_2)

Drug	Dopamine-sensitive adenylate cyclase, (A)	^{3}H-Haloperidol binding, (B)	Ratio, (A/B)
α-Flupenthixol	2.3×10^{-8}	2×10^{-8}	1.2
Chlorpromazine	1.1×10^{-6}	1.3×10^{-7}	8.5
(+)-Butaclamol	2×10^{-7}	1×10^{-8}	20
Haloperidol	7.5×10^{-7}	3.6×10^{-9}	208
Pimozide	1.5×10^{-5}	3.2×10^{-9}	4.687
Spiperone	2.2×10^{-6}	4.4×10^{-10}	5.500
Sulpiride	$>10^{-3}$	8×10^{-8}	>10.000
Halopemide	$>10^{-3}$	1×10^{-8}	>100.000
Domperidone	2.4×10^{-4}	1.4×10^{-9}	142.857

mone in the bovine parathyroid gland, which seems to be mediated through the D_1 sites (Brown et al., 1977), has been repeatedly presented as direct evidence for a physiological role of DA-stimulated adenylate cyclase (Stoof and Kebabian, 1984). Unfortunately, this effect remains purely anecdotal; it is restricted to one species and has not been demonstrated in vivo. Moreover, in humans the parathyroid gland is unresponsive to DA (Bansal et al., 1982) and to neuroleptic drugs, especially those, like thioxanthene and phenothiazine derivatives, that are potent antagonists of the D_1 sites. In retina, the presence of DA-sensitive adenylate cyclase has been clearly demonstrated, but its function there remains to be elucidated (Stoof and Kebabian, 1984). For these reasons, in contrast to other researchers (Stoof and Kebabian, 1984; Seeman, 1980; Creese et al., 1983), I find it premature to call them receptor sites. Since there is absolutely no doubt of the existence of the D_1 site, one may suspect that its role will be discovered, like that of adrenergic α_2 and histamine H_2 receptors. One cannot exclude the idea that the DA-stimulated adenylate cyclase should be involved in the phosphorylation of specific neuronal constituents like the DARPP-32 phosphoprotein (Nestler and Greengard, 1983) and perhaps the D_2 receptor, in the course of the neuronal growth and synaptogenesis. Whatever it is, such a long-term effect that represents more a metabolic process rather than a fast neuronal event may certainly have an important role in the brain dopaminergic system. It seems unlikely, however, that the D_1 site is directly implicated in the psychotic symptoms of schizophrenia, since the excitatory neuronal response to DA in the cerebral cortex is not mediated by the D_1 site, but by the D_2 receptor (Branshaw et al., 1985).

Recently, ^{3}H-flupenthixol and ^{3}H-piflutixol were reported to bind to the D_1 sites (Hyttel, 1981). This was somewhat surprising because of the lack of specificity of both drugs; they bind with high affinity to D_1 sites, but also to D_2, α_1, S_2, and H_1 receptors (Leysen, 1984). In fact, the correlation between the binding and the inhibition of DA-sensitive adenylate cyclase was not very satisfactory (Hyttel, 1981), a fact that is compatible with recent in vivo data showing a dissociation between the effects of chronic treatment with neuroleptics on the DA adenylate cyclase and the ^{3}H-piflutixol binding (Rupniak et al., 1985). Supporting this view is the fact that both the cyclase and the binding were found to represent different molecular entities (Nielsen et al., 1984). The recent introduction of the new ligand, ^{3}H-SCH 23390 (Schultz et al., 1985), does not seem to have resolved the specificity problems, since this drug is known to have a high affinity for serotonin S_2 receptors, a fact that is compatible with the relatively high number of ^{3}H-SCH 23390 binding sites found in the cortex (Schultz et al., 1985).

At the present time, one may conclude that the so-called D_1 sites identified by in vitro binding assays are probably the sum of various different binding sites that are not necessarily related to the D_1 site. As a consequence, the measurement of the increased cyclic AMP production elicited by DA, although laborious and sometimes difficult, remains probably the safest procedure when looking at the DA D_1 site.

3.2. Dopamine D_2 Receptors

Two ligands belonging to the butyrophenone series have played a prominent role in the identification of DA receptors (Fig. 2). As already stated, the first in vitro binding assay used ^{3}H-haloperidol as the ligand (Seeman et al., 1975; Creese et al., 1975). Later on ^{3}H-spiperone was introduced (Laduron and Leysen, 1977; Leysen et al., 1978a) and became one of the most popular ligands in neurobiology, probably because it can also bind to serotonin S_2

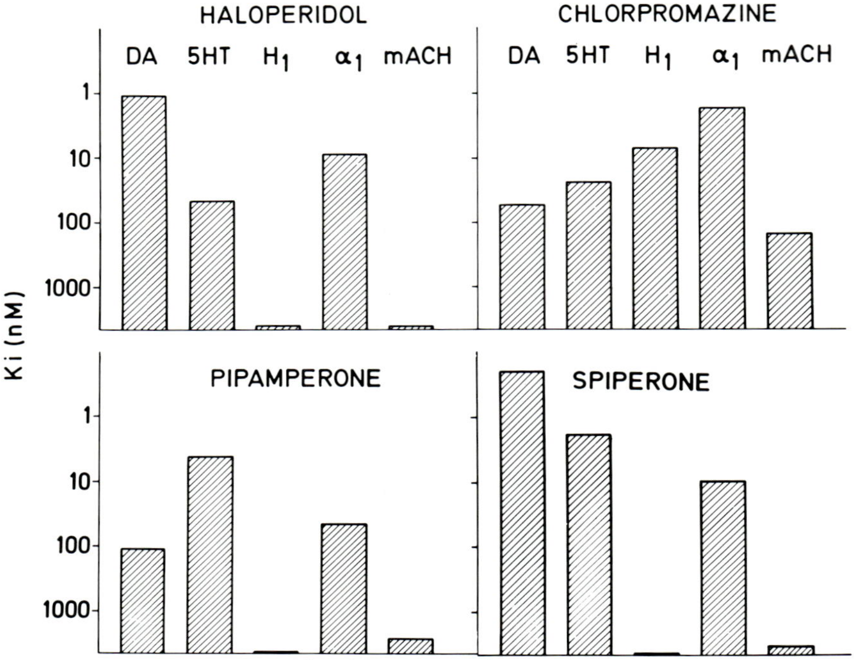

Fig. 3. Receptor profile of four neuroleptic drugs in various binding assays in vitro (DA, dopamine-D_2; 5-HT, serotonin-S_2; H_1, histamine-H_1; α_1, adrenergic-α_1; mACh, muscarinic). Note the selectivity of haloperidol as dopaminergic antagonist and pipamperone as serotonergic antagonist. Spiperone binds to both DA and serotonin receptors with high affinity (nanomolar).

receptors in the cortex (Leysen et al., 1978b) and works under in vivo conditions and in solubilized preparations. The great advantages of ^{3}H-spiperone are its high affinity, low nonspecific binding, and slow dissociation rate (Laduron and Leysen, 1977). Figure 3 clearly shows the reasons haloperidol and spiperone were chosen as the first ligands. They are more selective antidopamine drugs than chlorpromazine, a fact that has been known for many years from pharmacological data (Niemegeers and Janssen, 1979). Numerous other ligands for DA receptors were proposed—^{3}H-domperidone (Laduron and Leysen, 1979; Baudry et al., 1979; Lazareno and Nahorski, 1982), ^{3}H-thioproperazine (Blanchard et al., 1980), ^{3}H-sulpiride (Woodruff and Freeman, 1981), and other benzamides; the results obtained were essentially confirmation of those reported with ^{3}H-haloperidol and ^{3}H-spiperone. Recently, a promising ligand, ^{125}iodosulpride, was introduced; its high specific activity allowed for identification, for the first time, of DA receptors in the cortex (Martres et al., 1985). Sometimes it may be useful to use a more selective ligand than spiperone; however, it is always possible to differentiate ^{3}H-spiperone binding on dopamine D_2 (Leysen et al., 1978a) or serotonin S_2 receptors (Leysen et al., 1978b) by using selective displacers and different brain regions. The displacers of choice for DA receptors are domperidone, sulpiride, or ADTN (Quik et al., 1978), whereas methysergide and pipamperone are preferred for the serotonin S_2 receptors. As discussed several times before (Laduron, 1980, 1981, 1982, 1984a), the binding studies have generated many artifacts, perhaps simply because the binding assay is very easy to perform, but mostly because the interpretation is extremely difficult. The four main pitfalls that are responsible for numerous mistakes in this field are the following: (1) the belief that diplaceable binding means specific binding or binding on receptor sites, (2) the use of nonselective labeled ligand and unlabeled displacers, (3) the fact that the membrane preparations from brain or peripheral tissues are extremely heterogeneous, and finally (4) that the binding is rarely, if at all, in equilibrium, a fact that invalidates the mathematical models or computerized fitting methods (De Lean et al., 1978). Here the computer is a device to create the illusion of multiple receptor sites.

Hence, the so-called specific binding often represents the sum of binding on one or several receptor sites and binding on nonspecific displaceable sites. Spiperone, for instance, can bind to DA and serotonin receptors, but also to the nonspecific, but displaceable, spirodecanone sites (Howlett et al., 1979; Laduron, 1982, 1984a). A computer cannot make such a distinction; binding remains an operational approach. Unfortunately, many researchers do not realize that displaceable binding sometimes may only represent nonspecific binding. As discussed previously, numerous ligands, especially the labeled "classical" neurotransmitters, bind reversibly on nonspecific recognition sites. Therefore the best way to make a binding site a receptor site is to get functional correlates; the IC_{50} values of binding must correlate with the ED_{50} values in pharmacological tests performed in vitro or in vivo. In this regard, the IC_{50} values of numerous neuroleptic drugs obtained in in vitro binding assays correlate closely with their potency in numerous behavioral tests, in prolactin secretion, and even in clinical effectiveness to alleviate psychotic symptoms in schizophrenic patients (Creese et al., 1976; Seeman et al., 1976;

Leysen et al., 1977, 1978a,b; Caron et al., 1978; Laduron, 1980, 1982; Seeman, 1980).

Dopamine receptors were identified by in vitro binding assays in the striatum (Seeman et al., 1975; Creese et al., 1975), the pituitary (Creese et al., 1977; Caron et al., 1978), and the dopaminergic areas like olfactory tubercles, nucleus accumbens (Leysen et al., 1979), and area postrema (Stefanini and Clement-Cormier, 1981). Dopamine receptors were also reported in the rat spinal cord (Demenge et al., 1981), the cortex (Fields et al., 1977, Marchais et al., 1980) and the ventral tegmental area (Gundlack et al., 1982); however, for the latter two, the dopaminergic nature of the binding was not clearly established. The presence of DA receptors was also demonstrated by means of in vitro binding assays in human brain (Fields et al., 1978; Seeman, 1980; Ruberg et al., 1984; Maloteaux and Laduron, 1986), but it is not clear whether the number of these sites is increased in Parkinson's disease (Ruberg et al., 1984) and in schizophrenia (Lee et al., 1978; Owen et al., 1979; Mackay et al., 1980); for the latter, neuroleptic treatment seems to be the reason for an increase of DA receptors (Mackay et al., 1980). With the exception of degenerative diseases, a variation in the number of a given receptor in the brain is not necessarily expected to occur because the total number of sites may remain constant, whereas the turnover of receptor can be markedly enhanced or reduced. Therefore one should look only at the changes of functional synaptic receptors, which are only a very small part of the total number of receptors; unfortunately, up to now, there is no technical approach with which to make a distinction between functional and silent receptors.

^{3}H-Spiperone binding sites were also reported in mammalian lymphocytes, suggesting the presence of DA receptors on the blood cells (Le Fur et al., 1980); unfortunately these sites were found to be related to a trapping phenomenon of ^{3}H-spiperone within the intact lymphocytes (Maloteaux et al., 1983).

Although much work has been done on DA receptors during the last decade, the precise cellular localization of these sites in the brain remains unknown; what neuronal cells bear DA receptors in the striatum?

The fact that striatal acetylcholine release was enhanced by neuroleptic drugs and decreased by DA agonists (Stadler et al., 1973; Guyenet et al., 1975) suggests that striatal DA receptors might be located on cholinergic interneurons (Fig. 1). If the cellular localization of DA receptors remains to be proved, its subcellular localization is better known; fractionation studies have allowed clear differentiation of the DA-sensitive adenylate cyclase and ^{3}H-haloperidol binding (Leysen and Laduron, 1977) into two different entities, leading to the formulation of a two-receptor concept (Kebabian and Calne, 1979). In the striatum, DA receptors were found mainly in the microsomal P fraction (Laduron and Leysen, 1977; Laduron, 1981; Laduron et al., 1978a; Blanchard et al. 1980). In density gradient centrifugation, two peaks of DA receptors were found. The major peak, which equilibrated in a zone of lower density, corresponded to microsomal sites. The second smaller peak, which equilibrated in a zone of higher density, corresponded to synaptosomal receptors. Consequently, DA receptors like other receptors, possess at least a bimodal subcellular distribution that represents receptors located on nerve terminals, dendrites, and synaptic vesicles. Interestingly, DA receptors were not found to be associated with dopaminergic nerve terminals, which does not support the existence of DA autoreceptors (Laduron, 1981, 1984b).

3.3. What Is the Transducing System of Dopamine D_2 Receptors?

The DA site capable of stimulating adenylate cyclase is the D_1 site (Stoof and Kebabian, 1984). Recent studies suggest that the stimulation of the D_2 receptor inhibits adenylate cyclase activity in the pituitay gland (Cote et al., 1981). In these experiments DA inhibited the increase of cyclic AMP production induced by isoproterenol; this inhibitory effect was blocked by some neuroleptic drugs. Until now, there is no evidence for a negatively coupled adenylate cyclase in the striatum. It is even difficult to accept the idea that DA and adrenaline would be present in the same synapses; this represents a minimal condition to make the system operational. This remains unlikely because of the extremely low concentration of adrenaline in the striatum. Therefore the idea that the D_2 receptor should be negatively coupled to adenylate cyclase may be considered an interesting hypothesis that requires more experimental work.

3.4. Multiple Dopamine Sites

In the dopaminergic field, binding studies have generated a considerable number of controversies, in particular about the number of DA receptor subtypes. It is not my purpose to recall in detail all the adventures leading to multiple sites, because today this long period of controversy appears to be purely anecdotal. For several years, researchers ("binders") in the field were divided into two groups: the multiplicits and the unicists, the former being in the majority. The D_3 site was defined as a high-affinity site for DA (Seeman, 1982; Creese, 1982) and was believed, at least in the beginning, to be the DA autoreceptor (Creese, 1982). An apparent correlation between the IC_{50} values of various DA agonists for the D_3 site and the normalized doses that reversed the α-butyrolactone-induced elevation of Dopa by 50% was even obtained (Seeman, 1980). As previously discussed (Laduron, 1982), however, the correction factor used to normalize the ED_{50} values was much higher than the ED_{50} value itself, a fact that entirely invalidates such a type of correlation. Finally the D_4 site was reported to have high-affinity properties for agonists and antagonists (Seeman, 1980, 1982). In fact this site resulted from the analysis of anomalous displacement curves that may be expected when the binding is performed in a heterogeneous system like the membrane preparations used in the binding studies. In the meantime, Schwartz's group proposed three classes of dopamine receptors (D_2, D_3, D_4) identified by binding studies with ^{3}H-apomorphine and ^{3}H-domperidone (Sokoloff et al., 1980). Obviously all these studies were attempts to make binding reveal more than it could.

Five years ago, I wrote of the DA receptor ". . . the postulated existence of the subtypes D_1, D_2, D_3 . . . or even D_4, D_5, D_6 . . . is probably a short-lived fashion" (Laduron, 1980). This prediction seems to have come true, since even Seeman and Creese, who were in favor of multiple receptors and have explained, revisited, and reexplained many times the nomenclature of DA receptors, no longer believe the D_3 and D_4 sites to be separate entities (Leff and Cresse, 1983; Seeman et al., 1985). The fact that they now call D_4 the high-affinity site of D_2 and D_3 the high-affinity site of D_1 is simply a penultimate step before a more realistic unitary concept. As discussed briefly above, a unitary concept does not exclude the possibility of considering a second DA receptor if, in the future, the physiological role of DA-sensitive adenylate cyclase (D_1 site) is clearly demonstrated.

4. In Vivo Characterization of Dopamine Receptors

In vitro binding studies are undoubtedly the best way to identify brain receptors biochemically and even to assess the potential pharmacological effects of a drug through its receptor profile. This approach does not necessarily provide evidence that a drug will interact with brain receptors in vivo, however. A possible way is to determine the occupancy of the receptor in vitro after injection of a drug into the animal (ex vivo binding). A second more sensitive, although more laborious, way is to inject the labeled drug into the animal and then examine the disposition of labeling with or without displacement with unlabeled drug. ^{3}H-Spiperone was found to be the first suitable ligand to label DA receptors under in vivo conditions (Laduron and Leysen, 1977; Hollt et al., 1977; Laduron et al., 1978b,c). Figure 4 gives an example of in vivo binding in rat brain; first, the retention of ^{3}H-spiperone was much higher in the striatum, a dopaminergic area, than in the cerebellum. Moreover, the labeling could be displaced by unlabeled spiperone or other DA antagonists in the striatum, but not in the cerebellum. Therefore, the ratio of the amount of labeling between the striatum and cerebellum may be considered a good index for the existence of DA receptors in vivo. In addition to this, however, displacement experiments with selective drugs are necessary. Other ligands, such as ^{3}H-pimozide (Baudry et al., 1977; Laduron et al., 1978b) and ^{3}H-thioproperazine (Boireau et al., 1980), were also reported to label DA receptors in vivo, although the nonspecific binding was much higher than with ^{3}H-spiperone. Obviously, the problem of nonspecific binding is still more acute in vivo than in vitro; ^{3}H-haloperidol, for instance, works nicely in vitro, but cannot reveal specific binding in vivo (Laduron et al., 1978b). Another great advantage of ^{3}H-spiperone is that it can also label serotonin S_2 receptors in another brain region, the frontal cortex (Leysen et al., 1978b). When prevention experiments are performed (cold drug is given prior the ^{3}H-spiperone), one can measure whether a drug interacts in vivo with DA D_2 and serotonin S_2 receptors in the same animal (Laduron et al., 1982). Perfusion with ice-cold saline was reported to significantly improve the specific–nonspecific ratio of binding for both DA and serotonin receptors (Niehoff et al., 1979). In vivo binding with ^{3}H-spiperone was also revealed by autoradiography (Kuhar et al., 1978); DA receptors appear to be homogeneously distributed in the striatum from adult animals.

The studies of in vivo binding with ^{3}H-spiperone have been performed to visualize DA receptors in humans by means of positron emission tomography (PET). As was the case for in vivo binding in rat, spiperone is rapidly becoming the ligand of choice

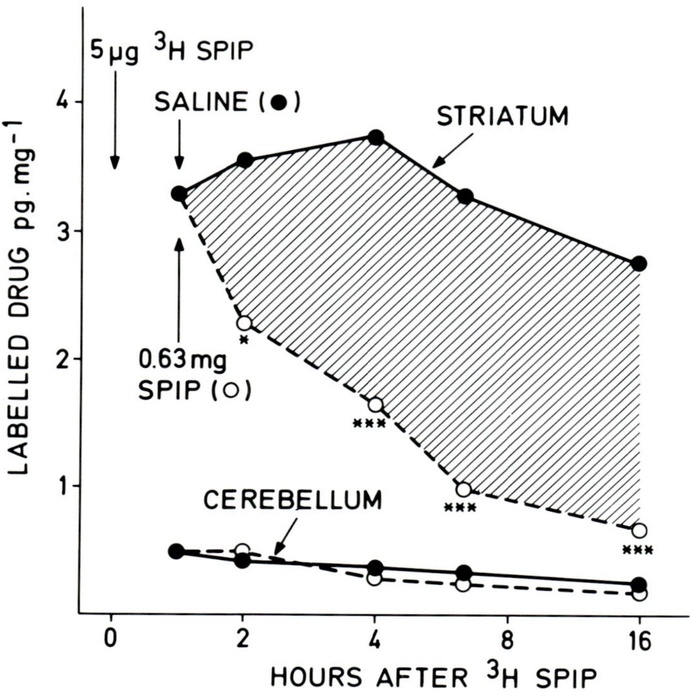

Fig. 4. In vivo ^{3}H-spiperone binding in rat brain. After iv injection of 5 μg ^{3}H-spiperone, the retention of labeling is much higher in the striatum than in the cerebellum and can be displaced, in the striatum but not in the cerebellum, by unlabeled spiperone (hatched part) injected 1 h after the labeled ligand.

for PET studies. In the first study, ^{11}C-methylspiperone was used (Wagner et al., 1983); then ^{77}Br-bromospiperone (Crawley et al., 1983) and ^{76}Br-bromospiperone (Maziere et al., 1985) were reported to reveal DA receptors in living human brain. It is not clear whether this will lead to clinical applications for the diagnosis of cerebral disorders. Nevertheless, the PET technique has already allowed for visualization in vivo of a loss of striatal DA receptors in patients with progressive supranuclear palsy (Baron et al., 1985a).

Another ligand, pimozide labeled with ^{11}C, was also reported to provide visualization of DA receptors in human brain (Baron et al., 1985b); the striatum–cerebellum ratio was not very marked, however, a fact that could be expected from the in vivo binding data obtained in rats (Laduron et al., 1978b). Although the study of DA receptors in the human brain in vivo seems feasible today,

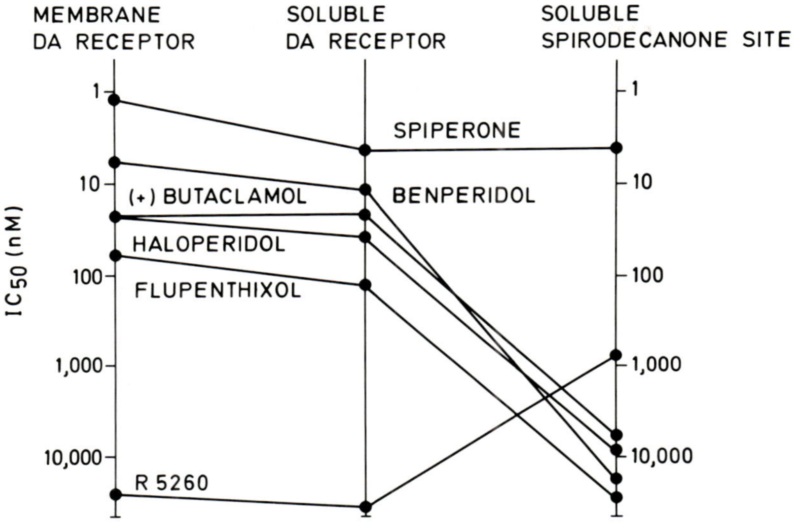

Fig. 5. IC_{50} values obtained with various drugs in ^{3}H-spiperone binding in a membrane preparation and in digitonin-solubilized extracts from dog striatum (soluble DA receptor) and from rat striatum (soluble spirodecanone site) (Gorissen and Laduron, 1978, 1979).

improvements may be still expected, especially in the fields of radiochemistry and detection devices.

5. Molecular Characterization of Dopamine Receptors

As a rule, the first binding studies for a given receptor are carried out in membrane preparations from brain tissue. Although the identification and characterization may be done on these membrane fractions, the receptor site must be solubilized before looking at its molecular structure.

5.1. Solubilization

The first attempts to obtain DA receptors in a molecularly dispersed form were not very encouraging (Gorissen and Laduron, 1978; Tam and Seeman, 1978); ^{3}H-spiperone binding sites solubilized from rat striatum by means of 1% digitonin had lost high affinity for all the DA antagonists except spiperone (Fig. 5). At that time, the reasons for such a failure were obscure. They have since been elucidated: only spirodecanone sites labeled with ^{3}H-spiperone (a nonspecific but displaceable site) were obtained in solution. Since they were present in excess, they masked the DA receptors. Figure 5 shows the characteristics of this recognition site for substances having the spirodecanone moiety; only spiperone and R 5260, a potent opiate, were found to bind to this site with high affinity. In contrast, haloperidol, benperidol, and (+)-butaclamol displayed a much lower affinity than in membrane preparations.

The choice of another animal species, apparently devoid of spirodecanone sites, was quite decisive; the first solubilization of high-affinity DA receptors was obtained from striatal membranes of dog brain using 1% digitonin (Gorissen and Laduron, 1979). Figure 5 shows that the IC_{50} values for DA antagonists belonging to different chemical classes of drug were practically the same when soluble extracts or membrane preparations were used for the binding assay. The ^{3}H-spiperone sites solubilized with digitonin retained the high-affinity properties of DA receptors (Gorissen et al., 1979). Thereafter, the concomitant use of a more sensitive technique to reveal soluble receptors, namely the charcoal procedure, and a spirodecanone site blocker (R 5260), enabled us to obtain DA receptors in molecularly dispersed form from rat striatum (Gorissen et al., 1980). The DA receptor can only be solubilized by means of mild detergents and not with high salt concentrations (Laduron and Ilien, 1982); therefore, the DA receptor must be considered an intrinsic protein. Brain-solubilized DA receptors have been obtained from various species, including humans, using different mild detergents (Davis et al., 1981; Goldstein and Lew, 1984; Leff and Creese, 1982; Lerner et al., 1981; Wheatley and Strange, 1983; Wouters et al., 1984a). Initially, the yield of soluble receptors was not very high; using digitonin, it was only 18% in dog (Gorissen and Laduron, 1979) and 20.25% in human striatum (Davis et al., 1981). More recently, a mixture of cholate and sodium chloride allowed a yield of about 50% (Wouters et al., 1984a). A value of 85% was also reported with 1% lysolecithin (Lerner et al., 1981); unfortunately, the characterization of solubilized ^{3}H-spiperone binding sites was not performed with this concentration of lysolecithin, but with 0.25%, which gave a much lower yield.

Various criteria must be fulfilled in order to identify soluble ^{3}H-spiperone binding sites as true DA receptors; among these, drug displacement and the correlation with the drug affinity

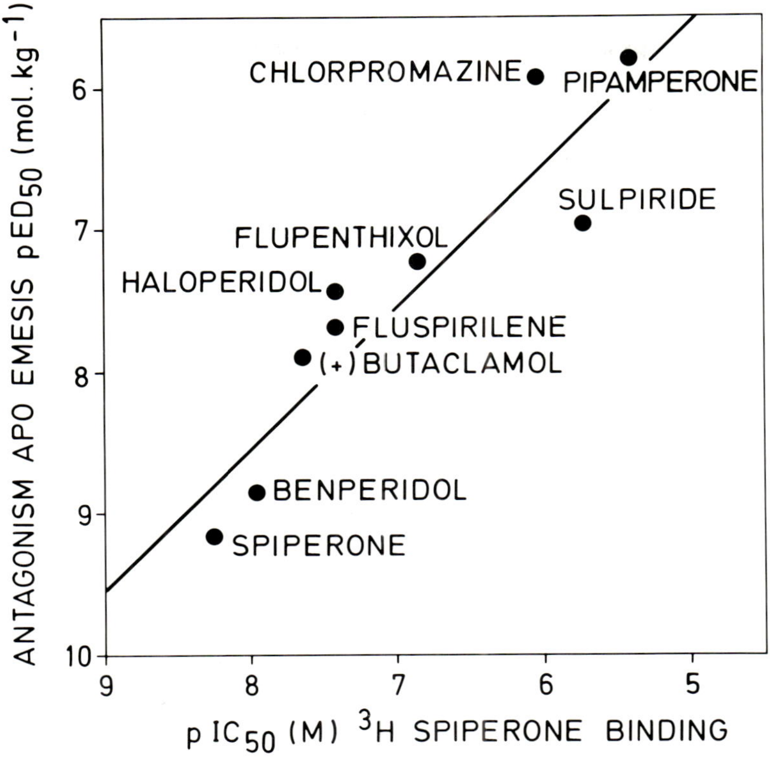

Fig. 6. Correlation between ^{3}H-spiperone binding in solubilized preparations from dog striatum and antagonism of apomorphine-induced emesis in the dog.

obtained on membrane preparations and with the drug potency obtained in vivo from pharmacological tests are the most important (Laduron and Ilien, 1982). Figure 6 shows the IC_{50} values of six compounds in membrane preparations and in two soluble extracts, one containing DA receptors and the other containing spirodecanone sites. The use of R 5260 and butyrophenones like haloperidol and benperidol, which do not possess a spirodecanone moiety, is, therefore, essential to exclude the presence of spirodecanone sites in the soluble extract. Moreover, the use of pipamperone, a potent serotonin antagonist, may exclude the presence of serotonergic receptors; indeed this drug competes at nanomolar concentrations on serotonin receptors and in the micromolar range on DA receptors (Fig. 5).

Finally, there was a good correlation (r = 0.89) between ^{3}H-spiperone binding in solubilized extracts from dog striatum and the antagonism of apomorphine-induced emesis in dog; therefore, one may conclude that the solubilized ^{3}H-spiperone sites displayed the properties of DA receptors.

In addition to the criteria of receptor specificity, one must establish that the DA receptor is really solubilized. Here, the criteria are mostly operational; they are based on the use of biochemical and morphological techniques. The sedimentation profile in sucrose gradients probably represents the most severe criterion; Table 4 shows that various sedimentation coefficients were obtained from 12.6 S to 2.5 S. Although this technique may be extremely useful to prove that the receptor is obtained in a molecular dispersed form, it cannot provide a good estimate of the molecular weight of protein; indeed, the sedimentation coefficient values are generally perturbed by an extensive binding of solubilizing detergent to the protein. By contrast, SDS-electrophoresis is not affected by detergent, but proteolysis problems may occur. As shown in Table 4, the sedimentation coefficient seems to be dependent not only on the nature of the detergent, but also on the presence or absence of salt or detergent in the gradient. For instance, when KCl was added in the sucrose, the S-value was generally lower than in its absence; the reasons for this are not clear, but a phenomenon of aggregation cannot be excluded. A sedimentation coefficient of 9 S was found many times; this corresponds to a molecular weight of 150,000–200,000, which is much higher than that obtained by radiation inactivation or SDS-electrophoresis.

5.2. Photoaffinity and Purification

The solubilization of a receptor is not a goal in itself, but a prerequisite to purifying the receptor protein and then analyzing its molecular structure. In 1987, eight years after the first report on the solubilization of high-affinity DA receptors (Gorissen and Laduron, 1979), one may ask why it took so much time to justify DA receptors. The reasons for this are not clear, but one may assume that all the attempts (except recently, by Senogles et al., 1986) to purify the receptor protein through conventional techniques led to changes of the receptor site such that it became undetectable in the binding assay; this was the case when using gel

RECEPTOR	PHOTOAFFINITY PROBE	
DOPAMINE D_2	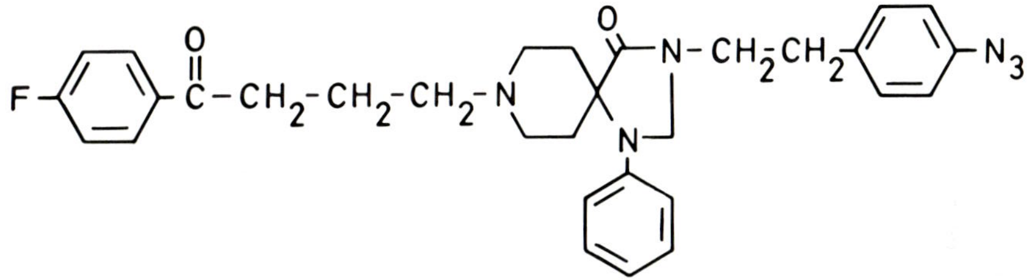 AZIDOPHENYLSPIPERONE	$Kd = 1.4\ nM$
DOPAMINE D_2	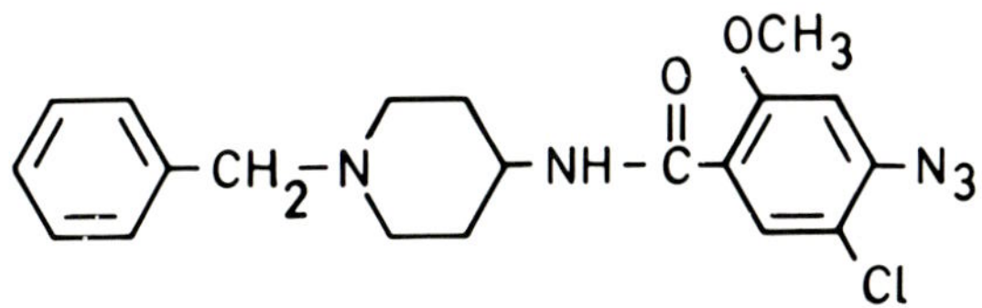 AZAPRIDE	$Kd = 480\ nM$

Fig. 7. Chemical structure of photoaffinity labels for dopamine receptors.

Table 4
Molecular Parameters of Dopamine Receptors

Source	Method	Sedimentation coefficient, S	Molecular weight, K	Stokes radius, nm	Reference
Dog	Digitonin sucrose gradient	~12	200–300	—	Gorissen and Laduron (1979)
	Digitonin sucrose gradient + KCl	9	150–200	—	Laduron and Wouters (1984a)
	Digitonin sucrose gradient				
	+ Triton X-100	5.5	136	—	Lilly et al. (1985)
	Gel filtration	—	—	5.8	Lilly et al. (1985)
	Cholate/salt sucrose gradient + KCl	12	200–300	—	Wouters et al. (1984a)
	Gel filtration	—	—	5.65	Wouters et al. (1984a)
	Lysophosphatidylcholine				
	Sucrose gradient	12.6	200–300	—	Strange et al. (1984)
	Sucrose gradient + KCl	9	150–200	—	Strange et al. (1984)
	Radiation inactivation	—	123	—	Lilly et al. (1983)

Human	Cholate/salt sucrose gradient + KCl	2.5	60	—	Wouters et al. (1984a)
	Gel filtration	—	—	3.42	Wouters et al. (1984a)
	Radiation inactivation	—	123	—	Lilly et al. (1983)
Bovine	Digitonin sucrose gradient	11.4	—	—	Strange et al. (1984)
	Gel filtration	—	—	8.0	Strange et al. (1984)
	Lysophosphatidylcholine				
	Sucrose gradient	11.9	—	—	Strange et al. (1984)
	Gel filtration	—	150–200	—	Lerner et al. (1981)
	Cholate/salt sucrose gradient	6	—	—	Strange et al. (1984)
	Gel filtration	—	—	6.6	Strange et al. (1984)
	Radiation inactivation	—	110 and 1500	—	Kuno and Tanaka (1983)
Rat	Cholate/salt sucrose gradient + KCl	9	150–200	—	Woulters et al. (1984a)
	SDS electrophoresis	—	94	—	Amlaiky and Caron (1985)
	Radiation inactivation	—	136	—	Nielsen et al. (1984)
	SDS electrophoresis	—	85	—	Redouane et al. (1985)

filtration techniques, affinity chromatography, and even gradient sedimentation. For the latter, for instance, DA receptors were only detectable when high sucrose concentrations (20–30%) were used in the gradient; in such conditions, the migration of protein only occurred over a short distance from the top of the gradient.

Photoaffinity is another approach that may be useful in attempts to purify the DA receptor. Two different probes azapride and NAPS (Fig. 7), have been developed to covalently label DA receptors. Azapride is an azido derivative of clebopride, a very selective DA antagonist that irreversibly blocked DA receptors from human (Wouters et al., 1984b) and rat (Niznik et al., 1984; Wouters et al., 1985) striatum after irradiation with ultraviolet light. The irradiation-induced inactivation by azapride was prevented by DA antagonists and remained unchanged in the presence of scavenger *p*-aminobenzoic acid, indicating a true photoaffinity mechanism (Wouters et al., 1985). Figure 7 shows, however, that the compound did not reveal a high affinity for DA receptors, which became a serious drawback for its use in a labeled form at nanomolar concentrations.

Another photoaffinity probe, N_3-NAPS (Amlaiky and Caron, 1985), was developed by starting from a potent spiperone analog, *n*-(*p*-aminophenethyl)-spiperone (Amlaiky et al., 1984). This azido derivative was found to bind to DA receptors with high affinity (Fig. 7). A radioiodinated form of this ligand was successfully used to identify a DA receptor subunit from crude membrane preparations (Amlaiky et al., 1985). Incubation of rat striatal membranes with $[^{125}I]N_3$-NAPS leads to the photo-dependent covalent incorporation of the probe into a peptide of $M_r = 94{,}000$ as assessed by gel electrophoresis. Labeling of this peptide has been blocked specifically and stereospecifically by dopaminergic antagonists such as (+)-and (–)-butaclamol, but not by nondopaminergic antagonists. It remains to be determined whether this peptide represents the DA receptor itself or a subunit of it. Very recently, another photoaffinity probe, azidosulpride, was found to label DA receptors covalently (Redouane et al., 1985). Analysis by gel electrophoresis revealed a peptide of 85,000 dalton, which is somewhat lower than that reported by the Caron's group.

Very recently, the partial purification of DA receptors was achieved by means of affinity chromatography (Senogles et al., 1986); earlier attempts through this approach had always been unsuccessful. Apparently, the use of a new affinity ligand, [(carboxymethylene)oximino]-spiperone, was a determining factor in obtaining purified DA receptors in an active form and with the expected dopaminergic specificity.

The molecular structure of the DA receptor can, now, be explored in detail.

6. Regulation of Dopamine Receptors

As a rule, the role of receptors is to transmit signals from one cell to another. A receptor is working when a neurotransmitter binds to it; this results in a physiological response. When a persistent and severe reduction of synaptic transmission occurs, however, the responsiveness of postsynaptic tissue to the deficient neurotransmitter is enhanced according to the well-known denervation rule. This phenomenon is also called postfunctional supersensitivity and represents a compensatory mechanism to the reduction or loss of signal molecules.

Dopaminergic supersensitivity was demonstrated after denervation with 6-hydroxydopamine (Ungerstedt, 1971; Feltz and de Champlain, 1972), after chronic treatment with haloperidol (Gianutsos et al., 1974) and after blockade of catecholamine synthesis with a α-methyl-*p*-tyrosine (Moore and Thornbury, 1975; Friedman et al., 1975). The supersensitivity phenomenon observed after denervation or drug treatment was always very marked in behavioral, electrophysiological, and pharmacological experiments. By contrast, when DA receptors were measured in in vitro binding studies, the increase of DA receptors after denervation or chronic treatment with neuroleptics remained relatively low (Burt et al., 1977; Thal et al., 1979; Owen et al., 1980; Feuerstein et al., 1981; Reches et al., 1983; Staunton et al., 1982; Theodorou et al., 1981). As a rule, the B_{max} ranged between 120 and 150% of control. There was an apparent discrepancy between the marked supersensitivity phenomenon detected in vivo and the low increase of DA receptor sites. One has to recall, however, that only a small percentage of the total number of DA receptors is functional at a given moment; we found, for instance, that about a 20% occupation of DA receptors in vivo by spiperone was sufficient to antagonize amphetamine-induced stereotypy (Laduron, 1981). This amount is probably still overestimated, however, because in vivo binding cannot make a clearcut distinction between functional and nonfunctional receptors. Consequently, one may assume that few sites are really involved in behavioral or pharmacological effects of DA antagonists. A consequence of this is that an increase of 20–30% of the total receptor sites should be of greater importance if they were expressed in terms of functional receptors. Unfortunately, these functional receptors cannot be separately detected by means of the methods used presently. This remains one of the main problems to be solved in the future.

7. Conclusion

Neuroleptic drugs are ligands that brain receptors are searching for; they are the guide that neurobiologists follow to try to understand the role of DA in the brain economy. Undoubtedly, neuroleptic drugs have been keystones in the identification (in vitro or in vivo), isolation, solubilization, and purification of DA receptors.

References

Amlaiky N. and Caron M. G. (1985) Photoaffinity labeling of the D_2 dopamine receptor using a novel high affinity radioiodinated probe. *J. Biol. Chem.* **260,** 1983–1986.

Amlaiky N., Kilpatrick B. F., and Caron M. G. (1984) A novel radioiodinated high affinity ligand for the D_2 dopamine receptor. *FEBS Lett.* **176,** 436–440.

Anden N.-E. (1972) Dopamine turnover in the corpus striatum and the limbic system after treatment with neuroleptic and anti-acetylcholine drugs. *J. Pharm. Pharmacol.* **24,** 905–906.

Ayd F. J. (1974) Do antiparkinson drugs interfere with the therapeutic effect of neuroleptics? *Int. Drug Ther. Newslett.* **9,** 29–31.

Bansal S., Woolf P. D., Fischer J. A., and Caro J. F. (1982) Dopamine does not affect parathyroid function in man. *J. Clin. Endocrinol. Metab.* **54,** 651–652.

Baron J. C., Maziere B., Loc'h C., Sojouropoulos P., Bonnet A. M., and Agid Y. (1985a) Progressive supranuclear palsy: Loss of striatal dopamine receptors demonstrated *in vivo* by position tomography. *Lancet* **i,** 1163–1164.

Baron J. C., Comar D., Zarifian E., Agid Y., Crouzel C., Loc'h C., Deniker P., and Kellershohn C. (1985b) Dopaminergic receptor sites in human brain: Positron emission tomography. *Neurology* **35,** 16–24.

Baudry M., Martres M. P., and Schwartz J. C. (1979) ^{3}H-Domperidone: A selective ligand for dopamine receptors. *Naunyn Schmiedebergs Arch. Pharmacol.* **308,** 231–237.

Baudry M., Martres M. P., and Schwartz J. C. (1977) *In vivo* binding of ^{3}H pimozide in mouse striatum: Effects of dopamine agonists and antagonists. *Life Sci.* **21,** 1163–1170.

Bertler A. and Rosengren E. (1959) Occurrence and distribution of dopamine in brain and other tissues. *Experientia* **15,** 10–11.

Bjorklund A. and Lindvall O. (1984) Dopamine-Containing Systems in the CNS, in *Handbook of Chemical Neuroanatomy* (Bjorklund A. and Hokfelt T., eds.) vol. 2, Elsevier, Amsterdam, New York.

Blanchard J.-C., Boireau A., Garret C., and Julou L. (1980) The use of thioproperazine, a phenothiazine derivative, as a ligand for neuroleptic receptors, I. *In vitro* studies. *Biochem. Pharmacol.* **29,** 2933–2938.

Boireau A., Blanchard J. C., Garret C., and Julou L. (1980) The use of thioproperazine, a phenothiazine derivative as a ligand for neuroleptic receptors. II. *In vivo* studies. *Biochem. Pharmacol.* **29,** 2939–2942.

Bradshaw C. M., Shendan R. D., and Szabadi E. (1985) Excitatory neuronal response to dopamine in the cerebral cortex: Involvement of D_2 but not D_1 dopamine receptors. *Br. J. Pharmacol.* **86,** 483–490.

Brown J. H. and Makman M. (1972) Stimulation by dopamine of adenylate cyclase in retinal homogenates and of adenosine 3'5' cyclic monophosphate formation in intact retina. *Proc. Natl. Acad. Sci. USA* **69,** 539–543.

Brown E. M., Carroll R. J., and Aurback G. D. (1977) Dopaminergic stimulation of cyclic AMP accumulation and parathyroid hormone release from disposed bovine parathyroid cells. *Proc. Natl. Acad. Sci. USA* **74,** 4210–4313.

Burt D. R., Creese I., and Snyder S. H. (1977) Antischizophrenic drugs: Chronic treatment elevates dopamine binding in brain. *Science* **196,** 326–328.

Carlsson A. and Lindqvist M. (1963) Effect of chlorpromazine or haloperidol on formation of 3 methoxytyramine and normetanephrine in mouse brain. *Acta Pharmacol.* **20,** 140–144.

Caron M. G., Beaulieu M., Raymond V., Sagne B., Drouin J., Lefkowitz R. J., and Labrie F. (1978) Dopaminergic receptors in the anterior pituitary gland. *J. Biol. Chem.* **253,** 2244–2253.

Changeux J. P. (1981) The acetylcholine receptor: An "allosteric" membrane protein. *Harvey Lect.* **75,** 85–254.

Clement-Cormier Y. C., Kebabian J. W., Petzold G. L., and Greengard P. (1974) Dopamine-sensitive adenylate cyclase in mammalian brain: A possible site of action of antipsychotic drugs. *Proc. Natl. Acad. Sci. USA* **71,** 1113–1117.

Cote T. E., Grewe C. W., and Kebabian J. W. (1981) Stimulation of a D_2-dopamine receptor in the intermediate lobe of the rat pituitary gland decreases the responsiveness of the β-adrenoceptor; biochemical mechanism. *Endocrinology* **108,** 420–426.

Crawley J. C. W., Smith T., Veall N., Zanelli G. D., Crow T. J., and Owen F. (1983) Dopamine receptors displayed in living human brain with ^{77}Br-*p*-bromo-spiperone. *Lancet* **ii.,** 975.

Creese I. (1982) Dopamine receptors explained. *Trends Neurosci.* **5,** 40–42.

Creese I., Burt D. R., and Snyder S. H. (1975) The dopamine receptor: Differential binding of *d*-LSD and related agents to agonist and antagonist states. *Life Sci.* **17,** 1715–1720.

Creese I., Hamblin M. W., Leff S. E., and Sibley D. R. (1983) CNS Dopamine Receptors, in *Handbook of Psychopharmacology* (Iversen L. L., Iversen S. D., and Snyder S. H., eds.) vol. 17 Plenum, New York.

Creese I., Burt D. R., and Snyder S. H. (1976) Dopamine receptor binding predicts clinical and pharmacological potencies of antischizophrenic drugs. *Science* **192,** 481–483.

Creese I., Schneider R., and Snyder S. H. (1977) ^{3}H Spiroperidol labels dopamine receptors in pituitary and brain. *Eur. J. Pharmacol.* **46,** 377–381.

Dahlstrom A. and Fuxe K. (1964) Evidence for the existence of monoamine-containing neurons in the central nervous system. I. Demonstration of monoamines in the cell bodies of brain stem neurons. *Acta physiol. Scand.* **62** (suppl. 232), 1–55.

Davis A., Madras B. K., and Seeman P. (1981) Solubilization of neuroleptic/dopamine receptors of human brain striatum. *Eur. J. Pharmacol.* **70,** 321–329.

Delay J., Deniker P., and Harl J. M. (1952) Utilisation en thérapeutique psychiatrique d'une phénothiazine d'action centrale élective (4500 RP). *Ann. Médicopsychol. (Paris)* **110,** 112–116.

De Lean A., Munson P. J., and Rodbard D. (1978) Simultaneous analysis of families of sigmoidal curves: Application to bioassay, radioligand assay and physiological dose–responses curves. *Am. J. Physiol.* **235** (2), E95–E102.

Demenge P., Mouchet P., Guerin B., and Feuerstein C. (1981) Identification and distribution of neuroleptic binding sites in the rat spinal cord. *J. Neurochem.* **37,** 53–59.

Ehringer R. H. and Hornykiewicz O. (1960) Verteilung von Noradrenalin und Dopamin (3-Hydroxytyramin) im Gehirn des Menschen und ihr Verhalten bei Erkrankungen des Extrapyramidalen Sytems. *Klin. Wschr.* **15,** 1236–1239.

Ernst A. M. and Smelik P. G. (1966) Site of action of dopamine and apomorphine on compulsive gnawing behavior in rats. *Experientia* **22,** 837–838.

Feltz P. and De Champlain J. (1972) Enhanced sensitivity of caudate neurones to microiontophoretic injections of dopamine in 6-hydroxydopamine treated cats. *Brain Res.* **43,** 601–605.

Feuerstein C., Demenge P., Caron P., Barrette G., Guerin B., and Mouchet P. (1981) Supersensitivity time course of dopamine antagonist binding after nigrostriatal denervation: Evidence for early and drastic changes in the rat corpus striatum. *Brain Res.* **226,** 221–234.

Fields J. Z., Reisine T. D., and Yamamura H. I. (1977) Biochemical demonstration of dopaminergic receptors in rat and human brain using [H^3] spiroperidol. *Brain Res.* **136,** 578–584.

Friedman E., Rotrosen J., Gurland M., Lambert G. A. and Gershon S. (1975) Enhancement of reserpine-elicited dopaminergic supersensitivity by repeated treatment with apomorphine and α-methyl-*p*-tyrosine. *Life Sci.* **17,** 867–874.

Gianutsos G., Drawbaugh R. B., Hynes M. D., and Lal H. (1974) Behavioral evidence for dopaminergic supersensitivity after chronic haloperidol. *Life Sci.* **14,** 887–898.

Goldstein M. and Lew J. Y. (1984) Studies on Solubilization and Characterization of Rat Striatal Dopamine Receptors, in *Catecholamines: Basic and Peripheral Mechanisms* (Usdin E., Carlsson A., Dahlstrom A., and Engel, J., eds.) Alan R. Liss, New York.

Gorissen H. and Laduron P. M. (1978) Solubilization of ^3H spiperone binding sites from rat brain. *Life Sci.* **23,** 575–580.

Gorissen H. and Laduron P. (1979) Solubilization of high affinity dopamine receptors. *Nature* **279,** 72–74.

Gorissen H., Aerts G., and Laduron P. M. (1979) Characterization of solubilized dopamine receptors from dog striatum. *FEBS Lett.* **100,** 281–285.

Gorissen H., Ilien B., Aerts G., and Laduron P. M. (1980) Differentiation of solubilized dopamine receptors from spirodecanone binding sites in rat striatum. *FEBS Lett.* **121,** 133–138.

Gundlack A. L., McDonald D., and Beart P. M. (1982) ^3H Spiperone labels non cyclase-linked dopamine receptors in the ventral segmental area of the rat brain. *J. Neurochem.* **39,** 890–894.

Guyenet P. G., Agid Y., Javoy F., Beaujouan J. C., Rossier J., and Glowinski J. (1975) Effects of dopaminergic receptor agonists and antagonists on the activity of the neo-striatal cholinergic system. *Brain Res.* **84,** 227–244.

Hollt V., Czlonkowski A., and Herz A. (1977) The demonstration *in vivo* of specific binding sites for neuroleptic drug in mouse brain. *Brain Res.* **130,** 176–183.

Howlett D. R., Morris H., and Nahorski S. R. (1979) Anomalous properties of ^{3}H spiperone binding sites in various areas of the rat limbic systems. *Mol. Pharmacol.* **15,** 505–514.

Hyttel J. (1981) Similarities between the binding of ^{3}H pifluthixol and ^{3}H flupenthixol to rat striatal dopamine receptors *in vitro. Life Sci.* **28,** 563–569.

Iversen L. L., Rogawski M. A., and Miller R. J. (1976) Comparison of the effects of neuroleptic drugs on pre- and postsynaptic dopaminergic mechanism in the rat striatum. *Mol. Pharmacol.* **12,** 251–262.

Janssen P. A. J. (1970) The Butyrophenone Story, in *Discoveries in Biological Psychiatry* (Ayd F. J. and Blackwell B., eds.) J. B. Lippincott, Philadelphia.

Janssen P. A. J., Niemegeers C. J. E. and Schellekens K. H. L. (1965) Is it possible to predict the clinical effects of neuroleptic drugs (major tranquilizers) from animal data? I. Neuroleptic activity spectra for rats. *Arzneimittelforsch./Drug Res.* **15,** 104–117.

Jenner P. and Marsden C. D. (1979) The substituted benzamides: A novel class of dopamine antagonists. *Life Sci.* **25,** 479–486.

Kebabian J. W. and Calne D. B. (1979) Multiple receptors for dopamine. *Nature* **277,** 93–96.

Kebabian J. W. and Greengard P. (1971) Dopamine-sensitive adenyl cyclase: Possible role in synaptic transmission. *Science* **174,** 1346–1349.

Kebabian J. W., Petzold G. L., and Greengard P. (1972) Dopamine-sensitive adenylate cyclase in caudate nucleus of rat brain, and its similarity to the "dopamine receptor". *Proc. Natl. Acad. Sci. USA* **69,** 2145–2149.

Kuhar M. J., Murrin C., Malouf A. T., and Klemm N. (1978) Dopamine receptor *in vivo:* The feasibility of autoradiographic studies. *Life Sci.* **22,** 203–210.

Kuno T. and Tanaka C. (1983) Correspondency between different affinity states and target size of the bovine striatal D_2 dopamine receptor. *Biochem. Biophys. Res. Commun.* **117,** 65–70.

Laduron P. M. (1976) Limiting factors in the antagonism of neuroleptics on dopamine-sensitive adenylate cyclase. *J. Pharm. Pharmacol.* **28,** 250–257.

Laduron P. M. (1980) Dopamine receptor: From an *in vivo* concept towards a molecular characterization. *Trends Pharmacol. Sci.* **1,** 471–474.

Laduron P. M. (1981) Dopamine Receptor: A unique Site with Multiple Postsynaptic Localization, in *Apomorphine and Other Dopaminomimetics* (Gessa G. L. and Corsini G. U., eds.) vol. 1, Raven, New York.

Laduron P. M. (1982) Brain Dopamine Receptor: Multiple Binding Sites or Physiological Receptor Site, in *Advances in Dopamine Research* (Kohsaka M., Kohsaka M., Sholmori T., Tsukada Y., and Woodruff G. N., eds.) vol. 37, Pergamon, Oxford, New York.

Laduron P. M. (1984a) Criteria for receptor sites in binding studies. *Biochem. Pharmacol.* **33,** 833–839.

Laduron P. M. (1984b) Lack of direct evidence for adrenergic and dopaminergic autoreceptors. *Trends Pharmacol. Sci.* **5,** 569–461.

Laduron P. M. and Ilien B. (1982) Solubilization of brain muscarinic, dopaminergic and serotonergic receptors: A critical analysis. *Biochem. Pharamacol.* **31,** 2145–2151.

Laduron P. and Leysen J. (1977) Specific in vivo binding of neuroleptic drugs in rat brain. *Biochem. Pharmacol.* **26,** 1003–1007.

Laduron P. and Leysen J. E. (1979) Domperidone, a specific in vitro dopamine antagonist devoid of in vivo central dopaminergic activity. *Biochem. Pharmacol.* **28,** 2161–2165.

Laduron P. M. and Wouters W. (1984) Isolation and Characterization of Dopamine Receptor Protein, in *Catecholamines: Basic and Peripheral Mechanisms* (Usdin E., Carlsson A., Dahlstrom A. and Engel J., eds.) Alan R. Liss, New York.

Laduron P., De Bie K., and Leysen J. (1977) Specific effect of haloperidol on dopamine turnover in the frontal cortex. *Naunyn Schmiedebergs Arch. Pharmacol.* **296,** 183–185.

Laduron P. M., Janssen P. F. M., and Leysen J. E. (1978a) Spiperone: A ligand of choice for neuroleptic receptors. 3. Subcellular distribution of neuroleptic drugs and their receptors in various rat brain areas. *Biochem. Pharmacol.* **27,** 323–328.

Laduron P. M., Janssen P. F. M., and Leysen J. E. (1978b) Spiperone: A ligand of choice for neuroleptic receptors. 2. Regional distribution and *in vivo* displacement of neuroleptic drugs. *Biochem. Pharmacol.* **27,** 317–321.

Laduron P. M., Janssen P. F. M., and Leysen J. E. (1978c) Characterization of specific *in vivo* binding of neuroleptic drugs in rat brain. *Life Sci.* **23,** 581–586.

Laduron P. M., Janssen P. F. M., and Leysen J. E. (1982) *In vivo* binding of ^3H ketanserin on serotonin S_2-receptors in rat brain. *Eur. J. Pharmacol.* **81,** 43–48.

Langley J. N. (1905) On the reaction of cells and of nerve endings to certain poisons, chiefly as regards the reaction of striated muscle to nicotine and curare. *J. Physiol.* **33,** 374–413.

Lazareno S. and Nahorski S. R. (1982) Selective labelling of dopamine (D_2) receptors in rat striatum by ^{3}H domperidone but not by ^{3}H spiperone. *Eur. J. Pharmacol.* **81,** 273–285.

Lee T., Seeman P., Tourtelotte W. W., Farley I. Y., and Hornykiewicz O. (1978) Binding of ^{3}H neuroleptics and ^{3}H apomorphine in schizophrenia brains. *Nature* **274,** 897–900.

Leff S. E. and Creese I. (1983) Dopamine receptors re-explained. *Trends Pharmacol. Sci.* **4,** 463–467.

Leff S. E. and Creese I. (1982) Solubilization of D-2 dopamine receptors from canine caudate: Agonist-occupation stabilizes guanine nucleotide sensitive receptor complexes. *Biochem. Biophys. Res. Commun.* **108,** 1150–1157.

Le Fur G., Phan T., and Uzan A. (1980) Identification of stereospecific [^{3}H]spiroperidol binding sites in mammalian lymphocytes. *Life Sci.* **26,** 1139–1148.

Lerner M. H., Rosengarten H., and Friedhoff A. (1981) Solubilization and characterization of ^{3}H spiroperidol binding sites of calf caudate. *Life Sci.* **29,** 2367–2374.

Leysen J. E. (1984) Receptors for Neuroleptic Drugs, in *Advances in Human Psychopharmacology* (Burrows G. B. and Werry J., eds.) vol. 3, JAI, Greenwich, Connecticut.

Leysen J. and Laduron P. (1977) Differential distribution of opiate acid neuroleptic receptors and dopamine-sensitive adenylate cyclase in rat brain. *Life Sci.* **20,** 281–288.

Leysen J. E., Gommeren W., and Laduron P. M. (1978a) Spiperone: A ligand of choice for neuroleptic receptors. 1. Kinetics and characteristics of in vitro binding. *Biochem. Pharmacol.* **27,** 307–316.

Leysen, J. E., Niemegeers C. J. E., Tollenaere J. P., and Laduron P. M. (1978b) Serotonergic component of neuroleptic receptors. *Nature* **272,** 168–171.

Leysen J. E., Tollenaere J. P., Koch M. H. J., and Laduron P. M. (1977) Differentiation of opiate and neuroleptic receptor binding in rat brain. *Eur. J. Pharmacol.* **43,** 253–267.

Leysen J., Gommeren W., and Laduron P. M. (1979) Distinction between dopaminergic and serotonergic components of neuroleptic binding sites in limbic brain areas. *Biochem. Pharmacol.* **28,** 447–448.

Lilly L., Davis A., Fraser C. M., Seeman P., and Venter J. C. (1985) Characterization of brain D_2 dopamine receptors. *Neurochem. Int.* **2,** 363–368.

Lilly L., Fraser C. M., Jung C., Seeman P., and Venter J. C. (1983) Molecular size of the canine and human D_2 dopamine receptor as determined by radiation inactivation. *Mol. Pharmacol.* **24,** 10–14.

Mackay A. V. P., Bird E. D., Spokes E. G., Rossor M., Iversen L. L., Creese I., and Snyder S. H. (1980) Dopamine receptors and schizophrenia: Drug effect or illness? *Lancet* **3**, 915–916.

Maloteaux J. M. and Laduron P. M. (1986) Receptors in Human Brain: From Regional Distribution to Molecular Characterization, in *The Role of Receptors in Biology and Medicine* (Gotto A. M. and O'Malley B. W., eds.) Raven, New York.

Maloteaux J. M., Gossuin A., Waterkein C., and Laduron P. M. (1983) Trapping of labelled ligands in intact cells: A pitfall in binding studies. *Biochem. Pharmacol.* **32,** 2545–2548.

Marchais D., Tassin J. P., and Bockaert J. (1980) Dopaminergic component of ^{3}H spiroperidol binding in the rat anterior cerebral cortex. *Brain Res.* **183,** 235–240.

Martres M. P., Sales N., Bouthenet M. L., and Schwartz J. C. (1985) Localisation and pharmacological characterization of D_2 dopamine receptors in rat cerebral neocortex and cerebellum using [^{125}I] iodosulpride. *Eur. J. Pharmacol.* **118,** 211–219.

Maziere B., Loc'h C., Baron J. C., Sojouropoulos P., Duquesnoy N., D'antona R., and Carnbon H. (1985) *In vivo* quantitative imaging of dopamine receptors in human brain using position emission tomography and [^{76}Br] bromospiperone. *Eur. J. Pharmacol.* **114,** 267–272.

Montagu K. R. (1957) Catechol compounds in rat tissues and in brain of different animals. *Nature* **180,** 244–245.

Moore K. E. and Thornbury J. E. (1975) Drug-induced dopaminergic supersensitivity. *Adv. Neurol.* **9,** 93–104.

Nestler E. J. and Greengard P. (1983) Protein phosphorylation in the brain. *Nature* **305,** 583–588.

Niehoff D. L., Palacios J. M., and Kuhar M. J. (1979) *In vivo* receptor binding: Attempts to improve specific/non specific ratios. *Life Sci.* **25,** 819–826.

Nielsen M., Klimek V., and Hyttel J. (1984) Distinct target of dopamine D_1 and D_2 receptors in rat striatum. *Life Sci.* **35,** 325–332.

Niemegeers C. J. E. and Janssen P. A. J. (1979) A systematic study of the pharmacological activities of dopamine antagonisits. *Life Sci.* **24,** 2201–2216.

Niznik H. B., Guan J. H., Neumeyer J. L., and Seeman P. (1984) A dopamine D_2 receptor photolabel: Azidoclebopride. *Eur. J. Pharmacol.* **104,** 389–390.

Owen F., Crow T. J., Poulter M., Cross A. J., Longden A., and Riley G. L. (1979) Increased dopamine receptor sensitivity in schizophrenia. *Lancet* **2,** 223–285.

Owen F., Cross A. J., Waddington J. L., Poulter M., Gamble S. J., and Crow T. J. (1980) Dopamine-mediated behaviour and ^{3}H spiperone binding to striatal membranes in rats after nine months haloperidol administration. *Life Sci.* **26,** 55–59.

Quik M., Iversen L. L., Larder A., and Mackay A. V. P. (1978) Use of ADTN to define specific ^{3}H spiperone binding to receptors in brain. *Nature* **274,** 513–514.

Randrup A. and Munkvad (1961) Pharmacology and physiology of stereotyped behavior. *J. Psychiat. Res.* **11,** 1–10.

Reches A., Wagner R. H., Jackson V., Yablonskaya-Alter, and Fahn S. (1983) Dopamine receptors in the denervated striatum: Further supersensitivity by chronic haloperidol treatment. *Brain Res.* **275,** 183–185.

Redouane K., Sokoloff P., Schwartz J. C., Hamdi P., Mann A., Wermuth C. G., Roy J., and Morgat J. L. (1985) Photoaffinity labeling of D_2 dopamine binding subunits from rat striatum, anterior pituitary and olfactory bulb with a new probe, ^{3}H azidosulpride. *Biochem. Biophys. Res. Commun.* **103,** 1086–1902.

Ruberg M., Bobobza B., Javoy-Agid F., Montfort J. C., and Agid Y. (1984) ^{3}H Spiperone binding in the nigrostriatal system in human brain. *Eur. J. Pharmacol.* **99,** 159–165.

Rupniak N. M. J., Hall M. D., Mann S., Fleminger S., Kilpatrick G., Jenner P., and Marsen C. D. (1985) Chronic treatment with clozapine, unlike haloperidol, does not induce changes in striatal D_2 receptor function in the rat. *Biochem. Pharmacol.* **34,** 2755–2763.

Scatton B., Glowinski J., and Julou L. (1976) Dopamine metabolism in the mesolimbic and mesocortical dopaminergic systems after single or repeated administrations of neuroleptics. *Brain Res.* **109,** 184–189.

Schultz D. W., Stanford E. J., Wyrick S. W., and Mailman R. B. (1985) Binding of [^{3}H] SCH 23390 in rat brain: Regional distributions and effects of assay conditions and GTP suggest interactions at a D_1 like dopamine receptor. *J. Neurochem.* **45,** 1601–1611.

Seeman P. (1980) Brain dopamine receptors. *Pharmacol. Rev.* **32,** 229–313.

Seeman P. (1982) Nomenclature of central and peripheral dopaminergic sites and receptors. *Biochem. Pharmacol.* **31,** 2563–2568.

Seeman P., Chau-Wong M., Tedesco J., and Wong K. (1975) Brain receptors for antipsychotic drugs and dopamine: Direct binding assays. *Proc. Natl. Acad. Sci. USA* **72,** 4376–4380.

Seeman P., Lee T., Chau-Wong M., and Wong K. (1976) Antipsychotic drug doses and neuroleptic/dopamine receptors. *Nature* **261,** 717–719.

Seeman P., Ulpian C., Grigoriadis D., Pri-Bar I., and Buckman O. (1985) Conversion of dopamine D_1 receptors from high to low affinity for dopamine. *Biochem. Pharmacol.* **34,** 151–154.

Senogles S. E., Amlaiky N., Johnson A. L., and Caron M. G. (1986) Affinity chromatography of the anterior pituitary D_2 dopamine receptor. *Biochemistry* **25,** 749–573.

Sokoloff P., Martres M. P., and Schwartz J. C. (1980) Three classes of dopamine receptor (D_2, D_3, D_4) identified by binding studies with ^3H apomorphine and ^3H domperidone. *Naunyn Schmiedebergs Arch. Pharmacol.* **315,** 89–102.

Stadler H., Lloyd K. G., Gadea-Ciria M., and Bartholini G. (1973) Enhanced striatal acetylcholine release by chlorpromazine and its reversal by apomorphine. *Brain Res.* **55,** 476–480.

Staunton D. A., Magistretti P. J., Koob G. F., Shoemaker W. J., and Bloom F. E. (1982) Dopaminergic supersensitivity induced by denervation and chronic receptor blockade is additive. *Nature* **299,** 72–74.

Stefanini E. and Clement-Cormier Y. (1981) Detection of dopamine receptors in the area postrema. *Eur. J. Pharmacol.* **74,** 257–260.

Stoof J. C. and Kebabian J. W. (1984) Two dopamine receptors: biochemistry, physiology and pharmacology, *Life Sci.* **35,** 2281–2296.

Strange P. G., Abbott W. M., and Wheatley M. (1984) Brain D_2-Dopamine Receptors: Isolation and Characterization, in *Catecholamines: Basic and Peripheral Mechanisms* (Usdin E., Carlsson A., Dahlstrom A., and Engel J., eds.) Alan R. Liss, New York.

Tam S. and Seeman P. (1978) Neuroleptic receptors in calf caudate. Solubilization by digitorium. *Eur. J. Pharmacol.* **52,** 151–152.

Thal L., Mishra R. K., Gardner E. L., Horowitz S. G., Varmuza S., and Makman M. H. (1979) Dopamine antagonist binding increases in two behaviorally distinct striatal denervation syndromes. *Brain Res.* **170,** 381–386.

Theodorou A., Gommeren W., Clow A., Leysen J., Jenner P., and Marsden C. D. (1981) Chronic neuroleptic treatment specifically alters the number of dopamine receptors in rat brain. *Life Sci.* **28,** 1621–1627.

Thierry A. M., Blanc C., Sobel A., Stinus L., and Glowinski J. (1973) Dopaminergic terminals in the rat cortex. *Science* **182,** 499–501.

Trabucchi M., Longoni R., Fresia P., and Spano P. F. (1975) Sulpiride: A study of the effects on dopamine receptors in rat neostriatum and limbic forebrain. *Life Sci.* **17,** 1551–1556.

Ungerstedt U. (1971) Postsynaptic sypersensitivity after 6-hydroxydopamine induced degeneration of the nigrostriatal dopamine system. *Acta Physiol. Scand.* **367** (suppl.), 69–93.

Van Rossum J. M. (1966) The significance of dopamine-receptor blockade for the mechanism of action of neuroleptic drugs. *Arch. Int. Pharmacodyn. Ther.* **160,** 492–494.

Wagner Jr. H. N., Burns D. H., Dannals R. F., Wong D. F., Langstrom B., Duelfer T., Frost J. J., Ravert H. T., Links J. M., Rosenbloom B. B.,

Lukas S. E., Kramer A. V., and Kuhar M. J. (1983) Imaging dopamine receptors in the human brain by position tomography. *Science* **221,** 1264–1266.

Weatley M. and Strange P. G. (1983) Characterization of brain D_2 dopamine receptors solubilised by lysophosphatidylcholine. *FEBS Lett.* **151,** 97–101.

Weil-Malherbe H. and Bone A. D. (1957) Intracellular distribution of catecholamines in the brain. *Nature* **180,** 1050–1051.

Woodruff G. N. and Freedman S. B. (1981) Binding of [^{3}H] sulpiride to purified rat striatal synaptic membranes. *Neuroscience* **6,** 407–410.

Wouters W., Van Dun J., and Laduron P. M. (1984a) Characterization of cholate-solubilized dopamine receptors from human, dog and rat brain. *Biochem. Pharmacol.* **33,** 4039–4044.

Wouters W., Van Dun J., and Laduron P. M. (1984b) Photoaffinity labeling of human brain dopamine receptors. *Biochem. Pharmacol.* **33,** 3517–3520.

Wouters W., Van Dun J., and Laduron P. M. (1985) Photoaffinity labelling of dopamine receptors. Synthesis and binding characteristics of azapride. *Eur. J. Biochem.* **145,** 273–278.

Use of 5-HT Receptor Agonists and Antagonists for the Characterization of Their Respective Receptor Sites

J. E. Leysen

1. Introduction

Serotonin (5-hydroxytryptamine, 5-HT), was discovered as a serum factor that exerts tonic effects on blood vessels; its chemical structure was elucidated in 1948. The compound gained wide interest among pharmacologists, and in a short time a great number of in vitro and in vivo pharmacological tests had been developed, and a variety of drugs interacting with the actions of serotonin were rapidly discovered. In 1966, Gyermek, in a review on "Drugs which antagonize 5-hydroxytryptamine and related indolealkylamines," compiled a table with 384 compounds with serotonin antagonistic properties; these compounds belonged to more than 18 different chemical and pharmacological classes. In addition, an impressive list was reported that consisted of 26 different tests for measuring serotonergic effects on isolated organs and 33 tests related to the action of serotonin in vivo, plus a number of indirect effects. At that time, the author wondered whether the vast interest paid to serotonin was fully deserved. He stated, "It turned out that besides the potent pharmacological actions of exogenous serotonin, the endogenous amine in mammals does not play many well-defined physiologic or pathologic roles." Existing drugs were found to be nonselective, and it was hoped that "specific antagonists" would be developed for analyzing the effects of indolealkylamines in various organisms (Gyermek, 1966).

In spite of the early-recognized variety of pharmacological actions of serotonin and the development of numerous tests in vitro and in vivo, these studies have not given rise to a thorough subclassification of serotonin receptors. Until the late 1970s, the only attempted classification was the one proposed by Gaddum and Picarelli (1957). They considered "M" and "D" receptors based on contractile effects of serotonin on guinea pig ileum preparations that were differentially antagonized by morphine (M) and dibenzyline (D), respectively. The antagonistic effect of morphine was found to be indirect, however, and caused by its inhibitory action on acetylcholine release. "M" receptors are now considered to be "neuronal receptors," i.e., presumably localized on nerve endings; they have been further classified into M_s (on sympathic neurons) and M_p (on parasympathic neurons) (Richardson et al., 1985).

Whereas originally attention was focused primarily on the peripheral actions of serotonin, particularly on the cardiovascular system, the discovery of serotonin-containing neurons in the brain (Dahlstrom and Fuxe, 1964) shifted the interest to the central nervous system. From studies of biosynthesis, metabolism, release, and turnover, a neurotransmitter role for serotonin has been postulated. But, here also many gaps remained, as pointed out in a recent critical review by Kuhn et al. (1986). The authors reflected how easily original assumptions and principles have become axiomatic.

The novelty of the past decade, unquestionably, has been the development of radioligand binding studies using brain tissue homogenates. They have revived the interest in serotonin, stimulated attempts toward the subclassification of serotonergic receptor sites, and led to the development of new classes of drugs. Radioligand binding studies also provide a readily accessible means for evaluating the selectivity of drugs. Originally, two different "serotonergic" binding sites were detected in mammalian brain tissue. Sites called 5-hydroxytryptamine$_1$, or 5-HT$_1$, sites (also called serotonin-S$_1$ sites) are labeled with high affinity by ^{3}H-serotonin and show high binding affinity for most serotonin agonists, but low binding affinity for known serotonin antagonist. Sites designated as 5-HT$_2$ or serotonin-S$_2$ sites were originally detected using ^{3}H-spiperone; they show high binding affinity for various serotonin antagonists and lower binding affinity for serotonin agonists (Bennett and Snyder, 1976; Leysen et al., 1978; Peroutka and Snyder, 1979; Leysen, 1981). The 5-HT$_1$ sites have been further subclassified into 5-HT$_{1A}$, 5-HT$_{1B}$, and 5-HT$_{1C}$ sites, based mainly on the analysis of binding curves and on distinct anatomical localization of binding sites (Pedigo et al., 1981; Pazos et al., 1985c). The challenge of recent research is to prove a role for the various binding site subtypes in functional effects of serotonin, a prerequisite before a binding site can be called a receptor. Characteristics of serotonergic binding sites have been extensively described in recent reviews (*see* Leysen, 1985; Peroutka, 1986).

A still distinct classification of serotonin receptors emerged from electrophysiological studies and refers to S_1 sites (excitatory receptors on motoneurons), S_2 sites (inhibitory autoreceptors on serotonergic neurons), and S_3 sites mediating depression of neuronal firing (Aghajanian, 1981). Since the electrophysiological classification is unrelated to the binding site classification, I will, in this chapter, use the 5-HT denomination for the binding sites and the S denomination for the electrophysiologically defined sites.

It appears that today we are faced with the same problems as pointed out by Gyermek (1966), i.e., a lack of selective drugs for antagonizing or mimicking the many different pharmacological actions of serotonin, and uncertainty about the physiological roles of the endogenous amine. The mountain of literature has grown exponentially, however, and has become incomprehensible. It surely is not a simple task to discuss "the use (or usefulness?) of serotonin agonists and antagonists for the characterization of their respective receptor sites." Trying to give an overview of all drugs acting on serotonergic receptor sites would perhaps be irrelevant. Yet it seemed important to refocus attention on the chemical and pharmacological diversity of drugs with alledged serotonergic agonist or antagonist activity. A chemical classification of serotonin agonists and antagonists of current interest is proposed. The use of these drugs for the identification of subclasses of serotonin receptor sites, in particular in radioligand binding studies, and the usefulness of the drugs to relate binding sites to functional effects is discussed. Finally, this chapter would perhaps gain most from extensive information on the receptor binding profile of the various drugs, not only regarding the interaction with subtypes of serotonergic binding sites, but also regarding interaction with various neurotransmitter receptor binding sites. Without extensive discussion, the information in itself should be a helpful guide to the selection of drugs for new studies, and it could be applied to interpretation and evaluation of new reported findings.

2. Chemical Classification of Serotonin Agonists and Antagonists

Based on the structural features of drugs, purported serotonin agonists and antagonists of current interest have been classified into eight major classes (Table 1). The classification is based pri-

Table 1
Chemical Classification of Currently Used Serotonin Agonists and Antagonists

I INDOLE DERIVATIVES

I.1. Tryptamine (T) derivatives

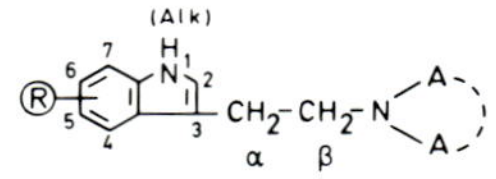

Ⓡ	$-N\langle A \rangle$	GENERIC OR TRIVIAL NAME	ALLEDGED SEROTONERGIC ACTIVITY	(ref)
5-OH	$-NH_2$	serotonin (5-HT)	agonist ligand 5-HT_1	 (1)
5-OCH_3	$-NH_2$	5-methoxy T	agonist	(2)
5-$CONH_2$	$-NH_2$	5-carbamoyl T	agonist 5-HT_1	(2)
5-OH, 2-CH_3	$-NH_2$	5-hydroxy-2-methyl T	agonist M_P	(3)
5-OH, α-CH_3	$-NH_2$	5-hydroxy-α-methyl T	agonist D	(3)
5-OH	$-NH-CH_3$	5-hydroxy-N-methyl T	agonist	(2)
5-OH Alk = CH_3	$-NH_2$	5-hydroxy-1-methyl T	agonist	(2)
5-OH	$-N(CH_3)_2$	bufotenine	agonist	(4)
5-OCH_3	$-N(CH_3)_2$	N,N-dimethyl-5-methoxy T	agonist 5-HT_1	(5)
4-OH	$-N(CH_3)_2$	psilocin	agonist	(6)
4-O-P(=O)(OH)-OH	$-N(CH_3)_2$	psilocybin	agonist	(6)
H	$-NH_2$	T	agonist	(4)
H	$-N(CH_3)_2$	N,N-dimethyl T	agonist	(6)

I. 2. Ergotamines

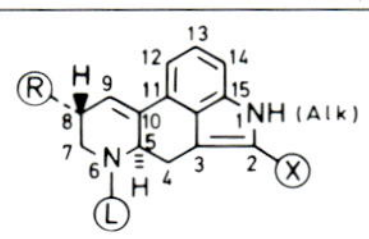

Ⓧ	Ⓛ	Ⓡ	GENERIC OR TRIVIAL NAME	ALLEDGED SEROTONERGIC ACTIVITY	(ref)
H	$-CH_3$	$-CO-N(C_2H_5)_2$	lysergide (LSD)	agonist-antagonist ligand 5-HT_1, 5-HT_2	(6) (1)
Br	$-CH_3$	$-CO-N(C_2H_5)_2$	2-bromo-LSD (BOL)	antagonist 5-HT_2	(7)

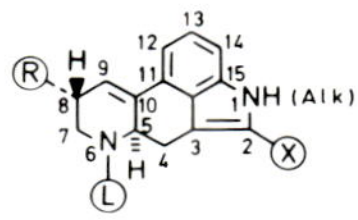

I. 2. Ergotamines

Ⓧ	Ⓛ	Ⓡ	GENERIC OR TRIVIAL NAME	ALLEDGED SEROTONERGIC ACTIVITY	(ref)
I	$-CH_3$	$-CO-N(C_2H_5)_2$	2-Iodo-LSD (IOL)	antagonist 5-HT_2 ligand 5-HT_2, 5-HT_{1C}	(8) (9)
I	$-CH_3$ Alk = CH_3	$-CO-N(C_2H_5)_2$	2-Iodo-1-methyl-LSD (IMIL)	antagonist 5-HT_2 ligand 5-HT_2	(10)
H	$-CH_3$	$-CO-NH-CH(CH_3)-CH_2OH$	ergometrine ergonovine	agonist-antagonist	(6)
H	$-CH_3$ Alk = CH_3	$-CO-NH-CH(C_2H_5)-CH_2OH$	methysergide	agonist-antagonist	(6) (4)
H	$-CH_3$	$-NH-CON(C_2H_5)_2$	lisuride	agonist-antagonist	(6)
H	$-CH_3$ Alk = CH_3	$-NH-SO_2-N(CH_3)_2$	mesulergine	antagonist 5-HT_2 ligand 5-HT_2, 5-HT_{1C}	(11) (12)

I. 3. 9,10-dihydroergotamines

Ⓧ	Ⓛ	Ⓡ	GENERIC OR TRIVIAL NAME	ALLEDGED SEROTONERGIC ACTIVITY	(ref)
H	$-CH_3$ Alk = iC_3H_7	$-COOCH(CH_3)-CH(CH_3)OH$	LY-53857	antagonist 5-HT_2	(13)
H	$-CH_3$ Alk = CH_3	$-CH_2-NH-COO-CH_2-$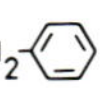	metergoline	antagonist ligand 5-HT_2	(14)

I. 4. Indolyltetrahydropyridine

	GENERIC OR TRIVIAL NAME	ALLEDGED SEROTONERGIC ACTIVITY	(ref)
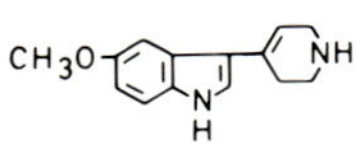	RU-24969	agonist 5-HT_{1B} 5-HT_{1A}	(5) (15)

(continued on next page)

Table 1 *(continued)*

I.5. Tropine derivatives

	GENERIC OR TRIVIAL NAME	ALLEDGED SEROTONERGIC ACTIVITY	(ref)
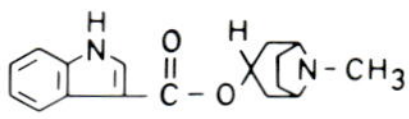	ICS-205930	antagonist M_S, M_P	(3)
Non-indole tropine derivative			
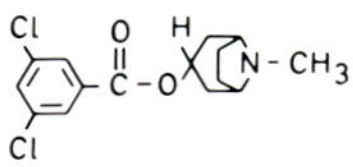	MDL-72222	antagonist M_S, M_P	(16)

II. ARYLPIPERAZINES (P)

Ar	(L)	GENERIC OR TRIVIAL NAME	ALLEDGED SEROTONERGIC ACTIVITY	(ref)
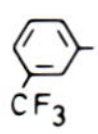	H	Trifluoromethylphenyl P (TFMPP)	agonist 5-HT_{1B} 5-HT_{1A+B}	(5) (15)
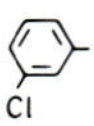	H	Metachlorophenyl P (mCPP)	agonist 5-HT_{1B} 5-HT_{1A+B}	(5) (15)
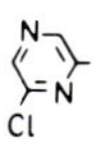	H	MK-212	agonist	
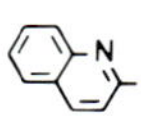	H	Quipazine	agonist-antagonist	
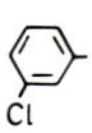	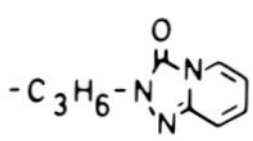	Trazodone	agonist-antagonist	
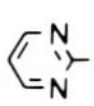	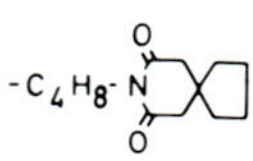	Buspirone	5-HT_{1A}-agonist	(17)
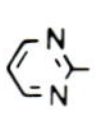	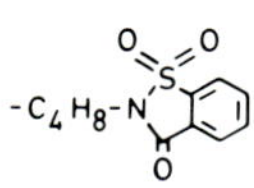	TVX Q-7821	5-HT_{1A}-agonist 5-HT_{1A}-ligand	(18) (19)
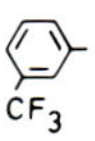	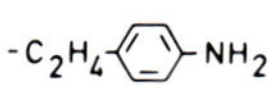	Aminophenylethyl-TFMPP (PAPP)	5-HT_{1A}-ligand	(20)

III. 2-AMINOTETRALINS

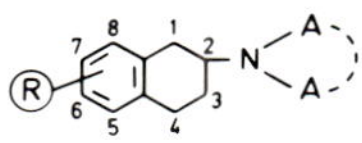

Ⓡ		GENERIC OR TRIVIAL NAME	ALLEDGED SEROTONERGIC ACTIVITY	(ref)
8-OH	$-N(CH_2-CH_2-CH_3)(CH_2-CH_2-CH_3)$	N, N-dipropyl-8-hydroxy-2-aminotetralin (8-OHDPAT or PAT)	agonist agonist 5-HT_{1A} ligand 5-HT_{1A}	(21) (22) (23)
8-OCH_3	$-N(CH_2-CH_2-CH_3)(CH_2-CH(Cl)-CH_3)$	2'-chloro-8-methoxy PAT	5-HT_{1A} alkylating ligand	(24)
8-OCH_3	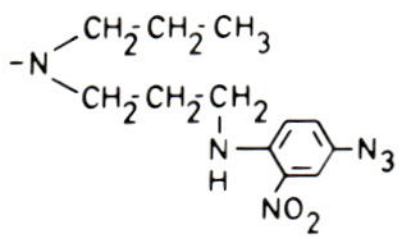	NAP-8-methoxy PAT	5-HT_{1A} photoaffinity ligand	(25)

IV. METHOXYPHENETYLAMINES

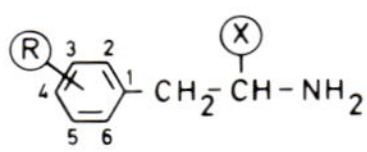

Ⓡ	Ⓧ	GENERIC OR TRIVIAL NAME	ALLEDGED SEROTONERGIC ACTIVITY	(ref)
2-OCH_3, 5-OCH_3, 4-Br	-CH_3	1-(4-bromo-2,5-dimethoxyphenyl)-2-aminopropane (DOB)	agonist 5-HT_2 ligand 5-HT_2	(26) (27)
2-OCH_3, 5-OCH_3, 4-CH_3	-CH_3	1-(2,5-dimethoxy-4-methylphenyl)-2-aminopropane (DOM)	agonist 5-HT_2	(26)
3-OCH_3, 4-OCH_3, 5-OCH_3	H	Mescaline	agonist	(6)

V. 6,7,6 MEMBERED RING TRICYCLICS

	GENERIC OR TRIVIAL NAME	ALLEDGED SEROTONERGIC ACTIVITY	(ref)
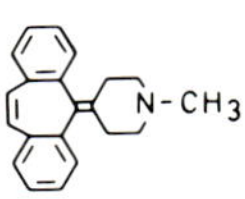	Cyproheptadine	antagonist 5-HT_2	(4) (28)
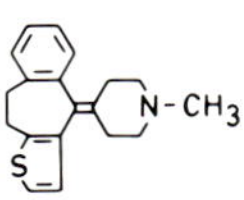	Pizotifen	antagonist 5-HT_2	(4) (28)

(continued on next page)

Table 1 *(continued)*

V. 6,7,6 MEMBERED RING TRICYCLICS

	GENERIC OR TRIVIAL NAME	ALLEDGED SEROTONERGIC ACTIVITY	(ref)
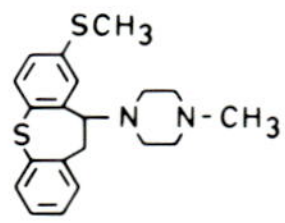	Metitepine (Methiotepine)	antagonist 5-HT_2 antagonist 5-HT_1	(4)(28) (29)
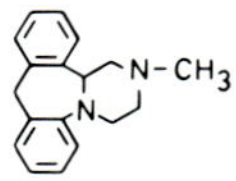	Mianserin	antagonist 5-HT_2 ligand 5-HT_2	(4) (28) (30)

VI. ARYLOXY(THIO)ALKYLAMINES

	GENERIC OR TRIVIAL NAME	ALLEDGED SEROTONERGIC ACTIVITY	(ref)
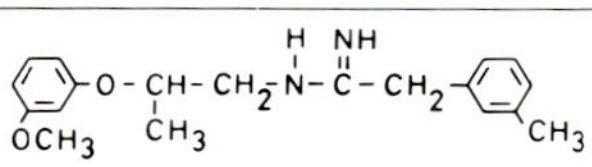	Xylamidine	antagonist 5-HT_2	(28)
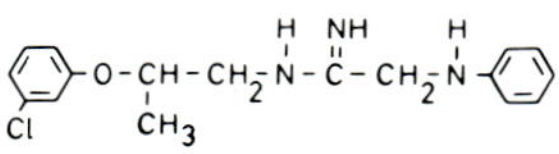	BW-501	antagonist 5-HT_2	
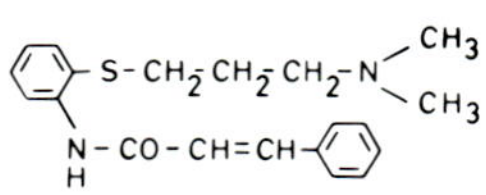	Cinanserin	antagonist 5-HT_2	(4) (28)

VII. 4-BENZOYLPIPERIDINES

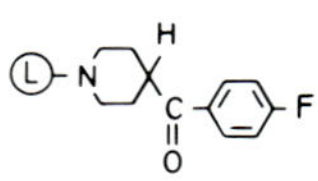

(L)	GENERAL OR TRIVIAL NAME	ALLEDGED SEROTONERGIC ACTIVITY	(ref)
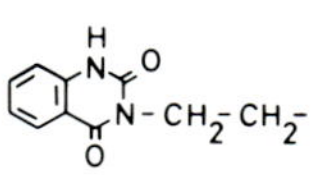	Ketanserin	antagonist 5-HT_2 ligand 5-HT_2	(4) (28)
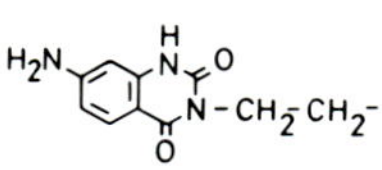	7-Aminoketanserin	antagonist 5-HT_2 ligand 5-HT_2	 (31)

VII. 4-BENZOYLPIPERIDINES

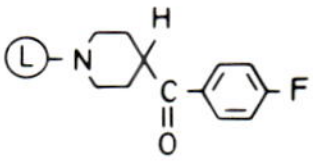

(L)	GENERAL OR TRIVIAL NAME	ALLEDGED SEROTONERGIC ACTIVITY	(ref)
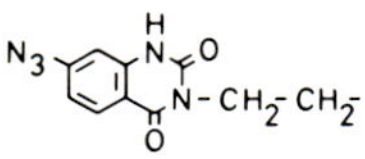	7-Azidoketanserin	Photoaffinity ligand 5-HT_2	(32)
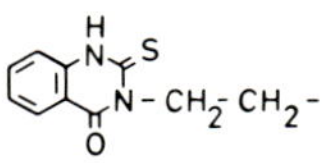	Altanserin	antagonist 5-HT_2	(33)
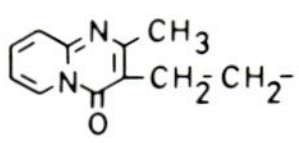	Pirenperone	antagonist 5-HT_2	(34)
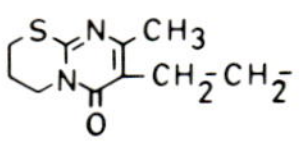	Setoperone	antagonist 5-HT_2	(33) (35)

VIII. 4-BENZHYDRILENEPIPERIDINES

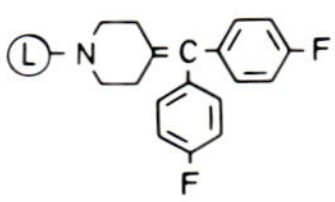

(L)	GENERAL OR TRIVIAL NAME	ALLEDGED SEROTONERGIC ACTIVITY	(ref)
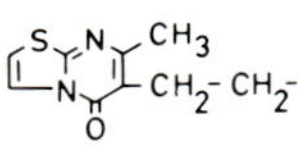	Ritanserin	antagonist 5-HT_2	(36) (33)
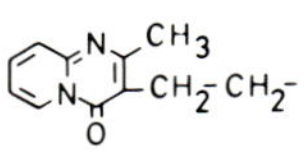	R 56413	antagonist 5-HT_2	(33)

[1]Peroutka and Snyder, 1979
[2]Engel et al., 1983
[3]Richardson et al., 1985
[4]Leysen et al., 1978
[5]Sills et al., 1984
[6]Gyermek, 1966
[7]Leysen and Laduron, 1977
[8]Engel et al., 1984b
[9]Yagaloff and Hartig, 1985

[10]Hoffman et al., 1985
[11]Pazos et al., 1985b
[12]Pazos et al., 1985c
[13]Cohen et al., 1985a
[14]Hamon et al., 1981
[15]Peroutka, 1985b
[16]Fozard, 1984
[17]Peroutka, 1985a
[18]Glaser and Traber, 1985
[19]Dompert et al., 1985
[20]Ransom et al., 1986
[21]Hjorth et al., 1982
[22]Middlemiss and Fozard, 1983
[23]Gozlan et al., 1983
[24]Emerit et al., 1985a
[25]Emerit et al., 1985b
[26]Glennon et al., 1984
[27]Titeler et al., 1985
[28]Leysen et al., 1982b
[29]Peroutka et al., 1981
[30]Peroutka and Snyder, 1981
[31]Wouters et al., 1986
[32]Wouters et al., 1985a
[33]Janssen, 1985
[34]Colpaert, 1982
[35]Leysen et al., 1986
[36]Leysen et al., 1985; further derivatives of ergotamines and dihydroergotamines were studied for their effects on 5-HT_1–5-HT_2 selectivity by Beart et al., 1986.

marily on chemical structure and does not distinguish between drugs with agonistic or antagonistic properties or between drugs with selectivity toward certain receptor subtypes. Groups have been arranged in such a way, however, that agonistic drugs (or drugs with mixed agonist–antagonist activity) are found mostly in classes I–IV and antagonists in classes V–VIII. Alleged activity (agonist, antagonist) or purported selectivity toward a serotonergic site subtype and references have been indicated.

By far the largest class, still containing five subclasses, consists of the indole derivatives, grouped in class I. It is composed of the earliest known agents such as the tryptamine derivatives and the ergotamines and dihydroergotamines. Tryptamine derivatives have predominantly agonistic activity; selectivity toward certain serotonin receptor subtypes has recently been ascribed to some of these agents (Table 1). The ergotamines and dihydroergotamines are a vast group of compounds showing a wide variety of activities, not only concerning agonist–antagonist action on the serotonergic system, but also concerning interactions with various different neurotransmitter systems. In this classification only a very restricted number of ergot derivatives that have recently been used as ligands or tools for receptor classification has been considered. More recent drugs in class I are the indoletetrahydropyridine with alleged agonist activity toward the 5-HT_1 sites, and the tropine derivative, which is claimed as a selective antagonist for "neuronal" (M_s, M_p) serotonin receptors.

Drugs belonging to class II, the arylpiperazines, are mostly agonistic compounds, although for some of them mixed agonist–antagonist properties have been reported. Recently, these agents have gained attention because of presumed selectivity toward the 5-HT_1 sites.

The most novel group of serotonin agonists consists of the 2-aminotetralins with an 8-hydroxy or 8-methoxy substituent, listed in class III. Whereas aminotetralins with hydroxyl substituents in the 4, 5, 6, or 7 positions were known as dopamine agonists, hydroxyl substitution in the 8 position yielded a compound with serotonin mimetic properties and lacking the dopaminergic effects (Hjorth et al., 1982). The compounds became particularly intriguing because of their high binding affinity and selectivity for the 5-HT_{1A} binding sites (Middlemiss and Fozard, 1983; Gozlan et al., 1983).

The fourth class of agonistic compounds consists of the methoxyphenethylamines, which may be more selective for the 5-HT_2 receptor sites (Glennon et al., 1984; Titeler et al., 1985).

Class V consists of 6, 7, 6-membered ring tricyclics, comprising the earlier-known serotonin antagonists. These drugs are nonselective, however, and potently interact with different neurotransmitter receptors (*see below*). Class VI is termed aryloxy-(thio)alkylamines, which can group the amidines and cinanserin, revealing structural similarities between these drugs. Also these drugs have been known for several years; they have turned out to be among the most selective antagonists at 5-HT_2 receptors (see below).

Novel serotonin antagonists are classified in groups VII and VIII. Class VII, the 4-benzoylpiperidines, has as a prototype, ketanserin, which is now widely regarded as "the" typical 5-HT_2 antagonist. The still more recent class VIII of 4-benzhydrilenepiperidines, has as a prototype ritanserin, which reveals extremely high affinity and very slow dissociation from 5-HT_2 receptor sites (Leysen et al., 1985).

Besides these drugs, which are known in the first place for their interaction with serotonergic receptors, many other drugs, belonging to various different pharmacological and chemical classes, have been found to have potent serotonin antagonistic properties or were recently found to bind with high affinity to a subtype of serotonin binding sites. Examples of such drugs are listed in Table 2.

In spite of the diversity of the chemical structures that can interact with serotonergic receptors, minor structural changes within a particular class sometimes totally abolish the affinity for the serotonergic sites. In the class of butyrophenones, pipamperone is primarily a potent 5-HT_2 antagonist, spiperone has both potent dopamine and serotonin antagonistic activity, but haloperidol has a much lower affinity for the 5-HT_2 receptor sites and is primarily a dopamine antagonist. Even more striking examples are found among the 6, 7, 6-membered ring tricyclics. Many compounds with this structural moiety, found among neuroleptics, histamine-H_1 antagonists, and monoamine uptake inhibitors, reveal potent serotonin antagonistic properties. Other congeners with only minor structural differences totally lack this activity. Pizotifen and ketotifen only differ in a keto substituent in the 10-position of the cycloheptane ring; pizotifen is a potent serotonin and histamine antagonist, yet ketotifen is a potent antihistamine devoid of serotonin antagonistic properties. The same is observed for azatadine, in which the thiophene ring in pizotifen is replaced by a pyridine ring, in spite of the fact that thiophene and pyridine rings can be considered as isosteric moieties. Thus, the question of structure–activity relationships for interaction with serotonin receptor sites is still enigmatic.

3. Serotonin Receptor Subclassification Based on Pharmacological Observations

As mentioned above, numerous pharmacological tests in vitro and in vivo related to the action of serotonin have been described, but they have not been the basis for a thorough serotonin receptor subclassification [*see* Gyermek (1966) and recent monographs edited by Vanhoutte (1985) and Green (1985)].

Table 2
Various Pharmacological Agents with High Binding Affinity for Serotonergic Binding Sites

Pharmacological and structural class	Serotonergic binding site	References
β-Adrenergic antagonist: indoles		
Pindolol	5-HT_{1A+B}	Hoyer et al., 1985b
Iodoisocyanopindolol	ligand 5-HT_{1B}	Hoyer et al., 1985a
(-) 21009	5-HT_{1A+B}	Hoyer et al., 1985a
α_1-Adrenergic antagonist: benzodioxan		
WB4101	5-HT_{1A} ligand	Norman et al., 1985
Neuroleptics		
Butyropherones		
Spiperone	5-HT_2 ligand	Leysen et al., 1978
Pipamperone	5-HT_2 antagonist	Leysen et al., 1978
6,6,6- and 6,7,6-Membered ring tricyclics		
Various agents	5-HT_2 antagonists	Leysen and Niemegeers, 1985
Benzodioxan-spirodecanone		
Spiroxatrine	5-HT_{1A}	Nelson and Taylor, 1985
Dopamine-D_1 antagonist: benzazepine		
SCH-23390	5-HT_2 antagonist	Leysen and Janssen, 1988

(continued on next page)

Table 2 *(continued)*

Pharmacological and structural class	Serotonergic binding site	References
Gastrokinetic: 4-(2-methoxybenzamide)piperidine		
Cisapride	5-HT_2 antagonist	Unpublished observation
Histamine-H_1 antagonists		
6,7,6-Membered ring tricyclics		
Doxepine	5-HT_2 antagonist	
4-piperidinyl,1H benzimidazol-2-amines		
Astemizole	5-HT_2 antagonist	Laduron et al. 1982a
R 43448	5-HT_2 antagonist	Leysen et al. 1982b
Diphenylmethyl-piperazine		
Oxatomide	5-HT_2 antagonist	Leysen and Gommeren, 1986
Monoamine-uptake inhibitors		
6,7,6-membered ring tricyclics		
Amitriptyline	5-HT_2 antagonist	Leysen and Tollenaere, 1982
Clomipramine	5-HT_2 antagonist	Leysen and Janssen, 1988

Tentative serotonin receptor subclassifications that have originated from pharmacological studies using isolated organs are summarized in Table 3. The observed effects and drugs that were found to be typical antagonists have been indicated. The earliest classification was proposed by Gaddum and Picarelli (1957) and considers "M" and "D" receptors. The antagonism by morphine of serotonin-induced contractions of the guinea pig ileum mediated by M receptors was found to be indirect, and morphine should not be considered a typical serotonin antagonist. M-receptor-mediated contraction of the guinea pig ileum by serotonin appeared to be a consequence of the stimulation of acetylcholine release probably via serotonergic receptors on cholinergic nerve terminals. Subclasses of "neuronal" serotonin receptors have been further distinguished by selective actions of new potent antagonists such as MDL-72222 (Fozard, 1984) and ICS-205930 (*see* Table 1, class I, 5) (Richardson et al., 1985). The sites have been called M_s and M_p according to their localization on sympathic and parasympathic neurons. The above-mentioned tropine derivatives do not interact with any other subtype of serotonergic receptors, and they reveal little affinity for various neurotransmitter receptor sites. Therefore the neuronal serotonin receptors have been considered as distinct receptor types, for which a corresponding radioligand binding site has not yet been found. D-type receptors, originally identified in the guinea pig ileum, were also claimed to be present in the fundus (Vane, 1957) and the uterus (Richardson et al., 1985). Dibenzyline was considered a typical antagonist at these sites. Engel et al. (1984a) demonstrated a relationship between the pharmacological properties of D-receptors in the ileum and the 5-HT_2 receptor binding sites: also, serotonin receptors in the uterus and trachea displayed features of 5-HT_2 binding sites (Cohen et al., 1985b; Van Nueten et al., 1982). A correlation did not exist, however, between 5-HT_2 binding site affinities of drugs and drug potencies in antagonizing serotonin-induced contractions in the fundus. The fundus receptor did not correspond to 5-HT_1 sites either (Leysen et al., 1982b; Leysen, 1983; Cohen et al., 1985b; Cohen and Wittenauer, 1985). Hence, serotonergic effects on nonvascular smooth muscle appear to be heterogenous, and differences between alleged D-type receptors apparently exist. Kaumann (1983, 1985) studied serotonergic effects on the heart atrium and ventricles, and these were found to be antagonized by phenoxybenzamine, an α-adrenergic antagonist. Because of the ineffectiveness of various purported serotonin antagonists, the effects could not be related to any of the previously defined serotonin receptor subtypes; the terminology 5-HT_3 and 5-HT_4 sites was proposed. One should not rush to such a conclusion based on the activity of only one, very nonselective agent such as phenoxybenzamine, however, particularly since interactions between α-adrenergic and serotonergic effects have been discribed (Curro et al., 1978).

Table 3
Serotonin Receptor Subclassification Based on Pharmacological or Electrophysiological Responses

Receptor subtype	Test system	Agonist	Antagonist	References
	Isolated organs			
M	Guinea pig ileum: activation of cholinergic neuron leading to smooth muscle contraction	5-HT	Morphine	Gaddum and Picarelli, 1957
D	Guinea pig ileum and uterus: smooth muscle contraction	5-HT S(+)-α-CH_3-5-HT	Dibenzyline	Gaddum and Picarelli, 1957 Richardson et al., 1985
M_S	Post ganglionic sympathic neurons: stimulation of transmitter release leading to chronotropic response of the heart	5-HT	MDL-72222 ICS-205930	Fozard, 1984 Richardson et al., 1985

M_P	Desheated rabbit vagus nerve	5-HT, 2-CH_3-5-HT	ICS-205930	Richardson et al., 1985
5-HT_3	Right and left atrium: enhanced sinoatrial beating rate (right), and contractile force (left)	5-HT	Phenoxybenzamine	Kaumann, 1983, 1985
5-HT_4	Kitten right ventricle	5-HT	Phenoxybenzamine	Kaumann, 1983, 1985
	Electrophysiological effects			
S_1	Motoneurons: amplification of response to excitatory amino acids	5-HT	Cyprohetadine Methysergide Metergoline	McCall and Aghajanian, 1979
S_2	Serotonergic neurons: autoreceptor in the nucleus raphé dorsalis (inhibition of firing)	5-HT	LSD	Aghajanian, 1981
S_3	Depression of neuronal firing	5-HT, LSD	Unknown	Aghajanian, 1981

Serotonin exerts different effects on vascular smooth muscle. Contractile effects on several blood vessels were shown to be mediated by 5-HT_2-type receptors, whereas serotonin-induced relaxation of blood vessels has tentatively been related to 5-HT_1 sites. Since in these cases the pharmacological effects were related *a posteriori* to binding site subtypes, these findings will be discussed below. The same pertains to in vivo pharmacological effects related to serotonin. In themselves they did not give rise to serotonin receptor subclassification, but following the identification of the binding sites, relationships have been investigated.

4. Serotonin Receptor Subclassification According to Electrophysiological Effects

In electrophysiological studies, multiple effects of serotonin on neuronal firing have been observed, leading to the proposition of three different receptor subtypes as described in Table 3 (Aghajanian, 1981). The excitatory effects of serotonin on motoneurons, classified as S_1 effects, were found to be antagonized by cyproheptadine, methysergide, and metergoline. Peroutka et al. (1981) suggested a relationship between these receptors and the 5-HT_2 binding sites. The correspondence has not been firmly established, however, because of a lack of sufficient quantitative data on drug potencies in the electrophysiological studies. Thus far, no antagonists have been found for serotonergic effects mediated by the electrophysiologically defined S_2 sites, i.e., inhibitory autoreceptors on serotonergic neurons and S_3 sites involved in depression of neuronal firing. As a consequence, attempts to relate these effects to 5-HT_1 binding sites are not substantiated (Peroutka et al. 1981).

5. Serotonergic Binding Sites

The classification and features of serotonergic binding sites, which have originated from radioligand binding studies, have been reviewed extensively on earlier occasions (*see* Leysen, 1985; Peroutka, 1986). The challenge of the binding studies is the search for selective drugs and, most importantly, the demonstration of a role for the binding sites in the functional effects of serotonin. This indeed is an ultimate prerequisite, since mere detection of binding sites, displaying the features of high binding affinity, saturability, and even typical distribution, is not a sufficient guarantee that they represent physiological or pharmacological receptor sites (Laduron, 1984; Leysen, 1984).

Very recently, a number of new radioligands and a number of purported selective inhibitors of radioligand binding have been introduced. I will comment on the selectivity of the various drugs, how their binding properties have led to the subclassification of binding sites, and to what extent they can contribute to the identification of the role of the binding sites. A summary of proposed binding site subtypes is presented in Table 4. Information is given on the localization of the sites, radioactive ligands used for their detection, and availability of selective drugs for the sites. The receptor binding profile of currently used serotonin agonists and antagonists with regard to interaction with various serotonergic binding sites is reported in Table 5; the binding affinities of the drugs for various neurotransmitter receptor sites is shown in Table 6.

5.1. 5-HT$_1$ Sites

Binding sites for ^{3}H-serotonin in brain homogenates were first described by Bennett and Snyder (1976). Even in the early studies, differences in binding features of sites labeled by ^{3}H-serotonin and ^{3}H-LSD were noted. It was proposed that serotonin and LSD-labeled the agonist and antagonist state of the serotonin receptor, respectively. In spite of a thorough characterization of the binding properties in the early study, no attempts were made to demonstrate a functional role for the binding sites. Shortly thereafter, binding sites in frontal cortex homogenates, labeled by the neuroleptic ^{3}H-spiperone, were found to be distinct from dopamine receptor sites detected with that ligand in the striatum. The sites labeled by ^{3}H-spiperone in the frontal cortex appeared to be identical to the sites labeled by ^{3}H-LSD in this area (Leysen and Laduron, 1977; Leysen et al., 1978). Peroutka and Snyder (1979) demonstrated that sites labeled by ^{3}H-serotonin and ^{3}H-spiperone in cortex tissue involved distinct molecular entities and introduced the terminology 5-HT$_1$ and 5-HT$_2$ sites. ^{3}H-LSD was found to interact with both types of sites.

5-HT$_1$ sites labeled by ^{3}H-serotonin occurred in all major brain areas, with the highest densities in the striatum and hippocampus.

Table 4
Subclassification of Serotonergic Binding Sites

Binding site	Enriched tissue	Labeled ligand and references		Specific drugs considering serotonergic sites[a]	Selective drugs[b]
5-HT_{1A}	Hippocampus Cortex	^{3}H-5-HT	Pedigo et al., 1981	8-OH-DPAT	8-OH-DPAT
		^{3}H-LSD	Peroutka and Snyder, 1979	Buspirone	
		^{3}H-8-OH-DPAT	Gozlan et al., 1983	TVX Q-7821	
		^{3}H-TVX Q-7821	Dompert et al., 1985	Spiroxatrine	
		^{3}H-WB4101	Norman et al., 1985		
		^{3}H-PAPP	Ransom et al., 1986		
5-HT_{1B}	Various brain areas	^{3}H-5-HT	Pedigo et al., 1981	None	None
		^{3}H-LSD	Peroutka and Snyder, 1979		
		^{125}I-Iodo-Cyanopindolol	Hoyer et al., 1985a		
5-HT_{1C}	Choroid plexus	^{3}H-5-HT	Pazos et al., 1985c	None	None
		^{125}I-LSD	Yagaloff and Hartig, 1985		
		^{3}H-Mesulergine	Hoyer et al., 1985b		

5-HT_3[c]	Striatum	^{3}H-8-OH-DPAT	Hall et al., 1985	None	None
5-HT_2	Frontal cortex Nucleus accumbens Tuberculum Olfactorium Platelets	^{3}H-Spiperone ^{3}H-LSD ^{3}H-Metergoline ^{3}H-Mianserin ^{3}H-Ketanserin ^{125}I-LSD ^{125}I-IMIL ^{3}H-Aminoketanserin ^{3}H-Azidoketanserin ^{3}H-DOB	Leysen and Laduron, 1977 Leysen et al., 1978 Hamon et al., 1981 Peroutka and Snyder, 1981 Leysen et al., 1982b Engel et al., 1984b Hoffman et al., 1985 Wouters et al., 1986 Wouters et al., 1985a Titeler et al., 1985	See compounds in table 5 under: Aryloxy(thio)-alkylamines, 4-Benzoylpiperidines, 4-Benzhydrilene-piperidines, Neuroleptics, SCH-23390, Antihistamines, Monoamine uptake, inhibitors	Xylamidine BW-501 Cinanserin Ritanserin R 56413 Pipamperone LY-53857

[a]Drugs that according to data in Table 5 reveal at least 50 times higher affinity for the corresponding serotonergic site subtype compared to the other serotonergic sites in Table 5.

[b]Drugs that, according to data in Tables 5 and 6 reveal at least 50 times higher affinity for the corresponding type of serotonergic site than for any other serotonergic site or various neurotransmitter receptor sites.

[c]These sites are distinct from 5-HT_3 sites defined by Kaumann in pharmacological experiments.

Table 5
Binding Affinity (K_i, nM) of Drugs for Serotonergic Binding Sites

	$5\text{-}HT_{1A}$, ^{3}H-8-OH-DPAT, rat hippocampus	$5\text{-}HT_{1B}$, ^{3}H-serotonin, + spiroxatrine, rat hippocampus	$5\text{-}HT_{1C}$[a], ^{3}H-mesulergine, pig choroid plexus	$5\text{-}HT_2$ ^{3}H-ketanserin, rat frontal cortex
I. *Indole derivatives*				
I.1. Serotonin	3.5	2.1	33	287
5-Carbamoyl T	0.2	0.7	617	9000
Bufotenine	5.6	50	71	135
N,N-Dimethyl-5-methoxy T	7.0	50	87	625
Tryptamine (T)	1320	1580		1530
I.2. LSD	4.2	12.5	7.9	2.4
Methysergide	23.5	133	2.0	1.3
Lisuride	1.3	12.5		3.3
Mesulergine	275[a]	>10000[a]	1.6[a]	3.8[a]
I.3. LY-53857	661	>10000	8.3	1.7
Metergoline	3.5	5.6	0.5	0.3
I.4. Ru-24969	3.9	0.4	398	590
I.5. ICS-205930	> 1000[b]	> 1000[b]		> 1000[b]
MDL-72222	>10000[c]	>10000[c]		>10000[c]

II. *Arylpiperazines*				
TFMPP	148	10.0	62	88
mCPP	263	75.0		118
MK-212	1860	1400		5570
Quipazine	1860	111	186	220
Trazodone	197	3330		7.8
Buspirone	10	>10000		443
TVX Q-7821	4.7	>10000		1620
III. *2-Aminotetralins*				
8-OH-DPAT	0.9	4980	7240	4870
IV. *Methoxyphenetylamines*				
DOM	6240	1770		154
Mescaline	5560	1180		6620
V. *6,7,6-Membered ring tricyclics*				
Cyproheptadine	74	>10000	8.3	0.57
Pizotifen	140	>10000	8.3	0.33
Metitepine	50	149	27.5	0.45
Mianserin	525	2360	10	1.43
VI. *Aryloxy(thio)alkylamines*				
Xylamidine	4960	>10000		1.54
BW-501	>10000	>10000		0.19
Cinanserin	331	1181	204	1.98

(continued)

Table 5 (*continued*)

	5-HT_{1A}, ^3H-8-OH-DPAT, rat hippocampus	5-HT_{1B}, ^3H-serotonin, + spiroxatrine, rat hippocampus	5-HT_{1C}[a], ^3H-mesulergine, pig choroid plexus	5-HT_2 ^3H-ketanserin, rat frontal cortex
VII. *4-Benzoylpiperidines*				
Ketanserin	4680	>10000	98	0.63
7-Aminoketanserin	3314	>10000		0.17
7-Azidoketanserin	371	>10000		0.35
Altanserin	1570	>10000		0.13
Pirenperone	1660	>10000	60	0.25
Setoperone	2633	>10000		0.37
VIII. *4-Benzhydrilenepiperi-dines*				
Ritanserin	1760	>10000	2.3	0.28
R 56413	1660	>10000		0.29
β-*Adrenergic antagonists*				
Pindolol	19	77 h[a]	53700	>10000
Iodoisocyanopindolol	4.1[a]	0.3[a]	9770[a]	
(-)-21009	7.4[a]	0.42[a]	5370	
α_1-*Adrenergic antagonist*				
WB4101	5.3	1050		>1000

Neuroleptics				
Spiperone	56	>10000	1148	0.64
Pipamperone	>10000	>10000		0.94
Spiroxatrine	2.7	>10000		86
D_1-*Antagonist*				
SCH-23390	331	994	5.26	
Gastrokinetic				
Cisapride	1660	>10000		2.49
Histamine-H_1 antagonists				
Doxepine	589	2225		4.12
Astemizole	1180	>10000		5.26
R 43448	4960	>10000		0.45
Oxatomide	882	>10000		2.85
Monoamine uptake inhibitors				
Amitriptyline	833	2800		6.2
Clomipramine	>10000	>10000		9.9

[a]Data taken from Hoyer et al., 1985b.
[b]Data from Richardson et al., 1985.
[c]Data from Fozard, 1984.

Table 6
Binding Affinity (K_i, nM) of Drugs for Various Neurotransmitter Receptor Sites[a]

	Histamine-H_1, ^{3}H-pyrilamine, guinea pig cerebellum	Dopamine-D_2, ^{3}H-haloperidol, rat striatum	Adrenergic-α_1, ^{3}H-WB4101, rat forebrain	Various
I. *Indole derivatives*				*α_2-adrenergic*
I.1. Serotonin	>100000	11100	18400	29200
5-Carbamoyl T	>100000	>10000	20600	5180
Bufotenine	2280	7420	4120	2320
N,N-Dimethyl-5-methoxy T	2700	>10000	3300	1400
Tryptamine	21500	39400	13000	14600
I.2. LSD	500	35	164	82
Methysergide	> 10000	140	2250	> 10000
Lisuride	15	0.88	6.5	1.3
I.3. LY-53857	> 10000	442	1840	412
Metergoline	420	26	35	292
I.4. RU-24969	8900	1250	1740	2320
I.5. ICS-205930	> 1000[b]	> 1000[b]	> 1000[b]	
MDL-72222	3800[c]		>10000[c]	> 10000[c]

II. *Arylpiperazines*				
TFMPP	1300	786	1300	887
mCPP	905	1760	653	436
MK-212	34000	>10000	1250	381
Quipazine	1400	1900	1160	4900
Trazodone	29	1110	73	3300
Buspirone	629	48	490	23200
TVX Q-7821	520	477	23	1740
III. *2-Aminotetralins*				
8-OH-DPAT	17000	1450	1550	224
IV. *Methoxyphenetylamines*				
DOM	43000	>10000	6200	1640
Mescaline	10800	10000	>10000	>100000
V. *6,7,6-Membered ring tricyclics*				*chol. musc.*
Cyproheptadine	1.08	74	90	10
Pizotifen	1.36	99	155	31
Metitepine	2.50	3.13	0.51	389
Mianserin	0.86	1400	85	806
VI. *Aryloxy(thio)alkylamines*				*α_2-adrenergic*
Xylamidine	4800	>10000	292	4600
BW-501	4050	3130	520	4120
Cinanserin	360	786	1704	> 10000

(continued)

Table 6 *(continued)*

	Histamine-H_1, ^{3}H-pyrilamine, guinea pig cerebellum	Dopamine-D_2, ^{3}H-haloperidol, rat striatum	Adrenergic-α_1, ^{3}H-WB4101, rat forebrain	Various
VII. *4-Benzoylpiperidines*				
Ketanserin	10	240	11	3270
7-Aminoketanserin	0.81	882	5.50	>10000
7-Azidoketanserin	0.48	183	7.76	>10000
Altanserin	20	62	4.55	141
Pirenperone	11	16	7.32	3.27
Setoperone	39	25	13	22
VIII. *4-Benzhydrilenepiperidines*				
Ritanserin	15	22	35	60
R 56413	15	125	73	56
β-*Adrenergic antagonists*				β-*adrenergic*
(-)-pindolol	>10000	>10000	1460	4.2
α_1-*Adrenergic antagonist*				
WB4101	>10000	64	0.46	—
Neuroleptics				μ-*opiate*
Spiperone	2030	0.26	10	174
Pipamperone	>10000	96	58	>10000
Spiroxatrine	>10000	1.4	78	13

D_1-*Antagonist*				*dopamine-D_1*
SCH-23390	572	442	732	1.3
Gastrokinetic				
Cisapride	1460	130	50	
Histamine-H_1antagonists				*chol. musc.*
Doxepine	0.37	394	22	62
Astemizole	13	849	6.7	—
R 43448	0.71	442	21	—
Oxatomide	22	29	150	—
Monoamine uptake inhibitors				
Amitriptyline	1.24	197	29	21
Clomipramine	68	106	29	90

[a] α_2-Adrenergic: ^{3}H-clonidine binding in rat cortex, chol. musc.: cholinergic muscarinic: ^{3}H-dexetimide binding in rat striatum, μ-opiate: ^{3}H-sufentanil binding in rat forebrain, dopamine-D_1: ^{3}H-SCH 23390 binding in striatum.

[b] Data from Richardson et al., 1985.

[c] Data from Fozard et al., 1984.

The main features of 5-HT_1 sites were that serotonin agonists, in particular the indole derivatives, displayed high-affinity binding. High-affinity binding was also observed for the ergot derivatives, but the nonindole antagonists bound only weakly to these sites.

5.1.1. Detection of 5-HT_{1A} Sites and Drug Selectivity

Heterogeneity of 5-HT_1 sites was pointed out by Pedigo et al. (1981), who showed that spiperone inhibited ^{3}H-serotonin binding in a biphasic manner. This was particularly evident in the frontal cortex, where about 60% of "specific" ^{3}H-serotonin binding (i.e., using cold serotonin as blank) was displaced at "low" concentration of spiperone (below $3 \times 10^{-7}M$) whereas the remainder of the binding became displaced only above $10^{-5}M$.

Hence the notation 5-HT_{1A} sites (showing high affinity for spiperone) and 5-HT_{1B} sites was proposed. Differences in the regional distribution of the 5-HT_{1A} sites (enriched in the frontal cortex) and 5-HT_{1B} sites (enriched in the striatum) were shown. Spiperone was not an ideal tool for identifying the 5-HT_{1A} sites, because of its high affinity for dopamine-D_2 and 5-HT_2 receptor sites (*see* Tables 5 and 6) (Leysen et al., 1978). More substantial proof for the existence of 5-HT_1 site subtypes was provided by the use of 8-hydroxy-2-(di-*n*propylamino)tetralin (8-OH-DPAT) (Table 1, class III). This compound, considered a simplified ergot alkaloid lacking an indole nucleus, was shown to have serotonin-mimetic properties (Hjorth et al., 1982). Middlemiss and Fozard (1983) demonstrated that the compound bound with a 3000-fold higher affinity to 5-HT_{1A} than to 5-HT_{1B} sites. The 8-OH-DPAT did not reveal appreciable binding for 5-HT_2 sites or for various other neurotransmitter receptor sites (*see* Tables 5 and 6); it can be considered the only presently known selective drug for 5-HT_{1A} sites (*see* Table 4).

Gozlan et al. (1983) used ^{3}H-8-OH-DPAT (which they called ^{3}H-PAT) to selectively label 5-HT_{1A} sites. A correspondence between ^{3}H-8-OH-DPAT-labeled sites and 5-HT_{1A} sites labeled by ^{3}H-serotonin was found in most brain areas, but not the striatum. The striatal sites labeled by ^{3}H-8-OH-DPAT showed somewhat different binding features and displayed a different sensitivity to neuronal lesions. Unlike cortical or hippocampal ^{3}H-8-OH-DPAT sites, striatal ^{3}H-8-OH-DPAT sites were reduced following destruction of serotonergic neurons by intraraphe administration of the neurotoxin 5,7-dihydroxytryptamine. Because of these anatomical features, the striatal ^{3}H-8-OH-DPAT binding sites were considered autoreceptors on serotonergic nerve terminals. Since the sites did not entirely show the same binding features as 5-HT_{1A} sites, the striatal 8-OH-DPAT binding sites were tentatively indicated as 5-HT_3 sites [Hall et al., 1985 (note that these sites are unrelated to the 5-HT_3 sites proposed by Kaumann, *see* Table 3)]. In functional studies, however, investigating the ability of drugs to regulate ^{3}H-serotonin release from brain slices, 8-OH-DPAT was ineffective. Hence functional autoreceptor properties of 8-OH-DPAT-labeled sites have been challenged (Middlemiss, 1984). The 8-methoxy-2-(*N*-chloropropyl,*N*-propyl)aminotetralin (2'-chloro-8-methoxy-PAT) has been described as an alkylating agent that irreversibly blocks 5-HT_{1A} and 5-HT_3 sites, but also seems to reduce the labeling by ^{3}H-serotonin of 5-HT_{1B} sites. The compound was presumed to be useful for irreversible blockade in vivo of central 5-HT binding sites and for studying the turnover rate of the various classes of sites (Emerit et al., 1985a). NAP-8-methoxy-PAT (*see* Table 1) was recently proposed as a photoaffinity ligand for the 5-HT_1 sites. This compound may represent a tool useful in the purification and investigation of the molecular properties of 5-HT_1 binding sites (Emerit et al., 1985b).

The arylpiperazine derivatives (Table 1, class II) buspirone and TVX Q-7821, which have been clinically investigated for purported anxiolytic properties, also revealed high binding affinity for 5-HT_{1A} sites (Peroutka, 1985a). The compounds are selective toward 5-HT_{1A} sites compared to the other serotonergic binding sites (*see* Table 5). Buspirone, however, previously known as a neuroleptic, binds with comparable affinity to 5-HT_{1A} sites and dopamine-D_2 receptors. The TVX Q-7821 binds to α_1-adrenergic receptor sites with approximately 10 times lower affinity than to the 5-HT_{1A} sites (*see* Table 6). ^{3}H-TVX Q-7821 has been reported to label sites with binding features and regional distribution corresponding to 5-HT_{1A} sites (Dompert et al., 1985; Glaser and Traber, 1985).

Recently another tritiated phenylpiperazine derivative, ^{3}H-1-[2-(4-aminophenyl)ethyl]-4-(3-trifluoromethylphenyl)piperazine, was claimed to selectively label 5-HT_{1A} sites. The ligand has been developed as a direct precursor for the corresponding azido derivative, which is potentially useful as a photoaffinity ligand for the 5-HT_{1A} sites (Ransom et al., 1986).

Finally ^{3}H-WB4101 [^{3}H-2-(2,6-dimethoxyphenoxy-ethyl)amino-methyl-1,4-benzodioxane], a known high-affinity ligand for α_1-adrenergic receptor sites, has been shown to label 5-HT_{1A} sites when used at concentrations above 2–3 n*M* and in the presence of 30 μM prazosin to preclude labeling of α_1-receptors (Norman et al., 1985). The advantages of this ligand to study 5-HT_{1A} sites are questionable, but it may be interesting to investigate which structural elements are responsible for its interaction with 5-HT_{1A} sites.

All above mentioned ligands for 5-HT_{1A} sites were thought to have agonistic properties either because of direct behavioral effects (8-OH-DPAT), or because of generalization with discriminative stimulus effects of serotonin agonists (buspirone, TVX Q-7821), or because of sensitivity of the binding to guanyl nucleotides (WB4101) (*see* ref. in Table 1).

No selective antagonistic agent that potently binds to 5-HT_{1A} sites is known. In screening for such drugs, Nelson and Taylor (1985) discovered that spiroxatrine selectively inhibited ^{3}H-serotonin and ^{3}H-8-OH-DPAT binding to 5-HT_{1A} sites. The compound was hypothesized to be a 5-HT_{1A} antagonist. Spiroxatrine,8[2-(1,4-benzodioxanyl)-methyl]-4-oxo-4-phenyl-1,2,8-triazo-spiro(4,5)decan], was described in 1964 as a compound that, like spiperone, belongs to the spirodecanone series of neuroleptics (Niemegeers et al., 1964). The purported selectivity of spiroxatrine only pertains to interaction with subtypes of serotonergic binding sites. The drug is a potent neuroleptic and binds with high affinity to dopamine-D_2 sites. In addition it has morphinomimetic properties with corresponding binding to μ-opiate receptor sites (*see* Table 6) (Niemegeers et al., 1964; Leysen et al., 1977). Therefore spiroxatrine is not a suitable tool for investigating functional effects mediated by 5-HT_{1A} sites. It is worth noting that, structurally, spiroxatrine has a spirodecanone moiety in common with spiperone and a benzodioxane moiety in common with WB4101. Both structural elements may be of importance for interaction with the 5-HT_{1A} sites.

5.1.2. Detection of 5-HT_{1B} Sites and Drug Selectivity

Labeling of 5-HT_{1B} sites in brain homogenates was achieved by using ^{3}H-serotonin in the presence of a selective blocker for the 5-HT_{1A} sites. Nonspecific binding was defined by displacement with excess unlabeled serotonin. Originally 1 μM spiperone was added to preclude labeling of 5-HT_{1A} sites (Pedigo et al., 1981). It now seems that 8-OH-DPAT or spiroxatrine, which show a higher binding affinity for 5-HT_{1A} sites and a wider dissociation between binding to 5-HT_{1A} and 5-HT_{1B} sites, are more suitable drugs to mask the 5-HT_{1A} sites. Recently Hoyer et al. (1985a) reported on the use of ^{125}I- iodocyanopindolol for detecting 5-HT_{1B} sites. Since this drug is in the first place a β-adrenergic receptor blocker, addition of 30 μ*M* isoprenaline together with the labeled ligand is necessary to prevent the labeling of β-adrenergic receptor sites. Iodocyanopindolol displayed a K_d value of 0.3 n*M* for 5-HT_{1B} sites, but its binding affinity for 5-HT_{1A} sites was only about 20-fold weaker. Hence the compound shows only poor selectivity and its usefulness as a ligand for 5-HT_{1B} sites may be limited. Sills et al. (1984) used a different approach to distinguish between drug binding to 5-HT_{1A} and 5-HT_{1B} subtypes. They investigated inhibition of ^{3}H-5-HT binding (used at the rather high concentration of 15 n*M*) in frontal cortex membranes in the presence of 1 m*M* GTP. The latter was claimed to be required to eliminate the high-affinity state of the receptor sites. Binding data were analyzed by computer-assisted modeling of inhibition curves. In this way RU-24969 (Table 1, class I, 4) and the arylpiperazines TFMP and mCPP (Table 1, class II) were claimed to bind more selectively to the 5-HT_{1B} sites, whereas tryptamine derivatives such as *N,N*-dimethyl-5-methoxytryptamine and *N,N*-dimethyltryptamine would show selectivity toward 5-HT_{1A} sites. Using assay conditions to selectively label 5-HT_{1A} and 5-HT_{1B} sites, respectively, these findings were not confirmed by Peroutka (1985b). In his assays, the arylpiperazines and tryptamine derivatives both bound somewhat more potently to 5-HT_{1A} than to 5-HT_{1B} sites. Our findings, using selective labeling conditions, are presented in Table 5 and are in line with the suggestions of Sills et al. (1984). The selectivity of the drugs is poor, however. It appears that, thus far, no markedly selective drugs for the 5-HT_{1B} sites have been discovered.

5.1.3. Detection of 5-HT_{1C} Sites

In a study on the autoradiographic localization of serotonergic binding sites using various ligands, binding sites for ^{3}H-mesulergine, ^{3}H-5-HT, and ^{3}H-LSD were found in the choroid plexus. That binding was insensitive to 8-OH-DPAT (a 5-HT_{1A}-selective drug), RU-24969 (used to block 5-HT_{1B} sites), and some β-blockers with high 5-HT_{1B} binding affinity, and ketanserin (a selective blocker of 5-HT_2 sites). The binding sites in the choroid plexus were termed 5-HT_{1C} sites (Cortes et al., 1984). In in vitro radioligand binding studies using ^{3}H-5-HT, ^{3}H-LSD, and ^{3}H-mesulergine and pig choroid plexus homogenates, the 5-HT_{1C} sites were further characterized. Serotonin antagonists such as mianserin, LSD, methysergide, and metitepine showed high binding affinity for both 5-HT_2 and 5-HT_{1C} sites. Yet, other antagonists such as ketanserin, cinanserin, and spiperone were found to differentiate between both types of sites. Among agonists, serotonin showed high binding affinity for the 5-HT_{1C} sites, but 5-carbamoyltryptamine, 8-OH-DPAT, and RU-24969 revealed only a weak binding affinity (Pazos et al., 1985c; Hoyer et al., 1985b). Yagaloff and Hartig (1985) detected a similar type of 5-HT_{1C} sites in the choroid plexus in autoradiographic studies in rat brain with ^{125}I-LSD. The 5-HT_{1C} sites are distinguished by their particular occurrence; the sites are enriched in the choroid plexus of different mammalian species, but they have also been detected in specific areas of the pig cortex (Pazos et al., 1985c). Selective drugs for 5-HT_{1C} sites have not yet been described.

5.1.4. Autoradiographic Studies of 5-HT_1 Sites

Several ligands have been used to study the detailed anatomical distribution of serotonergic binding sites by autoradiography. ^{3}H-serotonin was used to study specific distribution of 5-HT_1 sites in hippocampal regions of the rat (Kohler, 1984) and in human and canine basilar arteries (Peroutka and Kuhar, 1984). In these studies no distinction was made between 5-HT_1 site subtypes. In comparative studies using different labeled ligands (^{3}H-8-OH-DPAT, ^{3}H-TVX Q-7821, ^{3}H-mesulergine, ^{3}H-LSD, ^{125}I-LSD, ^{3}H-5-HT) in combination with various unlabeled displacers, distinctions in brain distribution of 5-HT_{1A}, 5-HT_{1B}, and 5-HT_{1C} sites were studied as well as sensitivity of the sites to neuronal lesions (Marcinkiewicz et al., 1984; Glaser et al., 1985; Yagaloff and Hartig, 1985; Pazos and Palacios, 1985; Weissman-Nanopoulos et al., 1985).

5.1.5. The Role of 5-HT_1 sites

Several roles have been claimed for 5-HT_1 sites (*see* Peroutka, 1986; Leysen, 1985). Among in vivo effects induced by serotoninmimetic agents, forepaw treading and reduction in body temperature were thought to be mediated by 5-HT_{1A} sites (Tricklebank, 1985; Goodwin and Green, 1985), and the decrease in locomotor activity induced by mCPP was thought to be mediated by 5-HT_{1B} sites (Sills et al., 1985). Since no specific antagonists for 5-HT_1 sites are known, however, the assumptions were based on the action of only a few serotonin agonists. Therefore substantial proof for these hypotheses is lacking. Similarly, roles of 5-HT_1 sites hypothesized from in vitro pharmacological studies were based mainly on the use of serotonin agonists, e.g., inhibition of serotonin and noradrenaline release by serotonin-mimetic agents (Engel et al., 1983) or serotonin agonist-induced relaxation of vascular smooth muscle (Cohen and Vanhoutte, 1985). Involvement of a particular subtype of 5-HT_1 sites was not specified. Serotonin-induced contractions in basilar artery segments were proposed to be mediated by 5-HT_1 sites because of observed antagonism of the effect by the nonspecific drug metitepine and an apparent lower potency of ketanserin (Peroutka et al., 1983). This assumption has been refuted, however, in independent studies showing that 5-HT_2 rather than 5-HT_1 sites are involved in serotonin-induced basilar artery contractions (Muller-Schweinitzer and Engel, 1983; Van Nueten et al., 1984). Because of their particular anatomical localization, it has tentatively been suggested that 5-HT_{1C} sites may have a role in the production of cerebrospinal fluid (Yagaloff and Hartig, 1985).

The only way to prove the role of a binding site is by careful examination of the correlation between the binding affinities of a series of drugs and the drug potencies in a functional test. Therefore selective drugs of widely varying potency are required. When in vivo functional effects are considered, it must be taken into account that some drugs may not fit the correlation line for pharmacokinetic or metabolic reasons. This should be verified, and the deviation of one point from a correlation line should not immediately be a reason for supposing receptor heterogeneity. When in vitro functional effects are compared to binding data, there should be a match in both rank order of potency of drugs and absolute potency. Establishment of a correspondence, however, between a pharmacologically defined pA_2-value and a binding affinity constant for only one or two drugs is not sufficient to prove the role of a binding site. This may be erratic since most drugs have multiple effects, and besides known interactions with receptors they may have as yet unknown effects. To prove the role of binding sites, investigation of a series of drugs is indispensable. Because of a lack of antagonists, this could not be accomplished for the 5-HT_1 sites.

5.1.6. Signal Transducing System Coupled to 5-HT_1 Sites

Binding of serotonin and serotonin agonists was found to be sensitive to GTP (Peroutka et al., 1979). This is generally considered an indication for involvement of an adenylate cyclase system, consisting of the receptor binding protein, a guanyl nucleotide binding protein and the catalytic adenylate cyclase unit (Lefkowitz et al., 1984). Peroutka et al. (1979) claimed a relationship between the weak 5-HT_1 binding affinity of some antagonists and their low potencies to inhibit serotonin-stimulated adenylate cyclase, leading to the suggestion of a coupling between both systems. This hypothesis has been refuted, however, by the fact that there was no correspondence between the regional and subcellular distribution of the 5-HT_1 binding sites and the serotonin-stimulated adenylate cyclase. Other authors showed a lack of correlation between drug potencies when larger series of drug were considered (Leysen and Tollenaere, 1982; Nelson et al., 1980a, b). Recently, a negative coupling between 5-HT_1 sites and adenylate cyclase was suggested by the observation of inhibition of cyclic AMP production by serotonin in primary cultures of neurons (Weiss et al., 1986). Observed effects were small, however, and only a very limited number of drugs was investigated (10–20% inhibition of cAMP production by 8-OH-DPAT, RU-24969, and serotonin; reversal of the effect by metergoline; ineffectiveness of ketanserin). No conclusions were drawn regarding the subtype of 5-HT_1 sites that should be involved. This study awaits further confirmation.

5.2. 5-HT_2 Sites

5.2.1. Detection of 5-HT_2 Sites and Drug Selectivity

5-HT_2 sites can be labeled by radioactive serotonin antagonists. Several ligands have been used, but only a few have displayed sufficient selectivity. ^{3}H-LSD labels 5-HT_2 and 5-HT_1 sites, and ^{3}H-spiperone labels 5-HT_2 and dopamine-D_2 receptor sites (Peroutka and Snyder, 1979; Leysen et al., 1978). ^{3}H-Metergoline has the disadvantage of high nonspecific binding (Hamon et al., 1981), ^{3}H-mianserin labels both 5-HT_2 and histamine-H_1 receptor sites (Peroutka and Snyder, 1981), and ^{3}H-metitepine interacts with multiple sites, as can be expected from its receptor binding profile (Tables 5 and 6). ^{3}H-Ketanserin was the first ligand that selectively labels 5-HT_2 sites (Leysen et al., 1982b). Although the drug reveals moderate binding affinity for α_1-adrenergic and histamine-H_1 receptor sites (K_i-values of 10 nM) (Leysen et al., 1981), these receptor sites are not labeled by the radioactive drug used at nanomolar concentrations. Using ^{3}H-ketanserin, the binding affinities of a large series of drugs for 5-HT_2 sites were investigated, which allowed for evaluation of the functional role of the binding sites.

Drugs that are known as serotonin antagonists show high binding affinity for 5-HT_2 sites. In the chemical classification of

Table 1, they are found in classes V–VIII. Also, certain drugs belonging to various pharmacological classes have revealed serotonin antagonistic properties and bind with high affinity to 5-HT_2 sites. The chemical and pharmacological diversity of these compounds is shown in Tables 1 and 2, and has been discussed in section 2. The receptor binding profile of the drugs is reported in Tables 5 and 6.

Regarding interaction with the serotonergic binding sites, the antagonists in Table 1(classes V–VIII) and in Table 2 all show higher binding affinity for 5-HT_2 than for 5-HT_1 site subtypes. Regarding interaction with various neurotransmitter receptor sites, however, only few drugs are selective (*see* Table 4). The most selective agents are found in class VI, the aryloxy(thio)alkylamines and in class VIII, the 4-benzhydrilenepiperidines. Ergot derivatives, which have mixed agonist–antagonist properties, also reveal high binding affinity for 5-HT_2 sites, but these drugs are not selective with regard to interaction with subtypes of serotonergic sites or interaction with various neurotransmitter receptor sites. Serotonin agonists with an indole structure (class I), arylpiperazines (class II), and methoxyphenethylamines (class IV) show low binding affinity (micromolar order) for 5-HT_2 sites; the 2-aminotetralins of class III are virtually inactive at 5-HT_2 sites. Indole derivatives and most arylpiperazines show higher binding affinity for the 5-HT_1 site subtypes than for 5-HT_2 sites. Although some methoxyphenethylamines have been claimed to show selectivity for 5-HT_2 sites, their binding affinity is still rather low. The first aonist with high binding affinity for 5-HT_2 sites may be DOB (*see* Table 1, class IV) (Titeler et al., 1985). This finding awaits confirmation, however.

The use of ^{3}H-ketanserin allowed for a thorough investigation of the regional distribution of 5-HT_2 sites in brains of several mammalian species, including humans (Leysen et al., 1982a,b, 1983b; Schotte et al., 1983). 5-HT_2 sites were enriched primarily in the frontal cortex and the mesolimbic areas (tuberculum olfactorium and nucleus accumbens). A high density was also found in the striatum. Neuronal lesions revealed that 5-HT_2 sites were not affected by destruction of serotonergic neurons or catecholaminergic neurons. This suggested that the sites were not localized on these types of neurons and that denervation supersensitivity did not seem to develop (Leysen et al., 1982a, 1983b). In autoradiographic studies, a very specific distribution of 5-HT_2 sites was found; highest density of 5-HT_2 sites occurred in the claustrum, olfactory tubercle, and layer IV of the neocortex, followed by the anterior olfactory nucleus, piriform cortex, and layer I of the neocortex. Caudate putamen and nucleus accumbens were among areas with intermediate concentration. Thalamus, hippocampus, brainstem, medulla, cerebellum, and spinal cord had low concentrations (Pazos et al., 1985a). Using ^{3}H-ketanserin,5-HT_2 sites were identified on cat platelet membranes (Leysen et al., 1983a). Furthermore, ^{3}H-ketanserin was found to be a suitable ligand for selective in vivo labeling of 5-HT_2 sites (Ladurion et al., 1982b), and was useful for labeling solubilized 5-HT_2 receptor sites (Ilien et al., 1982; Wouters et al., 1985b). ^{11}C-Ketanserin has been used successfully in PET scan studies (Baron et al., 1985).

In spite of the selectivity of ^{3}H-ketanserin for differential labeling of 5-HT_2 sites compared to other serotonergic binding sites or neurotransmitter receptors, one still has to be careful about the definition of specific binding. Indeed, particularly in striatal tissue and on cat platelet membranes, ^{3}H-ketanserin was found to bind, in addition to 5-HT_2 sites, to a saturable type of binding sites that was displaceable by unlabeled ketanserin (Leysen, 1984; Leysen et al., 1983a). In these tissues, binding of ^{3}H-ketanserin to 5-HT_2 sites has to be distinguished by using as a blank a 5-HT_2 antagonist, with a different chemical structure than ketanserin (e.g., methysergide, BW-501). The ketanserin-displaceable sites were found to be unrelated to a known neurotransmitter receptor site. The identity of these sites is currently being investigated in our laboratories.

Recently, derivatives of ketanserin that are promising tools for the study of the molecular properties of the 5-HT_2 sites have been synthetized. ^{3}H-Aminoketanserin shows high-affinity binding for 5-HT_2 sites and has a very low nonspecific binding. The compound has a low lipophilicity and may therefore be a candidate for the study of cell surface receptors. 7-Azidoketanserin is a suitable photoaffinity ligand for the irreversible blockade of 5-HT_2 sites. The azido derivative is being used in the study of the molecular properties of the 5-HT_2 receptor protein. Unfortunately, both 7-amino and 7-azidoketanserin show a higher affinity for histamine receptors than their parent compound, ketanserin (*see* Table 3). Hence precautions must be taken to distinguish 5-HT_2 site-labeling from possible histamine-H_1 site-labeling (Wouters et al., 1985a, Wouters et al., 1986). Iodinated ligands are 2-[^{125}Iodo]LSD and N_1-methyl-2-[^{125}I] LSD [(^{125}I)MIL] The iodinated derivatives of LSD, unlike ^{3}H-LSD, have been claimed to be more selective toward 5-HT_2 sites. They have less binding affinity for 5-HT_{1A} and

5-HT_{1B} sites (Engel et al., 1984b, Hoffman et al., 1985); Labeling of 5-HT_{1C} sites in the choroid plexus has been reported, however (Yagaloff and Hartig, 1985). The advantage of iodinated ligands is their high specific radioactivity, rendering detection of low receptor concentrations possible and speeding up autoradiographic detection of receptors. It appears, however, that background labeling (blank values) is proportionately increased with increasing specific radioactivity of the drug. This may be a drawback for their use in crude receptor preparations.

Recently, the first labeled 5-HT_2 agonist, ^{3}H-1-(4-bromo-2,5-dimethoxyphenyl)-2-amino propane (^{3}H-DOB) (*see* Table 1, class IV), has been proposed as a ligand for 5-HT_2 sites (Titeler et al., 1985). The available information is still scarce, but the use of an agonistic ligand may be interesting for the study of receptor regulation.

5.2.2. Role of 5-HT_2 Sites

By correlating binding affinities of drugs with drug potencies in antagonizing functional effects, various roles for 5-HT_2 sites in pharmacological effects induced by serotonergic agents in vivo and in vitro have been demonstrated. In contrast to 5-HT_1 sites, a wide variety of drugs acting at 5-HT_2 sites is available; therefore correlations between drug potencies in functional tests and drug binding affinities have been well established. Several components of the "serotonin syndrome" were shown to be mediated by 5-HT_2 sites. These include (1) behavioral excitation: tryptamine-induced clonic seizures, bilateral forepaw treading (Leysen et al., 1978), and head twitches induced by serotonin mimetics (Peroutka et al., 1981; Leysen et al., 1982b); and (2) impairment of blood circulation: tryptamine-induced cyanosis (Leysen, 1988), compound 48/80-induced gastric lesions (Awouters et al., 1982), and serotonin-induced paw edema (Ortmann et al., 1982). 5-HT_2 sites also have a role in discriminative stimulus effects elicited by LSD, DOM, and quipazine (Colpaert et al., 1982; Glennon et al., 1983; Friedman et al., 1984). Using isolated tissues, 5-HT_2 sites were shown to mediate serotonin-induced smooth muscle contractions in vascular tissue (Leysen, 1981; Leysen et al., 1982b; Muller-Schweinitzer and Engel, 1983; Van Nueten et al., 1984) and in certain nonvascular tissues such as the ileum, trachea, and uterus (Engel et al., 1984a; Van Nueten et al., 1982; Cohen et al., 1985b). 5-HT_2 sites detected on human and cat platelets showed similar binding properties to the sites in brain tissue, and a role of 5-HT_2 sites in serotonin-induced platelet function (shape change and aggregation) has been demonstrated (Geaney et al., 1984; Leysen et al., 1983a; De Clerck et al., 1984a,b).

Most evidence for demonstrating the various roles for 5-HT_2 sites was based on comparing activities of antagonistic drugs. The observation of a rather low 5-HT_2 binding affinity of serotonin and serotonin agonists has often been used as an argument against the physiological role of 5-HT_2 sites. It appears, however, that in order to elicit pharmacological effects with serotonin, high concentrations are required. This is particularly true in the pharmacological tests on isolated tissues, in which the active concentrations of serotonin (10^{-7} to 10^{-6} *M*) were found to match the 5-HT_2 site binding affinity of the biogenic amine (*see* Leysen, 1984).

5.2.3. Signal Transducing System Coupled to 5-HT_2 Sites

It appeared early on that 5-HT_2 sites were not likely to be coupled to adenylate cyclase (Peroutka et al., 1979). A study of phospholipid turnover in human platelets showed that serotonin induced a rapid increase in ^{32}P-phosphatidic acid formation at concentrations corresponding to the binding affinity of serotonin for 5-HT_2 sites. A concomitant transient increase in the phosphorylation of 20 and 40 kdalton proteins was observed. These serotonin-induced effects were found to be inhibited by various serotonin antagonists and the drug potencies corresponded closely to their binding affinities for 5-HT_2 sites (de Chaffoy de Courcelles et al., 1984, 1985). These findings provided evidence for a direct coupling between 5-HT_2 sites and phosphatidyl inositol turnover, a system in which inositol trisphosphate and diacylglycerol operate as second messengers (Berridge, 1984). Inositol trisphosphate is thought to mobilize Ca from intracellular stores through which Ca-dependent processes become activated. In the platelets, the role of inositol trisphosphate and consequently of Ca is reflected in the phosphorylation of the 20 kdalton protein (myosin light chain) by the activation of a Ca/calmodulin-dependent kinase. Diacylglycerol activates protein kinase C, for which the 40 kdalton protein in platelets is a substrate. An alternative method for demonstrating the coupling between a receptor and the phospholipid turnover system is the measurement of formation of ^{3}H-inositol phosphates. This technique was used in studies with rat brain slices. Serotonin-induced phosphorylation of ^{3}H-inositol has been observed in various brain areas, but antagonism by ketanserin was achieved only in the frontal cortex

(Conn and Sanders-Bush, 1985). Brown et al. (1984) observed serotonin-induced inositol phosphate formation in rat brain slices; but alterations in the magnitude of the effect following various drug treatments did not entirely parallel observed variations in the number of 5-HT_2 binding sites. Although it is likely that 5-HT_2 sites also operate through activation of phospholipid turnover in the brain, it could not firmly be established with the techniques using brain slices and water-soluble inositols.

5.2.4. Regulation of 5-HT_2 Sites

5-HT_2 sites in rat brain were found to be down-regulated following chronic treatment of the animals with tricyclic antidepressants (Peroutka and Snyder, 1980). This led to the suggestion of a role of 5-HT_2 sites in the etiology of depression (Snyder and Peroutka, 1982). According to the classical theory, the down-regulation of 5-HT_2 sites was thought to be a consequence of enhanced receptor stimulation because of the inhibition of serotonin uptake by the antidepressant drugs. Chronic treatment with potent and selective serotonin-uptake inhibitors, however, such as citalopram, did not affect 5-HT_2 receptor sites (Hyttel et al., 1984). In contrast, 5-HT_2 receptor down-regulation was observed with nonselective serotonin antagonists such as mianserin and cyproheptadine (Blackshear et al., 1983). The regulation of 5-HT_2 sites has recently been investigated using ritanserin, a potent, selective, and long-acting serotonin antagonist and setoperone, a potent but short-acting serotonin antagonist with concomitant dopamine antagonistic properties. It has been shown that down-regulation of 5-HT_2 sites, in contrast to current theories, occurs following occupation of the receptor sites by serotonin antagonists. The differences in the regulation of 5-HT_2 and dopamine-D_2 receptor sites was apparent by the use of setoperone, which when chronically administered caused a reduction in 5-HT_2 receptor site numbers in the frontal cortex and a significant increase in striatal dopamine-D_2 receptor sites (Leysen et al., 1986).

In view of the hypothesized role of 5-HT_2 sites in depressive disorders, it is interesting to note that ritanserin has revealed therapeutic efficacy in dysthymic disorders (Reyntjens et al., 1984). The complexity of 5-HT_2-receptor regulation has been further illustrated by the finding of up-regulation of 5-HT_2 sites following repeated electroconvulsive shocks, a phenomenon that required intact serotonergic neurons (Stockmeier and Kellar, 1986). The mechanism of 5-HT_2 receptor regulation is not yet understood.

Molecular investigation of the phenomenon will require the availability of cell lines rich in 5-HT_2 receptor sites, water-soluble ligands with which cell surface receptors can be labeled, and tools such as labeled irreversible ligands to identify the 5-HT_2 receptor proteins. By the rapid development of new drugs, these goals may be reached in the near future.

6. Conclusions

Forty years of serotonin research have continuously revealed the complexity of the serotonergic system: widespread occurrence in the central nervous system and in the periphery, many different central and peripheral pharmacological actions (probably involving different receptor subtypes), and interference of drugs with a wide variety of chemical structures. Although serotonin is thought to be involved in various pathologies such as affective disorders, migraine and cardiovascular diseases, the physiological role of the amine is still not clearly established. As has been stated over the years, a clue to the answer for many questions is probably to be found in the development of selective drugs and in their eventual clinical application. The recent introduction of various novel drugs may be a step in this direction.

Acknowledgments

Part of this work was supported by a grant from the Instituut voor Wetenschappelijk Onderzoek in Nijverheid en Landbouw. The author is indepted to C. Van der Eycken and Dr. M. Janssen for advice on the chemical classification of the drugs. Sincere thanks to Mady Verwimp for excellent technical assistance and to Chris Verellen, Suzy De Cauwer, and Jef Van Mierlo for typing and artwork.

References

Aghajanian G. K. (1981) The Modulatory Role of Serotonin of Multiple Receptors in Brain, in *Serotonin Neurotransmission and Behavior* (Jacobs B. L. and Gelperin A., eds.) MIT Press, Cambridge, Massachusetts.

Awouters F., Leysen J. E., De Clerck F., and Van Nueten J. M. (1982) General Pharmacological Profile of Ketanserin (R 41 468), A Selective 5-HT_2 Receptor Antagonist, in *5-Hydroxytryptamine in Peripheral Reactions* (De Clerck F. and Vanhoutte P. M., eds.) Raven, New York.

Baron J. C., Samson Y., Comar D., Crouzel C., Deniker P., and Agid, Y. (1985) Etude in vivo des récepteurs sérotoninergiques centraux chez l'homme par tomographie a positons. *Rev. Neurol.* (Paris) **141,** 537–545.

Beart P. M., McDonald D., Cincotta M., de Vries D. J., and Gundlach A. L. (1986) Selectivity of some ergot derivatives for 5-HT_1 and 5-HT_2 receptors of rat cerebral cortex. *Gen. Pharmacol.* **17,** 57–62.

Bennett, Jr., J. P. and Snyder S. H. (1976) Serotonin and lysergic acid diethylamide binding in rat brain membranes: Relationship to postsynaptic serotonin receptors. *Mol. Pharmacol.* **12,** 373–389.

Berridge M. J. (1984) Inositol trisphosphate and diacylglycerol as second messengers. *Biochem. J.* **220,** 345–360.

Blackshear M. A., Friedman R. L., and Sanders-Bush E. (1983) Acute and chronic effects of serotonin (5-HT) antagonists on serotonin binding sites. *Naunyn Schmiedebergs Arch. Pharmacol.* **324,** 125–129.

Brown E., Kendall D. A., and Nahorski S. R. (1984) Inositol phospholipid hydrolysis in rat cerebral cortical slices. I. Receptor characterization. *J. Neurochem.* **42,** 1379–1387.

Cohen M. L. and Wittenauer L. A. (1985) Further evidence that the serotonin receptor in the rat stomach fundus is not 5-HT_{1A} or 5-HT_{1B}. *Life Sci.* **38,** 1–5.

Cohen M. L., Kurz K. D., Mason N. R., Fuller R. W., Marzoni G. P., and Gabrecht W. L. (1985a) Pharmacological activity of the isomers of LY-53857, potent and selective 5-HT_2 receptor antagonists. *J. Pharmacol. Exp. Ther.* **235,** 319–323.

Cohen M. L., Schenck K. W., Colbert W., and Wittenauer L. (1985b) Role of 5-HT_2 receptors in serotonin-induced contractions of nonvascular smooth muscle. *J. Pharmacol. Exp. Ther.* **232,** 770–774.

Cohen R. A. and Vanhoutte P. M. (1985) Platelets, Serotonin and Endothelial Cells, in *Serotonin and the Cardiovascular System* (Vanhoutte P. M., ed.) Raven, New York.

Colpaert F. C., Niemegeers C. J. E., and Janssen P. A. J. (1982) A drug discrimination analysis of lysergic acid diethylamide (LSD): In vivo agonist and antagonist effects of purported 5-hydroxytryptamine antagonists and of pirenperone, an LSD antagonist. *J. Pharmacol. Exp. Ther.* **221,** 206–214.

Conn P. J. and Sanders-Bush E. (1985) Serotonin-stimulated phosphoinositide turnover: Mediation by the S_2 binding site in rat cerebral cortex but not in subcortical regions. *J. Pharmacol. Exp. Ther.* **234,** 195–203.

Cortes R., Palacios J. M., and Pazos A. (1984) Visualization of multiple serotonin receptors in the rat brain by autoradiography. *Br. J. Pharmacol.* **82,** 202P.

Curro F. A., Greenberg S., Verbeuren T. J., and Vanhoutte P. M. (1978) Interaction between alpha adrenergic and serotonergic activation of canine saphenous veins. *J. Pharmacol. Exp. Ther.* **207,** 936–949.

Dahlstrom A. and Fuxe K. (1964) Evidence for the existence of monoamine-containing neurons in the central nervous system. I. Demonstration of monoamines in cell bodies of brain neurons. *Acta Physiol. Scand.* **62** (suppl. 232), 1–55.

de Chaffoy de Courcelles D., Roevens P., and Van Belle H. (1984) Stimulation by serotonin of 40 kDa and 20 kDa protein phosphorylation in human platelets. FEBS Lett. **171,** 289–292.

de Chaffoy de Courcelles D., Leysen J. E., De Clerck F., Van Belle H., and Janssen P. A. J. (1985) Evidence that phospholipid turnover is the signal transducing system coupled to serotonin-S_2 receptor sites. *J. Biol. Chem.* **260,** 7603–7608.

De Clerck F., Xhonneux B., Leysen J., and Janssen P. A. J. (1984a) The involvement of 5-HT_2 receptor sites in the activation of cat platelets. *Thromb. Res.* **33,** 305–321.

De Clerck F., Xhonneux B., Leysen J. E., and Janssen P. A. J. (1984b) Evidence for functional 5-HT_2 receptor sites on human blood platelets. *Biochem. Pharmacol.* **33,** 2807–2811.

Dompert W. U., Glaser T., and Traber J. (1985) [^{3}H] TVX Q 7821: Identification of 5-HT_1 binding sites as target for a novel putative anxiolytic. *Naunyn Schmiedebergs Arch. Pharmacol.* 328, 467–470.

Emerit M. B., Gozlan H., Hall M. D., Hamon M., and Marquet A. (1985a) Irreversible blockade of central 5-HT binding sites by 8-methoxy-2'-chloro-PAT. *Biochem. Pharmacol.* **34,** 883–892.

Emerit M. B., Gozlan H., Hall M. D., Marquet A., and Hamon M. (1985b) Photoaffinity labelling of 5-HT binding sites. *Neurosci. Lett.* suppl. **22,** S552.

Engel G., Göthert M., Muller-Schweinitzer E., Schlicker E., Sistonen L., and Stadler P. A. (1983) Evidence for common pharmacological properties of [^{3}H] 5-hydroxytryptamine binding sites, presynaptic 5-hydroxytryptamine autoreceptors in CNS and inhibitory presynaptic 5-hydroxytryptamine receptors on sympathetic nerves. *Naunyn Schmiedebergs Arch. Pharmacol.* **324,** 116–124.

Engel G., Hoyer D., Kalkman H. O., and Wick M. B. (1984a) Identification of 5-HT_2-receptors on longitudinal muscle of the guinea pig ileum. *J. Recept. Res.* **4,** 113–126.

Engel G., Muller-Schweinitzer E., and Palacios J. M. (1984b) 2-[^{125}Iodo]LSD, a new ligand for the characterization and localisation of 5-HT_2 receptors. *Naunyn Schmiedebergs Arch. Pharmacol.* **325,** 328–336.

Fozard J. R. (1984) MDL 72222: A potent and highly selective antagonist at neuronal 5-hydroxytryptamine receptors. *Naunyn Schmiedebergs Arch. Pharmacol.* **326,** 36–44.

Friedman R., Barrett R. J., and Sanders-Bush E. (1984) Discriminative stimulus properties of quipazine: Mediation by $serotonin_2$ binding sites. *J. Pharmacol. Exp. Ther.* **228,** 628–635.

Gaddum J. H. and Picarelli Z. P. (1957) Two kinds of tryptamine receptors. *Br. J. Pharmacol.* **12,** 323–328.

Geaney D. P., Schachter M., Elliott J. M., and Graham-Smith D. G. (1984) Characterisation of [^{3}H] lysergic acid diethylamide binding to a 5-hydroxytryptamine receptor on human platelet membranes. *Eur. J. Pharmacol.* **97,** 87–94.

Glaser T. and Traber J. (1985) Binding of the putative anxiolytic TVX Q-7821 to hippocampal 5-hydroxytryptamine (5-HT) recognition sites. *Naunyn Schmiedebergs Arch. Pharmacol.* **329,** 211–215.

Glaser T., Rath M., Traber J., Zilles K., and Schleicher A. (1985) Autoradiographic identification and topographical analyses of high affinity serotonin receptor subtypes as a target for the novel putative anxiolytic TVX Q 7821. *Brain Res.* **358,** 129–136.

Glennon R. A., Young R., and Rosecrans J. A. (1983) Antagonism of the effects of the hallucinogen DOM and the purported 5-HT agonist quipazine by 5-HT_2 antagonists. *Eur. J. Pharmacol.* **91,** 197–206.

Glennon R. A., Titeler M., and McKenney J. D. (1984) Evidence for 5-HT_2 involvement in the mechanism of action of hallucinogenic agents. *Life Sci.* **35,** 2505–2511.

Goodwin G. M. and Green A. R. (1985) A behavioural and biochemical study in mice and rats of putative selective agonists and antagonists for 5-HT_1 and 5-HT_2 receptors. *Br. J. Pharmacol.* **84,** 743–753.

Gozlan H., El Mestikawy S., Pichat L., Glowinski J., and Hamon M. (1983) Identification of presynaptic serotonin autoreceptors using a new ligand: ^{3}H-PAT. *Nature* **305,** 140–142.

Green A. R., ed. (1985) *Neuropharmacology of Serotonin* Oxford University Press, Oxford, UK.

Gyermek J. (1966) Drugs Which Antagonize 5- Hydroxytryptamine and Related Indolealkylamines, in *Handbook of Experimental Pharmacology* (Eichler O. and Farah A., eds.; Erspamer V., sub-ed.) Springer-Verlag, Berlin-Heidelberg-New York.

Hall M. D., El Mestikawy S., Emerit M. B., Pichat L., Hamon M., and Gozlan H. (1985). [^{3}H] 8-Hydroxy-2-(di-*n*-propylamino)tetralin binding to pre- and postsynaptic 5-hydroxytryptamine sites in various regions of the rat brain. *J. Neurochem.* **44,** 1685–1696.

Hamon M., Mallat M., Herbet A., Nelson D. L., Audinot M., Pichat L., and Glowinski J. (1981) [^{3}H] Metergoline: A new ligand of serotonin receptors in the rat brain. *J. Neurochem.* **36,** 613–626.

Hjorth S., Carlsson A., Lindberg P., Sanchez D., Wikstrom H., Arvidsson L.-E., Hacksell U., and Nillsson J. L. G. (1982) 8-Hydroxy-2-(di-*n*-propylamino)tetralin, 8-OH-DPAT, a potent and selective simplified ergot congener with central 5-HT-receptor stimulating activity. *J. Neural Transm.* **55,** 169–188.

Hoffmann B. J., Karpa M. D., Lever J. R., and Hartig P. R. (1985) N_1-methyl-2-[^{125}I]LSD([I^{125}]MIL), a preferred ligand for serotonin 5-HT_2 receptors. *Eur. J. Pharmacol.* **110,** 147–148.

Hoyer D., Engel G., and Kalkman H. O. (1985a) Characterization of the 5-HT_{1B} recognition site in rat brain: Binding studies with (–) [^{125}I]iodocyanopindolol. *Eur. J. Pharmacol.* **118,** 1–12.

Hoyer D., Engel G., Kalkman H. O. (1985b) Molecular pharmacology of 5-HT_1 and 5-HT_2 recognition sites in rat and pig brain membranes: Radioligand binding studies with [^{3}H]5-HT, [^{3}H]8-OH-DPAT, (–) [^{125}I]iodocyanopindolol, [^{3}H]mesulergine and [^{3}H]ketanserin. *Eur. J. Pharmacol.* **118,** 13–23.

Hyttel J., Fredericson-Over K., and Arnt J. (1984) Biochemical effects and drug levels in rats after long-term treatment with the specific 5-HT-uptake inhibitor, citalopram. *Psychopharmacology* **83,** 20–27.

Ilien B., Schotte A., and Laduron P. M. (1982) Solubilized serotonin receptors from rat and dog brain. Selective labelling with [^{3}H] ketanserin. *FEBS Lett.* **138,** 311–315.

Janssen P. A. J. (1985) Pharmacology of potent and selective S_2-serotonergic antagonists. *J. Cardiovasc. Pharmacol.* **7** (suppl. 7), S2–S11.

Kaumann A. J. (1983) A classification of heart serotonin receptors. *Naunyn Schmiedebergs Arch. Pharmacol.* **322,** R42 (166), abstract.

Kaumann A. J. (1985) Two classes of myocardial 5-hydroxytryptamine receptors that are neither 5-HT_1 nor 5-HT_2. *J. Cardiovasc. Pharmacol.* **7** (suppl. 7), S76–S78.

Kohler C. (1984) The distribution of serotonin binding sites in the hippocampal region of the rat brain. An autoradiographic study. *Neuroscience* **13,** 667–680.

Kuhn D. M., Wolf W. A., and Youdim M. B. H. (1986) Serotonin neurochemistry revisited: A new look at some old axioms. *Neurochem. Int.* **8,** 141–154.

Laduron P. M. (1984) Criteria for receptor sites in binding studies. *Biochem. Pharmacol.* **33,** 833–839

Laduron P. M., Janssen P. F. M., Gommeren W., and Leysen J. E. (1982a) In vitro and in vivo binding characteristics of a new long-acting histamine-H_1 antagonist astemizole. *Mol. Pharmacol.* **21,** 294–300.

Laduron P. M., Janssen P. F. M., and Leysen J. E. (1982b) In vivo binding of [^{3}H] ketanserin on serotonin S_2-receptors in rat brain. *Eur. J. Pharmacol.* **81,** 43–48.

Lefkowitz R. J., Caron M. G., and Stiles G. L. (1984) Mechanisms of membrane-receptor regulation. Biochemical, physiological and clinical insights derived from studies of the adrenergic receptors. *N. Eng. J. Med.* **310,** 1570–1579.

Leysen J. E. (1981) Serotonergic receptors in brain tissue: Properties and identification of various [^{3}H]-ligand binding sites in vitro. *J. Physiol.* (Paris) **77,** 351–362.

Leysen J. E. (1983) Rationale and Importance of Binding Studies in Respect to Characterization of Serotonin Receptors, in *Vascular Neuroeffector Mechanisms* (Bevan G. A., Fujiwara M., Maxwell R. A., Mohri K., Shibata S., and Toda N., eds.) Raven, New York.

Leysen J. E. (1984) Problems in in vitro receptor binding studies and identification and role of serotonin receptor-sites. Neuropharmacology **23,** 247–254.

Leysen J. E. (1985) Characterization of Serotonin Receptor Binding Sites, in *Neuropharmacology of Serotonin* (Green A. R., ed.) Oxford University Press, Oxford, UK.

Leysen J. E. (1988) Caractérisation des récepteurs sérotoninergiques. La Revue Médicale de Liège *XLIII,* Suppl 1, 12–21.

Leysen J. E. and Gommeren W. (1986) The drug-receptor dissociation time: New tool for drug research. Receptor binding affinity and drug-receptor dissociation profiles of serotonin-S_2, dopamine-D_2, histamine-H_1 antagonists and opiates. *Drug. Dev. Res.* **8,** 119–131.

Leysen J. E. and Janssen P. A. J. (1988) Specificity of Ligands Used in Psychiatric Research, in *Receptors and Ligands in Psychiatry and Neurology* Vol. I (Sen A. K. and Lee T., eds.) Cambridge University Press, Cambridge, UK.

Leysen, J. E. and Laduron P. M. (1977) A serotonergic component of neuroleptic receptors. *Arch. Int. Pharmacodyn. Ther.* **230,** 337–339.

Leysen J. E. and Niemegeers C. J. E. (1985) Neuroleptics, in *Handbook of Neurochemistry* vol. 9 (Lajtha A., ed.) Plenum, New York.

Leysen J. E. and Tollenaere J. P. (1982) Biochemical models for serotonin receptors. *Ann. Rep. Med. Chem.* **17,** 1–10.

Leysen J. E., Tollenaere J. P., Koch M. H. J., and Laduron P. (1977) Differentiation of opiate and neuroleptic receptor binding in rat brain. *Eur. J. Pharmacol.* **43,** 253–267.

Leysen J. E., Niemegeers C. J. E., Tollenaere J. P., and Laduron P. M. (1978) Serotonergic component of neuroleptic receptors. *Nature* **272,** 168–171.

Leysen J. E., Awouters F., Kennis L., Laduron P. M., Vanderberk J., and Janssen, P. A. J. (1981) Receptor binding profile of R 41 468, a novel antagonist at 5-HT_2 receptors. *Life Sci* **28,** 1015–1022.

Leysen J. E., Geerts R., Gommeren W., Verwimp M., and Van Gompel P. (1982a) Regional distribution of serotonin-2 receptor binding sites in the brain and effects of neuronal lesions. *Arch. Int. Pharmacodyn. Ther.* **256,** 301–305.

Leysen J. E., Niemegeers C. J. E., Van Nueten J. M., and Laduron P. M. (1982b) [^{3}H] Ketanserin (R 41 468), a selective ^{3}H-ligand for serotonin$_2$ receptor binding sites. *Mol. Pharmacol.* **21,** 301–314.

Leysen J. E., Gommeren W., and De Clerck F. (1983a) Demonstration of S_2-receptor binding sites on cat blood platelets using [^{3}H] ketanserin. *Eur. J. Pharmacol.* **88,** 125–130.

Leysen J. E., Van Gompel P., Verwimp M., and Niemegeers C. J. E. (1983b) Role and Localization of Serotonin$_2$ (S_2)-Receptor Binding Sites: Effects of Neuronal Lesions, in *CNS Receptors. From Molecular Pharmacology to Behavior* (Mandel P. and De Feudis F. V., eds.) Raven, New York.

Leysen J. E., Gommeren W., Van Gompel P., Wynants J., Janssen P. F. M., and Laduron P. M. (1985). Receptor-binding properties *in vitro* and *in vivo* of ritanserin. A very potent and long acting serotonin-S_2 antagonist. *Mol. Pharmacol.* **27,** 600–611.

Leysen J. E., Van Gompel P., Gommeren W., Woestenborghs R., and Janssen P. A. J. (1986) Down regulation of serotonin-S_2 receptor sites in rat brain by chronic treatment with the serotonin-S_2 antagonists: Ritanserin and Setoperone. *Psychopharmacology* **88,** 434–444.

Marcinkiewicz M., Verge D., Gozlan H., Pichat L., and Hamon M. (1984) Autoradiographic evidence for the heterogeneity of 5-HT_1 sites in the brain. *Brain Res.* **291,** 159–163.

McCall R. B. and Aghajanian, G. K. (1979) Serotonergic facilitation of facial motoneuron excitation. *Brain Res.* **169,** 11–27.

Middlemiss D. N. (1984) 8-Hydroxy-2-(di-*n*-propylamino)tetralin is devoid of activity at the 5-hydroxytryptamine autoreceptor in rat brain. Implications for the proposed link between the autoreceptor and the [^{3}H] 5-HT recognition site. *Naunyn Schmiedebergs Arch. Pharmacol.* **327,** 18–22.

Middlemiss D. N. and Fozard J. R. (1983) 8-Hydroxy-2-(di-*n* propylamino) tetralin discriminates between subtypes of the 5-HT_1 recognition site. *Eur. J. Pharmacol.* **90,** 151–153.

Muller-Schweinitzer E. and Engel G. (1983) Evidence for mediation by 5-HT_2 receptors of 5-hydroxytryptamine-induced contraction of canine basilar artery. *Naunyn Schmiedebergs Arch. Pharmacol.* **324,** 287–292.

Nelson D. L. and Taylor E. W. (1985) Spiroxatrine: A selective antagonist of serotonin-1A receptors. *Soc. Neurosci. Abst.* part 2, 15th annual meeting, abstract 368.3.

Nelson D. L., Herbet A., Adrien J., Bockaert J., and Hamon M. (1980a) Serotonin-sensitive adenylate cyclase and [^{3}H] serotonin binding sites in the CNS of the rat. II. Respective regional and subcellular distributions and ontogenetic developments. *Biochem. Pharmacol.* **29,** 2455–2463.

Nelson D. L., Herbet A., Enjalbert A., Bockaert J., and Hamon M. (1980b) Serotonin-sensitive adenylate cyclase and [^{3}H] serotonin binding sites in the CNS of the rat. I. Kinetic parameters and pharmacological properties. *Biochem. Pharmacol.* **29,** 2445–2454.

Niemegeers C. J. E., Verbruggen F. J., Van Nueten J. M., and Janssen P. A. J. (1964) Spiroxamide (R 5188): A new compound producing morphine-like and chlorpromazine-like effects in animals. *Int. J. Neuropharmacol.* **2,** 349–354.

Norman A. B., Battaglia G., and Creese I. (1985) [^{3}H] WB4101 labels the 5-HT_{1A} serotonin receptor subtype in rat brain. Guanine nucleotide and divalent cation sensitivity. *Mol. Pharmacol.* **28,** 487–494.

Ortmann R., Bischoff S., Radeke E., Buech O., and Delini-Stula A. (1982) Correlations between different measures of antiserotonin activity of drugs. Study with neuroleptics and serotonin receptor blockers. *Naunyn Schmiedebergs Arch. Pharmacol.* **321,** 265–270.

Pazos A. and Palacios J. M. (1985) Quantitative autoradiographic mapping of serotonin receptors in the rat brain. I. Serotonin-1 receptors. *Brain Res.* **346,** 205–230.

Pazos A., Cortés R., and Palacios J. M. (1985a) Quantitative autoradiographic mapping of serotonin receptors in the rat brain. II. Serotonin-2 receptors. *Brain Res.* **346,** 231–249.

Pazos A., Hoyer D., and Palacios J. M. (1985b) Mesulergine, a selective serotonin-2 ligand in the rat cortex, does not label these receptors in porcine and human cortex: Evidence for species differences in brain serotonin-2 receptors. *Eur. J. Pharmacol.* **106,** 531–538.

Pazos A., Hoyer D., and Palacios J. M. (1985c) The binding of serotonergic ligands to the porcine choroid plexus: Characterization of a new type of serotonin recognition site. *Eur. J. Pharmacol.* **106,** 539–546.

Pedigo N. W., Yamamura H. I., and Nelson D. L. (1981) Discrimination of multiple [^{3}H] 5-hydroxytryptamine binding sites by the neuroleptic spiperone in rat brain. *J. Neurochem.* **36,** 220–226.

Peroutka S. J. (1985a) Selective interaction of novel anxiolytics with 5-hydroxytryptamine$_{1A}$ receptors. *Biol. Psychiat.* **20,** 971–979.

Peroutka S. J. (1985b) Selective labeling of 5-HT_{1A} and 5-HT_{1B} binding sites in bovine brain. *Brain Res.* **344,** 167–171.

Peroutka S. J. (1986) Serotonin Receptors, in *Neuromethods* vol. 4 *Receptor Binding* (Boulton A. A., Baker G. B., and Hrdina P., eds.) Humana, Clifton, New Jersey.

Peroutka S. J. and Kuhar M. J. (1984) Autoradiographic localization of 5-HT_1 receptors to human and canine basilar arteries. *Brain Res.* **310,** 193–196.

Peroutka S. J. and Snyder S. H. (1979) Multiple serotonin receptors: Differential binding of [^{3}H]-5-hydroxytryptamine, [^{3}H]-lysergic acid diethylamide, and [^{3}H]-spiroperidol. *Mol. Pharmacol.* **16,** 687–699.

Peroutka S. J. and Snyder S. H. (1980) Regulation of serotonin$_2$ (5-HT_2) receptors labeled with [^{3}H]spiroperidol by chronic treatment with the antidepressant amitriptyline. *J. Pharmacol. Exp. Ther.* **215,** 582–587.

Peroutka S. J. and Snyder S. H. (1981) [^{3}H] Mianserin: Differential labeling of serotonin$_2$ and histamine$_1$ receptors in rat brain. *J. Pharmacol. Exp. Ther.* **216,** 142–148.

Peroutka S. J., Lebovitz R. M., and Snyder S. H. (1979) Serotonin receptor binding sites affected differentially by guanine nucleotides. *Mol. Pharmacol.* **16,** 700–708.

Peroutka S. J., Lebovitz R. M., and Snyder S. H. (1981) Two distinct central serotonin receptors with different physiological functions. *Science* **121,** 827–829.

Peroutka S. J., Noguchi M., Tolner D. J., and Allen G. S. (1983) Serotonin-induced contraction of canine basilar artery: Mediation by 5-HT_1 receptors. *Brain Res.* **259,** 327–330.

Ransom R. W., Asarch K. B., and Shih J. C. (1986) [^{3}H]1-[2-(4-aminophenyl) ethyl]-4-(3-trifluoromethylphenyl)piperazine: A selective radioligand for 5-HT_{1A} receptors in rat brain. *Neurochem.* **46,** 68–75.

Reyntjens A., Waelkens J., Gelders Y., Ceulemans D., and Janssen P. A. J. (1984) A novel approach for psychosomatic normalization: A placebo-controlled pilot trial of R 55 667, a potent indolamine antagonist. Proc. 14th Congr. Collegium Internationale Neuro-Psychopharmacologicum.

Richardson B. P., Engel G., Donatsch P., and Stadler P. A. (1985) Identification of serotonin M-receptor subtypes and their specific blockade by a new class of drugs. *Nature* **316,** 126–131.

Schotte A., Maloteaux J. M., and Laduron P. M. (1983) Characterization and regional distribution of serotonin S_2-receptors in human brain. Brain Res. **276,** 231–235.

Sills M. A., Wolfe B. B., and Frazer A. (1984) Determination of selective and nonselective compounds for the $5\text{-}HT_{1A}$ and $5\text{-}HT_{1B}$ receptor subtypes in rat frontal cortex. *J. Pharmacol. Exp. Ther.* **231,** 480–487.

Sills M. A., Lucki I., and Frazer A. (1985) Development of selective tolerance to the serotonin behavioral syndrome and suppression of locomotor activity after repeated administration of either 5-MeODMT or mCPP. *Life Sci* **36,** 2463–2469.

Snyder S. H. and Peroutka S. J. (1982) A possible role of serotonin receptors in antidepressant drug action. *Pharmacopsychiatry* **15,** 131–134.

Stockmeier C. A. and Kellar K. J. (1986) In vivo regulation of serotonin-2 receptor in rat brain. *Life Sci.* **38,** 117–127.

Titeler M., Herrick K., and Lyon R. A. (1985) [^{3}H]DOB: A specific agonist radioligand for $5\text{-}HT_2$ serotonin receptors. *Eur. J. Pharmacol.* **117,** 145–146.

Tricklebank M. D. (1985) The behavioural response to 5-HT receptor agonists and subtypes of the central 5-HT receptor. *Trends Pharmacol. Sci.* **31,** 403–407.

Vane J. R. (1957) A sensitive method for the assay of 5-hydroxytryptamine. *Br. J. Pharmacol.* **12,** 344–349.

Vanhoutte P. M., ed. (1985) *Serotonin and the Cardiovascular System* Raven, New York.

Van Nueten J. M., Leysen J. E., Vanhoutte P. M., and Janssen P. A. J. (1982) Serotonergic responses in vascular and non-vascular tissues. *Arch. Int. Pharmacodyn. Ther.* **256,** 331–334.

Van Nueten J. M., Leysen J. E., De Clerck F., and Vanhoutte P. M. (1984) Serotonergic receptor subtypes and vascular reactivity. *J. Cardiovasc. Pharmacol.* **6** (suppl. 4), S564–S574.

Weiss S., Sebben M., Kemp D. E., and Bockaert J. (1986) Serotonin $5\text{-}HT_1$ receptors mediate inhibition of cyclic AMP production in neurons. *Eur. J. Pharmacol.* **120,** 227–230.

Weismann-Nanopoulos D., Mach E., Magre J., Demassey Y., and Pujol J.-F. (1985) Evidence for the localization of $5\text{-}HT_{1A}$ binding sites on serotonin containing neurons in the raphe dorsalis and raphe centralis nuclei of the rat brain. *Neurochem. Int.* **7,** 1061–1072.

Wouters W., Van Dun J., Leysen J. E., and Laduron P. M. (1985a) Differential photoaffinity labelling of serotonin-S_2 receptors and histamine-H_1 receptors using 7-azidoketanserin. *FEBS Lett.* **182,** 291–296.

Wouters, W., Van Dun J., Leysen J. E., and Laduron P. M. (1985b) Solubilization of rat brain serotonin-S_2 receptors using CHAPS/SALT. *Eur. J. Pharmacol.* **115,** 1–9.

Wouters W., Janssen C. G. M., Van Dun J., Thijssen J. B. A., and Laduron P. M. (1986) In vitro labelling of serotonin-S_2 receptors: Synthesis and binding characteristics of [^{3}H] 7-aminoketanserin. *J. Med. Chem.* **29,** 1663–1668.

Yagaloff K. A. and Hartig P. R. (1985) ^{125}I-Lysergic acid diethylamide binds to a novel serotonergic site on rat choroid plexus epithelial cells. *J. Neurosci.* **5,** 3178–3183.

Biochemical Models of Parkinson's Disease

R. E. Heikkila, P. K. Sonsalla, and R. C. Duvoisin

1. Introduction

There are several dopaminergic pathways present in mammalian brain (Ungerstedt, 1971a,b,c,d; Schultz, 1982). Perhaps the most studied of these is the nigrostriatal pathway, which arises from pigmented cell bodies in the zona compacta of the substantia nigra and projects rostrally to the neostriatum (corpus striatum). Parkinson's disease is a syndrome of diverse etiology that is characterized by a degeneration of the above-mentioned nigrostriatal pathway. It has been estimated that the incidence of parkinsonism in the general population ranges from 100 to 150 cases per 100,000, with the prevalence being considerably greater in an older population sample (Schoenberg, 1987; Yahr, 1977). Biochemical and pathological characteristics of this disease include a greatly diminished neostriatal content of dopamine (DA) and its major metabolites and a marked loss of nerve cells in the zona compacta of the substantia nigra. Additional neuronal systems involved include noradrenergic and cholinergic systems.

The clinical manifestations of parkinsonism comprise the classic triad of a resting tremor, bradykinesia, and rigidity; a stooped (kyphotic) posture, characteristic attitudes of the hands and feet, a short-stepped shuffling gait with festination and propulsion, and autonomic dysfunction. The disease often begins with a slight tremor in one hand, which slowly progresses to unilateral symptoms, and finally results in the typical symptoms bilaterally. It is generally accepted that the motor symptoms of parkinsonism primarily reflect dysfunction of the nigrostriatal dopaminergic system; it has been described as the syndrome of striatal DA deficiency (Hornykiewicz, 1966, 1973). The most effective therapy for Parkinson's disease and currently the drug of choice is L-3,4-dihydroxyphenylalanine (L-Dopa), the direct precursor of DA. L-Dopa is usually given in combination with a peripheral inhibitor of Dopa decarboxylase. The inhibition of the enzyme in extracerebral tissues makes it possible to achieve therapeutic blood levels of L-Dopa with smaller oral doses; in effect, circulating L-Dopa is protected from this ubiquitous enzyme, and a large proportion of the oral dose is able to enter the brain parenchyma. Incidentally, a major benefit of peripheral decarboxylase inhibitors is the virtual elimination of the side effects of nausea and vomiting because the formation of DA in the area postrema is blocked. Another therapeutic approach is the use of direct-acting DA receptor agonists. Unlike L-Dopa, these agents require no enzymatic transformation to be activated. None of the DA agonists developed thus far, however, approaches the therapeutic efficacy of L-Dopa; they can only serve as adjunctive therapy (Goldstein et al., 1985).

Since the nigrostriatal system may be involved in various pathophysiological processes, parkinsonism is a clinical syndrome of diverse etiology (Duvoisin, 1984). The most common disorder is Parkinson's disease, characterized by the presence of the intracytoplasmic Lewy body in the affected nuclei. Additional etiological groups include postencephalitic parkinsonism, now extremely rare; parkinsonism occurring as part of a more widespread neuronal system degeneration such as olivopontocerebellar atrophy, the Shy-Drager syndrome, and striatonigral degeneration; progressive supranuclear palsy; and parkinsonism resulting from various intoxications. Improved understanding of the dopaminergic system, the advent of L-Dopa therapy, and the introduction of the computerized tomography (CT) scan and, recently, magnetic resonance imaging (MRI), have led to more precise clinical diagnosis of parkinsonism in recent years, but the differential diagnosis of these different forms of parkinsonism remains quite difficult and occasional patients initially diagnosed by experienced neurologists as classical Parkinson's disease cases turn out with time, or on postmortem study, to have another disorder.

1.1. Animal Models of Parkinsonism

Investigators have long turned to animal models of parkinsonism in their quest for an understanding of its pathophysiology and to test new therapies. Animal models have proved to be a valuable supplement to the limited experimental work that can be accomplished in human patients. Research on human subjects is limited by the great effort and time required to collect data. The availability of tissue samples with which to do research is limited to that obtained at autopsy. It is desirable also to evaluate putative new therapies in animal models before undertaking clinical trials in patients. For these various reasons animal models of human disorders are extremely valuable.

The aim of the present study will be to discuss in depth several animal models of parkinsonism commonly used today. These include reserpinized rats or mice, rats with a unilateral 6-hydroxydopamine lesion of the nigrostriatal pathway, rats with bilateral lesions of the nigrostriatal pathway (Silbergeld and Calne, 1981), and rodents and primates treated with the recently discovered neurotoxin 1-methyl-4-phenyl-1,2,3,6-tetrahydropyridine (MPTP). This latter agent has generated considerable interest because of speculation that the etiology of idiopathic or classical parkinsonism might be related to an endogenous or exogenous "MPTP-like" substance.

2. Biochemical Manipulations Used in the Laboratory to Create Animal Models of Parkinsonism

2.1. Dopamine Depletion Caused by Storage and Synthesis Inhibitors

Reserpine is a rauwolfia alkaloid that has been used clinically as an antihypertensive and an antipsychotic agent. Most probably reserpine works by destroying the vesicular membrane of biogenic amine storage vesicles and thereby causing a concomitant depletion of biogenic amines. In the former clinical circumstance, reserpine most likely is effective because of its depletion of brain norepinephrine (noradrenaline) and, in the latter case, because of its depletion of brain DA. Reserpine also causes a depletion of brain serotonin (5-hydroxytryptamine, 5-HT). Since reserpine affects several biogenic amines and does not exhibit any marked regional specificity in the brain, it may be concluded that the actions of reserpine are nonselective.

Since parkinsonism is, by definition, a DA deficiency disease, it follows that reserpinization of experimental animals produces an animal model of parkinsonism. And, in fact, excessive therapy of humans with reserpine can lead to clinical features, including tremor, rigidity, and akinesia, that are virtually indistinguishable for those seen in Parkinson patients. In spite of this observation, the reserpinized rodent is deficient as an animal model for parkinsonism. Its response to various drugs is quite different from the response of the patient and the actions of reserpine are nonselective. Perhaps more importantly, although parkinsonism is a DA deficiency disease, it is also a degenerative disease. The Parkinson patient has low levels of DA in the nigrostriatal system because the entire system has undergone degenerative change. In contrast, reserpine causes a depletion of brain DA that is fully reversible.

After reserpinization, the laboratory mouse sits quietly with a posture that is considerably more curved than normal, even kyphotic, with its back sharply humped. It is generally immobile (i.e., bradykinetic) for long periods of time, although when prodded it can move relatively normally. When dropped or placed on its back, the reserpinized mouse can right itself, but does so quite slowly (Fig. 1). The slowness of movement in this animal has been described as a parallel to the akinesia or the bradykinesia seen in human Parkinson patients (Duvoisin, 1976). The curved posture of the reserpinized mouse may be analogous to the flexed posture characteristic of Parkinson patients. One of the consequences of reserpinization is a severe hypothermia; reserpinized mice often huddle in groups (probably to keep warm) and often shiver, which may be mistaken for tremor. Real tremor is, for the most part, not associated with reserpinized rodents. When given an injection of L-Dopa alone, or of L-Dopa combined with a peripheral Dopa decarboxylase inhibitor, activity returns to relatively normal levels. Depending upon the doses used, however, the return to normal activity is generally of relatively short duration. Thus the drug of choice used to treat Parkinson patients, namely L-Dopa, restores the motor deficit observed in the reserpinized rodent. This observation gives further credence to some that the reserpinized rodent is a good animal model for parkinsonism. As will be discussed in more detail below, however, amphetamine administration also restores locomotor activity in the reserpinized rodent. Since amphetamine treatment has no significant therapeutic effect in the parkinsonian patient, this observation points out a major deficiency of the reserpine model. It cannot be relied upon to predict the clinical efficacy of a new drug in Parkinson patients.

The drug α-methyl-*p*-tyrosine, an analog of the dietary amino acid tyrosine, is a potent inhibitor of the enzyme tyrosine hydroxylase. The biosynthesis of L-Dopa from L-tyrosine, catalyzed by tyrosine hydroxylase, is the rate-limiting step in the enzymatic formation of the catecholamines. When the enzyme is inhibited, the consequence is a marked diminution in the brain level of DA (and of norepinephrine) by a mechanism completely different from

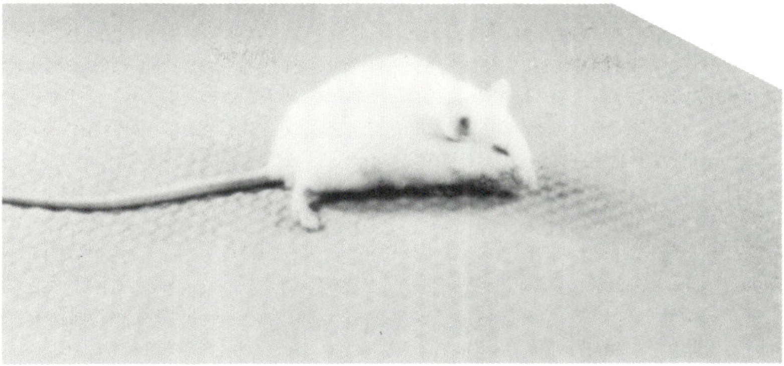

Fig. 1. A CF-W mouse treated with a single injection of reserpine 18 h previously.

that of reserpine. The laboratory animal treated with α-methyl-*p*-tyrosine, and depleted of brain DA, exhibits most of the behavioral deficits observed in the reserpinized animal. Since these animals do indeed have a DA deficiency, they represent a model of the DA deficiency disease parkinsonism. As was pointed out above for reserpine, however, this synthesis inhibitor is relatively nonselective in that it produces generalized depletions of DA and norepinephrine throughout the brain, leads to a DA depletion that is reversible, and does not cause degeneration of the nigrostriatal pathway.

In our laboratory, we routinely administer reserpine intraperitoneally at a dose of 5 mg/kg (Heikkila, 1981). The reserpine (Serpasil) is purchased in an injectable form from Ciba-Geigy (Summit, NJ). Shortly after they are administered reserpine, the mice show a temporary increase in locomotor activity, which is probably caused by the release of some DA from its stored vesicular sites that avoids inactivation by monoamine oxidase. Two hours after this dose of reserpine, the mice appear to be fully affected behaviorally and exhibit the behavioral characteristics described above. Higher doses of reserpine usually act somewhat more quickly, whereas lower doses (i.e., 0.5–4 mg/kg) take somewhat longer. Sixteen hours after administration of reserpine, the mice are fully affected and remain behaviorally affected for up to a month. In most behavioral experiments using reserpinized mice, we have chosen the 16-h time period. At this time the level of neostriatal DA is decreased by approximately 95% compared to the vehicle-injected control mice. Levels of the major metabolites of

DA, namely dihydroxyphenylacetic acid (DOPAC) and homovanillic acid (HVA), are affected to a somewhat lesser extent. This results in a higher ratio of metabolites to DA in reserpinized than in control mice. The level of neostriatal DA remains diminished for a long period of time after reserpine, taking about a month (depending on the dose) to recover to pretreatment values.

In our laboratory we also utilize α-methyl-*p*-tyrosine (AMPT, methyl ester hydrochloride salt, Sigma, St. Louis, MO) treatment to lower levels of neostriatal DA. The dose varies between 75 mg/kg and 200 mg/kg. We often give the AMPT at 75 mg/kg at zero time, repeat the injection 2 h later, and initiate behavioral observations 1 h after the last AMPT injection. At this time the DA content is approximately 20–40% of that observed in control animals. AMPT is rather difficult to work with because there is a narrow range between the effective dose and the lethal dose. In contrast, reserpine is a relatively safe drug with which to work, the range between its effective dose and its lethal dose being quite broad.

It should also be emphasized that although these two agents act to deplete brain DA, they do so by different mechanisms. Reserpine acts mainly on stored DA, and, in fact, levels of newly synthesized DA are higher in reserpinized animals than in corresponding controls. In contrast, AMPT acts on newly synthesized DA and has little direct effect on stored DA. These mechanistic differences can perhaps be best revealed by catecholamine-releasing agents or catecholamine-uptake inhibitors (Ross, 1979). For example, the releasing agent amphetamine, which acts on newly synthesized DA, is an effective stimulant in reserpinized mice, but has no effect in mice treated with AMPT. In contrast, amfonelic acid and other DA-uptake inhibitors are without stimulant effect in reserpinized mice, but are very active in mice depleted of DA by AMPT treatment.

2.2. Unilateral Destruction of the Nigrostriatal Pathway Caused by the Stereotaxic Administration of 6-Hydroxydopamine

6-Hydroxydopamine (2,4,5-trihydroxyphenylethylamine; 6-OHDA), a hydroxylated analog of DA, is widely used as a research tool by the neurobiologist. After its administration to experimental animals, it is accumulated by catecholamine-containing nerves and subsequently causes their degeneration. 6-OHDA possesses two important properties. First, it is transported by the catecholamine

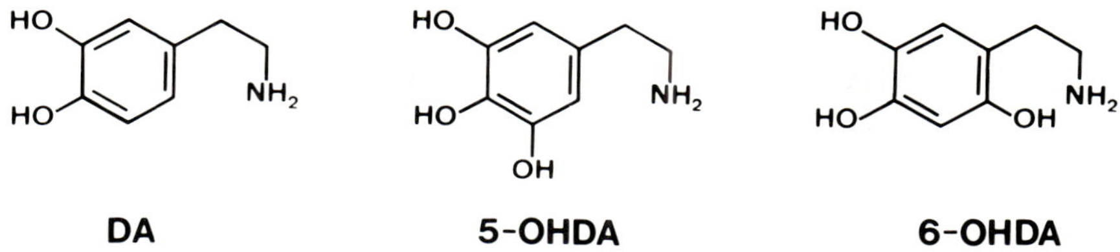

Fig. 2. The chemical structures of the neurotoxin 6-hydroxydopamine compared to dopamine and 5-hydroxydopamine.

transport (uptake) system and is concentrated within catecholamine-containing nerve cells. Second, because it possesses *ortho*- and *para*-dihydroxy groups, 6-OHDA is extremely unstable in aqueous solution and reacts with oxygen (autoxidizes) to form several highly reactive species. Among these are quinones of 6-OHDA, the superoxide radical, the hydroxyl radical, and hydrogen peroxide (Cohen and Heikkila, 1974; Graham, 1978). It has been suggested that one or a combination of these reactive species is responsible for 6-OHDA-induced neurotoxicity. The importance of the autoxidation of 6-OHDA as a determinant for neurotoxicity is clearly demonstrated by the fact that 5-hydroxydopamine (5-OHDA), which autoxidizes much more slowly than does 6-OHDA, is not neurotoxic. Note that the spatial arrangement of hydroxyl groups differs between 6-OHDA and 5-OHDA (Fig. 2). A number of excellent reports and review articles have discussed basic features of 6-OHDA neurotoxicity (e.g., Jonsson, 1980; Graham, 1984).

The rodent, and primarily the rat, with a unilateral lesion of the nigrostriatal pathway induced by 6-OHDA has been widely used as an animal model of parkinsonism. Interestingly, except for a relatively short time after the stereotaxic surgery, rats treated with 6-OHDA do not demonstrate any obvious behavioral deficits. They clearly do not exhibit the typical parkinsonian symptoms seen in reserpinized mice. They appear, even to the well-trained observer, to be normal. Then why are these lesioned rats used as an animal model for Parkinson's disease? It is because drugs used to treat Parkinson patients elicit a predictable behavioral response in these animals. They cause a contralateral (contraversive) circling or rotation. That is, they cause a circling away from the side of the lesion.

This animal model of parkinsonism was developed by the Swedish histologist Urban Ungerstedt in the late 1960s and early 1970s (e.g., Ungerstedt 1971a,b,c,d) and is often referred to as the

"Ungerstedt rat." As stated above, antiparkinsonian agents such as L-Dopa and other DA agonists cause a contralateral rotation in the 6-OHDA-lesioned rat. In contrast, indirectly acting agents, such as amphetamine, which acts by releasing DA, or cocaine, which acts by blocking the reuptake of synaptically released DA, cause an ipsilateral (ipsiversive) rotation. Ungerstedt first explained this puzzling phenomenon by suggesting that the direct-acting DA agonists act by stimulating supersensitive DA receptors on the lesioned side of the brain. In contrast, the indirectly acting agents act by increasing dopaminergic activity on the nonlesioned side of the brain.

It has been demonstrated by several groups of investigators (e.g., Creese and Snyder, 1979; Heikkila et al., 1981) that the 6-OHDA-induced lesion of the nigrostriatal pathway must be very extensive before the rats will rotate to dopaminergic agonists. It has been our experience, over the last decade or so, that the lesion must be almost complete before rats will rotate contralaterally after L-Dopa administration. For example, we have never had an animal that was a rotator have less than a 90% loss of neostriatal DA content (or less than a 90% loss of neostriatal tyrosine hydroxylase activity or in the capacity of neostriatal synaptosomes to take up ^{3}H-DA). In contrast, rats that exhibit only a marginal DA deficit will rotate ipsilaterally to amphetamine. A simple explanation for this observation is that a threshold deficit of about 90% must be attained before neostriatal DA receptors become supersensitive.

In our laboratory we routinely maintain a "colony" of rats with a unilateral lesion of the left nigrostriatal pathway induced by 6-OHDA. There is no particular reason to prefer the left side. Others, for example, routinely lesion on the right side. In fact, depending upon the nature of the experiment and the experimental objectives, it may be desirable to lesion some rats in the left and others in the right nigrostriatal pathway. The importance of this latter approach has perhaps been best pointed out by Robinson and Becker (1983).

In our laboratory, the colony of rotators is almost always female, for two specific and practical reasons. First, female rats grow at a slower rate than males, and consequently are of a more uniform size during the period of stereotaxic surgery. In many cases this issue is not relevant, but can become so if the period of surgery is done over several days. The second reason is that we routinely keep our rotators for several months (often up to 2 yr) and the smaller females can be more readily housed and maintained than the larger males. This latter point can be important if the space for animal maintenance is limited.

Specifically, female rats (e.g., Sprague-Dawley), weighing approximately 175 g are utilized. The amount of 6-OHDA injected, when the injection site is the medial forebrain bundle (MFB), is 8 μg of free base. The injection volume is 4 μL, and the 6-OHDA is kept on ice and dissolved in a 0.02% solution of ascorbic acid. The latter agent is utilized as an antioxidant, but depending upon the circumstances, it may not be necessary. Prior to surgery, the rats are anesthetized with sodium methohexital (Brevital, Eli Lilly and Co., Indianapolis, IN). The 6-OHDA is accumulated by both norepinephrine- and DA-containing nerves and causes destruction of both. We routinely pretreat rats with desmethylimipramine (25 mg/kg) 30 min prior to 6-OHDA. This agent, which selectively prevents the entry of 6-OHDA into norepinephrine-containing nerve terminals, prevents the destruction of the norepinephrine-containing nerves by the 6-OHDA. Theoretically, pretreatment of rats with a selective dopamine-uptake blocker would provide protection from 6-OHDA to dopaminergic nerves. To our knowledge, however, no such agent has been utilized in this way, although several recently developed uptake blockers possess the proper selectivity (i.e., are better blockers of uptake into dopaminergic nerves than of uptake into noradrenergic nerves) and thus have potential for such use (*see* Heikkila and Manzino, 1984, for a description of experiments with several uptake inhibitors that might have the appropriate pharmacological profile for this purpose). The surgery is done in a David Kopf model 900 stereotaxic apparatus. For injection into the MFB, lesion coordinates are AP + 4.9, L + 1.6, and DV – 2.4 (Konig and Klippel, 1963). If the injection site differs, obviously different injection coordinates and different amounts of 6-OHDA will be used. For example, we have injected 6-OHDA directly into the neostriatum itself or alternately into the substantia nigra. It has been our experience that the success rate of neostriatal surgery is very high (greater than 90%). In contrast, when we have attempted stereotaxic injections into smaller targets such as the substantia nigra, our success rate is considerably lower (often from 40 to 60%). Because the supersensitivity takes time to manifest itself, we routinely wait 3–4 wk postsurgery before initiating behavioral experiments.

At this time point animals are screened for rotation to L-Dopa combined with carbidopa, a peripheral inhibitor of the enzyme Dopa decarboxylase (10 mg/kg of each). The rotation is carried out in plastic bowls, 40 cm in diameter. The rats have a harness around their necks that is attached to a device for detecting circling behavior. Data are quantitated, both as to direction and number of turns, by a Coulbourn automated rotometer. The time period of observation can vary depending upon the experimental need, but 2 h is an appropriate time period. Animals are routinely retested about 1 wk later, and the colony developed. Only those animals that exhibit significant contralateral turns to L-Dopa (e.g., 500 turns in the 2-h period) are chosen for further experimentation. It has been our experience that all rats screened and selected in this manner will also rotate contralaterally to other directly acting dopaminergic agonists (e.g., apomorphine) and ipsilaterally to indirectly acting agents like amphetamine. Once the colony is developed, many different types of experiments can be carried out. We routinely use rats over and over, but generally wait several days (usually a week) before reusing the same animals. To avoid any potential experimental bias, we will randomize drug treatments in an experiment in which multiple drugs are being tested.

2.3. Bilateral Destruction of the Nigrostriatal Pathway by the Stereotaxic Administration of 6-Hydroxydopamine

Rats with bilateral 6-OHDA-induced lesions of the nigrostriatal pathway provide a different kind of animal model from the rat with a unilateral lesion. The rats with a sufficiently extensive bilateral 6-OHDA lesion clearly have behavioral deficits. These deficits can include a severe bradykinesia and, in fact, may closely resemble those seen in reserpinized rats. The rats can also demonstrate a marked aphagia and adipsia and can die if they are not administered food and water for the few days immediately following surgery (Schultz, 1982; Zigmond et al., 1984). After this initial time period, these animals usually recover motor function and eat and drink relatively normally, depending on the extent of the lesion (Schultz, 1982; Zigmond et al., 1984). After several days, the bilaterally lesioned rats exhibit an enhanced behavioral response compared to control animals when they are given L-Dopa or other DA agonists (Breese et al., 1985a,b). As with the unilaterally lesioned rats, all commonly used antiparkinson agents can elicit this enhanced locomotor response. In spite of this, the bilaterally lesioned rat is used considerably less as an animal model of parkinsonism. Perhaps this is because of the relative ease and reproducibility that can be obtained in the unilaterally lesioned rat.

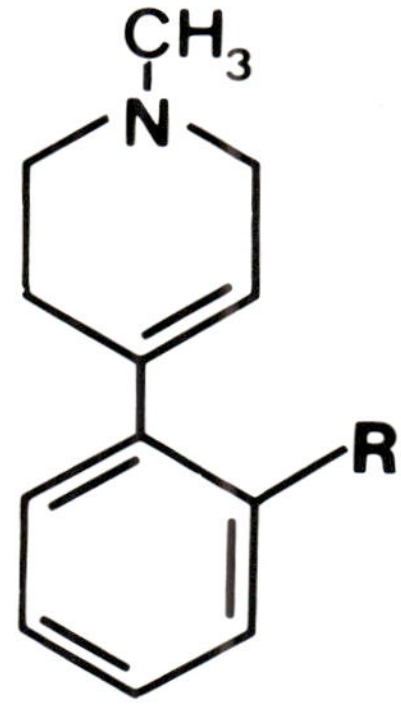

COMPOUND	R
MPTP	H
$2'CH_3$–MPTP	CH_3

Fig. 3. The structures of MPTP and a very toxic analog $2'CH_3$-MPTP.

This fact points out how an "animal model" can be used to great benefit even though the model does not possess the same clinical deficits that are present in the human disease state.

2.4. Nigrostriatal Degeneration in Experimental Animals Induced by the Peripheral Administration of MPTP

1-Methyl-4-phenyl-1,2,3,6-tetrahydropyridine (MPTP, *see* Fig. 3 for structure) has been extensively used as an intermediate in the synthesis of more complex molecules, in particular analgesic compounds related to meperidine (Demerol). Depending upon the reaction conditions, varying amounts of MPTP have been shown to be formed in the synthesis of the potent analgesic agent 1-methyl-4-phenyl-4-propionoxypiperidine (MPPP), which is the reverse ester of meperidine. The inadvertent self-administration of relatively small amounts of MPTP mixed with varying amounts of MPPP by several young drug abusers resulted in a severe parkinsonian syndrome (Davis et al., 1979; Langston et al., 1983). The rationale of this administration of MPPP was that it was a potent, but at the same time legal, alternative to heroin. The parkinsonism induced in these patients was similar to idiopathic Parkinson's disease with respect to its clinical, pathological, and biochemical features. It was subsequently shown that the intravenous injection of MPTP to monkeys caused behavioral, biochemical, and pathological changes similar to those seen in parkinsonian patients (Burns et al., 1983). This finding indicated that the agent responsible for causing parkinsonism in the human drug abusers was MPTP and not one of the other abused substances.

The observation that it produces the parkinsonian state in humans and nonhuman primates has stimulated considerable research with MPTP in a variety of in vivo and in vitro model systems. Moreover, a great deal of popular interest has been generated by these observations. Research with MPTP has, for example, been a topic of articles in major national newspapers and magazines, as well as the topic of several national television programs. Biochemical and pathological manifestations of MPTP toxicity have been characterized in several animal species. A considerable effort has been focused on the mechanism by which MPTP produces its selective destruction of nigrostriatal neurons, and there has been considerable speculation that a neurotoxin like MPTP could be involved in the pathogenesis of Parkinson's disease (Langston and Ballard, 1984; Snyder, 1984). This speculation is interesting when one considers that the low concordance rate for parkinsonism observed in monozygotic twins has appeared to rule out genetic factors as being causative (Duvoisin, 1984; Ward et al., 1983).

The first case of MPTP-induced parkinsonism was that of a 23-yr-old graduate student who inadvertently produced MPTP as a byproduct in the synthesis of MPPP (Davis et al., 1979). This individual apparently made and self-administered MPPP for several months with no problems, until he took certain synthetic shortcuts in one preparation. These shortcuts included a reduction in the reaction time, with a concomitant higher reaction temperature and a failure to isolate and properly recrystallize the product. Markey and colleagues at the National Institute of Mental Health analyzed this reaction mixture and found that it contained MPPP, 4-hydroxy-4-phenyl-*N*-methyl-piperidine, and MPTP. After several days of injection of this mixture, this patient developed severe motor disturbances. He was originally misdiagnosed as a catatonic schizophrenic, but a neurologist ultimately recognized the signs of parkinsonism. This was striking in that 23-yr-old Parkinson patients are quite rare. Treatment with L-Dopa/carbidopa or benztropine resulted in marked improvement. When treatment was discontinued, severe bradykinesia, generalized rigidity, and mild tremor developed. Resumption of drug therapy resulted in marked improvement. The diagnosis of parkinsonism was thus seemingly correct. The patient continued to abuse several drugs including the L-Dopa and ultimately died of an overdose of cocaine and codeine. Postmortem study revealed severe destruction of the pars compacta of the substantia nigra. A single rounded eosinophilic Lewy body similar to that seen in Parkinson patients was reported initially, but subsequent study has not confirmed that identification.

There subsequently have appeared several other published reports of MPTP-induced parkinsonism among a drug-abusing population in northern California who purchased MPTP-contaminated MPPP sold as new "synthetic heroin" (Langston et al., 1983; Ballard et al., 1983; Langston and Ballard, 1984; Ballard et al., 1985). All of these California patients became symptomatic within 1 wk after the use of the new drug. All patients responded to therapy with L-Dopa and carbidopa.

Parkinsonism following accidental MPTP exposure in humans has also been documented. For example, Langston and coworkers have reported MPTP-induced Parkinson's disease in a chemist who, while working at an east coast pharmaceutical firm, synthesized many compounds requiring repeated preparation of MPTP. A Danish chemist also developed parkinsonism, a short time after recrystallizing large amounts of the free base of MPTP, which is quite volatile. In both instances, opportunities for cutaneous and inhalation exposure to MPTP were present.

In 1983 it was reported that intravenous administration of MPTP to rhesus monkeys produced a Parkinson-like syndrome (Burns et al., 1983). Behavioral changes were initially observed after two or three doses of 0.33 mg/kg MPTP. At a cumulative dose of 1.7 mg/kg there were profound behavioral deficits, including persistent eyelid closure, decreased spontaneous motor activity, rigidity, postural tremor, and difficulty in swallowing. Levodopa treatment rapidly and dramatically corrected all of the above behavioral changes. The neurochemical changes induced by MPTP treatment in these primates included transient reductions in the brain and cerebrospinal fluid levels of 5-hydroxyindole-3-acetic acid (5-HIAA) and 3-methoxy-4-hydroxyphenylethylene glycol (MHPG). In contrast, concentrations of HVA in the cerebrospinal fluid and DA and HVA concentrations in the caudate and putamen were markedly diminished 3 mo after MPTP treatment. Histopathological examination of a single monkey brain indicated a severe loss of nerve cells in the zona compacta of the substantia nigra. In contrast, dopaminergic neurons of the ventral midbrain and median eminence were normal, as were noradrenergic neurons of the locus ceruleus and hypothalamus. Similar observations have also been made in other primates such as the squirrel monkey (Langston et al., 1984a), macaque monkey (Crossman et al., 1985), and marmoset (Jenner et al., 1984). MPTP treatment in all of these primates results in a neurological disorder with a clinical picture similar to that seen in humans in both MPTP-induced and idiopathic parkinsonism.

Several groups of investigators have attempted to establish a rodent model in which MPTP toxicity could be studied. Guinea pigs and rats were found, relative to primates, to be resistant to MPTP-induced nigrostriatal dopaminergic toxicity (Boyce et al., 1984; Chiueh et al., 1984; Enz et al., 1984; Perry et al., 1985). It has been shown by others, however, that with rigorous injection paradigms, MPTP can work to some extent in the rat (Fuller and Steranka, 1985; Jarvis and Wagner, 1985a).

It was subsequently discovered by several groups of investigators that the mouse, although less sensitive than the primate, was sensitive to MPTP-induced toxicity (Hallman et al., 1984; Heikkila et al., 1984a,b). For example, we found that 5–10 intraperitoneal injections of MPTP once a day at 30 mg/kg produced marked decreases in the neostriatal content of DA and its metabolites, whereas DA concentrations in the nucleus accumbens and hypothalamus were unaffected. Moreover, these MPTP-treated mice had a marked bilateral loss of neurons in the zona compacta of the substantia nigra (A9 region) on histological study, but there was no apparent effect in other brain areas including the A10 region, the locus ceruleus, the raphe nuclei, and the hypothalamus [*see* Fig. 4 for an example of a mouse treated with a very toxic structural analog of MPTP, namely 1-methyl-4-(2'methylphenyl)-1,2,3,6-tetrahydropyridine (2'CH_3-MPTP; structure in Fig. 3)].

The mechanism by which MPTP produces its profound pathological changes is at present only partially understood, although several important steps in the neurotoxic process have been identified. Markey et al. (1984), for example, first demonstrated that the corresponding pyridinium analog of MPTP, namely the 1-methyl-4-phenylpyridinium ion (MPP^+), was the major metabolite of MPTP, and that detectable levels of MPP^+ could be found in monkey brains many days after MPTP administration. Based on their findings, they reasoned that MPP^+ itself (or the process of

Fig. 4. A C57-bl mouse treated with two ip injections of 20 mg/kg 2'CH_3-MPTP at 6 h intervals. The photograph was taken 2 d after the last injection.

MPP^+ formation) was responsible for MPTP-induced toxicity. In contrast to the long half-life in primates, the half-life of MPP^+ in the mouse brain was found to be considerably shorter, approximately 2 h, whereas the half-life of MPP^+ in the rat was found to be even shorter, approximately 15 min. It seems to follow that if MPP^+ is indeed the toxin responsible for MPTP-induced neurotoxicity, its rapid clearance may explain why the mouse is less susceptible than the monkey to the neurotoxic action of MPTP.

Chiba et al. (1984) subsequently found that MPP^+ was formed in rat brain mitochondrial preparations in vitro and that this MPP^+ formation could be prevented by the presence of monoamine oxidase B (MAO-B) inhibitors. These investigators also found that deprenyl, a selective MAO-B inhibitor, was able to prevent the in vitro formation of MPP^+. In contrast, the selective MAO-A inhibitor clorgyline was relatively ineffective. These findings have been confirmed in additional studies with purified enzymes (Fritz et al., 1985). It was subsequently shown in vivo that MPTP-induced neurotoxicity in mice could be prevented by pretreatment of the animals with selective or nonselective inhibitors of MAO-B, but not inhibitors of MAO-A (Heikkila et al., 1984b, 1985a,b). These findings were subsequently confirmed in monkeys (Cohen et al., 1984; Langston et al., 1984b). It is evident that MAO-B plays a critical role in the oxidation and bioactivation of MPTP, and that MPP^+ formation is a critical part of the neurotoxic process.

A perplexing aspect of MPTP toxicity is its highly selective action on the nigrostriatal dopaminergic neurons and relative lack of effect in other brain areas. Although MAO-B activity is high in the substantia nigra and neostriatum, its activity is also high in other brain areas that are unaffected by MPTP treatment. MPP^+ formation, moreover, occurs rather ubiquitously throughout the brain as well as in many peripheral tissues (Johannessen et al., 1985). Although an answer is not available at this time, part of the answer must be related to the active transport of MPP^+ into the DA neurons by the DA transport system. This observation was initially made by Javitch et al. (1985), who found that MPP^+, but not MPTP, was readily accumulated by neostriatal synaptosomal preparations. In fact, the kinetics of MPP^+ uptake were found to be very similar to those for DA uptake. Consistent with these in vitro findings, it has been demonstrated that pretreatment of mice with compounds that block DA transport (e.g., mazindol or amfonelic acid) prevent MPTP-induced neurotoxicity to the nigrostriatal dopaminergic neurons (Fuller and Hemrick-Luecke, 1985; Javitch et al., 1985).

Based on the information thus far obtained, the following picture emerges. Since immunohistochemical studies have localized MAO-B to astrocytes and to serotonin-containing neurons, but not to catecholaminergic neurons (Westlund et al., 1985), MPP^+ formation probably occurs outside the dopaminergic neurons. It is likely that the lipophilic compound, MPTP, could be transported into astrocytes and be oxidized by MAO-B to $MPDP^+$, which in turn can be spontaneously oxidized to MPP^+. It is not clear how or if MPP^+ (or $MPDP^+$) is released into the extracellular space. Once in the extracellular space, $MPDP^+$ could be converted to MPP^+ and then accumulated into the DA neurons by the DA transport system. Inside the neuron, the MPP^+ is concentrated and reaches critical levels capable of producing damage.

The ultimate cellular mechanism of toxicity produced by MPTP on the nigrostriatal dopaminergic neurons is unknown. Several hypotheses have been proposed that invoke a role for DA in the neurotoxic process. Castagnoli et al. (1985), for example, proposed that MPTP may be oxidized by MAO-B within the DA neuron to $MPDP^+$, which could then oxidize the DA. The net effect of this oxidation–reduction reaction would be the reduction of $MPDP^+$ to MPTP and the oxidation of DA to its quinone. A byproduct would be the formation of the very toxic hydrogen peroxide and oxy-radicals (Castagnoli et al., 1985). A role for DA in the neurotoxic process appears unlikely, however, since it has recently been demonstrated that the effects of MPTP are not prevented in mice with a greatly diminished content of neostriatal DA (Fuller and Hemrick-Luecke, 1985; Schmidt et al., 1985). Moreover, it appears that MPP^+, which is much more difficult to reduce than $MPDP^+$, is the actual species transported into the dopaminergic neuron. We have recently reported that MPP^+ inhibits NADH-linked respiration in brain mitochondria (Nicklas et al., 1985). In these experiments, it was demonstrated that MPP^+ is capable of inhibiting the oxidation of NADH-linked substrates by brain mitochondrial preparations. Based on these findings we have suggested that MPP^+ may inhibit mitochondrial respiration within the dopaminergic neuron and thereby cause cell death.

In experiments with MPTP, we routinely use young male mice weighing at least 25 g. It has been our experience that results obtained with mice weighing less than 25 g tend to be somewhat more variable, perhaps because lighter animals are younger. Sershen et al. (1985) have studied the effects of MPTP in mice and showed a relationship between the age of the mice and the extent of neurotoxicity. Jarvis and Wagner (1985b), moreover, have shown that younger rats are less sensitive to the toxic actions of MPTP than are older rats. We generally inject the mice with one of two different experimental regimens. In the first regimen, which we used when we first started working with MPTP, and now use to a lesser extent, the mice receive an intraperitoneal injection of 30 mg/kg of free base of MPTP once a day for several consecutive days. The MPTP is the hydrochloride salt obtained from Research Biochemicals (Wayland, MA). This product is far preferable to the free base of MPTP for at least two important reasons. First, MPTP hydrochloride is much cheaper than the MPTP free base. Second, and very importantly, the hydrochloride salt of MPTP is much less volatile than the free base. Consequently, the chances of laboratory personnel inhaling MPTP during weighing and handling of the compound and handling of animals that receive the compound is greatly diminished. The MPTP is dissolved in distilled water at a concentration of 30 mg free base per 10 mL of solution. Each mouse receives an injection of 1 mL per 100 g of body weight. For example, a 25-g animal receives 0.25 mL. This injection sequence is repeated for several days.

The second injection paradigm, which we now favor, is based on the observation (Markey et al., 1984) that the half-life of MPP^+, the major and presumed toxic metabolite of MPTP, is approximately 2 h in the mouse. With this regimen, the mice are given four intraperitoneal injections of MPTP at 20 mg/kg of free base at 2-h intervals. As an example, the mice are injected at 9 A.M., 11 A.M., 1 P.M., and 3 P.M., with the total dose being 80 mg/kg. With this regimen, mice routinely show an 80% or greater decrease in the neostriatal content of DA and a somewhat lesser decrement in DOPAC and HVA. As many as 10 injections, once a day at 30 mg/kg, are required to obtain approximately the same effect (total dose of 300 mg/kg). In other experiments, we have found that 10 mg/kg of free base MPTP, given four times daily at 2-h intervals for 2 consecutive d (total dose of 80 mg/kg) is relatively ineffective. Thus the dosing paradigm is extremely important; not only the total dose administered, but also the frequency of administration, is critical.

There is a considerable difference in the sensitivity of various strains of mice to the neurotoxic actions of MPTP (e.g., Heikkila, 1985). For example, we have found that C57 black mice obtained from Charles River (Wilmington, MA) are very sensitive to the toxic actions of MPTP, much more so than CF-W (Swiss-Webster) mice obtained from the same supplier. Moreover, the CF-W mice from Charles River are much more sensitive to MPTP than are Swiss-Webster mice obtained from Perfection Breeders (Douglassville, PA). We have now worked with MPTP in approximately 15 different strains of mice, many of them pigmented like the C57 black and many of them albino like the CF-W. Many of the strains we have utilized are inbred, whereas many are outbred. No simple and clear generalization can be made about MPTP and its actions in mice, except perhaps that each of the 15 strains is affected. The effects in some are maximal (i.e., greater than 90%), whereas the effects in others are relatively minimal (i.e., less than 20%) with either of the two regimens described above. It should be emphasized that greater effects can be obtained in any of the various strains of mice with more rigorous injection paradigms. The approximate LD_{50} of MPTP in mice in our hands, however, is in the range of 50–80 mg/kg in a single injection. It should be emphasized that our goal is to keep as many mice alive as possible, so we have attempted to avoid a dosing regimen that kills more than a few animals.

2.5. Methods Used to Ascertain Dopaminergic Deficits in Animal Models of Parkinsonism

2.5.1. Measurement of the Neostriatal Content of Dopamine and Its Metabolites by HPLC with Electrochemical Detection

The experimental animals are sacrificed at various time periods after administration of the neurotoxin (or DA-depleting agent). Mice, for example, are stunned by a blow to the head, the brains removed, and the appropriate brain area (usually neostriatum) dissected free from the rest of the brain. In some cases, depending on the experiment, the left and right neostriata are handled separately. This would be the case in rats that had previously received a unilateral injection of 6-OHDA. In other experiments, the left and right neostriata may be pooled. In other instances, tissue from several animals might be pooled, but usually this is not necessary.

The tissue is homogenized in 0.1*M* perchloric acid containing 3,4-dihydroxybenzylamine as an internal standard. In most cases the tissue is homogenized at 10 mg of tissue per mL of medium. Care must be taken that the tissue is homogenized as quickly as possible after sacrifice of the experimental animals. We have noted that the levels of DA can fall markedly and the levels of DA metabolites increase markedly in tissue handled slowly, even that stored on ice. This is presumably because the catecholamine degradative enzymes, namely monoamine oxidase and catechol-*O*-methyltransferase, are very active. The perchloric acid tissue homogenate is centrifuged at 27,000*g* for 15 min. The supernatant, usually 20 μL (equivalent to 0.2 mg of tissue) is used for assays of DA, serotonin, and their metabolites. The method that we routinely use is an adaptation of that described by Mayer and Shoup (1983). The latter method is described in great detail in technical bulletins published by Bioanalytical Systems (W. Lafayette, IN).

For biogenic amine analysis, we use a LC-304T liquid chromatograph purchased from Bioanalytical Systems that is connected to a Hewlett-Packard model 3390A integrator. The operating potential is 750 mV and a Biophase ODS 5-μm column (Bioanalytical Systems) is routinely used. In certain cases we use columns obtained from other suppliers. The flow rate of the mobile phase ranges from 1.1 to 1.5 mL per min. The mobile phase is made up as follows: 35 mL of acetonitrile is added to 965 mL of 0.15 monochloroacetate buffer (pH 3.0) containing 193 mg of sodium octyl sulfate. This mixture is filtered and degassed, and 18 mL of tetrahydrofuran is added. This mobile phase is sealed until use. In some cases, usually when problems arise and separation is not ideal, the relative proportions of the organic solvents are altered. In other cases, EDTA (disodium salt) is added at 75 mg/L of mobile phase.

Data are calculated from standard curves that are run on the same day as the experiment. The standards consist of, in their order of elution: dihydroxybenzylamine (DHBA), DOPAC, DA, 5-HIAA, HVA, 3-*O*-methyl-dopamine (3-OMe-DA), and serotonin. Depending on the exact chromatography conditions chosen, a typical run will be completed in 7–15 min. We thus routinely handle as many as 30 tissue samples per day. It should be emphasized that the tryptamine derivatives (5-HIAA and serotonin) are somewhat unstable in the perchloric acid, so that the data obtained for these two compounds can be somewhat variable. And, in fact, in a "long day" there can be a systematic downward trend in their experimental values. It should also be emphasized that there is no such problem in the determination of DA and its metabolites, which is of primary interest. The data are routinely calculated as μg of amine or metabolite per g of tissue. In other cases, the data are based on the tissue protein content (i.e., ng of DA per mg of neostriatal protein).

2.5.2. Measurement of ^{3}H-Dopamine Uptake by a Neostriatal Synaptosomal Preparation

In situations in which there is a neurotoxic insult to the nigrostriatal pathway, for example after stereotaxic 6-OHDA administration or peripheral MPTP administration, a neostriatal synaptosomal preparation will show a diminished ^{3}H-DA uptake compared to an appropriate control. In the case of the rat with a unilateral 6-OHDA lesion of the nigrostriatal pathway, the contralateral or unoperated side can serve as control. In the case of the MPTP-treated animal, uninjected or sham-injected animals must serve as controls, since the neurotoxic lesion is obviously bilateral.

The appropriate tissue is dissected out after sacrifice of the animal and homogenized in 0.32*M* sucrose at 10 mg of tissue per mL of sucrose (Gershanik et al., 1979). The homogenate is placed into centrifuge tubes and centrifuged at 900*g* in a refrigerated Sorvall centrifuge for 7 min. The supernatant is transferred to a second centrifuge tube and centrifuged at 28,000*g* for 30 min. The supernatant is discarded and the pellet resuspended in a modified Krebs-Ringer phosphate (KRP) buffer at pH 7.4 at 8 mg tissue per mL of buffer. The composition of the buffer is as follows: 118 mM NaCl, 15.8 mM sodium phosphate, 4.7 mM KCl, 1.8 mM $CaCl_2$, 1.2 mM $MgSO_4$, 5.6 mM glucose, 1.3 mM EDTA, 0.08 mM pargyline hydrochloride, and 1.7 mM ascorbic acid. The pargyline is added to inhibit monoamine oxidase activity, whereas the EDTA and the ascorbate are added to prevent the decomposition of the ^{3}H-DA. Because the ascorbic acid is quite unstable, it is added to the buffer just prior to the experiment. Otherwise it can decompose to the brownish colored dehydroascorbic acid and hydrogen peroxide, which are clearly undesirable in uptake experiments.

The tissue preparation, 0.5–1 mL (4–8 mg of neostriatal tissue), is added to 9.5 mL or 9.0 mL of the KRP buffer in 25 mL Ehrlenmeyer flasks. The flasks are placed in a 37°C water bath with shaking for 5–10 min. Then ^{3}H-DA (New England Nuclear, Boston, MA, specific activity usually 25–50 Ci/mmol) is added to the flasks at timed intervals and the samples incubated with shaking for exactly 6 min. The final concentration of the ^{3}H-DA is routinely $10^{-8}M$, but can vary depending on the experimental circumstances. The samples are poured into centrifuge tubes and placed in an ice bath. The tubes are then centrifuged at 28,000*g* for 30 min. The supernatant (very radioactive) is carefully discarded into appropriate radioactive waste containers and the pellet is rinsed twice with 5 mL of isotonic saline. Absolute ethanol (3 mL) is added to the tubes, which are vortexed and allowed to stand at room temperature for 15 min. At this time, an aliquot (1–2 mL) of the extract containing the ^{3}H-DA is added to 9 mL of a scintillation fluid (e.g., Budgetsolve). The radioactivity is then determined by liquid scintillation spectrometry. Data are calculated and expressed in some convenient way (for example, as pmol of ^{3}H-DA per mg of tissue per min. The data from experimental animals can then be compared to data from controls. It has been our experience that rats that rotate contralaterally to an L-Dopa/carbidopa mixture (10 mg/kg of each) have a loss of neostriatal ^{3}H-DA uptake of at least 90% in a synaptosomal preparation from the neostriatum on the lesioned compared to the nonlesioned side.

2.5.3. Simultaneous Determination of Tissue ^{3}H-Dopamine Uptake and the Content of Dopamine and Its Metabolites

In many cases, it is desirable to measure both the capacity of a tissue preparation to accumulate ^{3}H-DA and the tissue content of DA. A decrement in neostriatal ^{3}H-DA uptake, for example, combined with a decrement in the neostriatal content of DA and its metabolites, indicates that a degenerative process has taken place. A decrement in DA content alone is not sufficient indication of an actual neuronal loss. Perhaps the best example of this point is that both reserpine and α-methyl-*p*-tyrosine administration lead to large decrements in the neostriatal DA content. Neither of these two pharmacological agents causes a decrement in neostriatal ^{3}H-DA uptake, however, which indicates that they do not cause a true neurotoxicity. In contrast, animals treated with 6-OHDA or MPTP, both of which are true neurotoxins, have a greatly diminished neostriatal content of DA and its metabolites and a decrement in the capacity of neostriatal synaptosomal preparations to take up ^{3}H-DA.

In these experiments we homogenize the neostriatal tissue in 0.32*M* sucrose. A small aliquot of this sucrose homogenate is added to an equal volume of 0.2*M* perchloric acid containing DHBA. The level of DA and its metabolites is then determined in this sample exactly as described above in section 2.5.1. In the remaining sucrose homogenate, ^{3}H-DA uptake is determined as described in section 2.5.2. It has been our experience that when a neurotoxic event has taken place, the decrement in neostriatal DA content is virtually identical in magnitude to the decrement in the capacity of a neostriatal synaptosomal preparation to take up ^{3}H-DA (*see,* for example, Heikkila et al., 1984b).

2.5.4. Determination of Tyrosine Hydroxylase Activity in Tissue Samples

It is well-known that the enzyme tyrosine hydroxylase is localized within catecholamine-containing nerve terminals. The destruction of catecholamine-containing nerve terminals by 6-OHDA or MPTP thus leads to a loss of tyrosine hydroxylase activity. For this reason, we measure tyrosine hydroxylase activity to assess the extent of damage after neurotoxin treatment (Mayer et al., 1986).

In our laboratory, we routinely measure tyrosine hydroxylase activity using modifications of the method of Nagatsu et al. (1964), as follows: The neostriatal tissue is homogenized in distilled water (5 μL of H_2O per mg of tissue) using a Sonifier cell disrupter (model W 185D, Heat Systems Ultrasonics Inc.) with converter (Branson Sonic Power Co.). Aliquots (50–100 μL) of the neostriatal homogenate (containing 10–20 mg tissue) are incubated in a 37°C shaking water bath for 15 min with 40 μL aliquots of buffer containing mercaptoethanol and $DMPH_4$. Aliquots (100 μL) of L-[^{14}C]tyrosine (New England Nuclear, 0.4–0.5 Ci/mmol, approximately 120,000 dpm) are added to each tissue solution. Blanks consist of boiled tissue or buffer alone. The final concentrations in the incubation mixture are 0.2*M* sodium acetate (pH 6.2), 0.1*M* mercaptoethanol, 2 m*M* 6,7-dimethyl-5,6,7,8-tetrahydropyridine ($DMPH_4$), and 50 μ*M* tyrosine. The enzymatic conversion of L-[^{14}C]tyrosine to L-[^{14}C]Dopa is terminated at exactly 10 min by the addition of 2.0 mL of 5% trichloroacetic acid containing 20 μg L-Dopa. This mixture is centrifuged for 15 min at 26,000*g*. The supernatant plus 10 mL of Tris (0.25*M*)-EDTA (0.01*M*), pH 9.4, is then swirled with 250 mg of acid alumina for 10 min at ambient temperature (final pH equals 8.5 ± 0.2). The aqueous phase is discarded and the alumina transferred to an AG-4 Bio-Rad column using two 5-mL rinses of distilled water. The alumina is rinsed two additional times with 10 mL of water to remove residual tyrosine. L-[^{14}C]Dopa is then eluted from the alumina with 2 × 2 mL vol of 0.3*N* acetic acid and collected into tubes. An aliquot is pipeted into a scintillation vial containing 15 mL of Aquasol (New England Nuclear) and radioactivity measured by liquid scintillation spectrometry. Data are expressed in some convenient measure (e.g., nmol L-Dopa formed per g of tissue per h).

2.5.5. Simultaneous Determination of Tyrosine Hydroxylase Activity and the Content of Dopamine and Its Metabolites

The tissue (e.g., mouse neostriatum) is homogenized in distilled water at 5 μL of distilled water per mg of tissue. An aliquot of this homogenate (e.g., 20 μL) is added into an aliquot of 0.1*M* perchloric acid containing DHBA (e.g., 310 μL). The neostriatal content of DA and its metabolites is determined exactly as described in section 2.5.1 above. The activity of tyrosine hydroxylase is then determined in the remaining water homogenate as described above in section 2.5.4. It has been our experience that neurotoxic damage, for example with MPTP, leads to a loss in neostriatal tyrosine hydroxylase activity that parallels the decrement in neostriatal DA content (*see,* e.g., Mayer et al., 1986).

2.5.6. Measurement of Dopamine Receptor Binding after Neurotoxin Treatment

Although none of the commonly used dopaminergic neurotoxins (i.e., MPTP, 6-OHDA) by themselves have any direct actions on DA receptors, the treatment with neurotoxins can lead to secondary changes in receptor function. For example, lesioning with 6-OHDA causes no short-term alterations in DA receptors, as judged by the lack of effect of 6-OHDA on ^{3}H-spiroperidol binding to neostriatal membrane preparations for the few days following the lesioning. With time, however, and if the lesion is extensive enough, 6-OHDA-treated animals show an increase in the neostriatal binding of ^{3}H-spiroperidol that develops slowly (*see*, e.g., Creese and Snyder, 1979; Heikkila et al., 1981; Staunton et al., 1981). This increase in ^{3}H-spiroperidol binding at about 3 wk–2 mo postsurgery has been shown to be caused by an increased B_{Max} (number of DA receptors); in most studies there has been little or no reported change in the K_d (affinity of the ^{3}H-spiroperidol for the DA receptor site). Although the magnitude of the effect observed after 6-OHDA lesioning has varied somewhat, an increase in receptor number of approximately 30% has been rather common.

In our laboratory (Heikkila and Cabbat, 1983), we routinely measure ^{3}H-spiroperidol binding with modifications of the procedure first described in detail by Bacopoulos (1981, 1982). We homogenize the neostriata (e.g., from male Sprague-Dawley rats) in 50 vol of 50 m*M* Tris buffer at pH 7.4. The homogenate is then centrifuged in a Sorvall RC-5B refrigerated centrifuge at 23,500*g* for 15 min. The resulting pellet is resuspended in 50 vol of the Tris buffer and recentrifuged at 23,500*g* for 15 minutes. The pellet is resuspended in 50 vol of the buffer and incubated at 37°C for 20 min. The sample is centrifuged at 23,500*g* for 15 min, and the pellet is rehomogenized and resuspended in a 50-m*M* Tris buffer at pH 7.4 containing 0.015 m*M* pargyline. The final tissue concentration is 5–6 mg/mL, based on original tissue weight. In the original technique described by Bacopoulos (1981, 1982), the final buffer used for binding contained 5 m*M* EDTA.

Aliquots (1 mL) of tissue are added to test tubes, which then receive the agent used to define specific binding or the appropriate vehicle. Other additions are made as necessary, the samples are preincubated for 10 min at 25°C, 0.8–3.0 n*M* ^{3}H-spiroperidol (25–50 Ci/mmol) is added, and the incubation carried out with shaking at 25°C for 30 min. The samples are filtered with suction through Gooch crucibles containing GFC filter papers, and the tubes and filters are rinsed twice with 5 mL cold Tris buffer. The filter paper is then placed in 3 mL absolute ethanol for extraction of the radioactivity. After 15 min, 10 mL Scintiverse is added, and radioactivity is measured by liquid scintillation spectrometry. The specific binding of ^{3}H-spiroperidol is defined as total binding minus binding in the presence of $10^{-6}M$ (+)-butaclamol. Data are expressed as specific ^{3}H-spiroperidol binding in convenient units, for example as fmol ^{3}H-spiroperidol bound per mg of neostriatal tissue. Depending upon the nature and design of the experiments, kinetic constants (K_d and B_{max}) can be determined. This may, however, require the pooling of tissue samples, which in the case of experiments in lesioned animals may call for relatively large numbers of very precious animals.

2.5.7. Histological Confirmation of Neurotoxicity: Diminished Cell Number in the Substantia Nigra

For most investigators, a large decrease in neostriatal DA and its metabolites, a large decrement in the capacity of neostriatal synaptosomal preparations to take up ^{3}H-DA, and a large decrease in neostriatal tyrosine hydroxylase activity, all taken together, would indicate neurotoxicity. From a personal case in point, however, this is not always true. For example, after we had found that MPTP induced the above changes in mice, we felt confident that MPTP caused a destruction of the dopaminergic nigrostriatal pathway in mice, much as it did in primates, although higher doses were clearly required in mice. To our surprise, more than a few investigators were unconvinced that MPTP was doing anything more than acting as a "super powerful, long-lasting pharmacological agent." This agent was perhaps an unknown drug that combined the properties of reserpine (DA depletion), α-methyl-*p*-tyrosine (inhibition of tyrosine hydroxylase), and cocaine (DA uptake blocker) with all effects lasting for a period of weeks. It was only after demonstrating to such individuals that MPTP caused a loss of nerve cells in the zona compacta of the substantia nigra that they were convinced. For such situations direct visualization by histological examination of the nerve cells in the zona compacta of the substantia nigra is invaluable.

This histological evaluation is briefly described as follows (*see* Heikkila et al., 1984a, for representative figures). Mice (or rats), both treated and control, are perfused with formalin. The brains are then excised and serial paraffin sections (10 μm in thickness) are cut through the superior colliculus containing the substantia nigra. The samples are stained with cresyl violet and viewed through a light microscope. Records can be made in either slide or photograph form. The extent of cell loss can be quantitated by an actual counting of the cells. As an example of this technique, we have recently found that the administration to mice of 2'CH_3-MPTP, a neurotoxic analog of MPTP, causes a very large loss of nerve cells in the zona compacta of the substantia nigra. At the same time 2'CH_3-MPTP caused approximately a 90% loss of neostriatal DA content and in the capacity of neostriatal synaptosomal preparations to accumulate ^{3}H-DA compared to controls (Youngster et al., 1986).

2.5.8. Locomotor Activity in Animal Models of Parkinsonism

As stated above, one of the classical features of parkinsonism is bradykinesia or a slowed movement. A true animal model of parkinsonism will also exhibit a decreased level of activity. In fact, rats with bilateral 6-OHDA lesions as well as mice with a severe dopaminergic deficit induced by MPTP or with 2'CH_3-MPTP all have greatly reduced levels of activity compared to normal untreated controls. Rats or mice, even those with a severe DA deficiency, however, tend to recover and may look relatively normal several days after lesioning. For this reason the lesioned rodent is not an ideal animal model of parkinsonism in which to carry out prolonged behavioral observations. For this purpose, the primate, which more closely resembles the human, is a better animal model. The difficulties and greater cost of working with primates, however, tend to make experiments in primates impractical for most investigators. Moreover, only a small number of animals can be studied.

In our laboratory, we routinely measure locomotor activity in mice as follows: Groups of four mice are put into each cage of a Varimex activity chamber (Columbus Instruments, Columbus, OH). The activity of the animals is monitored by an array of six proximity sensors located beneath the cages to detect movement. Each time a mouse passes over one of the sensors, it emits a signal. Activity is recorded over convenient time periods (*see* Heikkila, 1981). As an example of representative data, normal mice will have a level of activity of approximately 100 counts per 15-min period, mice treated with 2.5 mg/kg of the stimulant *d*-amphetamine will have approximately 400–500 counts in the 15-min period directly after injection, whereas mice treated 16 h previously with 5 mg/kg of reserpine will sit virtually motionless and exhibit almost no activity (i.e., less than 10 counts in 15 min). It has been our experience that MPTP-treated mice have a considerably lower 24-h level of locomotor activity compared to control mice for many days after MPTP treatment. This level of activity will gradually recover toward normal so that by 2 wk after MPTP the two groups of mice will have virtually the same level of activity. Thus far, in rather preliminary experiments, we are of the opinion that the recovery of activity after 2'CH_3-MPTP lesioning of mice may take a much longer period of time.

2.5.9. Rotational Studies in Rats with Unilateral Lesions of the Nigrostriatal Pathway

As stated above, the rat with a unilateral lesion of the nigrostriatal pathway, for example one induced by 6-OHDA, provides a very valuable model for research relevant to Parkinson's disease. This rat is a valuable animal model, not because it looks parkinsonian, but rather because it rotates contralaterally to drugs that are currently used as the primary therapy for the parkinsonian patient. These include L-Dopa as well as the direct-acting dopamine agonists.

The rotational response is measured as follows: Rats are placed in hemispheric, clear plastic containers 40 cm in diameter. A harness is placed around the neck of the rat and the rat is allowed to acclimate to its environment (for 15–30 min). The harness is connected to an axle on which a slotted disc is mounted. An array of four photo sensors spaced at 90° intervals detects each quarter circle of movement. A complete turn is recorded when the slot passes four consecutive sensors in the same direction. The sensors are connected to a Coulbourn Instruments automated rotometer (Lehigh, PA) which measures the total number of turns and the direction of turns. In most experiments the number of turns in a 10- or 15-min interval is recorded for a duration of 2 or 3 h. Longer time periods can be studied if necessary. The rats are given the drug to be studied and the data quantified for the desired time period. As an example, a rat given 10 mg/kg of L-Dopa will accomplish approximately 500 contralateral turns per hour for the first 2 or 3 h following the L-Dopa, after which time the rat will turn very little. It should be emphasized that the L-Dopa-treated rat will turn almost exclusively in the contralateral direction. It may randomly make a few ipsilateral turns during its acclimatization to the new environment and also after the L-Dopa has worn off. Rats treated with 10 mg/kg of L-Dopa plus 10 mg/kg of the peripheral Dopa decarboxylase inhibitor carbidopa will rotate faster (more turns per time interval) and longer than rats treated with L-Dopa alone at 10 mg/kg. A lower dose of L-Dopa combined with carbidopa is required to give the same number of turns as after L-Dopa alone. This is analogous to the situation in the Parkinson patient; carbidopa treatment diminishes the dose of L-Dopa required for therapeutic efficacy.

3. General Conclusions

In this review, we have focused on the use of rodents treated with various neurotoxins. We believe that these animals can provide very important models for research on parkinsonism. Depending upon the nature of the experimental question, the rodent model has differing degrees of suitability. It is up to each investigator to carefully select the model most appropriate to the experimental questions being asked. Depending on the goals of the investigator, other animals, for example primates, may be more suitable. Information gained from all of these diverse models of parkinsonism can, when carefully analyzed as a whole, lead to answers to the very difficult questions being posed. It is hoped that research conducted with these various animal models will lead to knowledge about the causes and ultimate prevention of this crippling disease.

References

Bacopoulos N. G. (1981) Acute changes in the state of dopamine receptors; *In vitro* monitoring with ^{3}H-dopamine. *Life Sci.* **29,** 2407–2414.

Bacopoulos N. G. (1982) Dopamine receptors in rat brain regions. Optimal conditions for ^{3}H-agonist binding, pH dependency and lack of inhibition by ascorbic acid. *Biochem. Pharmacol.* **31,** 3085–3091.

Ballard P. A., Langston J. W., Tetrud J. W., and Burns R. S. (1983) Chemically induced chronic Parkinsonism in young adults: Clinical and neuropharmacologic aspects. *Neurology* **33,** 90.

Ballard P. A., Tetrud J. W., and Langston J. W. (1985) Permanent human Parkinsonism due to 1-methyl-4-phenyl-1,2,3,6-tetrahydropyridine (MPTP): Seven cases. *Neurology* **35,** 949–956.

Boyce S., Kelly E., Reavill C., Jenner P., and Marsden C. D. (1984) Repeated administration of *N*-methyl-4-phenyl-1,2,5,6-tetrahydropyridine to rats is not toxic to striatal dopamine neurones. *Biochem. Pharmacol.* **33,** 1747–1752.

Breese G. R., Baumeister A., Napier T. C., Frye G. D., and Mueller R. A. (1985a) Evidence that D-1 dopamine receptors contribute to the supersensitive behavioral responses induced by L—dihydroxyphenylalanine in rats treated neonatally with 6-hydroxydopamine. *J. Pharmacol. Exp. Ther.* **235,** 287–295.

Breese G. R., Napier T. C., and Mueller R. A. (1985b) Dopamine agonist-induced locomotor activity in rats treated with 6-hydroxydopamine at differing ages: Functional supersensitivity of D-1 dopamine receptors in neonatally lesioned rats. *J. Pharmacol. Exp. Ther.* **234,** 447–455.

Burns R. S., Chiueh C. C., Markey S. P., Ebert M. H., Jacobowitz D. M., and Kopin I. J. (1983) A primate model of Parkinsonism: Selective destruction of dopaminergic neurons in the pars compacta of the substantia nigra by *N*-methyl-4-phenyl-1,2,3,6-tetrahydropyridine. *Proc. Natl. Acad. Sci. USA* **80,** 4546–4550.

Castagnoli Jr. N., Chiba K., and Trevor A. J. (1985) Potential bioactivation pathways for the neurotoxin 1-methyl-4-phenyl-1,2,3,6-tetrahydropyridine (MPTP). *Life Sci.* **36,** 225–230.

Chiba K., Trevor A., and Castagnoli Jr. N. (1984) Metabolism of the neurotoxic tertiary amine, MPTP, by brain monoamine oxidase. *Biochem. Biophys. Res. Commun.* **120,** 574–578.

Chiueh C. C., Markey S. P., Burns R. S., Johannessen J. N., Pert A., and Kopin I. J. (1984) Neurochemical and behavioral effects of systemic and intranigral administration of *N*-methyl-4-phenyl-1,2,3,6-tetrahydropyridine in the rat. *Eur. J. Pharmacol.* **100,** 189–194.

Cohen G. and Heikkila R. E. (1974) The generation of hydrogen peroxide, superoxide radical, and hydroxyl radical by 6-hydroxydopamine, dialuric acid, and related cytotoxic agents. *J. Biol. Chem.* **249,** 2447–2452.

Cohen G., Pasik P., Cohen B., Leist A., Mytilineou C., and Yahr M. D. (1984) Pargyline and deprenyl prevent the neurotoxicity of 1-methyl-4-phenyl-1,2,3,6-tetrahydropyridine (MPTP) in monkeys. *Eur. J. Pharmacol.* **106,** 209–210.

Creese I. and Snyder S. H. (1979) Nigrostriatal lesions enhance striatal ^{3}H-apomorphine and ^{3}H-spiroperidol binding. *Eur. J. Pharmacol.* **56,** 277–281.

Crossman A. R., Mitchell I. J., and Sambrook M. A. (1985) Regional brain uptake of 2-deoxyglucose in *N*-methyl-4-phenyl-1,2,3,6-tetrahydropyridine (MPTP)-induced Parkinsonism in the macaque monkey. *Neuropharmacology* **24,** 587–591.

Davis G. C., William A. C., Markey S. P., Ebert M. H., Caine E. D., Reichert C. M., and Kopin I. J. (1979) Chronic Parkinsonism secondary to intravenous injection of meperidine analogues. *Psychiatry Res.* **1,** 249–254.

Duvoisin R. C. (1976) Parkinsonism: Animal Analogues of the Human Disorder. in *The Basal Ganglia* (Yahr M. D., ed.) Raven, New York.

Duvoisin R. C. (1984) Is Parkinson's disease acquired or inherited? *Can. J. Neurol. Sci.* **11,** 151–155.

Enz A., Hefti F., and Frick W. (1984) Acute administration of 1-methyl-4-phenyl-1,2,3,6-tetrahydropyridine (MPTP) reduces dopamine and serotonin but accelerates norepinephrine metabolism in the rat brain. Effect of chronic pretreatment with MPTP. *Eur. J. Pharmacol.* **101,** 37–44.

Fritz R. R., Abell C. W., Patel N. T., Gessner W., and Brossi A. (1985) Metabolism of the neurotoxin MPTP by human liver monoamine oxidase B. *FEBS Lett.* **186,** 224–228.

Fuller R. W. and Hemrick-Luecke S. K. (1985) Effects of amfonelic acid, α-methyltyrosine, Ro 4-1284 and haloperidol pretreatment on the depletion of striatal dopamine by 1-methyl-4-phenyl-1,2,3,6-tetrahydropyridine in mice. *Res. Commun. Chem. Path. Pharmacol.* **48,** 17–25.

Fuller R. W. and Steranka L. R. (1985) Central and peripheral catecholamine depletion by 1-methyl-4-phenyl-tetrahydropyridine (MPTP) in rodents. *Life Sci.* **36,** 243–247.

Gershanik O., Heikkila R. E., and Duvoisin R. C. (1979) Asymmetric action of intraventricular monoamine neurotoxins. *Brain Res.* **174,** 345–350.

Goldstein M., Lieberman A., and Meller E. (1985) A possible molecular mechanism for the antiparkinsonian action of bromocriptine in combination with levodopa. *Trends Pharmacol. Sci.* **6,** 436–437.

Graham D. G. (1978) Oxidative pathways for catecholamines in the genesis of neuromelanin and cytotoxic quinones. *Mol. Pharmacol.* **14,** 633–643.

Graham D. G. (1984) Catecholamine toxicity: A proposal for the molecular pathogenesis of manganese neurotoxicity and Parkinson's disease. *Neurotoxicology* **5,** 83–96.

Hallman H., Olson L., and Jonsson G. (1984) Neurotoxicity of the meperidine analogue *N*-methyl-4-phenyl-1,2,3,6-tetrahydropyridine on brain catecholamine neurons in the mouse. *Eur. J. Pharmacol.* **97,** 133–136.

Heikkila R. E. (1981) Differential effects of several dopamine uptake inhibitors and releasing agents on locomotor activity in normal and in reserpinized mice. *Life Sci.* **28,** 1867–1873.

Heikkila R. E. (1985) Differential neurotoxicity of 1-methyl-4-phenyl-1,2,3,6-tetrahydropyridine (MPTP) in Swiss-Webster mice from different sources. *Eur. J. Pharmacol.* **117,** 131–133.

Heikkila R. E. and Cabbat F. S. (1983) Ascorbate-induced lipid peroxidation and inhibition of ^{3}H-spiroperidol binding in neostriatal membrane preparations. *J. Neurochem.* **41,** 1384–1392.

Heikkila R. E. and Manzino L. (1984) Behavioral properties of GBR 12,909, GBR 13,069 and GBR 13,098: specific inhibitors of dopamine uptake. *Eur. J. Pharmacol.* **103,** 241–248.

Heikkila R. E., Hess A., and Duvoisin R. C. (1985a) Dopaminergic neurotoxicity of 1-methyl-4-phenyl-1,2,5,6-tetrahydropyridine

(MPTP) in the mouse: Relationships between monoamine oxidase, MPTP metabolism and neurotoxicity. *Life Sci.* **36,** 231–236.

Heikkila R. E., Manzino L., Cabbat F. S., and Duvoisin R. C. (1985b) Studies on the oxidation of the dopaminergic neurotoxin 1-methyl-4-phenyl-1,2,5,6-tetrahydropyridine by monoamine oxidase B. *J. Neurochem.* **45,** 1049–1054.

Heikkila R. E., Hess A., and Duvoisin R. C. (1984a) Dopaminergic neurotoxicity of 1-methyl-4-phenyl-1,2,5,6-tetrahydropyridine (MPTP) in mice. *Science* **224,** 1451–1453.

Heikkila R. E., Manzino L., Cabbat F. S., and Duvoisin R. C. (1984b) Protection against the dopaminergic neurotoxicity of 1-methyl-4-phenyl-1,2,5,6-tetrahydropyridine by monoamine oxidase inhibitors. *Nature* **311,** 467–469.

Heikkila R. E., Shapiro B. S., and Duvoisin R. C. (1981) The relationship between loss of dopamine nerve terminals, striatal ^{3}H-spiroperidol binding and rotational behavior in unilaterally 6-hydroxydopamine lesioned rats. *Brain Res.* **211,** 285–292.

Hornykiewicz O. (1966) Dopamine (3-hydroxytyramine) and brain function. *Pharmacol. Rev.* **18,** 925–964.

Hornykiewicz O. (1973) Parkinson's disease: From brain homogenates to treatment. *Fed. Proc.* **32,** 183–190.

Jarvis M. F. and Wagner G. C. (1985a) Neurochemical and functional consequences following 1-methyl-4-phenyl-1,2,5,6-tetrahydropyridine (MPTP) and methamphetamine. *Life Sci.* **36,** 249–254.

Jarvis M. F. and Wagner G. C. (1985b) Age-dependent effects of 1-methyl-4-phenyl-1,2,5,6-tetrahydropyridine (MPTP). *Neuropharmacology* **24,** 581–583.

Javitch J. A., D'Amato R. J., Strittmatter S. M., and Snyder S. H. (1985) Parkinsonism-inducing neurotoxin, *N*-methyl-4-phenyl-1,2,3,5-tetrahydropyridine by dopamine neurons explains selective toxicity. *Proc. Natl. Acad. Sci. USA* **82,** 2173–2177.

Jenner P., Rupniak N. M. J., Rose S., Kelly E., Kilpatrick G., Lees A., and Marsden C. D. (1984) l-Methyl-4-phenyl-1,2,3,6-tetrahydropyridine-induced Parkinsonism in the common marmoset. *Neurosci. Lett.* **50,** 85–90.

Johannessen J. N., Kelner L., Hanselman D., Shih M. C., and Markey S. P. (1985) *In vitro* oxidation of MPTP by primate neural tissue: A potential model of MPTP neurotoxicity. *Neurochem. Int.* **7,** 169–176.

Jonsson G. (1980) Chemical Neurotoxins as Denervation Tools in Neurobiology, in *Annual Review of Neuroscience* vol. 6 (Cowan W. M., Hall Z. W., and Kandel E. R., eds.) Annual Reviews, Inc., Palo Alto, California.

Konig J. F. R. and Klippel R. A. (1963) *The Rat Brain. A Stereotaxic Atlas of the Forebrain and Lower Parts of the Brain Stem* Williams and Wilkins, Baltimore, Maryland.

Langston J. W. and Ballard P. (1984) Parkinsonism induced by l-methyl-4-phenyl-1,2,3,6-tetrahydropyridine (MPTP): Implications for treatment and the pathogenesis of Parkinson's disease. *Can. J. Neurol. Sci.* **11,** 160–165.

Langston J. W., Ballard P., Tetrud J. W., and Irwin I. (1983) Chronic parkinsonism in humans due to a product of meperidine-analog synthesis. *Science* **219,** 979–980.

Langston J. W., Forno L. S., Rebert C. S., and Irwin I. (1984a) Selective nigral toxicity after systemic administration of l-methyl-4-phenyl-1,2,5,6-tetrahydropyridine (MPTP) in the squirrel monkey. *Brain Res.* **292,** 390–394.

Langston J. W., Irwin I., Langston E. B., and Forno L. S. (1984b) Pargyline prevents MPTP-induced Parkinsonism in primates. *Science* **225,** 1480–1482.

Markey S. P., Johannessen J. N., Chiueh C. C., Burns R. S., and Herkenham M. A. (1984) Intraneuronal generation of a pyridinium metabolite may cause drug-induced Parkinsonism. *Natire* **311,** 464–467.

Mayer G. S. and Shoup R. E. (1983) Simultaneous multiple electrode liquid chromatographic-electrochemical assay for catecholamines, indoleamines and metabolites in brain tissue. *J. Chromatogr.* **255,** 533–544.

Mayer R. A., Walters A. S., and Heikkila R. E. (1986) 1-Methyl-4-phenyl-1,2,3,6-tetrahydropyridine (MPTP) administration to C57-black mice leads to parallel decrements in neostriatal dopamine content and tyrosine hydroxylase activity. *Eur. J. Pharmacol.* **120,** 375–377.

Nagatsu T., Levitt M., and Undenfriend S. (1964) Tyrosine hydroxylase: The initial step in norepinephrine biosynthesis. *J. Biol. Chem.* **239,** 2910–2917.

Nicklas W. J., Vyas I., and Heikkila R. E. (1985) Inhibition of NADH-lined oxidation in brain mitochondria by l-methyl-4-phenylpyridine, a metabolite of the neurotoxin, 1-methyl-4-phenyl-1,2,5,6-tetrahydropyridine. *Life Sci.* **36,** 2503–2508.

Perry T. L., Young V. W., Ito M., Jones K., Wall R. A., Foulks J. G., Wright J. M., and Kish S. J. (1985) l-Methyl-4-phenyl-1,2,3,6-tetrahydropyridine (MPTP) does not destroy nigrostriatal neurons in the scorbutic guinea pig. *Life Sci.* **36,** 1233–1238.

Robinson T. E. and Becker J. B. (1983) The rotational behavior model: Asymmetry in the effects of unilateral 6-OHDA lesions of the substantia nigra in rats. *Brain Res.* **264,** 127–131.

Ross S. B. (1979) The central stimulatory action of inhibitors of the dopamine uptake. *Life Sci.* **24,** 159–168.

Schmidt C. J., Bruckwick E., and Lovenberg W. (1985) Lack of evidence supporting a role for dopamine in l-methyl-4-phenyl-1,2,3,6-tetragydropyridine neurotoxicity. *Eur. J. Pharmacol.* **113,** 149–150.

Schoenberg B. S. (1987) Descriptive epidemiology of Parkinson's disease: Disease distribution and hypothesis formulation. *Adv. Neurol.* **45,** 277–283.

Schultz W. (1982) Depletion of dopamine in the striatum as an experimental model of Parkinsonism: Direct effects and adaptive mechanisms. *Prog. Neurobiol.* **18,** 121–166.

Sershen H., Mason M. F., Hashim, A., and Lajtha A. (1985) Effect of *N*-methyl-4-phenyl-1,2,3,6-tetrahydropyridine (MPTP) on age-related changes in dopamine turnover and transporter function in the mouse striatum. *Eur. J. Pharmacol.* **113,** 135–136.

Silbergeld E. K. and Calne D. B. (1981) Animal models of Parkinsonism. *Pharmacol. Ther.* **12,** 159–166.

Snyder S. H. (1984) Parkinson's disease: Clues to aetiology from a toxin. *Nature* **311,** 514.

Staunton D. A., Wolfe B. B., Groves P. M., and Molinoff P. B. (1981) Dopamine receptor changes following destruction of the nigrostriatal pathway: Lack of a relationship to rotational behavior. *Brain Res.* **211,** 325–327.

Ungerstedt U. (1971a) Stereotaxic mapping of the monoamine pathways in rat brain. *Acta Physiol. Scand.* **367,** (suppl.) 1–48.

Ungerstedt U. (1971b) Striatal dopamine release after amphetamine or nerve degeneration revealed by rotational behaviour. *Acta Physiol. Scand.* **367,** (suppl.), 49–68.

Ungerstedt U. (1971c) Postsynaptic supersensitivity after 6-hydroxydopamine induced degeneration of the nigrostriatal dopamine system. *Acta Physiol. Scand.* **367,** (suppl.), 69–93.

Ungerstedt U. (1971d) Adipsia and aphagia after 6-hydroxydopamine induced degeneration of the nigrostriatal dopamine system. *Acta Physiol. Scand.* **367,** (Suppl.), 95–117.

Ward C. D., Duvoisin R. C., Ince S. E., Nutt J. P., Eldridge R., and Calne D. B. (1983) Parkinson's disease in 65 pairs of twins and in a set of quadruplets. *Neurology* **33,** 815–824.

Westlund K. N., Denney R. M., Kochersperger L. M., Rose R. M., and Abell C. W. (1985) Distinct monoamine oxidase A and B populations in primate brain. *Science* **230,** 181–183.

Yahr M. D. (1977) Parkinson's disease—overview of its current status. *Mt. Sinai J. Med.* **44,** 183–191.

Youngster S. K., Duvoisin R. C., Hess A., Sonsalla P. K., Kindt M. V., and Heikkila R. E. (1986) l-Methyl-4-(2'methylphenyl)-1,2,3,6-tetrahydroxypyridine (2'CH_3-MPTP) is a more potent dopaminergic neurotoxin than MPTP in mice. *Eur. J. Pharmacol.* **122,** 283–287.

Zigmond M. J., Acheson A. L., Stachowiak M. K., and Stricker E. M. (1984) Neurochemical compensation after nigrostriatal bundle injury in an animal model of preclinical Parkinsonism. *Arch. Neurol.* **41,** 856–861.

Drug-Induced Behavioral Models of Central Disorders

David M. Jackson

1. Introduction

The purpose of the present chapter is not to review all drug-induced animal models of psychiatric and neurological diseases, but to discuss a number of models that highlight their usefulness and limitations. I have made my choice of models to discuss partly on the basis of their role (putative or actual) in modeling diseases, partly on their heuristic value in stimulating research, and partly because of my own personal interests. Some diseases will not be covered at all in this chapter because I do not think there are any adequate drug-induced models [e.g., depression; for models, *see* Everitt and Keverne (1979), Bollini et al. (1984), Willner (1984), Cooper et al. (1983), Telner and Singhal (1984), and Suomi (1983)]. Other models will only be discussed in part (e.g., anxiety and schizophrenia). The philosophical, psychological, and ethical considerations underlying animal models have been explored by many able authors in considerable depth, and the reader is referred to Bond (1984), Abramson and Seligman (1977), Bowd (1980), and Russell (1964) for detailed discussions.

A few points should be made about the requirements for model validity. In my view, a model should:

- produce anatomical, pathological, and biochemical lesions reminiscent of the disease,
- provide information on the etiology of the disease,
- mirror the clinical signs of the disease, whether physical and/or mental,
- be fairly specific to the disease,
- provide powerful predictive value for drugs potentially useful in the treatment of the disease,
- involve a genetic component, if such is part of the disease etiology, and
- be relatively easy to set up, economical, and ethically acceptable.

A detailed study of the literature makes clear that no animal model fulfills all these criteria. Perhaps the most important reason for this is that the etiology of most psychiatric and neurological disorders remains unknown—hence models are induced by means that are usually not related to the cause of the modeled disease. Another problem arises because some diseases may be uniquely human (e.g, schizophrenia), with only certain symptoms being reproducible in animal models. Since models generally serve a valuable heuristic role, however, these limitations should not deter the modeler. This has been particularly the case when only one aspect of the disease (rather than the whole disease) has been explored (e.g, the 6-aminonicotinamide model of parkinsonian rigidity). No matter how inaccurate models made be of particular diseases, most have provided a significant impetus for research into both basic and clinical aspects of the disease. They have also helped in the development of cleaner drugs; e.g., the reserpine hypokinesia model of Parkinson's disease (PDS) played a significant role in the development of the selective dopamine (DA) autoreceptor agonist, 3-PPP [3-(3-hydroxyphenyl)-*N*-(*n*-propyl)piperidine].

Animal models, whether drug-induced or otherwise, are absolutely necessary for research into central diseases to continue and progress. Because there are, and will remain, major ethical problems associated with the use of patients in clinical research, animal models are often the only alternative. Moreover, animal models provide a system that can be largely controlled for genetic and environmental variation. They also provide the possibility for near simultaneous behavioral, biochemical, and electrophysiological monitoring, which is generally impossible in humans.

It will be clear from this chapter that some of the best models use nonhuman primates. This is not surprising, given the evolutionary relationships between the primates. This need to use primates need not be a problem, because with well-planned experiments, wastage of animal resources can be kept to a minimum, and much more valuable data can often be obtained from one nonhuman primate than from hundreds of rodents.

2. Parkinson's Disease

2.1. General

Parkinson's disease (PDS) is the best-modeled of all the psychiatric and neurological diseases, partly because of a well understood psychopathology, a well-mapped and discrete nigrostriatal dopamine (DA) pathway with well-studied behavioral functions, fairly specific DA receptor agonists and antagonists, and, recently, a drug that produces in primates virtually all of the classical signs of the diseases. Thus the modeling of PDS can provide a guide to the general use of models in psychopathology, both in terms of their potential strengths and limitations (*see* Table 1 for a summary of the models).

2.2. The Disease

Parkinson's disease is characterized by a loss of the darkly pigmented cell bodies in the substantia nigra pars compacta (SN), resulting in the death of the nigrostriatal DA axons that innervate other parts of the basal ganglia, including the neostriatum. As a consequence, the levels of DA and the rate-limiting enzyme in its synthesis, tyrosine hydroxylase (TH), decrease to less than 20% of normal in these innervated areas. An important corollary is that the DA receptors (of the D_2 type) located postsynaptically to the degenerating terminals become supersensitive. Although it was earlier thought that the pathology was restricted to the nigrostriatal DA pathway, it is now clear that the mesolimbic DA pathway, with cell bodies mainly in the ventral tegmental area and terminals in several nuclei including the nucleus accumbens (NAC), is also compromised.

Not only DA is affected. The levels of many other neurotransmitters decrease; these include 5-hydroxytryptamine (5-HT) (Bernheimer et al., 1973; Scatton et al., 1983), norepinephrine (noradrenaline, NE) (Scatton et al., 1983), γ-aminobutyric acid (GABA) (Manyam, 1982; Rinne et al., 1978), and many peptides, including the enkephalins (Taquet et al., 1983). Although there is no doubt that the primary lesion of PDS involves a loss of DA, it is unclear whether the alterations in other neurotransmitters are also primary defects or are secondary to the loss of DA. In any case, the loss of the inhibitory transmitter DA leads to neurotransmitter imbalance in the basal ganglia, resulting in the characteristic rigidity and tremor. In contrast, the hypokinesia mainly results from the loss of DA in the mesolimbic system. Drug treatment involves replacing the lost DA [by L-3,4-dihydroxyphenylalanine (L-DOPA)], using DA receptor agonists [such as bromocriptine (BRC)] or damping down the excessive cholinergic activity (with, e.g., anti-muscarinics).

Table 1
Actions of Some Drugs in Some Biochemical Tests and in Possible Models of Parkinsons Disease

Compound	MPTP in monkeys	Rodent Stereotypy	Rodent locomotor stimulation	Rigidity reduction[a]	Rotation (contralateral)	DA D_2 binding	Stimulation of DA D_1 receptors	Reversal of reserpine catalepsy
BRC	+	+[b]	+[b]	+	+[b]	+	–	+
Lergotrile	?	+	+	+	+	+	–	+
Lisuride	?	+	+	+	+	+	–	+
L-DOPA	+	+	+	+	+	+[c]	+[c]	+

[a]Produced by either reserpine or 6-aminonicotinamide.

[b]BRC is not active in producing stereotyped behavior, locomotor stimulation, or contralateral rotation in rodents whose DA stores have been depleted.

[c]In the form of DA.

2.3. The Models

2.3.1. Rat Rotation Model

A unilateral lesion of the nigrostriatal pathway in rats either mechanically, electrolytically (Anden, 1966), or chemically with the neurotoxin 6-hydroxyDA (6-OHDA, Ungerstedt, 1971) produces an asymmetry of posture. In the case of 6-OHDA, the neurotoxin is usually injected into the SN (see below). Analogously to PDS, DA-containing neurons die, and the DA receptors postsynaptic to the degeneration become supersensitive. This supersensitivity is caused by an increase in the number of DA D_2 receptors rather than a change in affinity. There is no consistent change in the sensitivity of DA D_1 receptors linked to DA-stimulated adenyl cyclase (Creese et al., 1977). The DA receptors on the intact side remain normosensitive. The receptor asymmetry is the basis of the "Ungerstedt" rat rotation model. Directly acting drugs such as apomorphine cause rotation (involving components of an asymmetric posture and forward locomotion) away from the lesioned side (contralateral) because of the presence of supersensitive DA receptors on the lesioned side. In contrast, indirectly acting DA agonists such as *d*-amphetamine and methylphenidate, which release DA and have little or no affinity for the DA receptor *per se*, produce rotation toward the lesion (ipsilateral rotation) because there is no (or little) DA on the lesioned side available for release. Drugs with both direct and indirect properties will produce a rotation preference dependent upon their relative direct and indirect effects. Thus, behavior elicited by indirectly acting agonists is the result of stimulation of normosensitive DA receptors, whereas that elicited by direct agonists is via supersensitive receptors.

Neurotransmitters other than DA are involved in rotation controlled by basal ganglia structures (Iwamoto and Way, 1977; James and Starr, 1979; Hartgraves and Kelly, 1983). It is also clear that a pivotal role is played by the NAC, which seems to mediate most of the coordinated forwardly directly locomotor stimulant effects of DA agonists such as *d*-amphetamine and apomorphine (Jackson et al., 1975a). If, as now believed, the forward locomotion in this model is caused primarily by DA receptor stimulation in the NAC (Kelly and Moore, 1977), then the forward locomotion would involve the stimulation of normosensitive DA receptors in the NAC, whereas the postural asymmetry would be dependent upon both normosensitive and supersensitive receptors in the basal ganglia. An interesting variation of the model involves the destruction of the cell bodies unilaterally in the neostriatum with kainic acid. This lesion does not produce supersensitive DA receptors, but a loss of DA receptors on one side. DA agonist administration still produces an asymmetry and rotation, but of less intensity than in the supersensitive model (*see* section 4.2.1).

The use of the two models in parallel can provide a method of determining the relative activity of DA agonists at super- and normosensitive receptors. Rats are injected into one SN with 6-OHDA hydrobromide (8 μg/2 μL) dissolved freshly in 0.1% ascorbic acid solution. An equivalent volume of vehicle should be injected into the other side. The injections should take about 2 min and the cannulae should stay in place for 1 extra min to allow for equilibrium to be reached. Animals can be pretreated with the NE uptake inhibitor desipramine (25 mg/kg) 30 min before the injection to protect any NE neurons in the vicinity. After 2 wk, apomorphine (0.1–0.5 mg/kg sc) is administered, with successfully lesioned animals rotating vigorously away from the lesion. Biochemical estimation of the lesion is advantageous because it will enable quantitation of the lesion size, but also will enable determination of the amount of (unwanted) destruction in the mesolimbic DA system. The nonsupersensitive model can be prepared by injecting kainic acid (*see* section 4.2.1) into one of the nec striata and vehicle into the other, with testing being conducted ᶠter about 7–14 d.

2.3.1.1. Usefulness and Limitations of the Model. Rats are commonly used since they are generally plentiful and relatively cheap. Compounds effective or predictably effective in PDS are active in this model; those include apomorphine, L-DOPA, lergotrile, BRC, and pergolide. These drugs behave as directly acting agonists. The model also provides an indication of the duration of action of the drugs, with apomorphine being short-acting and BRC being long-acting. Indirectly acting drugs such as *d*-amphetamine are of limited use in PDS because of the decreased DA available for release. The model has also been useful in elucidating the role and interaction of the various neurotransmitters in the basal ganglia (*see* Hartgraves and Kelly, 1983). It has also provided a useful way of examining the chronic effects of DA receptor agonists such as L-DOPA, because both super- and normosensitive DA receptors are present, and there is some evidence that these may be differentially sensitive to down-regulation (Jackson et al., 1983; Schneider et al., 1984). The use of the model in this way is significant, since it has been suggested that some of the dyskinesias elicited in patients after the chronic use of L-DOPA may be caused by receptor sensitivity changes (Klawans et al., 1977b). One peculiarity of the Ungerstedt model is that BRC behaves as a directly acting postsynaptic DA receptor agonist (Fuxe et al., 1978), whereas in the reserpine hypokinesia model, it is inactive (Jackson and Jenkins, 1985). Currently, the involvement of DA D_1 and DA D_2 receptors in rotation is being investigated (Arnt and Hyttel, 1984), with evidence accumulating that the lesion and subsequent DA receptor supersensitivity and asymmetry may enable a functional role for DA D_1 receptors to be described.

A major problem of the rotating rat is that it does not really mirror the symptomatology of PDS, i.e., hypokinesia, rigidity, and tremor. The lesion is also much more specific than that seen in PDS, with no marked changes occurring, for example, in the NE levels of the locus ceruleus. In addition, the model does not reflect the cognitive deficits apparent in PDS, because it is used purely as a motor model.

In an endeavor to overcome this last criticism, Carli et al. (1985) attempted to see whether neostriatal lesions could affect sensory as well as motor functions. Each rat was trained to keep its head in a hole until an appropriate visual stimulus was presented, when it had to move its head either toward, or away from, the light. Rats unilaterally lesioned in the neostriatum with 6-OHDA displayed impairment of initiation (motor), but not of execution, which, in these authors' view, was compatible to findings in parkinsonian patients of changes in speed of reaction. These data suggested that manipulations of the basal ganglia produce motor rather than cognitive changes. An opposite conclusion was reached by Sabol et al. (1985), however, who examined the effect of unilateral 6-OHDA in the neostriatum on the accuracy of rats trained to extend their preferred forepaw into a narrow aperture to obtain small food pellets. The authors concluded that the lateral striatal DA system has a sensory role. The methodologies utilized by these two groups were quite different, so the different results are not surprising. These types of studies, using more refined lesions, represent a particularly valuable use of the basic model.

2.3.2. Reserpine-Induced Syndrome

When reserpine is administered to most animals, a characteristic behavioral syndrome occurs (Carlsson et al., 1957). Symptoms include hypokinesia, rigidity, catalepsy, a hunched posture, hypothermia, and an increase in muscle tone, particularly in the flexor muscles. Thus, at least two of the cardinal signs of PDS are produced by reserpine (hypokinesia and rigidity). In these experiments, rats or mice are pretreated ip with reserpine, 1–10 mg/kg, administered 4–24 h before experimentation. The duration of the reserpine pretreatment regulates the degree of DA receptor sensitivity, with longer (e.g., 24 h), but not shorter (4–6 h), pretreatment times producing enhanced behavioral responses. Although reserpine severely depletes the granular stores of DA, NE, and 5-HT, much of the behavioral syndrome is caused by the loss of DA, which causes a functional imbalance between DA and acetylcholine (ACh) in the basal ganglia. Support for the validity of the reserpine model comes from the observation that reserpine can exacerbate hypokinesia, rigidity, and tremor in PDS patients (Birkmayer and Hornykiewicz, 1964) (*see* Table 1).

2.3.2.1. Hypokinesia. The ability of DA receptor agonists to antagonize the reserpine-induced hypokinesia depends upon the stimulation of postsynaptic DA D_2 receptors in the neostriatum (producing stereotypies) and in the NAC (producing forward locomotion). DA autoreceptors play little or no role in the reserpine model. The receptor stimulation can be monitored behaviorally in activity cages as an increase in locomotor activity, or subjectively as an increase in stereotyped behavior. The reversal of the hypokinesia produced by apomorphine, a relatively pure DA receptor agonist, is characterized by a mixture of stereotypies (such as sniffing, biting, and rearing) and somewhat bizarre, uncoordinated, and sometimes explosive locomotion. Neither clonidine, an α_1 and α_2 agonist, nor the 5-HT precursor, 5-hydroxytryptophan, are able to reverse the hypokinesia. Alpha receptors play an important role, however, because the noncoordinated locomotion produced by apomorphine can be "normalized" to some extent by concomitantly stimulating the α_1 receptors with clonidine (Strombom, 1976).

If the apomorphine effect is measured in automated activity cages (Jackson, 1972; Strombom, 1976), the "activity" measured will consist of stereotypies as well as forward locomotion. Consequently, it is good practice to assess *both* locomotion and stereotyped behavior. The latter cannot be assessed adequately with automated equipment, and needs to be assessed subjectively. There are several ways of doing this. A graded scoring system from 0 to 5 can be used, where 0 is the absence of all behavior and 5 is, e.g., continuous sniffing, rearing, and biting (*see* Costall et al., 1972). Alternatively, the incidence of individual behaviors can be assessed as present or absent throughout the observation period (for the use of such a scoring system, *see* Dourish, 1982). This method is superior to the 0 to 5 system because it generates more precise data and can accommodate the different types of stereotypies produced in various species and strains by different DA agonists. Since all the assessment systems are subjective, they must be performed by an observer blind to the treatment schedule and shown to be reliable and to produce dose-related responses. Reserpinized rodents are quite sensitive to DA agonists, with apomorphine producing stereotypies at doses as low as 25–100 μg/kg SC and locomotion at about 200 μg/kg ip. Stereotypy, like locomotor activity, is not a pure DA-mediated behavior, but is also subjected to influences from a variety of neurotransmitters, including NE (Braestrup and Randrup, 1978).

Directly acting DA receptor agonists such as L-DOPA and apomorphine produce the expected postsynaptic stimulation (increased locomotion and stereotypy). The situation is slightly more complex with indirectly acting DA receptor agonists. Some (for example, methylphenidate), but not others (for example, *d*-amphetamine), appear to require reserpine-sensitive stores of DA (Scheel-Kruger, 1971). In reserpine-treated rodents, DA synthesis still occurs, and the resultant DA can be released by *d*-amphetamine, but not, apparently, by methylphenidate. A modification of the reserpine model takes this into account by also pretreating the animals with the TH inhibitor, alpha-methyl-*p*-tyrosine (AMPT). With these two pretreatments the granular stores of DA are depleted and its synthesis blocked. In this modified model, apomorphine, lisuride, and L-DOPA are still active (Wachtel, 1983), but *d*-amphetamine is not. Moreover, BRC is completely inactive in this model (Jackson and Jenkins, 1985; Fuxe et al., 1978). Thus, the rotation model clearly places BRC with the directly acting agonists, whereas the reserpine-hypokinesia model places it with the drugs that require endogenous DA. This discrepancy may be caused by a requirement by BRC for endogenous DA and/or upon a requirement for concomitant DA D_1 and D_2 receptor stimulation (Jackson and Jenkins, 1985; Jackson and Hashizume, 1986). Drugs that are selective for DA autoreceptors such as 3-PPP are inactive (Hjorth et al., 1983).

2.3.2.2. Catalepsy. Reserpine-induced catalepsy, in which a rat or mouse will stay in an abnormal position for a period of time (e.g., placed in a rearing position with its forepaws on a raised bar), can be reversed by DA receptor agonists. Like some of the symptoms of PDS, catalepsy involves a DA–ACh interaction since catalepsy produced, for example, by neuroleptics can be reversed with anticholinergic drugs (Costall and Olley, 1971). It is of interest that BRC is quite effective in this model (Johnson et al., 1976) in contrast to its inability to substantially reverse reserpine-induced akinesia. Catalepsy reflecting that seen in PDS can also be produced in rodents by high doses of any of the classical neuroleptics (Morpurgo, 1962).

2.3.2.3 RIGIDITY. Reserpine-induced rigidity is caused by an increase in both spontaneous and reflex alpha motoneuron activity (Steg, 1964), but there is still some argument about the involvement of gamma motor neurons. It seems that the basal ganglia play a dominant role because apomorphine applied directly into the neostriatum, but not into the NAC, can antagonize reserpine-induced rigidity (Anden and Johnels, 1977). A relatively simple method for quantitating changes in rat rigidity has been reported by Johnels et al. (1978). Reserpine-induced rigidity can be antagonized by antiparkinsonian drugs, including L-DOPA, apomorphine, BRC, amantadine, and atropine (Roos and Steg, 1964; Jurna et al., 1972; Jurna, 1976; Johnels et al., 1978). Data suggest that alpha and gamma motor neuron activity is influenced by GABA systems because the application of GABA into the caudate nucleus normalized the alpha reflex activity (Jurna et al., 1978).

Other drugs can also produce muscular rigidity, which can be used as a model of PDS rigidity. For example, the antimetabolite 6-aminonicotinamide produces a degenerative disorder in rats with long-lasting parkinsonian-like rigidity. The rigidity is similar to that produced by reserpine and is reversed by L-DOPA, lisuride, or BRC (Halbhubner et al., 1978). This model has the advantage over the reserpine model in that the rigidity lasts for many months, enabling different drugs to be tested in the same animal, but has the disadvantage of being less well-documented than the reserpine model.

2.3.3. Manganese Model

It has been proposed that manganese intoxication may produce a useful experimental model of PDS. Although many parkinsonian-like signs are produced in humans by manganese poisoning, other symptoms also occur. The reader is referred to a comprehensive and critical review of manganese and parkinsonism by Barbeau (1984).

2.3.4. MPTP (1-Methyl-4-phenyl-1,2,3,6-tetrahydropyridine) Model

2.3.4.1. General. Perhaps the most exciting breakthrough in modeling PDS was the presentation of a number of young people at a clinic in California with many symptoms of rapid-onset PDS (Langston et al., 1983). Inspired detective work demonstrated that the disease had been caused by the intravenous injection of MPTP, a contaminant of an alphaprodine synthetic street narcotic. Virtually all the symptoms of PDS were evident. Postmortem studies of one of these patients indicated lesions typical of PDS, with cell loss in the SN. A vast amount of work has since been done (*see* Langston, 1985a, for review) to determine the usefulness of MPTP in PDS models (*see* Table 2). The parent compound itself, MPTP, does not seem to be neurotoxic, but requires uptake by glial cells with subsequent conversion by monoamine oxidase B to the putative active neurotoxin, MPP^+ (Heikkila et al., 1984a; Castagnoli et al., 1985; Sundstrom and Jonsson, 1985). After several doses, MPTP is neurotoxic to the nigrostriatal DA pathway of most species, including humans (Langston, 1985a), monkeys (Burns et al., 1983; Chiueh et al., 1985), rats (Steranka et al., 1983; Jarvis and Wagner, 1985; but *see* Boyce et al., 1984), mice (Sundstrom and Jonsson, 1985; Heikkila et al., 1985), and amphibia (Barbeau et al., 1985). The quantity and quality of damage may not be the same in all species, however (*see below*). Primates are the most sensitive, perhaps because MPP^+ is trapped intraneuronally in primates, whereas rodents clear the compound fairly rapidly (Johannessen et al., 1985; Markey et al., 1984). Another explanation for the species differences may hinge upon the the involvement of neuromelanin located in the SN in the toxicity produced by MPTP. It seems likely that the active metabolite of MPTP, MPP^+, may bind with high affinity to neuromelanin (D'Amato et al., 1986). Interestingly, the DA terminals in the dorsolateral part of the neostriatum are the most vulnerable (Chiueh et al., 1985), corresponding to a recent finding that the lateral part of striatal DA disappears in early hemiparkinsonism (Garnett et al., 1984). The common laboratory mouse seems quite sensitive to MPTP if a dose of 30 mg/kg ip is given once a day for several days (Heikkila et al., 1984b), with lesions so produced lasting for at least 40 d (Heikkila et al., 1985). DA pathways other than the nigrostriatal pathway are also affected (Gupta et al., 1985). There are also reports, however, that the neostriatal DA supersensitivity so produced in mice may not last

Table 2
Comparison of the Behavioral Effects of MPTP in Rat, Guinea Pig, Monkey, and Human[a]

Extrapyramidal Syndrome	Rat	Guinea Pig	Monkey	Human
Rigidity	–	+	+	+
Hypokinesia	–	+	+	+
Flexed posture	–	+	+	+
Freezing episodes	+–	+	+	+
Difficulty in vocalization	+–	+	+	+
Persistence	–	–	+	+
Duration	15–30 min	4–6 h	?	9 yr

[a]After Chiueh et al. (1984).

long (Peroutka et al., 1985). In any event, as a result of MPTP administration to primates, the pigmented cells of the SN die, with a resultant loss of DA from the innervated areas. At least 80% of the DA must disappear before clear clinical signs appear in monkeys (Chiueh et al., 1985). Such a loss produces the characteristic cardinal signs of PDS (*see* above references and Table 2). In contrast to the situation seen in idiopathic PDS, the lesion appears to be restricted to the nigrostriatal DA pathway, with the mesolimbic pathway remaining relatively unscathed (Chiueh et al., 1985) (however, see below.) MPTP is usually injected sc, ip, or iv. A schedule of 45 mg/kg twice daily for 2 d in mice produces a decrease in forebrain DA of 68% at 7 d and 63% at 20 d (Wallace et al., 1984). An alternative treatment schedule in mice was one dose of 20 mg/kg (Sundstrom and Jonsson, 1985). One regimen for squirrel monkeys utilized 2 mg/kg ip at 2-h intervals for four doses (Langston, 1985b), whereas in macacque monkeys, Chiueh and coworkers (Chiueh et al., 1984) used 0.33 mg/kg iv daily for 5 or 6 d. In monkeys, three phases of the MPTP syndrome have been identified:

- An acute phase characterized by myoclonus, salivation, motor disorders, and "nodding off," in which the monkey's head nods,

- A subacute phase, in which there is a gradual increase in hypokinesia and rigidity,
- A chronic phase, in which the classical signs of PDS are manifested (Langston, 1985b).

It is interesting that the occurrence of the chronic phase does not always take place. Note that mortality during the treatment schedules is often quite high—in mice it may be 10–30 % (Wallace et al., 1984), and in rats it may be as high as 50 % (Jarvis and Wagner, 1985).

2.3.4.2. USEFULNESS AND LIMITATIONS OF THE MODEL. In many respects, the nonhuman primate model appears to mimic idiopathic PDS from the behavioral symptoms (see above) through to the pathology. Therapeutically, the primate model responds to treatment with L-Dopa or BRC (Chiueh et al., 1985; Close et al., 1985a), but not to the DA D_1 agonist SKF 38393 (Close et al., 1985b). The MPTP primate model does not apparently reveal significant decrements in NE levels, in contrast to PDS, although recent studies indicate that NE systems can, in fact, be affected by MPTP treatment (Fuller and Hemrick-Luecke, 1986; Mitchell et al., 1985). Finally, there is now some evidence indicating that MPTP-lesioned primates may spontaneously recover their motor function after several months (*see,* e.g., Jenner and Matsden, 1988), which does not occur in true PDS. Further similarities and differences between the model and true PDS will be revealed by more research. Since there is some dispute about the response of nonprimates such as mice, extrapolation of biochemical changes in this species to PDS must be made cautiously. I think it is important, therefore, whatever species is studied, that the lesion produced be shown to be permanent and not transitory. Nevertheless, since the use of primates will be beyond the financial scope of most workers, it is necessary for extensive species comparisons to be made. The use of MPTP in amphibians looks promising (Barbeau et al., 1985) because the species incorporates a change in pigment-containing cells.

The most significant contribution of the work with MPTP is that it has focused attention once more on the causes of PDS and has provided significant circumstantial evidence that at least some cases of PDS may be caused by environmental factors. The MPTP model has already provided the impetus toward an investigation of the possibility that MAO-B inhibitors, such as deprenyl, may be able to prevent PDS, by virtue of their ability to prevent the conversion of MPTP to the active neurotoxin, MPP^+. The nonhuman primate model will also provide a significant tool with which to examine potential treatment regimens, such as novel D_1 agonists, administered alone or in combination with other DA agonists, such as BRC and L-Dopa (*see* Hinzen et al., 1986). Since there is some evidence that MPTP and/or analogs may produce PDS by inhalation or skin contact (Langston and Ballard, 1983), researchers must consider the ethical problems of staff working with such a potentially biohazardous material.

3. Tardive Dyskinesia

3.1. The Disease

Tardive dyskinesia (TD) is an iatrogenic disease associated with the chronic intake of neuroleptic drugs (Faurbye et al., 1964; Ayd, 1967; Crane, 1973; Tarsy and Baldessarini, 1976). Symptomatically, it is a choreiform disorder characterized by involuntary and often bizarre dyskinesias of the mouth region, face, limbs, and trunk. The hypothesis that has underpinned a massive amount of basic and clinical research is that TD is caused by DA receptor supersensitivity in the neostriatum (Klawans, 1973a,b) produced as an homeostatic response to the prolonged receptor blockade. There are a number of clinical observations that indirectly support this hypothesis. Thus, TD is exacerbated if the neuroleptic is withdrawn, but ameliorated if the neuroleptic dose is increased (Kazamatsuri et al., 1973). It is worsened by L-DOPA and other DA agonists (Gerlach et al., 1974) and by muscarinic antagonists, but is alleviated by blocking DA synthesis, depleting the granule stores of DA (Sato et al., 1971; Kazamatsuri et al., 1972; Fahn, 1983), or using low, autoreceptor-selective doses of apomorphine (Carroll et al., 1977). These observations suggest that TD is caused in some way by excessive DA activity. One cause of increased DA activity could be the presence of supersensitivity. There have been several attempts to demonstrate supersensitive DA receptors in TD patients, but the data are conflicting and unconvincing (Ettigi et al., 1976; Tamminga et al., 1977). Thus, the major problem with the supersensitivity hypothesis is that it has only circumstantial and indirect evidence to support it. This should be remembered in the following discussion of the chief animal model for TD.

3.2. Neuroleptic-Induced DA Receptor Supersensitivity Model

Most species respond to chronic neuroleptic treatment with the development of DA receptor supersensitivity. In rodents, this has been demonstrated behaviorally (Tarsy and Baldessarini, 1973; Jackson et al., 1975b), electrophysiologically (Yarbrough, 1975), with receptor binding techniques (Burt et al., 1977; Muller and Seeman, 1977) and biochemically (Gianutsos et al., 1975; Stock et al., 1980). These data indicate that pre- (autoreceptors) and postsynaptic DA receptors become supersensitive in various parts of the brain (*see,* e.g., Jackson et al., 1975b). The supersensitivity is caused by an increase in DA D_2 receptor number, rather than a change in affinity, although very long periods of treatment (>6 mo) may produce changes in affinity as well as in receptor number (*see* Owen et al., 1980). There has been dispute as to whether the DA D_1 receptor as measured by adenyl cyclase and its product cyclic AMP is also supersensitized, with some workers reporting no change and others reporting an increase in sensitivity (Marshall and Mishra, 1980; Rotrosen et al., 1975). This uncertainty will probably remain until the relationship between DA D_1 and D_2 receptors is clarified, although the latest studies indicate that chronic treatment of animals with D_2 antagonists produces changes only in DA D_2 receptors, whereas chronic treatment with DA D_1 antagonists produces changes only in DA D_1 receptors (as measured by ligand binding studies) (*see* Hyttel, 1986). Neuroleptic-induced changes are not restricted to changes in DA receptor sensitivity. There are marked increases in DA turnover after neuroleptic treatment, and it is interesting that tolerance to these changes occurs sooner in the basal ganglia than in the limbic system (Scatton, 1977).

Neurotransmitters and modulators other than DA are affected, however, with various workers reporting changes in calmodulin levels (Gnegy et al., 1977), ACh function (Gianutsos and Lal, 1976), GABA function (Gunne and Haggstrom, 1985; Marco et al., 1978; Freed et al., 1980), 5-HT and α_1 receptor function (Muller and Seeman, 1977; Dunstan and Jackson, 1977), and peptides, including cholecystokinin (Chang et al., 1983) and neurotensin (Govoni et al., 1980). It has also been proposed by various workers that neuroleptic-induced supersensitivity may be modulated by hormones such as prolactin, although this has been disputed (Rupniak et al., 1983).

Part of the wide profile of changes is because of the use of different neuroleptics. Each has a unique profile with varying affinities for DA D_2 and D_1, α_1 and α_2, 5-HT2, and muscarinic receptors, and the relative potency of each neuroleptic at each receptor will influence the chronic effects produced. It is also clear that the atypical neuroleptics such as clozapine are biochemically distinct from the traditional ones such as haloperidol (see, e.g., Friedman et al., 1983a; Montanaro et al., 1982). Pertinent to this is a recent report that chronic treatment of rats with haloperidol reduced the number of spontaneously active DA cells detectable in both the basal ganglia and in the mesolimbic system. Clozapine, which seems to be devoid of extrapyramidal side effects and which might not induce TD, reduced the number only in the mesolimbic system. This selectivity of clozapine for the mesolimbic system may be an electrophysiological correlate to the observation that clozapine does not produce DA receptor supersensitivity in the basal ganglia (Sayers et al., 1975). These observations also may be correlates of the ability of traditional neuroleptics, but not clozapine, to produce severe extrapyramidal side effects. In all the vast literature, however, there are only a few papers that have reported spontaneous, rather than drug-elicited, changes after such treatments. This is peculiar, since TD is, after all, a spontaneous motor disorder.

Nevertheless, there are a few studies in this area. Thus, rodents may exhibit a hyperkinesia on withdrawal from the neuroleptic that can be blocked by neuroleptics or by inhibiting DA synthesis (Engel et al., 1976; Jackson et al., 1979). Rats may also display vacuous chewing after many months of neuroleptic treatment (Waddington et al., 1983). Motor changes in nonhuman primates have been reported (Weiss and Santelli, 1978; Haggstrom et al., 1983; Paulson, 1973; Bedard et al., 1977). Two separate syndromes are seen in monkeys and it is important to distinguish them. First, monkeys may exhibit dyskinesias when acutely challenged with neuroleptic, even if this challenge occurs months after withdrawal from the chronic neuroleptic treatment (Weiss and Santelli, 1978; Neale et al., 1984; Bedard et al., 1977). This "induced" syndrome can also be precipitated by tetrabenazine and reversed by amantadine, L-DOPA, and anticholinergics, although not by 3-PPP (Neale et al., 1984). Thus, this syndrome seems to model the acute extrapyramidal syndrome seen in humans and primates. The second syndrome was reported by workers who, in addition to noting the "induced" syndrome, also noted spontaneous TD-like dyskinesias (Gunne and Barany, 1976; Haggstrom et al., 1983). These dyskinesias, in contrast to those precipitated by an acute neuroleptic challenge, were reduced by acute neuroleptic treatment and exacerbated by an antimuscarinic agent, as well as being alleviated by 3-PPP (Haggstrom et al., 1983). The second syndrome more truly mirrors TD.

In practical terms, the neuroleptic is given daily either by injection, minipump, orally as part of the diet, or in the drinking water. Oral administration is quite suitable and is certainly more economical than using minipumps and less stressful to the subjects than systemic administration. Some studies have utilized depot injections of esters of neuroleptics (*see,* e.g., Gunne and Haggstrom, 1985). The dose required is dependent upon the neuroleptic, with haloperidol, one of the best studied, being given in the range of 0.1–5 mg/kg/d. Nonhuman primates are given doses at the lower end of the range, whereas rodents require higher doses. The duration of treatment is important. Most studies have used treatment schedules in rodents of less than 1 mo. Since TD only appears after a few years of neuroleptic treatment, longer pretreatment times (perhaps $>$6 mo) may be desirable in order to produce spontaneous dyskinesias. Note that there are a few reports that even one dose of a neuroleptic may produce some signs of supersensitivity in rodents (Hyttel, 1974), but this phenomenon probably bears no relationship to TD. The most usual way to assess the presence of supersensitive DA receptors in rodents is a challenge with a DA receptor agonist or antagonist. If the receptors are supersensitive, there will be an increased response to agonists and a decreased response to antagonists. Any model that is sensitive to DA drugs will be suitable. For example, DA agonist-induced locomotor stimulation and stereotypy in mice and rats (Dunstan and Jackson, 1977), orofacial dyskinesias ("vacuous chewing) (Waddington and Gamble, 1980; Waddington et al., 1983), and induction of rotation in Ungerstedt rats (Siever et al., 1981) have been used. An alternative behavioral test is to assess the animal's sensitivity to neuroleptic-induced catalepsy (Allikmets et al., 1981). These methods assess DA D_2 receptor sensitivity. Biochemical models include receptor binding assays in DA-rich nuclei for DA D_2 receptors (*see* Seeman, 1980, for references), the inhibition by apomorphine of the gamma-butyrolactone (GBL)-induced rise in DA or L-DOPA (a measure of DA autoreceptor sensitivity) (*see* Stock et al., 1980), or DA-sensitive adenyl cyclase activity (DA D_1 receptors) (*see* references above). It is important that the biochemical measure reflect the reports that DA-rich areas of the brain may be differentially affected by chronic neuroleptics (Chiodo and Bunney, 1983; Sayers et al., 1975). Whatever biochemical measures of DA receptor function are used, they will bear little relationship to TD unless it can be demonstrated that spontaneous dyskinesias, which are reduced by neuroleptics and exacerbated by antimuscarinics, are also present.

3.2.1. Usefulness and Limitations of the Supersensitivity Model

There are many problems with the supersensitivity hypothesis of TD and consequently with the models that depend upon this. First, it would be predicted that if supersensitive DA receptors are responsible for TD, then patients who exhibit TD would be those who exhibited the most drug-induced parkinsonism. In other words, there should be a significant positive correlation between sensitivity to the parkinsonism-producing effects of neuroleptics and the occurrence of TD. This does not appear to be true (Klawans, 1973a). In this context, it would also be predicted that there should be an exacerbation of schizophrenia on withdrawal of patients from neuroleptics, because DA receptors in the limbic system (where the antischizophrenic effects of neuroleptics may reside) also become supersensitive. There is no convincing evidence of this occurring. Second, DA receptors become supersensitive fairly rapidly (see above), whereas TD may take years to develop. Conversely, whereas TD may last for many years, supersensitivity is a relatively short-lived phenomenon, resolving in a few weeks. It is interesting, therefore, that some of the spontaneous behaviors discussed above may last for several months. Although TD appears during treatment and worsens when neuroleptic medication is stopped, most workers have not looked for spontaneous dyskinesias in their models (at least not in the rodent models) and have used DA agonists to demonstrate supersensitivity. The nonhuman primate model would seem to be the best species to use, with its major problem being one of very high cost. The caveat is that although there is little doubt that TD-like dyskinesias can be produced in nonhuman primates, there is nothing substantial to support a connection between these dyskinesias and DA receptor supersensitivity. One other problem with the supersensitivity hypothesis of TD is that not all patients develop TD, whereas all animals become supersensitive. Finally, there are now a number of studies that have suggested that D_2 binding in the putamen of schizophrenic patients with and without involuntary movements is similar, although binding assay sensitivity may not be great enough to register small, but potentially significant, changes (*see* Waddington, 1985).

An alternative hypothesis of TD that will not be discussed here is that chronic neuroleptic treatment produces a loss of neurons in the neostriatum (Pakkenberg et al., 1973; Nielsen and Lyon, 1978; Benes et al., 1983). This fact, together with the other limitations discussed above, has led to a number of alternative suggestions, including a role for GABA (*see* Fibiger and Lloyd, 1984, for discussion), NE, and β-phenylethylamine (Wolf and Mosnaim, 1984).

In conclusion, the supersensitivity model should be divided into two: one in which the chronic treatment is of short duration and the endpoint is a reflection of DA receptor supersensitivity, and another in which the treatment is of many months duration and an attempt is made to observe spontaneous, rather than drug-precipitated, changes. Although there is justifiably much dispute about the supersensitivity hypothesis, there is no doubt that it has contributed significantly to our understanding of homeostasis in the DA system, together with the primary and secondary effects of sustained receptor blockade. It has also provided a substantial impetus for the investigation of the effects of DA receptor agonists at normo- and supersensitive DA receptors.

3.3. Other Methods of Producing Dyskinesias

Although it is logical to model TD by administering neuroleptics long-term, dyskinesias can be produced in a number of other ways, and these may also be useful in examining the etiology of TD. A high dose of L-DOPA (normally combined with a peripheral decarboxylase inhibitor), apomorphine, or piribedil produces a variety of involuntary movements in monkeys, as well as intense stereotyped behavior (Goldstein et al., 1973; Sassin et al., 1971; Sassin, 1975). These movements are blocked by haloperidol and AMPT, reduced by the GABA transaminase (GABA-T) inhibitor aminooxyacetic acid, but potentiated by bicuculline and picrotoxin (Goldstein et al., 1975). Interference with the GABA system by injecting the GABA antagonist bicuculline into the lentiform nucleus of monkeys produced contralateral choreoathetoid dyskinesia (Jackson and Crossman, 1984). Finally, when 3,3-iminodipropionitrile is administered to rats for 7–10 d (150 mg/kg/d), motor abnormalities occur, including increased locomotor activity, vertical and lateral head movements, and other postural changes (Schneider et al., 1981; Schneider, 1984). The abnormalities were potentiated by amphetamine and alleviated by chlorpromazine and physostigmine, but not by pentobarbitone. The syndrome seems to be permanent and, from its responsiveness to drugs, may mirror not only the dyskinesias of TD, but those of HDS and Gilles de la Tourette's syndrome.

4. Huntington's Disease

4.1. *The Disease*

Huntington's disease (HDS) is a progressive, irreversible, genetic (autosomal dominant) disorder involving neuronal degeneration that is especially marked in the basal ganglia. Phenomenologically, the disease includes both mental and physical signs, either of which may appear first. The mental signs include impaired judgment, irritability, depression, violent behavior, impaired memory, and other schizophrenic-like symptoms. Physical signs include gross motor abnormalities characterized by bizarre choreiform movements. Pathologically, there is a loss in brain weight, with marked atrophy of the neostriatum and increased ventricular volume. Biochemically, there are marked changes in the neostriatum and associated structures (Table 3; Coyle et al., 1978; Sanberg and Johnston, 1981). Thus, the neurotransmitter systems in the basal ganglia become imbalanced, with an effective deficit of GABA and ACh function and a functional dominance of DA. This is evident also in the ability of anticholinergics, for example, to exacerbate HDS. The imbalance is the reverse of that seen in PDS, but is similar to that in TD. Interestingly, it may sometimes be difficult to distinguish the choreiform movements of TD from those of HDS (Buxton, 1976).

4.2. *The Models*

4.2.1. The Kainic Acid Model

Kainic acid is a neuroexcitatory analog of the transmitter glutamate, which, on microinjection into the neostriatum, destroys local neurons without damaging axons of passage or axons terminating in the area (McGeer and McGeer, 1976). Biochemically, the changes are very similar to those in HDS (Table 3), although there

Table 3
Biochemical Changes in the Neostriatum in Huntington's Disease and in the Kainate Rat[a]

Biochemical parameter	HDS	Kainate rat
GABA	Decreased	Decreased
Glutamic acid decarboxylase	Decreased	Decreased
Choline uptake	Decreased	Decreased
ACh	Decreased	Decreased
Muscarinic receptors	Decreased	Decreased
DA	No change or increased	No change or increased
Angiotensin-converting enzyme	Decreased	Decreased
Enkephalin	Decreased	Decreased
Cholecystokinin	Decreased	Decreased

[a]Modified after Coyle et al. (1978) and Sanberg and Johnston (1981).

are some differences. For example, GABA-T levels are normal in HDS, but decreased in the kainate model. Qualitative differences also exist that are probably a function of the injection technique and of species differences. On the whole, the pathological and biochemical changes in HDS and in the kainate model are remarkably similar (*see* Table 3).

The parallelism of the behavioral changes in HDS and in the kainate model are more difficult to assess. Whereas kainate-treated rats do not become choreic, there are some spontaneous changes in motor function. Thus, Hruska and Silbergeld (1979) noted abnormal placement of their limbs, whereas Mason and Fibiger (1979) reported that the activity of kainate-treated rats was markedly potentiated by arousal, an observation also seen in HDS. The latter authors also noted that treated animals were much more active during the night (but not during the day) than control rats, and that the hyperactivity was not specifically suppressed by the DA antagonist pimozide. Fibiger's group (Fibiger, 1978; Mason and Fibiger, 1979) likened this nocturnal hyperactivity to the choreiform movements of HDS that abate during rest and sleep, but worsen during movement (Duvoisin, 1972).

There are also changes in drug sensitivity, with treated animals being more sensitive to the stereotypy-producing effects of *d*-amphetamine, but not to apomorphine (Mason et al., 1978). This is interesting, because the choreic movements in HDS are exacerbated by amphetamine and the DA precursor L-DOPA (Gerstenbrand et al., 1963; Klawans et al., 1972), but not by apomorphine (Lal et al., 1973; Tolosa, 1974; Carroll et al., 1977). This increased sensitivity to *d*-amphetamine is apparently not related to a change in DA receptor sensitivity, but may be related to the dependence of some of the actions of amphetamine on the striatonigral pathway or on the release of DA within the SN (Bunney and Aghajanian, 1978). It is almost impossible in animal models to mirror psychiatric changes, but an attempt has been made to show that there are learning problems in kainate-treated animals. Such animals take longer to learn a passive avoidance task (Sanberg et al., 1979), but not a continuously reinforced bar press response, than control animals. Once the latter response is learned, however, it is more difficult to extinguish than in control animals, which may be related to the increased arousal in the treated animals. One major problem with these learning deficits is that they may merely be a reflection of motor deficits produced in the neostriatum. Thus, it is well documented that interference with DA and thus motor function in either the basal ganglia or NAC produces changes in a variety of learning paradigms (Bracs et al., 1982).

For a full discussion on how to set up the model, the reader is referred to McGeer et al. (1978a,b). In brief, kainic acid is dissolved in a phosphate buffer at pH 7.2, and 3 nmol (in 0.5 μL) injected into each neostriatum over about 3 min. The injection cannulae should be left in place for a few minutes longer to allow equilibrium to be established. The death rate can be quite high in the initial few days, sometimes because of kainate-induced convulsions. These can be reduced by treating the rats at the time of surgery with diazepam (1 mg/kg ip). Mortality can also be reduced by orally feeding the animals for a few days postlesion. The rate of injection, dose, and injection volume influence the extent of the lesion (McGeer et al., 1978).

4.2.1.1. Usefulness and Limitations of the Model. Many of the parallels between the kainate model and HDS have been discussed above. Whereas many of the mental symptoms of HDS are thought to be caused by cortical degeneration, however, this is not seen in the kainate animal. The major usefulness of the model has been to show that a drug can induce many of the signs of a genetic disease. This has led to the hypothesis that a causative factor in HDS may be an endogenous kainate-like material (Skerritt and Johnston, 1981; Coyle et al., 1983). An interesting use of the model was a partially successful attempt to reverse some of the motor abnormalities associated with the kainate lesion by fetal tissue striatal implants (Deckel et al., 1983).

L-Pyroglutamate (Rieke et al., 1984) has been proposed as an alternative to kainic acid, but few data are available as yet.

4.2.2. The AF64A Model

A series of choline analogs have been tested as potential neurotoxins of cholinergic neurons, and one of them (AF64A) displays some interesting properties. The drug's neurotoxic properties seem to be caused by irreversible inhibition of choline acetyl transferase (CAT) and perhaps of other choline-recognizing enzymes (Sandberg et al., 1985; also *see* section 7.2.3) On injection into the neostriatum, AF64A produces a selective loss of cholinergic neurons (Sandberg et al., 1984a). Lesioned rats display deficits in the acquisition and retention of a passive avoidance task (Sandberg et al., 1984c) and increased spontaneous nocturnal locomotor activity (Sandberg et al., 1984b). This latter finding corresponds to that seen in the kainate model (see above). AF64A has the advantage over kainate in that the lesion is restricted to cholinergic neurons. AF64A has recently been reviewed by Mantione et al. (1984).

5. Lesch-Nyhan Syndrome

5.1. The Disease and Some of Its Models

This is a rare, sex-linked recessive disorder of purine metabolism, caused by a deficiency in the enzyme hypoxanthine-guanine phosphoribosyltransferase and associated with hyperuricemia. Symptoms include physical and mental retardation, compulsive self-mutilation, dystonias, and choreoathetosis. There are many biochemical changes, and one of these appears to be a deficit of basal ganglia DA function (Lloyd et al., 1981). This is not, of course, the whole story, since parkinsonian patients who also suffer from a basal ganglia DA deficit do not self-multilate. Noting that the Lesch-Nyhan syndrome presents at birth, Breese et al. (1984) theorized that a neonatal injection of 6-OHDA into rats might reproduce some of the symptoms. When these neonatally 6-OHDA-treated rats were adult, they responded to L-Dopa challenge with self-mutilation behavior and self-biting, an effect that was not blocked by a DA-beta-hydroxylase inhibitor or a high dose of haloperidol, suggesting that the effect of L-Dopa might not have been mediated by DA receptors. Since 5-hydroxytryptophan has been used to suppress self-mutilation in patients, future studies should examine the possibility that the L-Dopa in this rat model is being converted to DA in 5-HT neurons and exerting some effect there. Another model depends upon the administration of high doses of caffeine to rats, which produces self-destructive behavior. The caffeine model is accompanied by an increase in hypoxanthine-guanine phosphoribosyltransferase, however, rather than the decrease seen in the disease (Minana et al., 1984). Caffeine-induced elevations in DA may be the cause of the self-destructive behavior, which can also be produced by excessive doses of many DA agonists.

6. Schizophrenia

6.1. The Disease

Schizophrenia is a group of disorders characterized by disintegration of the personality. Symptoms include a retreat from reality, autism, thought disorder, delusions, hallucinations, and inappropriate affect. Other symptoms may include anxiety, neurosis, prominent mood disorders, lack of initiative, and indifference. It is not a single disease, and its causes appear to involve both genetic and environmental factors. Certainly, many drugs have been reported to induce a form of psychosis. These include the hallucinogens (e.g., mescaline), amphetamines (e.g., *d*-amphetamine), and sedative hypnotics (e.g., bromides and paraldehyde), as well as certain drug combinations (e.g., diphenhydramine plus methaqualone). Over the years, a number of histopathological changes have been reported in schizophrenic brains, but many of them are nonspecific. Computerized axial tomographic scans do, however, indicate that the lateral ventricles are enlarged in a proportion of schizophrenics (Andreasen et al., 1981), which has been interpreted as indicating cerebral atrophy.

Although the cause of schizophrenia is unknown, there have been many hypotheses to explain it, and the models to be discussed are based upon these hypotheses. The most influential is the DA hypothesis. Broadly, this states that there is excessive DA function in certain parts of the brain. It is based upon several observations. First, the clinically used antipsychotics (with the exception of clozapine and a number of other "new generation" antipsychotics) block DA D_2 receptors in the brain. Their clinical potency is highly correlated with their ability to block DA D_2 receptor-mediated biochemical and behavioral effects, including *d*-amphetamine- and apomorphine-induced stereotypy and locomotion in rodents, apomorphine-induced vomiting in dogs, and displacing potency in binding to the DA D_2 receptor (but not to the DA D_1 receptor). There is no significant correlation between ability to block alpha receptors and clinical efficacy. Second, it was observed quite early in humans that *d*-amphetamine abuse can produce what appears to be a frank paranoid psychosis (Connell, 1958; Snyder et al., 1974), as well as activating preexisting psychotic symptoms (Angrist and Van Kammen, 1984). *d*-Amphetamine can release both DA and NE in the human brain, but DA seems to be most important because the euphoria is blocked by blocking DA synthesis (Jonsson et al., 1971) and the psychosis by neuroleptics. High doses of other DA agonists such as BRC and L-DOPA can also produce schizophrenic-like symptoms. Third, the stimulation of DA receptors in the basal ganglia produces various stereotypies. Schizophrenics may also perform repetitious, compulsive, bizarre behaviors (Rylander, 1972). Fourth, there is some evidence from postmortem studies in schizophrenic brains that there may be an increase in the number of DA D_2 receptors, although it has been difficult to eliminate drug-induced changes as causative factors in these changes. Other biochemical studies looking for changes in DA function have produced conflicting results that are difficult to interpret and also may be related to medication (Crow et al., 1979). One recent report noted that there was a significant increase in DA (but not NE) levels in the amygdala of a group of schizophrenics, and that this increase was asymmetric (Reynolds, 1983).

Closely related to the DA hypothesis is the β-phenylethylamine (PE) hypothesis (Sandler and Reynolds, 1976), which suggests that schizophrenia may be associated with an increase in activity of this naturally occurring endogenous amine. Certainly, it shares many of the behavioral and biochemical actions of *d*-amphetamine (Jackson, 1978). The evidence in support of the hypothesis is indirect, but includes its presence in many human tissues (Mosnaim et al., 1973), including brain (Karoum et al.,

1979), its release of DA and NE in the CNS, its ability to cause locomotor stimulation and stereotypy (Jackson, 1978; Borison et al., 1979; Dourish, 1982), and its specific metabolism by monoamine oxidase-B (Phillips and Boulton, 1979), an enzyme that may be markedly decreased in platelets during acute schizophrenia (Murphy and Wyatt, 1972). There are also some reports that urinary excretion of PE may be elevated in schizophrenia (Potkin et al., 1979), although nonspecific factors such as diet may play an important role here.

Other authors have proposed that NE metabolism may be altered in schizophrenia. The NE hypothesis of schizophrenia is based partly upon the observation that there may be an increased level of NE in limbic ("emotional") brain and in parts of the basal ganglia of brains from schizophrenics (Hornykiewicz, 1982). There are, however, many reports that have failed to note any such changes (*see,* e.g., Crow et al., 1979). There is also substantial evidence that NE is involved in stress, which occurs in schizophrenia, but is not, of course, restricted to this disease.

5-HT has also been examined in some detail, but results are again unclear. Perhaps the most interesting aspect of 5-HT and schizophrenia is the hypothesis that a metabolite of 5-HT, *N-N*-dimethyltryptamine [which has been reported to be excreted in schizophrenia (Murray et al., 1979)], can produce a short-lived model of schizophrenia (*see* Gillin et al., 1976). Interestingly, 5-HT is involved in the behavioral effects of high doses of *d*-amphetamine and PE (Dourish, 1981; Sloviter et al., 1978). The possibility of 5-HT involvement has also gained considerable support from the structural relationships between 5-HT and the psychotomimetics such as LSD and mescaline and the chronic, undifferentiated, schizophrenia-like syndrome produced by these drugs (Glass and Bowers, 1970).

The formation of an endogenous schizotoxin in schizophrenia has been proposed by a number of workers. Some workers, for example, have proposed that *d*-amphetamine produces its psychoses via the formation of a toxic metabolite, *p*-methoxyamphetamine (Angrist and Gershon, 1970; Anggard et al., 1973). Seiden and Vosmer (1984) showed that 6-OHDA is formed in rat brain after one dose of methylamphetamine, and thus it is interesting that methylamphetamine causes irreversable damage to nigrostriatal DA pathways (Seiden et al., 1975). Other workers have proposed that 6-OHDA may play a role in schizophrenia (Stein and Wise, 1971).

Many other putative neurotransmitters and neuromodulators have been examined in schizophrenia (GABA, opiates, prostaglandins), and the reader is referred to discussions by Horrobin (1979) and Rodnight (1983).

6.2. The Models

6.2.1. General

One of the difficulties in modeling schizophrenia is that it is a uniquely human disease, and one for which the etiology is unknown. Therefore, models generally aim to mirror one or more aspects of schizophrenia. It is important to separate animal models of schizophrenia from those used to screen neuroleptics. With the dominance of the DA hypothesis of schizophrenia, many workers have fallen into the trap of confusing the two. There is, nevertheless, a close relationship between the two. Although all the models used to screen neuroleptics assume that the DA hypothesis of schizophrenia is true, however, and therefore depend upon the ability of putative neuroleptics to block DA receptors, this is not true of models of schizophrenia. Thus neuroleptics will: inhibit conditioned avoidance responding in rodents and cause hypothermia, catalepsy, and reduced locomotor exploration. They will also block DA-mediated behaviors such as apomorphine- or *d*-amphetamine-induced stereotypy or locomotor stimulation in many species and apomorphine-induced vomiting in dogs. These are acute screening methods for drugs active at DA receptors. These rigid DA-oriented screening procedures have produced many new drugs with essentially similar properties but have also contributed to an enormous amount of valuable research into the pharmacology of DA. They have also, however, restricted the possibility of finding drugs active in schizophrenia that may have a novel action mediated by mechanisms other than DA receptor blockade.

6.2.2. *d*-Amphetamine-Induced Changes

d-Amphetamine can be administered to animals acutely or chronically. When given acutely, it (at, e.g.) 5 mg/kg ip or sc) produces a mixture of stereotyped behavior and forward locomotion. Both behaviors are caused by the release of newly synthesised DA, since they can be blocked by AMPT. Both behaviors are readily and accurately assessed (*see* section 2.3.2.1) and are blocked by neuroleptics, whose potency parallels both their affinity for the DA

D_2 receptor and their clinical efficacy (Seeman, 1980). Note that stereotypy emanating from the neostriatum cannot reflect schizophrenia, since the latter disorder is believed to derive from a disorder in the mesolimbic/mesocortical DA system. Stereotypy reflects, in my view, nothing more or less than a behavior that is produced by the stimulation of a relatively well-defined population of neostriatal DA receptors, and is, therefore, suitable for screening drugs that might block DA receptors. Even in this regard, its limitation is reinforced by the observation that the clinically effective antipsychotic clozapine does not readily antagonize DA-mediated behaviors from the neostriatum (Burki et al., 1977) and does not produce catalepsy.

The observation that high-dose *d*-amphetamine abuse produces some of the signs of schizophrenia provides one of the strongest supports for the DA hypothesis. Thus it seems logical to administer the drug chronically. In most of the chronic studies with *d*-amphetamine, the drug was administered once or several times per day. The doses were usually high and tended to produce transient high plasma levels that may be neurotoxic (Fuller and Heinrick-Luecke, 1980). Such treatments certainly produce behavioral and biochemical changes, with DA systems being the most commonly studied. Changes observed include tolerance to its anorectic (Magour et al., 1974), stimulus perseveration (Kokkinidis and Anisman, 1981), and self-stimulation (Leith and Barrett, 1976) effects. The intensity of other behaviors may, in contrast, be increased; these include its locomotor stimulant and stereotypy effects (Segal and Mandell, 1974; Bailey and Jackson, 1978), and it has been proposed that this increased response may "contribute to the development of amphetamine psychosis" (Segal et al., 1980). It seems, despite much controversy, that the "supersensitivity" is not mediated by a change in postsynaptic DA receptor sensitivity (Jackson et al., 1981), but in part by conditioning factors (Ellison and Morris, 1981) and by the down-regulation of DA autoreceptors (Muller and Seeman, 1979). It is hard to relate such changes to the human situation. All these studies used relatively short, high-dose treatment schedules, and, in animals, the chronic drug regimen appears to be critical to the chronic effects produced (Nelson and Ellison, 1979). Thus it is hard to relate the changes seen in animal studies to the human situation, in which abusers seem to aim to maintain plasma levels at measurable levels for several days (Kramer et al., 1967). This observation led a number of workers to more closely reproduce the social situation by administering the

Table 4
The Behavioral Effects of Chronically Administered Amphetamine in Humans, Monkey, and Rat[a]

Intoxication stage	Humans	Monkey	Rat
Initial	Euphoria, exhilaration except in low doses to addicts	Increased activity, staring, and stereotypies	Increased activity and motor stereotypies
Crash	Increased irritability, intellectual flattening, anhedonia, depression	General inactivity, hunched posture, dyskinesias	Retreat to burrows
Late	Paranoid delusions, parasitosis, hallucinations	"Hallucinations," wet dog shakes	Parasitosis, "hallucinations" ("limb flicks, wet dog Shakes"), abnormal social behavior, fight and flight

[a]After Ellison and Eison (1983).

drug continually to animals via implanted pumps. For example, Ellison and colleagues (Ellison and Eison, 1983) reviewed the parallelism between the effects of *d*-amphetamine in humans and its effects in monkeys and rats when administered to these two species by implanted minipumps (Table 4). There seemed to be a very close concurrence between the behavioral effects seen in these species when *d*-amphetamine was administered by implanted minipumps, although this needs to be replicated independently. There are also extensive biochemical changes, including a loss of DA in the basal ganglia and gross damage to the DA axons, indicating a neurotoxic effect. Ellison and Eison concluded that the damage to the neostriatum inactivated it, so to speak, and, by allowing the mesolimbic system to dominate, produced psychoses. This is perhaps the weakest part of Ellison's arguments, because there is no evidence of any such simplistic division between the functions of the basal ganglia and the mesolimbic systems.

There can, of course, be a "motor antagonism" between the two regions, with, for example, excessive stimulation of the DA receptors in the neostriatum producing stereotypies that override any coordinated locomotion derived from stimulation of DA receptors in the NAC. I suspect that the model is an excellent way to study *d*-amphetamine toxicity, rather than a model of schizophrenia, but the idea is provocative and the model warrants further study.

6.2.3. The Opioid Model of Schizophrenia

The opioid model of schizophrenia derives from the hypothesis that there is an over- or underproduction of an endogenous opioid in schizophrenia and is based on the ability of opiates to cause catalepsy, the alleviation by naloxone of some of the symptoms of schizophrenia, the elevation in certain types of schizophrenia of CSF opioids, and the ability of endorphins in animals to alter learning. The last forms the basis of the animal model. Gamma-endorphin markedly facilitates the extinction of active avoidance behavior and attenuates passive avoidance in rats (De Wied and Jolles, 1982). In contrast, alpha- and beta-endorphin delay extinction of active and facilitated passive avoidance behavior. Thus, different endorphins produced diametrically opposite effects, with gamma-endorphin facilitating what can be viewed as a normal adaptive response to new environmental stimuli. This lack of adaptation is one facet of schizophrenia, so gamma-endorphin can be viewed in some ways as a neuroleptic, whereas the other two endorphins can be viewed as schizogens. This type of approach is interesting, but requires more study.

6.2.4. The Norepinephrine Model of Schizophrenia

Few battles in pharmacology have been more marked than that between the proponents of the DA and the NE hypotheses of schizophrenia. This is a pity, because the proponents of both sometimes seem to forget that schizophrenia is a disease with many putative causes, and that many neurotransmitters (DA, NE, opioids, and so on) may be involved. The NE animal model really began with the observation that animals will actively work to obtain an electrical stimulation from electrodes implanted in the medial forebrain bundle. This releases NE in the forebrain (Stein,

1964; Wise and Stein, 1969). Since NE in this situation may mediate "reward," and since one of the symptoms of schizophrenia is often a lack of "reward"-oriented behavior, it was proposed that schizophrenia might result from the degeneration of NE pathways. There is substantial evidence, however, that DA is also involved in self-stimulation behavior. Mason and Lin (1980) proposed that lesions (with, e.g, 6-OHDA) of the dorsal NE bundle in rats induced behavioral changes reminscent of some schizophrenic symptoms. The changes indicated attention deficits (*see* Robbins et al., 1982, for discussion). Although there is no doubt that such global damage can markedly affect learning, the model's relevance to schizophrenia remains obscure. For example, the dorsal lesion caused NE to disappear from all forebrain regions, including the amygdala (Mason and Iversen, 1979), whereas in schizophrenia there is no change in amygdala NE levels (Bird et al., 1979). Another problem is that the clinical efficacy of neuroleptics is unrelated to their affinity at alpha or beta receptors (see discussion of DA hypothesis of schizophrenia above). Reynolds (1984) suggested that the dorsal NE model might be a useful model for so-called type II schizophrenia in which negative symptoms predominate and a poor response to neuroleptics is apparent.

7. Alzheimer's Disease

7.1. The Disease

Alzheimer's Disease (AD) is the most common dementia of the elderly. Symptomatically, it is characterized by severe memory disorder, learning deficits, and personality disintegration. Pathologically, brains at autopsy reveal senile plaques, neurofibrillary tangles, and vacuolar degeneration, especially in the hippocampus and parts of the cortex. Biochemically, the most striking changes are decreases in the terminal cholinergic markers CAT and ACh esterase, with no marked change in the number of muscarinic receptors. There are no alterations in other neurotransmitters such as cholecystokinin, vasopressin, or vasoactive intestinal peptide. Thus, dementia seems to be strongly associated with cholinergic hypofunction in the brain (Coyle et al., 1983), and this is supported by a variety of pharmacological, biochemical, and behavioral studies (Bartus et al., 1982; Drachman, 1977). The cause remains un-

Table 5
Comparison of the Biochemical Changes in AD with Those Seen in the Kainate and AF64A Models[a]

Parameter	AD	Kainate lesion	AF64A Lesion
CAT	Decreased	Decreased	Decreased
ACh esterase	Decreased	Decreased	Decreased
ACh	Decreased	Decreased	Decreased
Muscarinic receptors	No change	No change	No change
Somatostatin	Decreased	No change	—
NE	Variably decreased	No change	No change
5-HT	No change	No change	No change
Glutamic acid decarboxylase	No change	No change	—
GABA	No change	No change	—
Glucose utilization	Decreased	No change	—

[a]For references, *see* the text.

known, although there is the possibility of genetic involvement (Corkin et al., 1982).

Most models have concentrated upon the cholinergic hypofunction in AD.

7.2. The Models

7.2.1. Cholinergic Drugs and Memory

A well-documented way to interfere with memory is with anticholinergic drugs such as hyoscine (Pepeu et al., 1980; Drachman, 1977). The memory deficits produced seem to be fairly specific, in that the effects of anticholinergics can be antagonized by physostigmine, but not by stimulants such as methylphenidate or amphetamine (Drachman, 1977; Bartus, 1978; Bartus et al., 1982). Although such studies have been very important in testing the acute effects of drugs on a variety of behaviors, they do not mirror the long-term deficits seen in AD. A more promising way is to reproduce the most obvious lesion seen in AD, permanent cholinergic hypofunction. There are at least two ways of doing this with drugs (*see* Table 5).

7.2.2. *Kainate-Induced Cholinergic Hypofunction Model*

A few years ago, Johnston et al. (1979) reported that a bilateral kainate or ibotenic acid lesion of the ventromedial globus pallidus in rats produced marked changes in CAT and ACh esterase levels in the ipsilateral cerebral cortex. This was confirmed by Lehmann et al. (1980). The lesion was restricted to the cholinergic neurons because there was no change in the number of muscarinic receptors. These changes closely but not completely parallelled the biochemical changes seen in AD (Table 5).

Rats lesioned bilaterally with kainic acid reveal changes in pharmacological responsiveness as well as learning deficits. Thus, there is an increased sensitivity to nicotine (Ksir and Benson, 1983), as well as a marked retention deficit 24 h after exposure to a step-through passive avoidance task (Friedman et al., 1983b). In a T-maze task, lesioned rats show an impaired choice performance relative to controls (Salamone et al., 1984). Thus the hypocholinergic state produces marked behavioral changes that may be analogous to those seen in AD. This model is supported by the AF64A model.

7.2.3. *The Cholinergic Neurotoxin (AF64A) Model*

After intraventricular injection of AF64A into mice (about 60 nmol) or rats (15–30 nmol), animals displayed weight loss, hypokinesia, and an unsteady gait (Fisher et al., 1982a,b). Mortality in rats was 41% 7 d after the application of 30 nmol (Walsh et al., 1984). Biochemically, there was a relatively long-lasting (more than 7 d) decrease in ACh levels in the hippocampus and cortex, with a shorter duration in the striatum. There was a concomitant decline in CAT and AChE and in high-affinity choline transport, but no change in the number of muscarinic receptors. GABA, 5-HT, and DA neurons seemed to be unaffected (Coyle et al., 1982; Walsh et al., 1984). Thus, AF64A behaves as a selective presynaptic cholinergic neurotoxin. Behaviorally, although treated rats had levels of locomotor activity comparable to control rats, they displayed poor retention of a step-through passive avoidance task (compare the kainate model). In addition, they exhibited deficits in radial-arm maze performance, with more selections being required to complete the task (Walsh et al., 1984). It is interesting that lesioned rats were more sensitive to the hot plate test, which, however, was not correlated with any learning deficits. Nevertheless, this change in heat sensitivity requires detailed study. (The use of the neurotoxin AF64A is also discussed in section 4.2.2 above.)

The potential problem with both these types of models (and others that are similar) is that they are single-factor paradigms and therefore neglect the role of other neurotransmitters and modulators in the behaviors being studied; e.g, a role for vasopressin (De Wied, 1984). Both these models do, however, provide a fascinating insight into the role of forebrain cholinergic projections in various types of learning.

8. Anxiety

8.1. The Disease

Anxiety is one of the most widespread psychiatric illnesses affecting most people at one or more times during life. It is also the one disease for which there are possibly more human than animal models available. Since this section does not intend to be comprehensive, the reader is referred to a number of excellent reviews on anxiety and its numerous animal and human models (Lal and Emmett-Ogleby, 1983; Rainey and Nesse, 1985; Guttmacher et al., 1983; Pitts and Allen, 1979). Anxiety is both a clinical and social concept, closely related to fear, and apparently serves to mobilize a person to both action and defense. Unfortunately, at its most pathological, it is crippling to action. Symptoms are broad, including signs that are psychological [loss of self-esteem, social humiliation, apprehension, fear without obvious stimuli (the doomsday syndrome), and panic], physical (dyspnoea, faintness, gastrointestinal effects, sweating, tachycardia, and palpitations), and biochemical (increased plasma ACTH, cortisol, and catecholamines). Although the disease manifests itself through many organ systems, there is some pharmacological specificity in that the treatment is relatively specific (for example, benzodiazepines, but not neuroleptics or antidepressants). The causes include genetic and environmental factors, disease (e.g., hyperthyroidism), or challenge with a variety of drugs (e.g., pentylenetetrazol, caffeine, carbon dioxide, lactate, and some newer benzodiazepine receptor antagonists; see below). Many of these agents have been administered to volunteers in human models of anxiety (*see* review references above). Most recent work in drug-induced animal models has concentrated on the role of benzodiazepine receptors in anxiety.

8.2. The Models

8.2.1. Benzodiazepine-Related Models

Specific high-affinity and saturable binding sites for benzodiazepines were first reported in brain in 1977 (Squires and Braestrup, 1977; Mohler and Okada, 1977). All benzodiazepines interact with these sites in a competitive and reversible way. Moreover, a good correlation exists between the potency of benzodiazepines at these sites and their relative potencies in vivo. As soon as these sites were identified, researchers began looking for an endogenous ligand. Many have been proposed, including nicotinamide (Mohler et al., 1979), inosine and hypoxanthine (Skolnick et al., 1978), ethyl-beta-carboline-3-carboxylate (BCC, Braestrup et al., 1980), and diazepam binding factor (DBI, Guidotti et al., 1983). The last two are perhaps the most interesting, with DBI approaching final purification and BCC currently providing one of the more interesting animal models of anxiety. BCC (usually used as the ethyl or methyl esters, BCCE and BCCM, respectively) was isolated from urine (Braestrup et al., 1980) and displays high-affinity antagonist binding to the benzodiazepine receptor. Despite the fact that the isolated compound may well have been an artifact of the isolation procedure, it seems likely that a carboline type of compound may be one of the endogenous benzodiazepine ligands. Apart from this carboline, two main classes of benzodiazepine receptor antagonists have been described; the imidazobenzodiazepines (e.g., Ro 15-1788) and the pyrazoquinolinones (e.g., CGS 8216).

There is some evidence that benzodiazepine receptors are involved in animal (and human) models of anxiety. Chemo- or electroshock (Paul and Skolnick, 1978), cold stress (Soubrie et al., 1980), or swim stress (Skerritt et al., 1981) in rodents causes a rapid rise in the number of benzodiazepine receptors. When the receptors are blocked, behavioral and biochemical effects are sometimes produced that provide the basis of the nonhuman primate model of anxiety described below. Benzodiazepine antagonists act as anxiety-producing agents in a variety of behavioral models such as the social interaction test (File and Lister, 1983), the thirsty lick conflict test in rats, and interoceptive discriminative stimulation in rats (Lal and Shearman, 1980), as well as producing signs of anxiety in both human and nonhuman primates. It is interesting that an antagonist (e.g., Ro 15-1788) can sometimes antagonize the behavioral effects of another antagonist (BCC), leading to the suggestion that there are differences between the various antagonists, with some, such as BCC, having much more intrinsic activity than others, such as Ro 15-1788. In any event, in vivo, BCC can antagonize the anticonvulsant, anxiolytic, and sedative effects of a variety of benzodiazepines.

8.2.2. BCC in Nonhuman Primates

In monkeys, intravenous administration of BCC in doses between 100 and 5000 μg/kg produces, within minutes, marked physical and biochemical signs of anxiety (Ninan et al., 1982). For example, after a dose of 5 mg/kg, animals exhibit agitation such as struggling, a general increase in movement with, at the same time, excessive immobility, bleeding caused by scratches, a reduction in food and water intake, excessive defecation and/or urination, and penile erection, as well as elevated plasma levels of catecholamines, ACTH, and cortisol, and an elevated heart rate. In addition, there is a marked change in the interaction between the monkeys and the rater, with the animals displaying either fear or aggression. All of these effects were antagonized by diazepam or Ro 15-1788 (Ninan et al., 1982), most of them by clonidine, and some of them by the 5-HT blocker cyproheptadine, the beta-blocker propranolol, and the GABA agonist THIP (Crawley et al., 1984). Moreover, the doses of the drugs that are also used in the treatment of the BCC-induced anxiety (diazepam, cyproheptadine, propranolol, and clonidine) were similar to their clinically effective doses. These data strongly support the involvement of multiple neurotransmitter systems in the syndrome, as well as indicating the closeness of the model to human anxiety.

Further support for the model comes from some preliminary studies in humans with the carboline derivative, FG7142 (beta-carboline monoethylamide). Two volunteers administered this drug experienced a number of symptoms including uneasiness, motor tension, and biochemical alterations indicative of anxiety. An iv benzodiazepine relieved the symptoms in one of the patients (Dorow et al., 1983). Validation of this human model will depend on the assessment of a wide range of anxiolytic compounds from a variety of chemical classes, as well as on the comparison of the potency of the compounds in this model with those already determined in behavioral models of anxiety. FG7142 promises to be a particularly interesting compound, however, because a recent report has indicated that it can produce a behavioral syndrome in rats reminiscent of "learned helplessness," a well-studied behavioral model of depression (Drugan et al., 1985). When this drug was administered to rats, it blocked their ability to acquire an escape response in a shuttle box, an effect that was potently antagonized by the benzodiazepine antagonist, Ro 15-1788. This confirms that there may be a close causal relationship between elements of anxiety and depression, and suggests that FG7142 or an analog may provide the first valid (in my opinion) drug-induced model of depression.

9. Conclusions

Drug-induced behavioral models of various psychiatric and neurological diseases have made an indispensable contribution to our understanding and treatment of them. They have also contributed to our basic knowledge of synaptic homeostasis and neurotransmission. With the development of new drugs such as the anxiogenic beta-carbolines and our increasing understanding of the biological bases of many of these diseases, it seems that drug-induced models may continue to make a significant contribution.

Acknowledgments

Research supported by Grants from the National Health and Medical Research Council and the Australian Research Grants Scheme. Ms. Mayko Hashizume is thanked for checking the references, and Ms. Jenny Watson is thanked for her helpful comments.

References

Abramson L. Y. and Seligman M. E. P. (1977) Modelling Psychopathology in the Laboratory, in *Psychopathology: Experimental Models* (Maser J. D. and Seligman M. E. P, eds.) Freeman, San Francisco.

Allikmets L. H., Zarkovsky A. M., and Nurk A. M. (1981) Changes in catalepsy and receptor sensitivity following chronic neuroleptic treatment. *Eur. J. Pharmacol.* **75,** 145–147.

Anden N.-E. (1966) On the Function of the Nigro-neostriatal Dopamine Pathway, in *Mechanisms of Release of Biogenic Amines* (von Euler, U. S., Rosell, S., and Uvnas, B., eds.) Pergamon, Oxford.

Anden N.-E. and Johnels B. (1977) Effect of local application of apomorphine to the corpus striatum and to the nucleus accumbens on the reserpine-induced rigidity in rats. *Brain Res.* **133,** 386–389.

Andreasen N. C., Smith M. R., Jacoby C. G., Dennert J. W., and Olsen S. A. (1981) Ventricular enlargment in schizophrenia: Definition and prevalence. *Am. J. Psychiat.* **139,** 292–296.

Anggard E., Jonsson L.-E., Hogmark A.-L., and Gunne L.-M. (1973) Amphetamine metabolism in amphetamine psychosis. *Clin. Pharmacol. Ther.* **14,** 870–880.

Angrist B. M. and Gershon S. (1970) The phenomenology of experimentally induced amphetamine psychosis—Preliminary observations. *Biol. Psychiat.* **2,** 95–107.

Angrist B. and Van Kammen D. P. (1984) CNS stimulants as tools in the study of schizophrenia. *Trends Neurosci.* **7,** 388–390.

Arnt J. and Hyttel J. (1984) Differential inhibition by dopamine D-1 and D-2 antagonists of circling behavior induced by dopamine agonists in rats with unilateral 6-hydroxydopamine lesions. *Eur. J. Pharmacol.* **102,** 349–354.

Ayd F. J. (1967) Persistant dyskinesia: A neurologic complication of major tranquillizers. *Med. Sci.* **18,** 32–40.

Bailey R. and Jackson D. M. (1978) A pharmacological study of changes in central nervous system responsiveness after long-term dexamphetamine and apomorphine administration. *Psychopharmacology* **56,** 317–326.

Barbeau A. (1984) Manganese and extrapyramidal disorders. *Neurotoxicology* **5,** 13–36.

Barbeau A., Dallaire L., Buu N. T., Veilleux F., Boyer H., de Lanney L. E., Irwin I., Langston E. B., and Langston J. W. (1985) New amphibian models for the study of 1-methyl-4-phenyl-1,2,3,6-tetrahydropyrinine (MPTP). *Life Sci.* **18,** 1125–1134.

Bartus R. T. (1978) Evidence for a direct cholinergic involvement in the scopolamine-induced amnesia in monkeys: Effects of concurrent administration of physostigmine and methylphenidate with scopolamine. *Pharmacol. Biochem. Behav.* **9,** 833–836.

Bartus R. T., Dean R. L., Beer B., and Lippa A. S. (1982) The cholinergic hypothesis of geriatric memory dysfunction. *Science* **217,** 408–414.

Bedard P., DeLean J., Lafleur J., and Larochelle L. (1977) Haloperidol-induced dyskinesias in the monkey. *Can. J. Neurol. Sci.* **4,** 197–201.

Benes F. M., Paskevich P. A., and Domesick V. B. (1983) Haloperidol-induced plasticity of axon terminals in rat substantia nigra. *Science* **221,** 969–971.

Bernheimer H., Birkmeyer W., Hornykiewicz O., Jellinger K., and Seitelberger F. (1973) Brain dopamine and the syndromes of Parkinson and Huntington: Clinical, morphological and neurochemical correlates. *J. Neurol. Sci.* **20,** 415–455.

Bird E. D., Spokes E. G., and Iversen L. L. (1979) Increased dopamine concentration in limbic areas of brain from patients dying with schizophrenia. *Brain* **102,** 347–360.

Birkmayer W. and Hornykiewicz O. (1964) Weitere experimentelle untersuchungen uber L-DOPA beim Parkinson-Syndrom und Reserpin-Parkinsonismus. *Arch. Psychiat. Nervenk.* **206,** 367–381.

Bollini P., Cotecchia S., De Blasi A., Romandini S., and Tognoni G. (1984) Drugs: Guide and caveats to explanatory and descriptive approaches. II. Drugs in psychiatric research. *J. Psychiat. Res.* **18,** 391–400.

Bond N. W. (1984) Animal Models in Psychopathology: An Introduction, in *Animal Models in Psychopathology* (Bond N. W., ed.) Academic, Australia, Sydney.

Borison R. L., Hitri A., Klawans H. L., and Diamond B. I. (1979) A New Animal Model for Schizophrenia: Behavioural and Receptor Binding Studies, in *Catecholamines. Basic and clinical Frontiers* (Koplin F. J. and Barchas T., eds.) Pergamon, New York.

Bowd A. D (1980) Ethical reservations about psychological research with animals. *Psychol. Record* **30,** 201–210.

Boyce S., Kelly E., Reavill C., Jenner P., and Marsden C. D. (1984) Repeated administration of 1-methyl-4-phenyl-1,2,5,6-tetrahydropyridine to rats is not toxic to striatal dopamine neurones. *Biochem. Pharmacol.* **33,** 1747–1752.

Bracs P. U., Jackson D. M., and Gregory P. (1982) Alpha-methyl-*p*-tyrosine inhibition of a conditioned avoidance response: Reversal by dopamine applied to the nucleus accumbens. *Psychopharmacology* **77,** 159–163.

Braestrup C. and Randrup A. (1978) Stereotyped Behaviour in Rats Induced by Phenylethylamine, Dependence on Dopamine and Noradrenaline, and Possible Relation to Psychosis?, in *Non-catecholic Phenylethylamines* part 1 (Mosnaim A. D. and Wolf M. E., eds.) Dekker, New York.

Braestrup C., Nielsen M., and Olsen C. F. (1980) Urinary and brain beta-carboline-3-carboxylates as potent inhibitors of brain benzodiazepine receptors. *Proc. Natl. Acad. Sci. USA* **77,** 2288–2292.

Breese G. R., Baumeister A. A., McCown T. J., Emerick S. G., Frye G. D., and Mueller R. A. (1984) Neonatal-6-hydroxydopamine treatment: Model of susceptibility for self mutilation in the Lesch-Nyhan syndrome. *Pharmacol. Biochem. Behav.* **21,** 459–461.

Bunney B. S. and Aghajanian G. K. (1978) D-Amphetamine-induced depression of central dopamine neurons: Evidence for mediation by both autoreceptors and a striato-nigral feedback pathway. *Naunyn Schmiedebergs Arch. Pharmacol.* **304,** 255–261.

Burki H. R., Sayers A. C., Ruch W., and Asper H. (1977) Effects of clozapine and other dibenzo-epines on central dopaminergic and cholinergic systems. *Arzneimittel forsch.* **27,** 1561–1566.

Burns R. S., Chiueh C. C., Markey S. P., Ebert M. H., Jacobowitz D. M., and Kopin I. J. (1983) A primate model of parkinsonism: Selective destruction of dopaminergic neurons in the pars compacta of the substantia nigra by *N*-methyl-4-phenyl-1,2,3,6 tetrahydropyridine. *Proc. Natl. Acad. Sci. USA* **80,** 4546–4550.

Burt D. R., Creese I., and Snyder S. H. (1977) Antischizophrenic drugs: Chronic treatment elevates dopamine receptor binding in brain. *Science* **197,** 326–328.

Buxton M. (1976) Diagnostic problems in Huntington's chorea and tardive dyskinesia. *Comp. Psychiat.* **17,** 325–333.

Carli M., Evenden J. L., and Robbins T. W. (1985) Depletion of unilateral striatal dopamine impairs initiation of contralateral actions and not sensory attention. *Nature* **313,** 679–682.

Carlsson A., Lindqvist M., and Magnusson T. (1957) 3,4-Dihydroxyphenylalanine and 5-hydroxytryptophan as reserpine antagonists. *Nature* **180,** 1200.

Carroll B. J., Curtis G. C., and Kokmen E. (1977) Paradoxical responses to dopamine agonists in tardive dyskinesia. *Am. J. Psychiat.* **134,** 785–789.

Castagnoli Jr. N., Chiba K., and Trevor A. J. (1985) Potential bioactivation pathways for the neurotoxin 1-methyl-4-phenyl-1,2,5,6-tetrahydropyridine (MPTP). *Life Sci.* **36,** 225–230.

Chang R. S. L., Lotti V. J., Martin G. E., and Chen T. B. (1983) Increase in brain 125I-cholecystokinin (CCK) receptor binding following chronic haloperidol treatment, intracisternal 6-hydroxydopamine or ventral tegmental lesions. *Life Sci.* **32,** 871–878.

Chiodo L. A. and Bunney B. S. (1983) Typical and atypical neuroleptics: Differential effects of chronic administration on the activity of A9 and A10 midbrain dopaminergic neurons. *J. Neurosci.* **3,** 1607–1619.

Chiueh C. C., Burns R. S., Markey S. P., Jacobowitz D. M., and Kopin I. J. (1985) Primate model of parkinsonism: Selective lesion of nigrostriatal neurons by 1-methyl-4-phenyl-1,2,3,6-tetrahydropyridine produces an extrapyramidal syndrome in rhesus monkeys. *Life Sci.* **36,** 213–218.

Chiueh C. C., Markey S. P., Burns R. S., Johannessen J. N., Jacobowitz D. M., and Kopin I. J. (1984) Neurochemical and behavioural effects of 1-methyl-4-phenyl-1,2,3,6-tetrahydropyridine (MPTP) in rat, guinea pig, and monkey. *Psychopharmacol. Bull.* **20,** 548–553.

Close S. P., Marriott A. S., and Pay S. (1985a) Parkinsonism Induced by 1-Methyl-4-phenyl-1,2,3,6-tetrahydropyridine (MPTP) in the Marmoset, in *Dopaminergic Systems and Their Regulation* (Woodruff G. N., Poat J. A., and Roberts P. J., eds.) Basingstoke, Macmillan.

Close S. P., Marriott A. S., and Pay S. (1985b) Failure of SKF38393-A to relieve parkinsonian symptoms induced by 1 methyl-4 phenyl-1,2,3,6-tetrahydropyridine in the marmoset. *Br. J. Pharmacol.* **85,** 320–322.

Connell P. H. (1958) *Amphetamine Psychosis. Maudsley Monographs* 1st Edn., no. 5. Oxford University Press, Oxford.

Cooper B. R., Howard J. L., and Soroko F. E. (1983) Animal models used in prediction of antidepressant effects in man. *Clin. Psychiat.* **44,** 63–66.

Corkin S., Davis K. L., Growdon J. H., Usdin E., and Wurtman R. J. (1982) *Alzheimer's Disease: A Report of Progress in Research* Raven, New York.

Costall B. and Olley J. E. (1971) Cholinergic and neuroleptic-induced catalepsy: Modification by lesions in the caudate putamen. *Neuropharmacology* **10,** 297–306.

Costall B., Naylor R. J., and Wright T. (1972) The use of amphetamine induced stereotyped behaviour as a model for the experimental evaluation of antiparkinson agents. *Arzneimittel forsch.* **22,** 1178–1183.

Coyle J. T., Ferkany J. W., and Zaczek R. (1983) Kainic acid: Insights from a neurotoxin into the pathophysiology of Huntington's disease. *Neurobehav. Toxicol. Teratol.* **5,** 617–624.

Coyle J. T., McGeer E. G., McGeer P. L., and Schwarcz R. (1978) Neostriatal Injections: A Model for Huntington's Chorea, in *Kainic Acid as a Tool in Neurobiology* (McGeer E. G., Olney J. W., and McGeer P. L., eds.) Raven, New York.

Coyle J. T., Singer H., McKinney M., and Price D. (1984) Neurotransmitter specific alterations in dementing disorders: Insights from animal models. *J. Psychiat. Res.* **18,** 501–512.

Crane G. E. (1973) Persistant dyskinesia. *Br. J. Psychiat.* **122,** 395–405.

Crawley J. N., Ninan P. T., Pickar D., Chrousos G. P., Skolnick P., and Paul S. M. (1984) Behavioural and electrophysiological responses to benzodiazepine receptor antagonists. *Psychopharmacol. Bull.* **20,** 403–407.

Creese I., Burt D. R., and Snyder S. H. (1977) Dopamine receptor binding enhancement accompanies lesion-induced behavioural supersensitivity. *Science* **197,** 596–598.

Crow T. J., Baker H. F., Cross A. J., Joseph M. H., Lofthouse R., Longden A., Owen F., Riley G. J., Glover V., and Killpack W. S. (1979) Monoamine mechanisms in chronic schizophrenia: Post mortem neurochemical findings. *Br. J. Psychiat.* **134,** 249–256.

D'Amato R. J., Lipman Z. P., and Snyder S. H. (1986) Selectivity of the Parkinsonian neurotoxin MPTP: Toxic metabolite MPP+ binds to neuromelanin. *Science* **231,** 987–989.

Deckel A. W., Robinson R. G., Coyle J. T., and Sanberg P. R. (1983) Reversal of long-term locomotor abnormalities in the kainic acid model of Huntington's disease by day 18 fetal striatal implants. *Eur. J. Pharmacol.* **93,** 287–288.

De Wied D., (1984) The importance of vasopressin in memory. *Trends Neurosci.* **7,** 62–64 (for discussion) and 64–66 (for reply).

De Wied D. and Jolles J. (1982) Neuropeptides derived from pro-opiocortin: Behavioural, physiological and neurochemical effects. *Physiol. Rev.* **62,** 976–1059.

Dorow R., Horowski R., Paschelke G., Amin M., and Braestrup C. (1983) Severe anxiety induced by FG 7142, a beta-carboline ligand for benzodiazepine receptors. *Lancet* **2,** 98–99.

Dourish C. T. (1981) Behavioural effects of acute and chronic beta-phenylethylamine administration in the rat: Evidence for the involvement of 5-hydroxytryptamine. *Neuropharmacology* **20,** 1067–1072.

Dourish C. T. (1982) A pharmacological analysis of the hyperactivity syndrome induced by beta-phenylethylamine in the mouse. *Br. J. Pharmacol.* **77,** 129–139.

Drachman D. A. (1977) Memory and cognitive function in man: Does the cholinergic system have a specific role? *Neurology* **27,** 783–790.

Drugan R. C., Maier S. F., Skolnick P., Paul S. M., and Crawley J. N. (1985) An anxiogenic benzodiazepine receptor ligand induces learned helplessness. *Eur. J. Pharmacol.* **113,** 453–457.

Dunstan R. and Jackson D. M. (1977) The effect of apomorphine and clonidine on locomotor activity in mice after long term treatment with haloperidol. *Clin. Exp. Physiol. Pharmacol.* **4,** 131–141.

Duvoisin R. (1972) Clinical diagnosis of the dyskinesias. *Med. Clin. North Am.* **56,** 1321–1341.

Ellison G. D. and Eison M. S. (1983) Continuous amphetamine intoxication: An animal model of the acute psychotic episode. *Psychol. Med.* **13,** 751–761.

Ellison G. and Morris W. (1981) Opposed changes of continuous amphetamine administration: Parallel alterations in motor stereotypies and in vivo spiroperidol accumulation. *Eur. J. Pharmacol.* **74,** 207–214.

Engel J., Liljequist S., and Johannessen K. (1976) Behavioural Effects of Long-Term Treatment with Antipsychotic Drugs, in *Antipsychotic Drugs, Pharmacodynamics and Pharmacokinetics* (Sedvall G., Uvnas B., and Zotterman G. Y., eds.) Pergamon. Oxford.

Ettigi P., Nair N. P. V., Lal S., and Cervantes P. (1976) Effect of apomorphine on growth hormone and prolactin secretion in schizophrenic patients with or without oral dyskinesia, withdrawn from chronic neuroleptic therapy. *J. Neurol. Neurosurg. Psychiat.* **39,** 870–876.

Everitt B. J. and Keverne E. B. (1979) Models of Depression Based on Behavioural Observations of Experimental Animals, in *Psychopharmacology of Affective Disorders* (Paykell E. S. and Coppen A., eds.) Oxford University Press, Oxford.

Fahn S. (1983) Treatment of tardive dyskinesia: Use of dopamine-depleting agents. *Clin. Neuropharmacol.* **6,** 151–158.

Faurbye A., Rasch P. J., Peterson P. B., Brandborg G., and Pakkenberg H. (1964) Neurological symptoms in pharmacotherapy of psychoses. *Acta Psychiatr. Scand.* **40,** 10–27.

Fibiger H. C. (1978) Kainic Acid Lesions of the Striatum: A Pharmacological and Behavioural Model of Huntington's Disease, in *Kainic Acid as a Tool in Neurobiology* (McGeer E. G., Olney J. W., and McGeer P. L., eds.) Raven, New York.

Fibiger H. C. and Lloyd K. G. (1984) Neurobiological substrates of tardive dyskinesia: The GABA hypothesis. *Trends Neurosci.* **12,** 462–464.

File S. E. and Lister R. G. (1983) Interactions of ethyl-beta-carboline-3-carboxylate and Ro15-1788 with CGS 8216 in an animal model of anxiety. *Neurosci. Lett.* **39,** 91–94.

Fisher A., Mantione C. R., Abraham D. J., and Hanin I. (1982a) Long-term central cholinergic hypofunction induced in mice by ethylcholine aziridinium ion (AF64A) in vivo. *J. Pharmacol. Exp. Ther.* **222,** 140–145.

Fisher A., Mantione C. R., Grauer E., Levy A., and Hanin I. (1982b) Manipulation of Brain Cholinergic Mechanisms by Ethylcholineaziridinium Ion (AF64A), A Promising Animal Model for Alzheimer's Disease, in *Behavioural Models and the Analysis of Drug Action* (Spiegelstein M. Y. and Levy A., eds.) Elsevier, Amsterdam.

Freed W. J., Gillin J. C., and Wyatt R. J. (1980) Anomalous behavioural response to imidazoleacetic acid a GABA agonist, in animals treated chronically with haloperidol. *Biol. Psychiat.* **15,** 21–35.

Friedman E., Gianutsos G., and Kuster J. (1983a) Chronic fluphenazine and clozapine elicit opposite changes in brain muscarinic receptor binding: Implications for understanding tardive dyskinesia. *J. Pharmacol. Exp. Ther.* **226,** 7–12.

Friedman E., Lerer B., and Kuster J. (1983b) Loss of cholinergic neurons in the rat neocortex produces deficits in passive avoidance learning. *Pharmacol. Biochem. Behav.* **19,** 309–312.

Fuller R. W. and Hemrick-Luecke S. (1980) Long-lasting depletion of striatal dopamine by a single injection of amphetamine in iprindole-treated rat. *Science* **209,** 395–397.

Fuller R. W. and Hemrick-Luecke S. K. (1986) Depletion of norepinephrine in mouse heart by l-methyl-4-phenyl-1,2,3,6-tetrahydropyridine (MPTP) mimicked by l-methyl-4-phenylpyridinium (MPP+) and not blocked by deprenyl. *Life Sci.* **39,** 1645–1650.

Fuxe K., Fredholm B. B., Agnati L. F., Ogren S.-O., Everitt B. J., Jonsson G., and Gustafsson J.-A. (1978) Interaction of ergot drugs with central monoamine systems—evidence for a high potential in the treatment of mental and neurological disorders. *Pharmacology* **16** (suppl.1), 99–134.

Fuxe K., Nystrom M., Tovi M., Smith R., and Ogren S.-O. (1974) Central catecholamine neurons, behaviour and neuroleptic drugs: An analysis to understand the involvement of catecholamines in schizophrenia. *J. Psychiat. Res.* **11,** 151–161.

Garnett E. S., Nahmias C., and Firnau G. (1984) Central dopaminergic pathways in hemiparkinsonism examined by positron emission tomography. *Can. J. Neurol. Sci.* **11,** 174–179.

Gerstenbrand F., Pateisky K., and Prosenz P. (1963) Erfahrungen mit L-DOPA in der therapie des parkinsonismus. *Psychiat. Neurol.* **146,** 246–261.

Gerlach J., Reisby N., and Randrup A. (1974) Dopaminergic hypersensitivity and cholinergic hypofunction in the pathophysiology of tardive dyskinesia. *Psychopharmacology* **34,** 21–35.

Gianutsos G., and Lal H. (1976) Alteration in the action of cholinergic and anticholinergic drugs after chronic haloperidol: Indirect evidence for cholinergic hyposensitivity. *Life Sci.* **18,** 515–520.

Gianutsos G., Hynes M. D., and Lal H. (1975) Enhancement of apomorphine-induced inhibition of striatal dopamine-turnover following chronic haloperidol. *Biochem. Pharmacol.* **24,** 581–582.

Gillin J. C., Kaplan J., Stillman R., and Wyatt R. J. (1976) The psychodelic model of schizophrenia—the case of *N,N*-dimethyl tryptamine. *Am. J. Psychiat* **133,** 203–208.

Glass G. S. and Bowers M. B. (1970) Chronic psychosis associated with long-term psychotomimetic drug abuse. *Arch. Gen. Psychiat.* **23,** 97–103.

Gnegy M. E., Lucchelli A., and Costa E. (1977) Correlation between drug-induced supersensitivity of dopamine dependent striatal mechanisms and the increase in striatal content of the Ca^{++} regulated protein activator of c-AMP phosphodiesterases. *Naunyn Schmiedebergs Arch. Pharmacol.* **301,** 121–127.

Goldstein M., Battista A. F., and Miyamoto T. (1975) Modification of Involuntary Movements by Centrally Acting Drugs, in *Advances in Neurology* vol. 9 (Calne D. B., Chase T. N., and Barbeau A., eds.) Raven, New York.

Goldstein M., Battista A. F., Ohmoto T., Anagoste B., and Fuxe K. (1973) Tremor and involuntary movements in monkeys: Effect of L-DOPA and of a dopamine receptor stimulating agent. *Science* **179,** 816–817.

Govoni S., Hong J. S., Wang H.-Y. T., and Costa E. (1980) Increase of neurotensin content elicited by neuroleptics in nucleus accumbens. *J. Pharmacol. Exp. Ther.* **215,** 413–417.

Guidotti A., Forchetti C. M., Corda M. G., Konkel D., Bennett C. D., and Costa E. (1983) Isolation, characterization and purification to homogeneity of an endogenous polypeptide with agonist action on benzodiazepine receptors. *Proc. Natl. Acad. Sci. U.S.A.* **80,** 3531–3535.

Gunne L.-M. and Barany S. (1976) Haloperidol-induced tardive dyskinesia in monkeys. *Psychopharmacology* **50,** 237–240.

Gunne L.-M. and Haggstrom J.-E. (1985) Experimental tardive dyskinesia. *J. Clin. Psychiat.* **46,** 48–49.

Gupta M., Felten S. Y., and Felten D. L. (1985) MPTP alters both mesolimbic and nigrostriatal dopaminergic systems in mice. *Soc. Neurosci. Abst.* **11,** 427.

Guttmacher L. B., Murphy D. L., and Insel T. R. (1983) Pharmacologic models of anxiety. *Comp. Psychiat.* **24,** 312–326.

Haggstrom J.-E., Gunne L. M., Carlsson A., and Wikstrom H. (1983) Antidyskinetic action of 3-PPP, a selective dopaminergic autoreceptor agonist in Cebus monkeys with persistent neuroleptic-induced dyskinesias. *J. Neural Transm.* **58,** 135–142.

Halbhubner K., Herken H., and Loos D. (1978) Experimental neuropathy with Parkinson-like muscular rigidity. *Arzneimittel forsch.* **28,** 1743–1752.

Hartgraves S. L. and Kelly P. H. (1983) Inhibition of drug-induced circling by GABA-ergic activity in the nucleus accumbens. *Brain Res. Bull.* **11,** 315–320.

Heikkila R. E., Manzino L., Cabbat F. S., and Duvoisin R. C. (1984a) Protection against the dopaminergic neurotoxicity of 1-methyl-4-phenyl-1,2,5,6-tetrahydropyridine by monoamine oxidase inhibitors. *Nature* **311,** 467–469.

Heikkila R. E., Hess A., and Duvoisin R. C. (1984b) Dopaminergic neurotoxicity of 1-methyl-4-phenyl-1,2,5,6-tetrahydropyridine in mice. *Science* **224,** 1451–1453.

Heikkila R. E., Hess A., and Duvoisin R. C. (1985) Dopaminergic neurotoxicity of 1-methyl-4-phenyl-1,2,5,6-tetrahydropyridine (MPTP) in the mouse: Relationships between monoamine oxidase, MPTP metabolism and neurotoxicity. *Life Sci.* **36,** 231–236.

Hinzen D., Hornykiewicz O., Kobinger W., Pichler L., Pifl C., and Schingnitz C. (1986) The dopamine autoreceptor agonist B-HT920 stimulates denervated postsynaptic brain DA receptors in rodent and primate models of Parkinson's disease: A novel approach to treatment. *Eur. J. Pharmacol.* **131,** 75–86.

Hjorth S., Carlsson A., Wikstrom H., Lundberg P., Sanchez D., Hacksell U., Arvidsson L. E., Svensson U., and Nilsson J. L. G. (1983) 3-PPP. A new centrally acting dopamine-receptor agonist with selectivity for autoreceptors. *Life Sci.* **28,** 1225–1238.

Hornykiewicz O. (1982) Brain catecholamines in schizophrenia—a good case for noradrenaline. *Nature* **299,** 484–486.

Horrobin D. F. (1979) Schizophrenia: Reconciliation of the dopamine, prostaglandin, and opioid concepts and the role of the pineal. *Lancet* **1,** 529–531.

Hruska R. E. and Silbergeld E. K. (1979) Abnormal locomotion in rats after bilateral intrastriatal injection of kainic acid. *Life Sci.* **25,** 181–193.

Hyttel J. (1974) Endogenous levels and turnover of catecholamines in mouse brain after repeated administration of haloperidol. *Psychopharmacology* **36,** 237–241.

Hyttel J. (1986) Effect of prolonged treatment with neuroleptics on dopamine D-1 and D-2 receptor density in corpus striatum of mice. *Acta Pharmacol. Toxicol.* **59,** 387–391.

Iwamoto E. T. and Way E. L. (1977) Circling behaviour and stereotypy induced by intranigral opiate microinjection. *J. Pharmacol. Exp. Ther.* **203,** 347–359.

Jackson D. M. (1972) The effect of beta-phenethylamine upon spontaneous motor activity in mice: A dual effect on locomotor activity. *J. Pharm. Pharmacol.* **24,** 383–389.

Jackson D. M. (1978) Beta-phenylethylamine: Studies on the Mechanism of Its Stimulant Effects, in *Non Catechol Phenylethylamines* part 1 *Phenylethylamine: Biological Mechanisms and Clinical Aspects* (Mosnaim A. D. and Wolf M. E., eds.) Dekker, New York.

Jackson A. and Crossman A. R. (1984) Experimental choreoathetosis produced by injection of a gamma-aminobutyric acid antagonist into the lentiform nucelus in the monkey. *Neurosci. Lett.* **46,** 41–45.

Jackson D. M. and Hashizume M. (1986) Bromocriptine induces marked locomotor stimulation in dopamine-depleted mice when DI dopamine receptors are stimulated with SKF38393. *Psychopharmacology* **90,** 147–149.

Jackson D. M. and Jenkins O. F. (1985) Hypothesis: Bromocriptine lacks intrinsic dopamine receptor stimulating properties. *J. Neurol. Transm.* **62,** 219–230.

Jackson D. M., Anden N.-E., and Dahlstrom A. (1975a) A functional effect of dopamine in the nucleus accumbens and in some other dopamine rich parts of the rat brain. *Psychopharmacology* **45,** 139–149.

Jackson D. M., Anden N.-E., Engel J., and Liljequist S. (1975b) The effect of long-term penfluridol treatment on the sensitivity of the dopamine receptors in the nucleus accumbens and in the corpus striatum. *Psychopharmacology* **45,** 151–155.

Jackson D. M., Bailey R. C., Christie M. J., Crisp E. A., and Skerritt J. H. (1981) Long-term *d*-amphetamine in rats: Lack of change in post-synaptic dopamine receptor sensitivity. *Psychopharmacology* **73,** 276–280.

Jackson D. M., Dunstan R., and Perrington A. (1979) The hyperkinetic syndrome following long term haloperidol treatment: Involvement of dopamine and noradrenaline. *J. Neural Transm.* **44,** 175–186.

Jackson D. M., Jenkins O. F., Malor R., Christie M. J., and Gregory P. (1983) Chronic L-DOPA treatment of rats and mice does not change the sensitivity of post-synaptic dopamine receptors. *Naunyn Schmiedebergs Arch. Pharmacol.* **324,** 271–274.

James T. A. and Starr M. S. (1979) Effects of substance P injected into the substantia nigra. *Br. J. Pharmacol.* **65,** 423–429.

Jarvis M. F. and Wagner G. C. (1985) Neurochemical and functional consequences following 1-methyl-4-phenyl-1,2,5,6-tetrahydropyridine (MPTP) and methamphetamine *Life Sci.* **36,** 249–254.

Jenner P. and Marsden C. D. (1988) Neuronal compensation and age as factors affecting MPTP-induced parkinsonism in the common marmoset, in *Pharmacology and Functional Regulation of Dopamine Neurons* (Beart P. B., Woodruff G. N., and Jackson D. M., eds.) Macmillan, London, in press.

Johannessen J. N., Chiueh C. C., Burns R. S., and Markey S. P. (1985) Differences in the metabolism of MPTP in the rodent and primate parallel differences in sensitivity to its neurotoxic effects. *Life Sci.* **36,** 219–224.

Johnels B., Steg G., and Ungerstedt U. (1978) A method for mechanographical recording of muscle tone in the rat: The effect of some antiparkinsonian drugs on rigidity induced by reserpine. *Brain Res.* **140,** 177–181.

Johnson A. M., Loew D. M., and Vigouret J. M. (1976) Stimulant properties of bromocriptine on central dopamine receptors in comparison to apomorphine, (+) amphetamine and L-DOPA. *Br. J. Pharmacol.* **56,** 59–68.

Johnston M. V., McKinney M., and Coyle J. T. (1979) Evidence for a cholinergic projection to neocortex from neurons in basal forebrain. *Proc. Natl. Acad Sci. USA* **76,** 5392–5396.

Jonsson L.-E., Anggard E., and Gunne L.-M. (1971) Blockade of intravenous amphetamine euphoria in man. *Clin. Pharmacol. Ther.* **12,** 889–896.

Jurna I. (1976) Striatal monoamines and reserpine and chlorpromazine rigidity. *Pharmacol. Ther.* **2,** 113–128.

Jurna I., Brenner M., and Drum P. (1978) Abolition of spinal motor disturbance by injections of dopamine receptor agonists, atropine and GABA into the caudate nucleus. *Neuropharmacology* **17,** 35–44.

Jurna I., Grossmann W., and Nell T. (1972) Depression by amantadine of drug-induced rigidity in the rat. *Neuropharmacology* **11,** 559–564.

Karoum F., Nasrallah H., Potkin S., Chuang L., Moyer-Schwing J., Phillips I., and Wyatt R. J. (1979) Mass fragmentography of phenylethylamine, *m*- and *p*-tyramine and related amines in plasma, cerebrospinal fluid, urine, and brain. *J. Neurochem.* **33,** 201–212.

Kazamatsuri H., Chen C., and Cole J. O. (1972) Treatment of tardive dyskinesia. I. Clinical efficacy of a dopamine-depleting agent, tetrabenazine. *Arch. Gen. Psychiat.* **27,** 95–99.

Kazamatsuri H., Chen C.-P., and Cole J. O. (1973) Long-term treatment of tardive dyskinesia with haloperidol and tetrabenazine. *Am. J. Psychiat.* **130,** 479–483.

Kelly P. H. and Moore K. E. (1977) Mesolimbic dopamine neurons: Effects of 6-hydroxydopamine-induced destruction and receptor blockade on drug-induced rotation in rats. *Psychopharmacology* **55,** 35–41.

Klawans H. L. (1973a) The pharmacology of tardive dyskinesias. *Am. J. Psychiat.* **130,** 82–86.

Klawans H. L. (1973b) The Pharmacology of Extrapyramidal Movement Disorders, in *Monographs in Neural Science* Karger, Basel.

Klawans H. L., Goetz C., Nausieda P. A., and Weiner W. J. (1977b) Levo-dopa induced dopamine receptor hypersensitivity. *Trans. Am. Neurol. Assoc.* **102,** 80–83.

Klawans H. L., Paulson G. W., Ringel S. P., and Barbeau A. (1972) Use of L-DOPA in the detection of presymptomatic Huntington's chorea. *N. Engl. J. Med.* **286,** 1332–1334.

Kokkinidis L. and Anisman H. (1981) Amphetamine psychosis and schizophrenia: A dual model. *Neurosci. Biobehav. Rev.* **5,** 449–461.

Kramer J. C., Fischman V. S., and Littlefeld D. C. (1967) Amphetamine abuse. Pattern and effects of high doses taken intravenously. *J. Am. Med. Assoc.* **201,** 305–309.

Ksir C. and Benson D. M. (1983) Enhanced behavioural response to nicotine in an animal model of Alzheimer's disease. *Psychopharmacology* **81,** 272–273.

Lal H. and Emmett-Ogleby M. W. (1983) Behavioural analogues of anxiety-animal models. *Neuropharmacology* **22,** 1423–1441.

Lal H. and Shearman G. T. (1980) Interoceptive discriminative stimuli in the development of CNS drugs and a case of an animal model of anxiety. *Ann. Rep. Med. Chem.* **15,** 51–58.

Lal S., de la Vega C. E., Garelis E., and Sourkes T. L. (1973) Apomorphine, pimozide, L-DOPA and the probenecid test in Huntington's chorea. *Psychiat. Neurol. Neurochir.* **76,** 113–117.

Langston J. W. (1985a) MPTP and parkinson's disease. *Trends Neurosci.* **8,** 79–83.

Langston J. W. (1985b) MPTP neurotoxicity: An overview and characterization of phases of toxicity. *Life Sci.* **36,** 201–206.

Langston J. W. and Ballard P. A. (1983) Parkinson's disease in a chemist working with 1-methyl-4-phenyl-1,2,5,6-tetrahydropyridine. *N. Engl. J. Med.* **5,** 309–310.

Langston J. W., Ballard P., Tetrud J. W., and Kopin I. J. (1983) Chronic parkinsonism in humans due to a product of meperidine-analog synthesis. *Science* **219,** 979–980.

Lehmann J., Nagy J. I., Atmadya S., and Fibiger H. C. (1980) The nucleus basalis magocellularis: The origon of a cholinergic projection to the neocortex of the rat. *Neuroscience* **5,** 1161–1174.

Leith N. J. and Barrett R. J. (1976) Amphetamine and the reward system: Evidence for tolerance and post-drug depression. *Psychopharmacology* **46,** 19–25.

Lloyd K. G., Hornykiewicz O., Davidson L., Shannak K., Farley I., Goldstein M., Shibuya M., Kelley W. N., and Fox I. H. (1981) Biochemical evidence of dysfunction of brain neurotransmitters in the Lesch-Nyhan syndrome. *N. Engl. J. Med.* **305,** 1106–1111.

Magour S., Coper H., and Fahndrich C. G. (1974) The effects of chronic treatment with *d*-amphetamine on food intake, body weight, locomotor activity and subcellular distribution of the drug in rat brain. *Psychopharmacology* **34,** 45–54.

Mantione C. R., Fisher A., and Hanin I. (1984) iii Possible mechanisms involved in the presynaptic cholinotoxicity due to ethylcholine aziridinium (AF64A) in vivo. *Life Sci.* **35,** 33–41.

Manyam B. V. (1982) Low CSF gamma-aminobutyric acid levels in Parkinson's disease. Effect of levodopa and carbidopa. *Arch. Neurol.* **39,** 391–392.

Marco E., Mao C. C., Revuelta A., Peralta E., and Costa E. (1978) Turnover rates of gamma-aminobutyric acid in substantia nigra, N. caudatus, globus pallidus and N. accumbens of rats injected with cataleptic and non-cataleptic antipsychotics. *Neuropharmacology* **17,** 589–596.

Markey S. P., Johannessen J. N., Chiueh C. C., Burns R. S., and Herkenham M. A. (1984) Intraneuronal generation of a pyridinium metabolite may cause drug-induced parkinsonism. *Nature* **311,** 464–467.

Marshall A. M. and Mishra R. K. (1980) Dopamine-Sensitive Adenylate Cyclase and Dopamine/Neuroleptic Receptor Binding: Effect of Neuroleptic Drugs, in *Long-Term Effects of Neuroleptics* vol. 24 (Cattabeni F., Racagni G., Spano P. F., and Costa E., eds.) Raven, New York.

Mason S. T. and Fibiger H. C. (1979) Kainic acid lesions of the striatum in rats mimic the spontaneous motor abnormalities of Huntington's disease. *Neuropharmacology* **18,** 403–407.

Mason S. T. and Iversen S. D. (1979) Theories of the dorsal bundle extinction effect. *Brain Res.* **180,** 107–137.

Mason S. T. and Lin D. (1980) Dorsal noradrenergic bundle and selective attention in the rat. *J. Comp. Physiol. Psychol.* **94,** 819–832.

Mason S. T., Sanberg P. R., and Fibiger H. C. (1978) Kainic acid lesions of the striatum dissociate amphetamine and apomorphine stereotypy: Similarities to Huntington's disease. *Science* **201,** 352–355.

McGeer E. G. and McGeer P. L. (1976) Duplication of biochemical changes of Huntington's chorea by intrastriatal injections of glutamic and kainic acids. *Nature* **263,** 517–519.

McGeer P. L., McGeer E. G., and Hattori T. (1978a) Kainic Acid as a Tool in Neurobiology, in *Kainic Acid as a Tool in Neurobiology* (McGeer E. G., Olney J. W., and McGeer P. L., eds.) Raven, New York.

McGeer E. G., Olney J. W., and McGeer P. L. (1978b) *Kainic Acid as a Tool in Neurobiology,* Raven, New York.

Minana M. D., Portoles M., Jorda A., and Grisolia S. (1984) Lesch-Nyhan syndrome, caffeine model: Increase of purine and pyrimidine enzymes in rat brain. *J. Neurochem.* **43,** 1556–1560.

Mitchell I. J., Cross A. J., Sambrook M. A., and Crossman A. R. (1985) Sites of the neurotoxic action of 1-methyl-4-phenyl-1,2,3,6-tetrahydropyridine in the macaque monkey include the ventral tegmental area and the locus coeruleus. *Neurosci. Lett.* **61,** 195–200.

Mohler H. and Okada T. (1977) Benzodiazepine receptor: Demonstration in the central nervous system. *Science* **198,** 849–851.

Mohler H., Polc P., Cumin R., Pieri L., and Kettler R. (1979) Nicotinamide is a brain constituent with benzodiazepine-like actions. *Nature* **278,** 563–565.

Montanaro N., Dall'olio R., Gandolfi O., and Vaccheri A. (1982) Differential enhancement of behavioural sensitivity to apomorphine following chronic treatment of rats with (–) sulpiride and haloperidol. *Eur. J. Pharmacol.* **81,** 1–9.

Morpurgo C. (1962) Effects of anti-parkinson drugs on a phenothiazine induced catatonic reaction. *Arch. Int. Pharmacodyn. Ther.* **137,** 84–90.

Mosnaim A. D., Inwang E. E., Sugerman J. H., and Sabelli H. C. (1973) Identification of 2-phenylethylamine in human urine by infrared and mass spectroscopy and its quantification in normal subjects and cardiovascular patients. *Clin. Chim. Acta* **46,** 407–413.

Muller P. and Seeman P. (1977) Brain neurotransmitter receptors after long-term haloperidol: Dopamine, acetylcholine, serotonin, alpha noradrenergic and naloxone receptors. *Life Sci.* **21,** 1751–1758.

Muller P. and Seeman P. (1979) Presynaptic subsensitivity as a possible basis for sensitization by long-term dopamine mimetics. *Eur. J. Pharmacol.* **55,** 149–157.

Murphy D. L. and Wyatt R. J. (1972) Reduced monoamine oxidase activity in blood platelets from schizophrenic subjects. *Nature* **238,** 225–226.

Murray R. M., Oon M. C. H., Rodnight R., Birley J. L. T., and Smith A. (1979) Increased excretion of dimethyltryptamine and certain features of psychosis: A possible association. *Arch. Gen. Psychiat.* **36,** 644–649.

Neale R., Gerhardt S., and Liebman J. M. (1984) Effect of dopamine agonists, catecholamine depletors and cholinergic and GABAergic drugs on acute dyskinesias in squirrel monkeys. *Psychopharmacology* **82,** 20–26.

Nelson L. R. and Ellison G. (1979) Enhanced stereotypies after repeated injections but not continuous amphetamines. *Neuropharmacology* **17,** 1041–1084.

Nielsen E. B. and Lyon M. (1978) Evidence for cell loss in corpus striatum after long term treatment with a neuroleptic drug (flupenthixol) in rats. *Psychopharmacology* **59,** 85–89.

Ninan P. T., Insel T. M., Cohen R. M., Cook J. M., Skolnick P., and Paul S. M. (1982) Benzodiazepine receptor mediated experimental "anxiety" in primates. *Science* **218,** 1332–1334.

Owen F., Cross A. J., Waddington J. L., Poulter M., Gamble S. J., and Crow T. J. (1980) Dopamine-mediated behaviour and 3H-spiperone binding to striatal membranes in rats after nine months haloperidol administration. *Life Sci.* **26,** 55–59.

Pakkenberg H., Fog R., and Nilakantan B. (1973) The long-term effect of perphenazine enanthate on the rat brain. Some metabolic and anatomical observations. *Psychopharmacology* **29,** 329–336.

Paul S. M. and Skolnick P. (1978) Rapid changes in brain benzodiazepine receptors after experimental seizures. *Science* **202,** 892–894.

Paulson G. (1973) Dyskinesias in Monkeys in *Advances in Neurology* vol. 1 (Barbeau A., Chase T. N., and Paulson G. W., eds.) Raven, New York.

Pepeu G., Gori G., and Bartholini L. (1980) Pharmacologic and Therapeutic Perspective on Dementia: An Experimental Approach, in *Aging of the Brain and Dementia* (Armaducci L., Davison A. N., and Antuono P., eds.) Raven, New York.

Peroutka S. J., DeLanney L., Irwin I., Ison P. J., Ricaurte G., Schlegel J. R., and Langston J. W. (1985) MPTP induced dopamine D2 receptor hypersensitivity in the mouse is transient. *Res. Comm. Chem. Path. Pharmacol.* **48,** 163–168.

Philips S. R. and Boulton A. A. (1979) The effect of monoamine oxidase inhibitors on some arylalkylamines in rat striatum. *J. Neurochem.* **33,** 159–167.

Pitts F. N. and Allen R. E. (1979) Biochemical Indicators of Anxiety, in *Phenomenology and Treatment of Anxiety* (Fann W. E., ed.) Spectrum, New York.

Potkin S. G., Karoum F., Chuang L., Cannon-Spoor H. E., Phillips I., and Wyatt R. J. (1979) Phenylethylamine in paranoid chronic schizophrenia. *Science* **206,** 470–471.

Rainey J. M. and Nesse R. M. (1985) Psychobiology of anxiety and anxiety disorders. *Psychiatr. Clin. North Am.* **8,** 133–144.

Reynolds G. P. (1983) Increased concentrations and lateral assymetry of amygdala dopamine in schizophrenia. *Nature* **305,** 527–528.

Reynolds G. P. (1984) Noradrenaline and schizophrenia. *Trends Pharmacol. Sci.* **5,** 138.

Rieke G. K., Scarfe A. D., and Hunter J. F. (1984) L-pyroglutamate: An alternative neurotoxin for a rodent model of Huntington's disease. *Brain Res. Bull.* **13,** 443–456.

Rinne U. K., Koskinen V., Laaksonen H., Lonnberg P., and Sonninen V. (1978) GABA receptor binding in the parkinsonian brain. *Life Sci.* **22,** 2225–2228.

Robbins T. W., Everitt B. J., Fray P. J., Gaskin M., Carli M., and de la Riva C. (1982) The Role of the Central Catecholamines in Attention and Learning, in *Behavioural Models and the Analysis of Drug Action,* Proceedings of the 27th OHOLO conference (Spegelstein M. Y. and Levy A., eds.) Elsevier, Amsterdam.

Rodnight R. (1983) Schizophrenia: Some current neurochemical approaches. *J. Neurochem.* **41,** 12–21.

Roos B.-E. and Steg G. (1964) The effect of L-3,4-dihydroxyphenylalanine and D,L-5-hydroxytryptophan on rigidity and tremor induced by reserpine, chlorpromazine and phenoxybenzamine. *Life Sci.* **3,** 351–360.

Rotrosen J., Friedman E., and Gershon S. (1975) Striatal adenylate cyclase activity following reserpine and chronic chlorpromazine administration. *Life Sci.* **17,** 563–568.

Rupniak N., Hong M., Mansfield S., Fleminger S., Dyer R., Jenner P., and Marsden C. D. (1983) Elevation of circulating prolactin concentrations may not cause striatal dopamine receptor supersensitivity. *Eur. J. Pharmacol.* **93,** 195–200.

Russell R. W. (1964) Extrapolation from Animals to Man, in *Animal Behaviour and Drug Action* (Steinberg H., ed.) Churchill, London.

Rylander G. (1972) Psychoses and the punding and choreiform syndromes in addiction to central stimulant drugs. *Psychiat. Neurol. Neurochir.* **75,** 203–212.

Sabol K. E., Neill D. B., Wages S. A., Church W. H., and Justice J. B. (1985) Dopamine depletion in a striatal subregion disrupts performance of a skilled motor task in the rat. *Brain Res.* **335,** 33–43.

Salamone J. D., Beart P. M., Alpert J. E., and Iversen S. D. (1984) Impairment in T-maze reinforced alternation performance following nucleus basalis magnocellularis lesions in rats. *Behav. Brain Res.* **13,** 63–70.

Sanberg P. R. and Johnston G. A. R. (1981) Glutamate and Huntington's disease. *Med. J. Aust.* **2,** 460–465.

Sanberg P. R., Pisa M., and Fibiger H. C. (1979) Avoidance, operant and locomotor behaviour in rats with neostriatal injections of kainic acid. *Pharmacol. Biochem. Behav.* **10,** 137–144.

Sandberg K., Hanin I., Fisher A., and Coyle J. T. (1984a) Selective cholinergic neurotoxin: AF64A's effects in rat striatum. *Brain Res.* **293,** 49–55.

Sandberg K., Sanberg P. R., and Coyle J. T. (1984b) Effects of intrastriatal injections of the cholinergic neurotoxin AF64A on spontaneous nocturnal locomotor activity in the rat. *Brain Res.* **299,** 339–343.

Sandberg K., Sanberg P. R., Hanin I., Fisher A., and Coyle J. T. (1984c) Cholinergic lesion of the striatum impairs acquisition and retention of a passive avoidance response. *Behav. Neurosci.* **98,** 162–165.

Sandberg K., Schnaar R. L., McKinney M., Hanin I., Fisher A., and Coyle J. T. (1985) AF64A: An active site directed irreversible inhibitor of choline acetyltransferase. *J. Neurochem.* **44,** 439–445.

Sandler M. and Reynolds G. P. (1976) Does phenylethylamine cause schizophrenia? *Lancet* **1,** 70–71.

Sassin J. F. (1975) Drug-Induced Dyskinesia in Monkeys, in *Advances in Neurology* vol 10 (Meldrum B. S. and Marsden C. D., eds.) Raven, New York.

Sassin J., Taub S., and Weitzman E. (1971) Abnormal behaviour patterns produced in normal monkeys by L-DOPA. *Neurology* **21,** 403.

Sato S., Daly R., and Peters H. (1971) Reserpine therapy of phenothiazine induced dyskinesia. *Dis. Nerv. Sys.* **32,** 680–685.

Sayers A. C., Burki H. R., Ruch W., and Asper H. (1975) Neuroleptic-induced hypersensitivity of striatal dopamine receptors in the rat as a model of tardive dyskinesias. Effects of clozapine, haloperidol, loxapine and chlorpromazine. *Psychopharmacology* **41,** 97–104.

Scatton B. (1977) Differential regional development of tolerance to increase in dopamine turnover upon repeated neuroleptic administration. *Eur. J. Pharmacol.* **46,** 363–369.

Scatton B., Javoy-Agid F., Rouquier L., Dubois B., and Agid Y. (1983) Reduction of cortical dopamine, noradrenaline, serotonin and their metabolites in Parkinson's disease. *Brain Res.* **275,** 321–328.

Scheel-Kruger J. (1971) Comparative studies of various amphetamine analogues demonstrating different interactions with the metabolism of the catecholamines in the brain. *Eur. J. Pharmacol.* **14,** 47–59.

Schneider J. S. (1984) Role of the basal ganglia in a chemically induced dyskinesia in rat. *Exp. Neurol.* **84,** 524–532.

Schneider G., Oepen H., and Klapproth A., (1981) The effect of the neuro-lathyrogenic substance beta, beta iminodipropionitrile (IDPN) on some biological parameters in rats and mice. *Gen. Pharmacol.* **12,** 109–114.

Schneider M. B., Murrin L. C., Pfeiffer R. F., and Deupree J. D. (1984) Dopamine receptors: Effects of chronic L-DOPA and bromocriptine treatment in an animal model of Parkinson's disease. *Clin. Neuropharmacol.* **7,** 247–257.

Seeman P. (1980) Brain dopamine receptors. *Pharmacol. Rev.* **32,** 229–313.

Segal D. S. and Mandell A. J. (1974) Long-term administration of *d*-amphetamine: Progressive augmentation of motor activity and stereotypy. *Pharmacol. Biochem. Behav.* **2,** 249–255.

Segal D. S., Weinberger S. B., Cahill J., and McCunney S. J. (1980) Multiple daily amphetamine administration: Behavioural and neurochemical alterations. *Science* **207,** 904–907.

Seiden L. S. and Vosmer G. (1984) Formation of 6-hydroxydopamine in caudate nucleus of the rat brain after a single large dose of methylamphetamine. *Pharmacol. Biochem. Behav.* **21,** 29–31.

Seiden L. S., Fischman M. W., and Schuster C. R. (1975) Long-term methamphetamine induced changes in brain catecholamines in tolerant rhesus monkeys. *Drug Alcohol Depend.* **1,** 215–219.

Siever L., Cohen R., and Pert A. (1981) Assessing pharmacologically induced dopamine receptor sensitivity changes with the Ungerstedt turning model. *Psychopharmacology* **75,** 212–213.

Skerritt J. H. and Johnston G. A. R. (1981) Postnatal development of GABA and kainate binding sites and their endogenous inhibitors in rat brain. *Aust Neurosci Proc.* **1,** 65C.

Skerritt J. H., Trisdikoon P., and Johnston G. A. R. (1981) Increased GABA binding in mouse brain following acute swim stress. *Brain Res.* **215,** 398–403.

Skolnick P., Marangos P. J., Goodwin F. K., Edwards M., and Paul S. M. (1978) Identification of inosine and hypoxanthine as endogenous inhibitors of 3H-diazepam binding in the central nervous system. *Life Sci.* **23,** 1473–1480.

Sloviter R. S., Drust E. G., and Connor J. D. (1978) Evidence that serotonin mediates some behavioural effects of amphetamine. *J. Pharmacol. Exp. Ther.* **206,** 348–352.

Snyder S. H., Banerjee S. P., Yamamura H. I., and Greenberg D. (1974) Drugs, neurotransmitters and schizophrenia. *Science* **184,** 1243–1253.

Soubrie P., Thiebot M., Jobert A., Montastruc J. L., Hery F., and Hamon M. (1980) Decreased convulsant potency of picrotoxin and pentylenetetrazol and increased [3H]-flunitrazepam cortical binding following stressful manipulation in the rat. *Brain Res.* **189,** 505–517.

Squires R. F. and Braestrup C. (1977) Benzodiazepine receptors in rat brain. *Nature* **266,** 732–734.

Steg G. (1964) Efferent muscle innervation and rigidity. *Acta Physiol Scand.* **61,** (suppl. 225), 1–53.

Stein L. (1964) Self stimulation of the brain and the central stimulant action of amphetamine. *Fed. Proc.* **23,** 836–850.

Stein L. and Wise C. D. (1971) Possible etiology of schizophrenia: Progressive damage to the noradrenergic reward system by 6-hydroxydopamine. *Science* **171,** 1032–1036.

Steranka L. R., Polite L. N., Perry K. W., and Fuller R. W. (1983) Dopamine depletion in rat brain by MPTP. *Res. Comm. Subs. Abuse* **4,** 315–323.

Stock G., Steinbrenner J., and Kummer P. (1980) Supersensitivity of dopamine-autoreceptors. The effect of gamma butyrolactone in long-term haloperidol treated rats. *J. Neural Transm.* **47,** 145–151.

Strombom U. (1976) On the functional role of pre- and postsynaptic catecholamine receptors in brain. *Acta Physiol. Scand.* **431,** (suppl.), 1–43.

Sundstrom E. and Jonsson G. (1985) Pharmacological interference with the neurotoxic action of 1-methyl-4-phenyl-1,2,5,6-tetrahydropyridine (MPTP) on central catecholamine neurons in the mouse. *Eur. J. Pharmacol.* **110,** 293–299.

Suomi S. J. (1983) Models of depression in primates. *Psychol. Med.* **13,** 465–468.

Tamminga C. A., Smith R. C., Pandey G., Frohman L. A., and Davis J. M. (1977) A neuroendocrine study of supersensitivity in tardive dyskinesia. *Arch. Gen. Psychiat.* **34,** 1199–1203.

Taquet H., Javoy-Agid F., Hamon M., Legrand J. C., Agid Y., and Cesselin F. (1983) Parkinson's disease affects differently met5- and leu5 enkephalin in the human brain. *Brain Res.* **280,** 379–382.

Tarsy D. and Baldessarini R. J. (1973) Pharmacologically induced behavioural supersensitivity to apomorphine. *Nature* **245,** 262–263.

Tarsy D. and Baldessarini R. J. (1976) The tardive dyskinesia syndrome. *Clin. Neuropharmacol.* **1,** 29–61.

Telner J. J. and Singhal R. L. (1984) The learned helplessness model of depression. *J. Psychiat. Res.* **18,** 207–215.

Tolosa E. S. (1974) Paradoxical suppression of chorea by apomorphine. *J. Am. Med. Assoc.* **229,** 1579–1580.

Ungerstedt U. (1971) Postsynaptic supersensitivity after 6-hydroxydopamine induced degeneration of the nigro-striatal dopamine system. *Acta Physiol. Scand.* **367** (suppl.), 69–93.

Wachtel H. (1983) Central Dopaminergic and Antidopaminergic Effects of Ergot Derivatives Structurally Related to Lisuride, in *Lisuride and Other Dopamine Agonists* (Calne D. B., Horowski R., McDonald R. J., and Wuttke W.) Raven, New York.

Waddington J. L. (1985) Further anomalies in the dopamine receptor supersensitivity hypothesis of tardive dyskinesia. *Trends Neurosci.* **8,** 200.

Waddington J. L. and Gamble S. J. (1980) Emergence of apomorphine-induced vacuous chewing during 6 months continuous treatment with fluphenazine decanoate. *Eur. J. Pharmacol.* **68,** 387–388.

Waddington J. L., Cross A. J., Gamble S. J., and Bourne R. C. (1983) Spontaneous orofacial dyskinesia and dopaminergic function in rats after 6 months of neuroleptic treatment. *Science* **220,** 530–532.

Wallace R. A., Boldry R., Schmittgen T., Miller D., and Uretsky N. (1984) Effect of 1-methyl-4-phenyl-1,2,5,6-tetrahydropyridine (MPTP) on monoamine neurotransmitters in mouse brain and heart. *Life Sci.* **35,** 285–291.

Walsh T. J., Tilson H. A., De Haven D. L., Mailman R. B., Fisher A., and Hanin I. (1984) AF64A, a cholinergic neurotoxin, selectively depletes acetylcholine in hippocampus and cortex, and produces long-term passive avoidance and radial-arm deficits in the rat. *Brain Res.* **321,** 91–102.

Weiss B. and Santelli S. (1978) Dyskinesias evoked in monkeys by weekly administration of haloperidol. *Science* **200,** 799–801.

Willner P. (1984) The validity of animal models of depression. *Psychopharmacology* **83,** 1–16.

Wise C. D. and Stein L. (1969) Facilitation of brain self-stimulation by central administration of norepinephrine. *Science* **163,** 299–301.

Wolf M. E. and Mosnaim A. D. (1984) Phenylethylamine and Tardive Dyskinesia, in *Neurobiology of the Trace Amines* (Boulton A. A., Baker G. B., Dewhurst W. G., and Sandler M., eds.) Humana, Clifton, New Jersey.

Yarbrough G. G. (1975) Supersensitivity of caudate neurones after repeated administration of haloperidol. *Eur. J. Pharmacol.* **31,** 367–369.

Determination of Acetylcholine Dynamics

Darwin L. Cheney, John Lehmann, Cristina Cosi, and Paul L. Wood

1. Introduction

During the last 15 years, the methods proposed to study the regulation of cholinergic transmission in vivo have required the labeling of the acetylcholine (ACh) and choline stored in neurons by injecting labeled precursors of ACh, and by measuring simultaneously the changes with time in the specific activities of choline and ACh (Cheney et al., 1974, 1975a; Costa et al. 1975; Dross and Kewitz, 1966; Hanin et al., 1973; Jenden et al., 1974; Racagni et al., 1974; Saelens et al., 1973; Schuberth et al., 1969; Zsilla et al., 1977). In addition to these isotopic methods, various nonisotopic approaches have been suggested to estimate the fractional rate constant of ACh efflux (Bartholini et al., 1976; Celesia and Jasper, 1966; Hemsworth and Neal, 1968; Mitchell, 1960; Pepeu et al., 1975; Smith, 1972; Szerb, 1975; Yaksh and Yamamura, 1975). Some of the nonisotopic methods have used drugs to instantaneously inhibit synthesis, metabolism, or transport of ACh, its precursor, or metabolites (Costa and Neff, 1966). A flaw common to all the methods involving the use of inhibitors derives from the widely spread assumption that drugs are so specific that, even in high doses, they selectively inhibit a given metabolic step. Obviously, such specificity is exceptional; more often, drugs that inhibit ACh metabolism exert collateral actions on other processes that directly or indirectly regulate a number of other metabolic steps. The isotopic methods that use tracer doses of a precursor of ACh are not based on this assumption, but may be plagued by other conceptual flaws. Therefore, they are not devoid of problems; in fact, they include a number of tacit assumptions concerning the dynamics and compartmentation of ACh and its precursors that are open to critique. The assumption implicit in all of the techniques, whether isotopic or nonisotopic, is that at steady-state the newly taken up pool of choline is in rapid equilibrium with the pool of choline that is the precursor of ACh. Moreover, it is assumed that the rates of synthesis and degradation of ACh are in equilibrium, and, therefore, principles of steady-state kinetics can be used to calculate ACh "turnover rate." "Turnover rate" has been defined as the dynamics whereby a tissue constituent is renewed, whereas "turnover" has been classically defined as the continuous renewal of a substance at steady-state (Zilversmit, 1960).

The action of drugs on cholinergic systems in the brain can be inferred by measuring the ACh turnover rate in vivo (Cheney et al., 1974; Racagni et al., 1974), the V_{max} of choline uptake (Simon et al., 1976), or the rate of ACh release using a push–pull cannula (Bartholini et al., 1976). The use of ACh turnover rate measurements to estimate the cholinergic neuronal activity stems from the belief that the turnover rate of ACh increases or decreases when the activity of cholinergic neurons is enhanced or depressed, respectively. Historically, this belief developed as a result of the demonstration that, in cat superior cervical ganglion, the turnover rate of ACh at rest never exceeded 20% of the maximum rate that could be attained during electrical stimulation of preganglionic nerves (MacIntosh, 1963). Since the ACh stored in cat sympathetic ganglia is located in terminal axons, the ganglion was a good model to correlate the neuronal activity with the ACh turnover rate. Unfortunately, the whole brain cannot be used as a model for this correlation because of the intricate network of cholinergic neurons.

It is highly unlikely that drugs would simultaneously change the activity of every cholinergic neuron in the brain in the same direction, unless they act directly on choline acetyltransferase or acetylcholinesterase (AChE). Hence, the nonuniformity and intricacy of the brain cholinergic network make measurements of ACh turnover in the whole brain devoid of any significance in the study of the action of analgesics, neuroleptics, hypnotics, and so on. In contrast, measurements of ACh turnover rates in specific brain nuclei containing small, intrinsic cholinergic neurons or important cholinergic projections are of value in understanding how drugs change the cholinergic function at the level of specific synapses. Furthermore, when such studies are carried out in several nuclei of the brain simultaneously, one might be able to elucidate whether a drug given chronically or acutely acts primarily on cholinergic mechanisms or whether the involvement of cholinergic neurons is secondary to a drug-induced modification of the synaptic input mediated by other transmitters, i.e., serotonin, dopamine, gamma-aminobutyric acid, substance P, and so on. These results allow drug actions to be described in terms of their effects on the integration of different synaptic mechanisms. Thus, a given drug effect may be expressed in terms of the changes in the biochemical correlate of various transmitter systems in selected brain nuclei.

The following is a discussion of various approaches to the pharmacological analysis of cholinergic systems. It will be the specific scientific problem posed that will determine which experimental paradigm will be utilized in any given experiment.

2. Superfusion Methodologies

2.1. General Considerations

For catecholamines, some differences in interpretation have arisen when comparing turnover measurements with neurotransmitter release measured directly with superfusion techniques. When comparing these approaches in the study of cholinergic systems, few differences have arisen. For the monoamines, enzymes capable of catabolizing the neurotransmitters are present within the cytosol of the presynaptic terminal (Fig. 1, top panel). In contrast, it is thought that AChE, the chief enzyme responsible for the catabolism of ACh (Quastel et al., 1936), is not found within the interior of the cholinergic nerve terminal (Koelle et al., 1974; Fig. 1, lower panel).

Superfusion techniques have been employed in vitro and in vivo to measure neurotransmitters released into the extracellular space. In vitro methods have significant advantages when an effect of a drug or compound is directly on the cholinergic neuron, as opposed to involving multisynaptic communication. In vivo methods are better suited to those circumstances in which the influence of a drug is relayed from one brain nucleus to another, thereby having an effect on cholinergic neurons and their release of ACh.

The choice between using perfusion cup methods or cannula methods (either push–pull or dialysis) is dictated mainly by anatomical considerations. The neocortex, because of its superficial orientation, is an excellent tissue for perfusion cups, but a more difficult tissue for the techniques of push–pull perfusion and intracerebral dialysis. The substantia nigra, on the other hand, is completely inaccessible to perfusion cups, but accessible to push–pull cannulae.

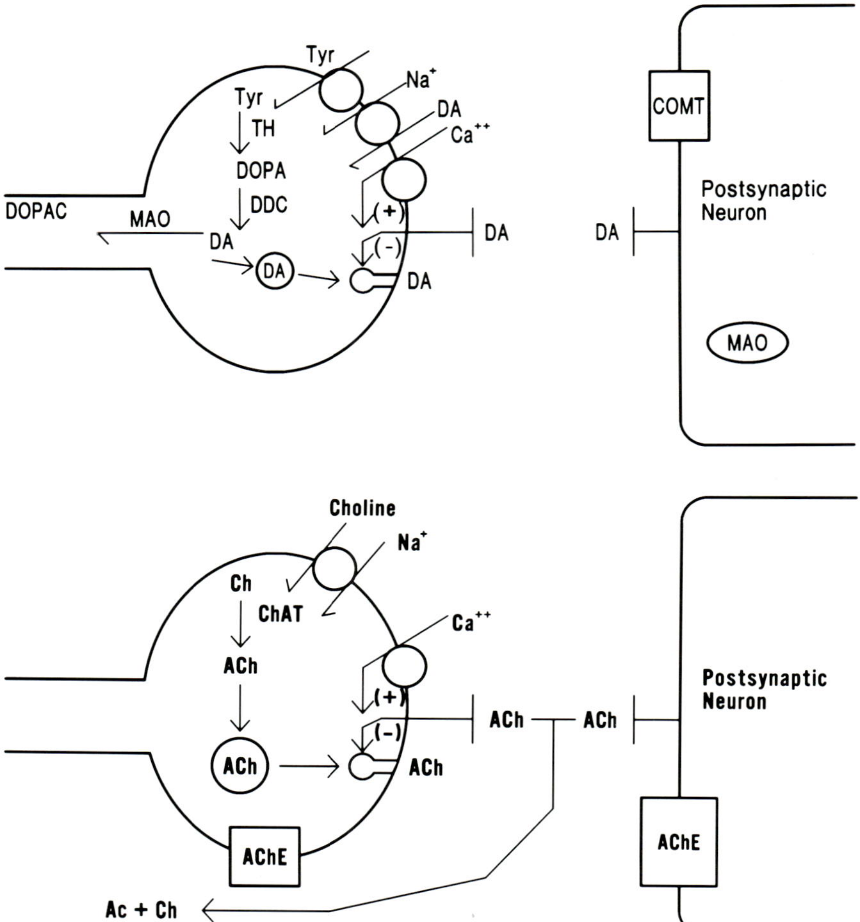

Fig. 1. Comparison of a model dopaminergic (top panel) and cholinergic (lower panel), synapse.

Altered release of neurotransmitter is often interpreted implicitly to have occurred if there are changes in parameters of turnover. Even though arguments may be made that other indices reflect the neuronal release of ACh, the direct measurement of ACh release itself is often the simplest approach.

The use of superfusion techniques to measure neurotransmitter release may be further subdivided into isotopic or nonisotopic methods. In the majority of cases, isotopic methods are technically simpler, but they involve several assumptions in order to extrapolate from release of radioactivity to the release of endogenous neurotransmitter.

2.2. In Vitro Measurement of ACh Release

The release of endogenous ACh from in vitro brain preparations following inhibition of AChE, to prevent degradation of release ACh, has been a well-utilized model. A major advantage to this method is that it measures, in a straightforward fashion, release of the endogenous pool of ACh. The only major disadvantage is the presence of an AChE inhibitor that interferes with the dynamics of, and pharmacological control mechanisms normally exerted over, ACh and its release.

The bioassays previously used to measure ACh are being replaced by analytical methods that are more specific, convenient, and rapid. In addition to the high-pressure liquid chromatography (HPLC) methods employing electrochemical detection (section 3.2.1.1.), the chemiluminescent method (Israel and Lesbats, 1981) is being employed. This approach circumvents the problem of potential interference of test compounds at the level of the bioassay. An obvious example would be the use of muscarinic agonists and antagonists to modulate the release of ACh via presynaptic muscarinic autoreceptors. Since the receptor contained in the guinea pig ileum (a standard tissue for the bioassay of ACh) is muscarinic, there would be an obvious interference by muscarinic agents not only with the release, but also with the bioassay. When the pharmacology of tested compounds becomes more complex (i.e., adenosine modulation of cholinergic function), the difficulty of interpreting the results multiplies substantially. Adenosine may: 1. Affect the release of ACh from the brain tissue 2. Affect the release of other substances from the brain, which may also affect the contracility of the ileum (or other tissue employed) or 3. Interact directly with its own class(es) of receptors in the ileum.

A major advance in the study of ACh release was made in 1972 when Somogyi and Szerb showed that neuronal pools of ACh could be labeled by using [^{3}H]choline and exploiting the high-affinity choline uptake system that is characteristic of cholinergic neurons (Rommelspacher and Kuhar, 1975b). However, unlike the dopaminergic system (Westfall et al., 1976) or other biogenic amine systems studied with analogous methods, it is not necessary to perform a chromatographic separation of the labeled neurotransmitter from its labeled precursor and other products.

This is primarily because of the very high levels of AChE, which hydrolyze the released neurotransmitter [^{3}H]ACh to [^{3}H]choline. Therefore, when reuptake of this extracellularlly formed [^{3}H]choline is blocked with hemicholinium, the efflux of radioactivity (i.e., [^{3}H]choline) from superfused brain slices is a direct index of ACh release (Somogyi and Szerb, 1972). The most rigorous methodolgical comparison between the release of endogenous ACh and tritium derived from [^{3}H]choline is that of Beani et al. (1984), which explores the effects of numerous pharmacological conditions and regional variables systematically.

2.2.1. Brain Slices

The brain slice is simple compared to the intact system. Most pharmacological effects on [^{3}H]ACh release can be assumed to be caused by an effect of the compounds on the cholinergic neuron. However, to be more rigorous, one may adopt the study of [^{3}H]ACh release from synaptosomes. Technically, this approach is substantially more challenging (Marchi et al., 1981), and the cholinergic neuron deprived of its dendrites may show a reduced repertoire of pharmacological responses (Lehmann and Scatton, 1982).

2.2.1.1. Detailed Protocol. 2.2.1.1.1. Apparatus. Many types of stimuli have been employed to measure [^{3}H]ACh release. Electrical field stimulation is the closest to physiological stimulation wherein the effects of paramaters, such as frequency, pulse width, and voltage gradient may be altered and studied systematically. Electrical stimulation may be applied continuously for long periods without compromising the preparation. For most pharmacological applications, however, depolarization evoked by elevated potassium chloride produces equivalent results and is easier technically to accomplish, particularly when dealing with large numbers of chambers.

The system shown in Fig. 2 has been designed for stimulation by elevated KCl and other chemical depolarizing agents, including aconitine, veratridine, and particularly excitatory amino acids. The most important consideration in this case is to deliver the depolarizing stimulus with as little mixing as possible, in a pulse of defined duration. To this end, it is best to position the peristaltic pump to pull the superfusion solution through the perfusion chambers. Conically shaped chambers produce less turbulence than those that have abrupt changes in diameter. Void volumes throughout the system should be kept low to minimize mixing. Finally, the ideal chamber should be disposable, since certain

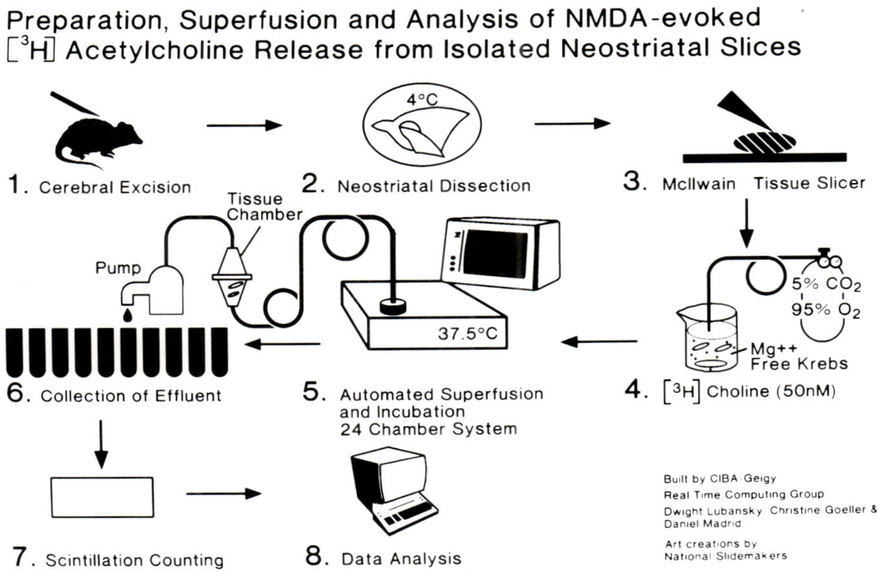

Fig. 2. Methodology for studying ACh release from striatal slices.

drugs such as neuroleptics, benzodiazepines, and phencyclidine analogs will be absorbed by virtually any plastic, including Teflon, confounding results as the apparatus continues to be used. An adequate chamber can be simply made by trimming two 200-μL Eppendorf pipet tips, which then fit easily together forming an airtight and water-tight seal with virtually perfect conical shape. Polyethylene tubing serves to connect the chambers with the reservoirs of Krebs medium. For the peristaltic pump, a recently developed tubing called C-flex is superior to silastic tubing, and is also used for connecting the polyethylene tubing to chambers and other joints.

As shown in Fig. 2, quasiphysiological Krebs medium is maintained slightly above 37°C to compensate for the loss of temperature produced by moving through the tubing. Following passage through the chambers and then through the pump, the superfusate is collected by a fraction collector. Commercially available fraction collectors are designed to collect one sample at a time. It is possible to modify such collectors to receive up to 20 channels at a time, but the alignment is a constant problem. For this reason, a custom-designed fraction collector, which advances a tray containing vials in a fixed alignment, can reduce many of the manual manipulations. Moreover, the tray should hold vials in the cassettes that can be directly inserted into a scintillation counter. These modifications result in significant savings of time in transferring the vials used in experiments involving release of radioactive substances and reduce the possibility of confusing the order of vials, since each rack's position is labeled. Such automation allows for the simultaneous operation of 24 chambers. It is therefore possible to perform extensive concentration–response studies under strictly parallel conditions, which is important, for instance, in demonstrating competitive interactions vs noncompetitive interactions (Lehmann et al., 1987).

Scintillation counting represents an additional challenge, inasmuch as typically 1.2 mL of Krebs medium (0.2 mL/min for 6 min) is collected in each fraction, in a minivial with 4.5 mL of total volume. The oxygen content produces major chemiluminescence and must be scavenged by the addition of 50 μL of 30% ascorbate to each vial. Subsequently, the vials receive a highly water-accepting scintillation cocktail (e.g., Beckman Readi-Solv MP) and are counted as a gel. For calculations of fractional release, it is also necessary to know the radioactivity content of the slice at the end of the experiment. Once recovered from the chamber, the slice is dissolved in a tissue solubilizer and neutralized with acetic acid, and the radioactivity is determined. A systematic error in the difference in counting efficiency for the tissues and superfusates may be overcome by calculating dpm for each vial.

Each time an experiment (consisting of 10 fractions and 24 chambers) is run, 264 numbers consisting of four or five decimal places are generated by the scintillation counter. To reduce time lost in transcribing all of those numbers, it is advisable to transmit data via the RS232 port of the scintillation counter to a buffer box, which can then upload the data to a microcomputer, which calculates the results and prints them out in a readily understandable form.

2.2.1.1.2. Execution of Experiments. The most important "art" in the execution of an experiment is the preparation and handling of slices. After sacrifice of the animals, the brain should be chilled as rapidly as possible on a cold plate (e.g., petri dish inverted over crushed ice). The dissection of tissue must be done quite carefully to avoid distorting, crushing, or stretching of the brain tissue to be used. Once the brain region of interest is isolated, it may be sliced in one dimension mechanically on a McIlwain chopper. A thickness of 0.4 mm seems easy to handle and yields reliable results. The slices are then added to a previously prepared

Krebs medium (composition in m*M:* NaCl, 113; KCl, 4.2; $NaHCO_3$, 25; NaH_2PO_4, 1.0; $CaCl_2$, 1.3; glucose, 11), which has been bubbled with 95% O_2/5% CO_2 to bring the pH to neutrality. The pH must be carefully monitored with this medium, as it is quite unstable, and bubbling must be continuous at either room temperature or at 37°C. Bubbling at 4°C results in an overly acidic pH. Magnesium is routinely left out of the medium with NMDA-type receptor stimulation, since this ion blocks NMDA-induced responses non-competitively (Lehmann and Scatton, 1982). Once the slice is added to a container of Krebs (with a minimum of 10 mL Krebs for each 50 mg tissue), [^{3}H]choline is added to the medium. The uptake of [^{3}H]choline is extremely rapid, and nearly all of the radioactivity is removed from the medium within five min. The incubation is carried out for a total of 30 min, however, to achieve attainment of equilibrium. Subsequently, the radioactive medium is discarded, and chilled Krebs is added to the vessel containing the slices. The chilling allows the slices to be inserted into their chambers with minimal trauma. Empirically, two slices per chamber offers the best precision and reproducibility. As soon as all chambers are filled (and this should be done as rapidly as possible), the pumps are turned on and the slices superfused for 45 min to enable the system to equilibrate.

After the 45-min "washing" time has elapsed, the superfusate is collected in 6-min intervals. Two fractions are first collected to give evidence of a stable baseline, after which depolarizing substances may be added. It is advisable to correct for the time it requires from changing of solution, until the depolarizing solution (together with evoked neurotransmitter) arrives at the level of the collection vials.

2.3. *In Vivo* Measurement of ACh Release

2.3.1. Cortical Cup Technique

The cortical cup technique is a tool that can be used to investigate neurotransmitter release from the cerebral cortex. MacIntosh and Oborin (1953) were the first to demonstrate successfully the release of ACh from the intact cerebral cortex by using cortical cups. They also showed that ACh release was related to the spontaneous electrical activity of the cortex. Mitchell (1960, 1961, 1963) has extended these studies by describing in more detail the method for collecting and measuring ACh output from the cerebral cortex, its advantages, and its limitations. By using anesthetized cats and sheep, he has observed that:

1. ACh is cortical in origin, since no ACh can be recovered from white matter underlying the cortex
2. The ACh collected does not originate from the blood flowing beneath the cup, since ACh is not detected in the blood collected from the sagittal sinus or from cortical veins during the collection of ACh from the somatosensory cortex
3. The ACh is not derived from the cerebrospinal fluid, since no measurable ACh is found in this fluid
4. The ACh leaves the cortex by diffusion, since during the first 15 min of collection, the ACh liberated into the cups increases in a linear manner. Over longer periods of collection, an equilibrium between ACh in the cup and the brain is reached.

A modification of this technique has been used by some authors (Smith, 1972; Dudar, 1975) to investigate the release of ACh from the hippocampus of the rabbit. ACh release from cerebellum has also been studied (Mitchell, 1961; Phillis and Chong, 1965). A correlation between cortical ACh release and nervous electrical activity has been shown. Peripheral stimulation of the sciatic nerve causes an increase of ACh output from the contralateral somatosensory cortex (Mitchell, 1963). Stimulation of the medial geniculate nucleus is followed by an increase of ACh release from the ipsilateral auditory cortex (Hemsworth and Mitchell, 1969). Stimulation of the reticular formation also increases ACh release in various cortical areas (Phillis, 1968). Direct stimulation of the cerebral cortex only causes an ipsilateral increase of ACh release (Mitchell, 1963). ACh release appears to be correlated with behavioral and electrocortical activity: Collier and Mitchell (1966) have demonstrated that ACh release increases in freely moving conscious animals during periods of activity; Jasper and Tessier (1971) have correlated the release with electroencephalographic potentials recorded from the cortex of chronically implanted freely moving cats. They have observed that the increase of the release is correlated with EEG desynchronization, shown by the animals both during active wakefulness and during paradoxical (REM) sleep.

It has been repeatedly shown during the past few years that the major cholinergic innervation of the cortical mantle is provided

Table 1
Drugs That Decrease ACh Release from Cerebral Cortex of Rats

Drug	Dose, mg/kg	Route of administration	Refs.
Morphine[a,b,c]	10	—	Domino et al. (1973)
Nalorphine[a]	1–10	iv	Domino et al. (1973)
Cholecystokinin[a,b]	0.01	ip	Magnani et al. (1984)
Muscimol[d,e]	0.001*	Loc. inj. NBM[f]	Casamenti et al. (1986)
Haloperidol[a]	1	ip	Casamenti et al. (1980)

* In a volume of 0.5 μL.
[a] Anesthetized animals.
[b] Reversed by naloxone.
[c] Antagonized by calcium in anesthetized animals (Sanfacon and Labreque, 1982).
[d] Prevented by picrotoxin.
[e] Antagonized by amphetamine.
[f] Nucleus basalis magnocellularis.

in the rat and in other animal species by the nucleus basalis magnocellularis (Johnston et al., 1979; Kelly and Moore, 1978; Lehmann et al., 1980; Wenk et al., 1980). The cup technique, in correlation with lesions of this necleus, has become a useful tool to investigate the psysiological function and neurotransmitter modulation of the N. basalis–cortical cholinergic pathway (Casamanti et al., 1986; Lo Conte et al., 1982a; Lo Conte et al., 1982b).

2.3.1.1. Drug Effects on ACh Release. The principal aims of the research concerning drug effects on ACh release are: (1) to investigate the effects of neurotropic drugs on central cholinergic pathways, and (2) to use the drugs as tools to define transsynaptic modulation (decreasing or increasing) of ACh release in cortical and subcortical regions. The cup technique has been extensively employed in studies of ACh release; Tables 1 and 2 summarize some results of such investigations.

2.3.1.2. Detailed Protocol. The cortical cup can be applied in different animal species: sheep, cat, rabbit, guinea pig, rat. Its

Table 2
Drugs That Increase ACh Release from Cerebral Cortex of Rats

Drug	Dose, mg/kg	Route of administration	References
Pentylenetetrazol[a]	150	ip	Hemsworth & Mitchel (1969)
Picrotoxin	12	iv	Hemsworth & Mitchel (1969)
4-Aminopyridine[a]	1.3	ip, iv	Casamenti et al. (1982)
Amphetamine[a,b]	1	ip	Casamenti et al., (1980)
Pimozide[b]	1	ip	Casamenti et al., (1980)
Apomorphine[b]	10	ip	Casamenti et al., (1986)
Scopolamine[b]	2.5	ip	Lo Conte et al. (1982a)
Atropine[b]	0.75–1.3	ip	Aquilonius et al. (1972)
Cholecystokinin[a,c]	0.0015	ip	Magnani et al. (1984)
Phosphatidylserine[a]	150	iv	Casamenti et al. (1979)

[a] Anesthetized animals.
[b] Unanesthetized animals.
[c] Prevented by proglumide.

size varies according to the animal species used. However, it is preferable to use as small a cortical cup diameter as possible in order to investigate ACh release from more discrete cortical zones and to have a collection volume that allows a high neurotransmitter concentration. In general, the cortical cup is formed by a cylinder held in tight contact with the cortical surface, after removing part of the skull.

2.3.1.2.1. Acute, Anesthetized Rats. The type (Moroni et al., 1981) and the depth of anesthesia (Mitchell, 1963) can affect neurotrasmitter release. It is important to obtain a long-lasting, stable light surgical plane of anesthesia.

In the rat, urethane (1 mg/kg ip) maintains a relatively stable ACh output with only a slight decrease over a few hours (Casamenti et al., 1979). The animal is placed in a stereotaxic apparatus. Part of the parietal bone and the underlying dura mater are removed from both sides. This avoids different pressures on each hemisphere. The Perspex cup used in this model has a shape that allows its adhesion to the somatosensory cortical surface. A gentle pressure on the cylinder, which is held by an electrode carrier, is sufficient to avoid the leakage of the Ringer solution out of the cup. The cylinder covers an area of 0.2 cm^2 (Mulas et al., 1974). It is filled with 0.4 mL of Ringer solution of the following composition (g/L): NaCl, 9; KCl, 0.42; $CaCl_2$, 0.24; $NaHCO_3$, l; glucose, l; and physostigmine sulfate, 0.1.

Some authors add $MgCl_2$ (0.05 g/L) to the Ringer solution. The cortex is maintained in contact with the eserinized Ringer solution for 1 h, and is then washed with fresh eserinized solution. The presence of eserine is necessary to inhibit the activity of AChE in order to obtain a quantifiable amount of ACh released. The Ringer solution, maintained at 37°C throughout the experiment, is removed by suction every 10 min and stored at –20°C. An electric heating pad maintains the body temperature of the animal at 37°C.

2.3.1.2.2. Midpontine-Pretrigeminal Transected Rats. With this procedure (Sanfacon and Labreque, 1982), the influence of anesthetics on ACh release can be avoided. The animal, anesthetized with ketamine (100 mg/kg ip), is placed in a stereotaxic apparatus in the flat-skull position. After removing the occipital bone and dura mater, the neuroaxis is transected at the midpontine-pretrigeminal level by a spatula inserted into the cerebellum approximately 1 mm posterior to the occipital suture and forward at an angle of 45° to the horizontal plane of the stereotaxic apparatus. The cerebrum now has only the first four cranial nerves intact: olfactory, trochlear, oculomotor, and visual. All the other motor and sensory pathway are disconnected. This model presents a desynchronized electroencephalographic pattern typical of active wakefulness. After this surgery, a hole 6 mm in diameter is placed in both parietal bones by a hand drill. Two Perspex cylinders are screwed into the holes until they exert a slight pressure on the dura mater. The cups are filled with eserinized Ringer solution (0.4 mL, 37°C). The animal is utilized after about two h following the transection, to allow the recovery from the anesthesia. The Ringer solution is collected and replaced every 20 min.

2.3.1.2.3. Freely Moving Chronically Implanted Rats. No differences between midpontine-pretrigeminal transected rats and freely moving rats have been found in spontaneous ACh output from the cerebral cortex (Casamenti et al., 1986). It may be assumed, therefore, that the activity of the cholinergic neurons ascending to the cortex is similar in the two preperations. On the other hand, some differences in cortical ACh release have been found between midpontine transected and chronically implanted, freely moving cats (Pepeu, 1973). The use of freely moving preparations makes it possible not only to avoid the influence of the anesthetic, but also to investigate the relationship between behavioral changes and ACh output (Casamenti et al., 1980). The animal is initially anesthetized with ketamine (100 mg/kg ip), and placed in a stereotaxic apparatus. One Perspex cup is inserted into one side of the cranium, as described above, and is fixed to the skull with dental cement (Beani and Bianchi, 1970). A Perpex plug is inserted in the cylinder, which is then closed with a screw cap so that the bottom of the plug exerts a pressure on the dura mater to prevent them from bending outward because of the intracranial pressure. For 3 d prior to using the animal, a terramycin solution (100 mg/mL) is placed in the cup and replaced daily. For the release experiments, a Ringer solution is collected as described for the previous models. In freely moving and midpontine-pretrigeminal transected rats, the cup is applied on the dura mater; this avoids damage to the cortical surface, but could also decrease the ACh output. In rabbits with intact dura (Beani et al., 1968), the ACh output is approximately 70% of that measured after removing the dura. However, the ability of drugs to increase or decrease ACh release is not affected by the presence or absence of dura mater (Beani et al., 1968; Casamenti et al., 1980).

2.3.1.2.4. Quantification of ACh Released. The spontaneous release, in the presence of eserine, from the cerebral cortex in rats, as well as in all the species investigated, is between 0.5–2 ng/min/cm^2. Until recently, this small amount of ACh has been routinely quantified by bioassay using the dorsal muscle of the leech or the guinea-pig ileum. This kind of assay has a sensitivity for ACh at least 5×10^{-9}g/L (Pepeu, 1973). However, radioenzymatic (Goldberg and McCaman, 1973) GC-MS (Wood and Peloguin, 1982) and HPLC (Potter et al., 1984) methods developed in the past few years have picomolar sensitivity, thus making them also suitable for release studies from the brain in vivo.

2.3.2. Push–Pull Cannulae

The use of push–pull cannulae to monitor ACh release has been limited as a result of the problems associated with local tissue damage (Yaksh and Yamamura, 1974) and the use of AChE inhibitors. Regional release studies have been performed in the rat, with significant ACh release being monitored in the striatum, thalamus, and substantia nigra (Oelszner et al., 1973). High rates of ACh release have also been monitored in the monkey thalamus (Myers and Beleslin, 1970). In pharmacological studies, morphine has been reported to decrease striatal ACh release in the cat (Yaksh and Yamamura, 1975) and in the rabbit cortex (Mullin, 1974). Neuroleptics have also been reported to enhance ACh release in the cat striatum (Stadler et al., 1973).

3. Biochemical Methodologies

3.1. Nonsteady-State Methods

3.1.1. Inhibition of ACh Synthesis by Hemicholinium-3 (HC-3)

The studies of Slater (1969) have clearly demonstrated the ability of intraventricular (ivt) HC-3, as well as other quaternary compounds, to decrease whole rat brain ACh levels, an action that can be prevented by the simultaneous administration of choline. Subsequently, the laboratories of Domino have investigated the actions of drugs on this rate of decline in whole brain ACh after intraventricular HC-3 (20 μg) or acetylseco HC-3 (5 μg). The decline in ACh levels, resulting from decreased choline availability for ACh synthesis after ivt. HC-3, is linear for 45–60 min, such that investigators generally measure ACh levels 30 min after HC-3 administration, in control and drug-treated rats. With drug/HC-3 combinations, enhanced ACh utilization is characterized by lowered ACh levels at 30 min, and decreased ACh utilization is characterized by increased levels, relative to HC-3 alone.

Using this approach along with bioassay of whole brain ACh, morphine and methadone have been demonstrated to decrease ACh turnover as evidenced by a slower rate of decline of ACh levels after HC-3 (Domino and Wilson, 1973). These data have been subsequently verified by a gas chromatographic assay of ACh in opiate-treated rats (Wilson and Domino, 1973). LSD-25, methotrimepazine, naloxone, nalorphine, cyclazocine, and pentazocine were found not to alter the decline rate of ACh (Domino and Wilson, 1973). These data clearly demonstrate the limitations of using whole brain analyses for such studies, since cyclazocine and pentazocine were subsequently shown to selectively depress ACh turnover in the parietal cortex (Wood and Stotland, 1980), an action masked by whole-brain analyses. Tolerance to the actions of morphine have also been demonstrated (Domino and Wilson, 1975a), as well as cross-tolerance between methadone and heroin (Domino and Wilson, 1975b).

Using the HC-3 method to examine the regional actions of morphine on ACh utilization, Schmidt and Buxbaum (1978) have found that morphine decreases ACh utilization in the midbrain, hypothalamus, and hippocampus, but not the striatum of the rat. These data agree with assessment of the actions of morphine using isotopic methods to monitor TR_{ACh} (Zsilla et al., 1977; Wood and Stotland, 1980). Similiar regional analyses have demonstrated a selective decrease in hippocampal ACh utilization by delta-9-tetrahydrocannabinol (Domino et al., 1977), with no effect in the striatum, midbrain, cortex, hypothalamus, or thalamus. The thiamine antagonist, pyrithiamine, has also been shown to decrease ACh utilization in the rat hippocampus and striatum (Vorhees et al., 1978).

3.1.2. Postmortem Changes in ACh Content

In animals sacrificed by decapitation, the enzymes involved in ACh metabolism are not inactivated instantaneously. The ACh content of brain structures may change considerably because of the rapid postmortem metabolism of ACh and the increase of brain choline content. Thus, it may be hypothesized that the slope of the decline of brain ACh content might increase under conditions where the ACh turnover rate has been increased, and decrease under conditions where the ACh turnover rate has been decreased. Unfortunately, there are no reports that would validate such an hypothesis.

Some investigators have proposed that, in rats sacrificed by decapitation, brain ACh content may vary inversely proportional to the ACh turnover rate (Guyenet et al., 1975; Stadler et al., 1973). It has been suggested that an increase in striatal ACh content in rats that have been decapitated at a certain time after the injection of dopamine receptor agonists (Consolo et al., 1974; Guyenet et al., 1974; Ladinsky et al., 1974; McGeer et al., 1974; Rommelspacher and Kuhar, 1975b; Sethy and Van Woert, 1974) indicates a decrease in the turnover rate of striatal ACh. Conversely, it has been reported that a decrease of striatal ACh content following the injection of dopamine receptor antagonists (Guyenet et al., 1974; McGeer et al., 1974; Rommelspacher and Kuhar, 1975b; Sethy and Van Woert, 1974) or after cerebral hemisection (Rommelspacher and Kuhar, 1975a) indicates an increase in ACh turnover rate. However, when the rats are sacrificed by microwave irradiation (Guidotti et al., 1974), which prevents the postmortem accumulation of free choline in the brain and blocks ACh destruction, the injection of dopamine receptor agonists or antagonists fails to change striatal ACh content. Since the stability of ACh levels depends on the method used to sacrifice the rats, predictions made concerning the turnover rate of ACh based on ACh content may or may not be correct. For example, a decrease of striatal ACh content occurs after injection of ACh receptor blockers, although they do not reduce the ACh turnover rate (Cheney et al., 1976; Racagni et al., 1975a).

3.1.3. Changes in V_{max} of Sodium-Dependent High Affinity Choline Uptake

Several laboratories have independently discovered the modifications in high-affinity choline uptake brought about by activation of cholinergic nerve terminals. Barker and Mittag (1975) have shown that, after depolarization of rat brain synaptosomes in vitro by elevated K^+, the V_{max} of [^{3}H]choline uptake is increased, whereas the K_m is not altered. Jenden et al. (1976) have found an inverse correlation between alterations in choline uptake and alterations in ACh levels following in-vivo pharmacological treatments. Finally, Simon et al. (1976) have demonstrated that increases or decreases in neuronal activity in different regions of the brain correlate with the regional changes in sodium- and temperature-dependent choline uptake.

It has been shown that the increase in high-affinity choline uptake caused by depolarizing stimuli in vitro is dependent upon the presence of calcium in the medium (Higgins and Neal, 1982a,b). Thus, it is not the depolarization *per se* or the movement of sodium ions that is responsible for the activation of the choline uptake. Rather, the voltage-dependent inward flux of calcium or the consequent release of ACh from the terminal would appear to be key in the mechanism of choline uptake activation. According to the results of Murrin et al. (1977), the entry of divalent cation, rather than the release of ACh, is the rate-controlling step. In fact, these workers have found that magnesium entry can support activation of choline uptake, but of course, this divalent cation does not support neurotransmitter release. Since uptake of the neurotransmitter precursor choline appears to be rate-limiting for ACh biosynthesis, the regulation of high-affinity choline uptake is a logical control point for ACh turnover regulation.

3.1.3.1. APPLICATION OF TECHNIQUE. The most common exploitation of the phenomenon of activation of high-affinity choline uptake is in the "ex vivo" design of experiments. Animals are treated with given drugs, and at a carefully chosen time point, sacrificed to allow measurement of the high-affinity choline uptake. Although the key variable thought to be altered is the V_{max} of the uptake, the uptake is typically measured at a single (low) concentration of [^{3}H]choline, and the measured parameter of velocity of high-affinity choline uptake is taken to reflect the V_{max}.

3.1.3.2. DETAILED PROTOCOL. The measurement of uptake enhancement requires the preparation of a crude "synaptosome," or P_2 fraction that contains resealed nerve terminals responsive to depolarizing stimuli. Following sacrifice of the animals, the brain is removed, chilled to 4°C, and homogenized in 10–20 vol of 0.32*M* sucrose/10 m*M* sodium phosphate, pH 7.4, using a Teflon pestle fitted to a glass homogenizing tube (125 μm clearance) at a speed of approximately 200 rpm. During the homogenization, all conditions are controlled such that the 4°C temperature is maintained, i.e., holding the homogenizing tube in a beaker containing ice and water. Ten slow strokes up and down result in a creamy suspension homogeneous to the naked eye.

The homogenate is centrifuged at 1000*g* for 10 min at 4°C. The pellet (P_1), containing nuclei, myelin, and large cell fragments, as well as blood vessels, is discarded, whereas the supernatant (S_1) is subjected to a second centrifugation at 11,000*g* for 15 min at 4°C (Note: Variations include centrifugation at speeds up to 18,000*g*, resulting in a higher percentage recovery of tissue, but a pellet that is harder and more difficult to resuspend homogeneously.) The supernatant (S_2) is discarded, and the pellet resuspended in 0.32*M* sucrose.

One hundred μL of the P_2 suspension is then added to tubes containing 900 μL of incubation medium. The final concentrations are: NaCl 122 m*M*; KCl 4.9 m*M*; $CaCl_2$ 1.3 m*M*; $MgSO_4$ 1.2 m*M*; NaH_2PO_4 15 m*M* (adjusted to pH 7.4 with NaOH); dextrose 11 m*M*; and [^{3}H]choline 50 n*M*. Since the calcium and phosphate will precipitate on storage, it is best to prepare fresh medium. The frequent user of Krebs medium finds it more economical to make it fresh daily from salts, with the exception of the sodium phosphate solution, which must be titrated to neutrality. No bicarbonate or oxygen is usually found in the Krebs medium used for synaptosome work such as this, since it is not easy to maintain constant pH and oxygen levels in multiple tubes.

The definition of net uptake is important for the final expression of results, and such uptake may be determined operationally by the several methods used to generate a "blank":

1. Temperature-dependent uptake (blank is held on ice)
2. Sodium-dependent uptake (blank has potassium- or other cation-replacing sodium)
3. Hemicholinium-3-sensitive uptake (blank contains hemicholinium-3 10 μ*M*)
4. Neuronal uptake (blank contains 100 μ*M* veratridine or aconitine).

By far the most popular are methods 1 and 2, although the most simple to use is method 3. All methods theoretically yield the same results. In general, no more than 10% of substrate should be consumed, and no more than 200 μg of protein should be added to each tube. The incubation should not last for more than 4 min.

The standard method of measuring uptake for neurotransmitters involves separating the label in the medium from that accumulated in the synaptosomes. This may involve filtration or centrifugation at 4500g for 15 min at 4°C. Once the conditions for measuring an activation have been established, it is important to adhere to them rigorously. It is also important to run all appropriate groups for statistical comparison in the same experiment and to arrange these appropriately within the assay so as not to create a systematic bias.

3.2. Steady-State Methods

3.2.1. General Considerations

Although ACh was first synthesized in 1867, more than 100 years had to pass before technological advances made it possible to study the turnover rate of this biologically active compound. The lengthy delay was the result of the instability of ACh in tissues and the lack of sensitive methods to measure it. Assays that were developed were time-consuming, lacked absolute specificity, and produced accurate data only in the hands of skilled investigators.

In the 1960s during the time when tremendous advances were being made in the understanding of the dynamics of catecholaminergic neurons (Costa and Neff, 1966; Costa et al., 1966), studies of the cholinergic neurons in the central nervous system lagged substantially behind, while investigators struggled to develop sensitive and specific chemical methods for the measurement of ACh and its precursor choline. Introduction of these methods in the late 1960s is generally believed to be the single most important stimulus to research into cholinergic mechanisms. Although they varied widely in sensitivity and specificity, numerous radiochemical and physical methods were developed to measure the concentration of ACh (Hanin, 1974). However, the problem of the instability of the ACh molecule in tissues continued to hinder progress in this field. Furthermore, choline was found to increase in postmortem brain tissues at the rate of 21 nmol/g/min (Dross and Kewitz, 1972). It was not until the development of focused microwave techniques to instantly inactivate the brain enzymes responsible for the synthesis and degradation of ACh and choline that the estimation of the tissue content of both of these compounds became reliable (Guidotti et al., 1974; Stavinoha et al., 1973). The union of these two techniques provided the catalyst for the investigation of the dynamic state of the neuronal stores of ACh.

Several attempts have been made to describe neurotransmitter compartmentation by suitable mathematical models (Costa, 1970; Sedvall et al., 1968). However, as discussed by Bergman (1969), it is difficult to formulate an exact model that appropriately weighs the numerous factors involved in controlling the steady state of endogenous chemicals functioning in a biological system. Usually, the biological system is not fully understood; the measurements include a number of compartments, rather than a single compartment; the number of measurements that can be performed in the same sample is often limited to one; and the uncertainty involved in these measurements is usually quite large. These complications have frustrated attempts to describe the regulation of transmitter turnover with a satisfactory mathematical model. A number of assumptions have been made that have led to the formulation of simple, but probably inexact, models of neurotransmitter function (Costa, 1970; Costa and Neff, 1966; Costa and Neff, 1970; Costa et al., 1966) (Fig. 5). Simple models are important as a basis from which to make hypotheses and design experiments. The results can then be used to modify and improve the model. One way to test the validity of a model and give relevance to the basic assumptions is the coincidence of experimental results obtained using different methods to measure neurotransmitter turnover. An operational device often followed is that of parallel experimentation with various methods to measure neurotransmiter turnover rates, i.e., isotopic methods versus nonisotopic methods (Costa and Neff, 1966). Kinetically, neurotransmitters have been viewed as if they were stored in a compartment open at both ends (Neff et al., 1969; Neff et al., 1971).

The rate of ACh efflux from endogenous stores, including the efflux from neuronal stores, can be measured by labeling the ACh at steady-state (Dross and Kewitz, 1966). The rate-limiting step of ACh synthesis is the uptake of Ch (Barker and Mittag, 1975; Guyenet et al., 1973; Yamamura and Snyder, 1973). This choline, in part, is transported from plasma and, in part, is generated from the hydrolysis of ACh. Since this recylcing of choline could cause a reutilization of the label that would interfere with the estimation of ACh turnover based on principles of steady-state kinetics, the interference created by the recycling of labeled choline must be minimized (Cheney et al., 1974; Racagni et al., 1974).

Calculations of the efflux rates of ACh are made by applying principles of steady-state kinetics. Since the labeling procedures do not alter the steady-state of the precursor, the rate of synthesis and degradation of ACh can be monitored by measuring the change with time of the abundance of label in choline and ACh. It is imperative with this approach that the total amounts of choline and ACh do not change while the relative abundance of a given isotope in the chemical composition of Ch and ACh does change. Furthermore, this approach requires that the enzymes responsible for the synthesis and degradation of ACh and choline be instantaneously inactivated at the time of sacrifice, preferably with microwave irradiation. Even the 20 s that are required to freeze the brain following the immersion of rats in liquid nitrogen could cause a variability in the measurement of choline concentrations that would not be tolerable for the measurement of ACh turnover with steady-state methods. Although these methods eliminate some of the drawbacks associated with the nonsteady-state methods, the steady-state methods are based on the following assumptions, which are not yet entirely documented by experimental evidence (Cheney et al., 1974; Cheney et al., 1975a,b,c; Hanin et al., 1973; Racagni et al., 1974):

1. Plasma choline is assumed to be preferentially transferred to a pool of free brain choline that, in turn, equilibrates with the various metabolic pools of brain Ch at a rate proportional to their intrinsic metabolic rates. The pool size of choline that functions as a precursor for ACh is very small. Since the rate of ACh synthesis is fast, the specific activity of free brain choline would mirror with a certain accuracy the specific activity of choline serving as ACh precursor.
2. Simple kinetic models based on a single open metabolic compartment of ACh store are assumed. Some equations are derived from this assumption and used to calculate turnover.
3. With the constant rate of label infusion and the time limits proposed in this model, the feedback resulting from recycling of labeled choline formed from the hydrolysis of ACh is insignificant and fails to exceed the noise of the methods available to measure isotopic abundance.
4. The rate of conversion of free brain choline into brain ACh is assumed to be much more rapid than the turnover rates that describe the other choline metabolic pathways.

There are three methodologies to measure the concentrations of ACh and choline and to measure the incorporation of label into ACh and choline that will be discussed: High-pressure liquid chromatography (HPLC), gas chromatography (GC), and combined gas chromatography–mass fragmentography (GC–MF).

3.2.1.1. High-Pressure Liquid Chromatography. HPLC is a recent addition to the armamentarium of the cholinergic pharmacologist. The HPLC measurement of ACh depends upon the separation of ACh and choline by reverse-phase HPLC with subsequent mixing of the effluent with AChE to hydrolyze the ACh to choline, which is next oxidized to betaine and hydrogen peroxide by choline oxidase (Potter et al., 1983, 1984). The production of hydrogen peroxide is continuously monitored by electrochemical detection (Bymaster et al., 1985; Potter et al., 1984). A more recent modification has been to immobilize the two enzymes on a postcolumn anion-exchange resin (Damsma et al., 1985; Eva et al., 1984), thereby reducing the quantities of expensive enzymes required to maintain the assay.

This HPLC methodology, along with a pulse injection of [^{3}H]choline, has been used to measure ACh turnover in the mouse cortex (Potter et al., 1984). However, no drug studies have been reported using this approach. A nonisotopic method for measurement of acetylcholinesterase activity has also been developed using the HPLC methodology (Kaneda et al., 1985). The recent report of direct thermospray ionization of ACh offers the new potential for an LC-MS assay of ACh and choline in biological samples (Liberato and Yergey, 1986).

3.2.1.2. Gas Chromatography Methodology. A sample containing derivatized ACh, choline, and appropriate internal standards dissolved in either chloroform or ethylacetate is injected into a gas chromatograph fitted with a fraction collector. The separation of the various compounds is achieved in the gas chromatographic portion of the instrument in a silanized Pennwalt 223 amine packing (80–100 mesh) that has been conditioned at 214°C for at least 72 h. Analysis conditions are: column temperature, 160°C; injector temperature, 230°C. Ninety percent of the effluent from the gas chromatographic column is diverted via a heated transfer line (95°C) to the fraction collector, which is maintained at room temperature. The carrier gas passes through the trapping medium (molecular sieve 5A, 45–60 mesh), whereas the radiolabeled teritiary amines are trapped by the highly reactive molecular sieve. The flame ionization detector senses the remaining 10% of the effluent from the gas chromatographic column. By monitoring the flame ionization detector response, the tertiary amines are selectively trapped as they are seen to emerge from the gas chromatographic column. The trapped radioactivity of each sample is counted by transferring the entire contents of each trap to a scintillation counting vial containing 20 mL of scintillation fluid (Hanin et al., 1973). The concentration of choline and ACh is calculated from the areas under the corresponding peaks on the flame ionization detector record. Specific activities of each compound are then determined and reported as disintegrations/min/nmol of substance.

3.2.1.3. Gas Chromatography–Mass Fragmentography Methodology. The coupling of a gas chromatograph to a mass spectrometer results in the marriage of two powerful technologies. The tremendous separative power of the gas chromatograph is coupled to the direct identification and quantitation capabilities of the mass spectrometer. The mass spectrometer produces ions by electron bombardment of eluted compounds, and the subsequent decomposition of these ions is monitored. The observed fragmentation patterns of "mass spectra" have characteristic ions that can be selectively monitored to obtain enhanced sensitivity for quantitative studies.

Aliquots of ethylacetate containing derivitized ACh, choline, their deuterated analogs, and deuterated internal standards are injected into the gas chromatograph–mass spectrometer. The separation of the various compounds is achieved in the gas chromatographic portion of the instrument in a silanized glass column, 1.8 m long and 2 mm inside diameter, packed with Pennwalt 223 amine packing (80–100 mesh), which has been conditioned at 215°C for at least 72 h. Analysis conditions are: column temperature: 160°C; injector temperature, 230°C; separator temperature, 200°C; carrier gas, helium at 12 mL/min; the beam current, 100 mA, and electron energy, 70 eV. Using gas fragmentographic analysis, concentrations of choline, ACh, and their deuterated analogs are calculated from peak height ratios with respect to the internal standards after comparing these ratios with those obtained using various known concentrations of external standards (Zsilla et al., 1977; Fig. 3).

Figure 4 shows the mass spectra of derivatized choline and its esters [base peak mass-to-charge (*m*/*e*) 58], [$^2H_4$]choline and its esters (*m*/*e* 60), and [$^2H_9$]choline and its esters (base peak *m*/*e* 64). The mass spectrometer is focused on *m*/*e* values 58, 60, and 64, respectively (Zsilla et al., 1977).

3.2.2. Pulse Injection of Isotopic Label

3.2.2.1. Application of the Technique. This approach utilizes the measurement of the specific radioactivity of ACh and its precursor choline, or the percentage of incorporation of deuterium into ACh and choline at various times after the pulse injection of labeled choline or a labeled precursor of choline (Cheney et al., 1975a; Dross and Kewitz, 1966; Hanin et al., 1973; Jenden et al., 1974; Saelens et al., 1973; Schuberth et al., 1969). Both radioactive and stable isotopes of choline have been used to pulse label brain ACh and measure the turnover rate of ACh in whole mouse brain (Jenden et al., 1974; Sparf, 1973), whole rat brain (Dross and Kewitz, 1966), or in large brain regions (Choi et al., 1973).

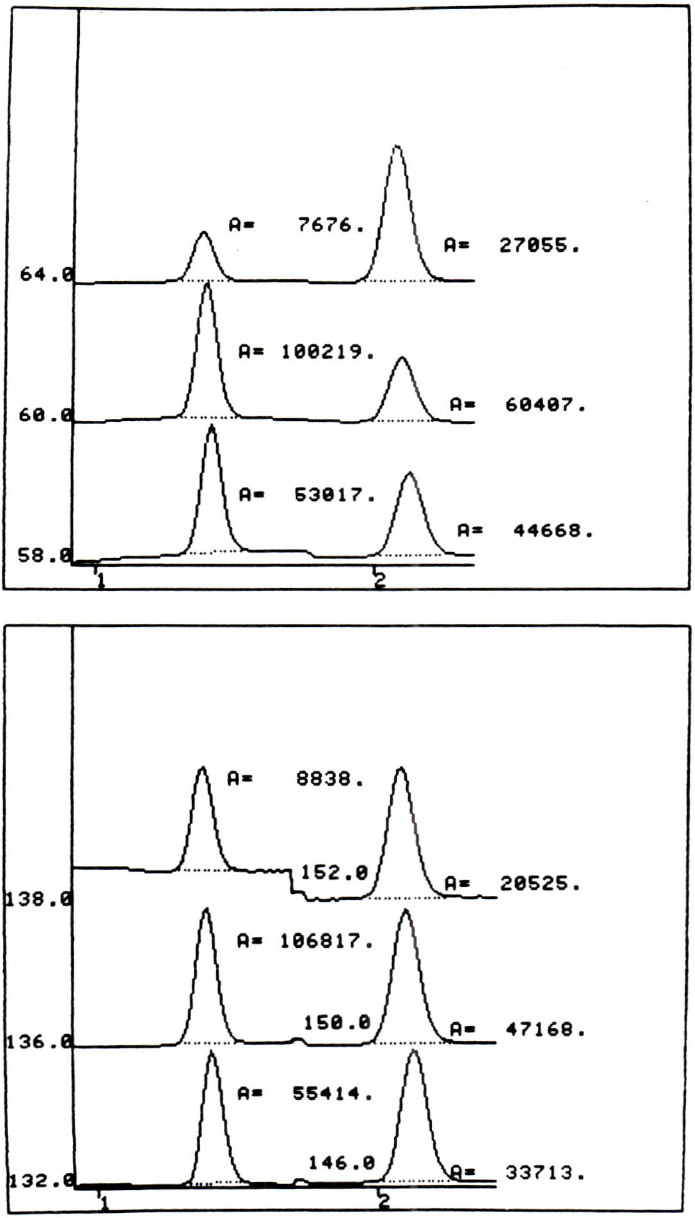

Fig. 3. SIM traces of ACh and choline both in electron impact and positive chemical ionization modes.

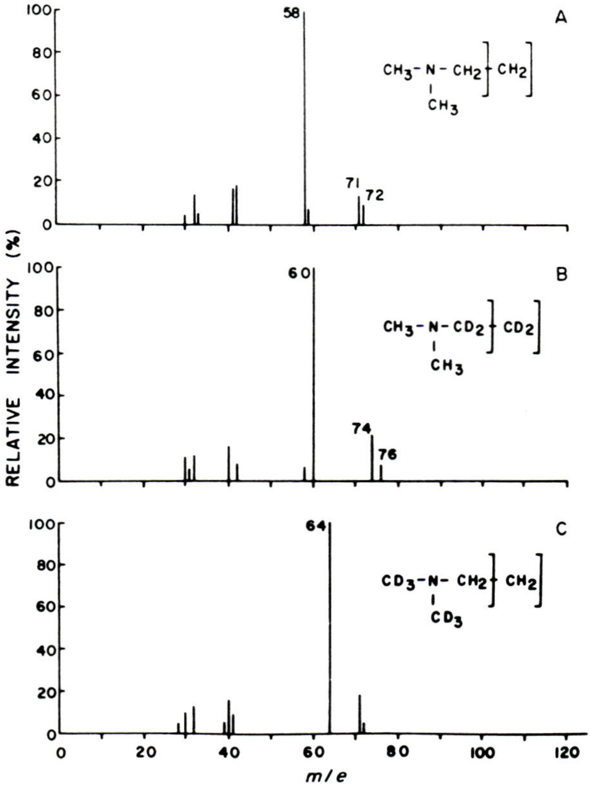

Fig. 4. Electron impact mass spectra of derivatized choline and its esters (a). [$^2H_4$]choline (B); [$^2H_9$]choline (C). (From Zsilla et al., 1977.)

The injections of radiolabeled choline necessary to label the ACh content of small brain areas may change the steady-state of choline; hence, phosphorylcholine has been used to bypass this difficulty (Cheney et al., 1974; Cheney et al, 1975a; Cheney et al., 1975c; Costa et al., 1975; Hanin et al., 1973; Racagni et al., 1974; Trabucchi et al., 1975). In fact, in plasma and brain there is a large pool of phosphorylcholine (Ansell and Spanner, 1971; Ansell and Spanner, 1975). Perhaps in brain tissue, phosphorylcholine may be even a natural precursor of brain choline (Ansell and Spanner, 1975). Radioactive phosphorylcholine efficiently labels brain choline (Hanin et al., 1973).

With pulse injections of labeled phosphorylcholine one can measure the ACh turnover rate in samples of brain tissue weighing about 100–300 mg (Cheney et al., 1975a). However, it is impossible

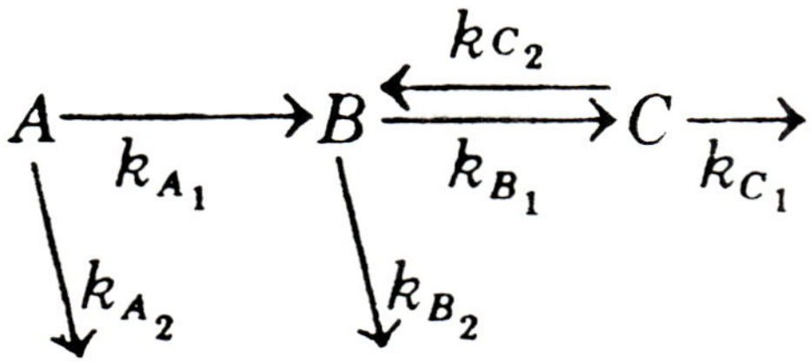

$$k_{A_1} + k_{A_2} = k_A$$
$$k_{B_1} + k_{B_2} = k_B$$
$$k_{C_1} + k_{C_2} = k_C$$

Fig. 5. Model for studying ACh and choline dynamics by constant infusion labeling.

to label ACh in brain nuclei with pulse injections of isotopically labeled phosphorylcholine, because the amount of ^{14}C or deuterium that can be incorporated into ACh at steady-state is too small to be detectable.

3.2.2.2. Description of the Model. Based on the assumption that: (a) isotopic-labeled ACh and choline are each in one homogenous pool; (b) endogenous ACh and choline are each in one homogenous pool; and (c) labeled and endogenous ACh and choline are in the same pool, and model shown in Fig. 5 has been postulated (Cheney et al., 1975a). In Fig. 5, A is the plasma choline compartment, B is the brain choline compartment, and C is the brain ACh compartment. k_A, k_B, and k_C are the fractional rate constants describing the efflux of choline or ACh from the various compartments. Specifically, k_{A1} and k_{B1} are the fractions of the total, which are involved in the biosynthesis of brain ACh, whereas k_{A2} and k_{B2} represent all other pathways of choline metabolism. At steady-state, the fraction involved in ACh biosynthesis is a constant proportion of the rates indicated by $k_{A1} + k_{A1}$ and $k_{B1} + k_{B2}$. The metabolism of C includes two components, k_{C2}, which feeds back on B, and k_{C1}, which is the efflux of C from the system.

Mathematically, the equilibrium conditions of the precursor–product relationship can be stated by

$$\frac{dB}{dt} = Ak_{A1} - Bk_B + Ck_{C2} = 0 \tag{1}$$

and

$$\frac{dC}{dt} = Bk_{B1} - Ck_C = 0 \tag{2}$$

Similarly, equations describing radiolabeled compounds (labeled compounds and denoted by an asterisk) following injection of the labeled precursor are:

$$\frac{dB^*}{d_t} = A^*k_{A1} - B^*k_B + C^*k_{C2} \tag{3}$$

and

$$\frac{dC^*}{dt} = B^*k_{B1} - C^*k_c \tag{4}$$

If Eq. (3) is divided by B and if the specific radioactivity is denoted by $S_B = B^*/B$, then using the relationship of Eq. (1), Eq. (5) is obtained.

$$\frac{dS_B}{d_t} = k_B\left[\left(\frac{A^*k_{A1} + C^*k_{C2}}{Ak_{A1} + Ck_{C2}}\right) - S_B\right] \tag{5}$$

Since the increase of radioactivity in C is much slower than that in B, it can be assumed that $Ck_{C2} << Ak_{A1}$. Thus, if one used data obtained during the first few minutes before C* increases to a much higher value than A*, when $C^*k_{C2} << A^*k_{A1}$, then Eq. (5) can be approximated by:

$$\frac{dS_B}{d_t} \cong k_B(S_A - S_B) \tag{6}$$

If Eq. (4) is divided by C and if the specific radioactivity is denoted by $S_C = C^*/C$, then using the relationship of Eq. (2), one obtains:

$$\frac{dS_C}{d_t} = k_C(S_B - S_C) \tag{7}$$

In integral form, Eqs. (6) and (7) become:

$$S_B = k_B \exp^{-}{}_{k}{}^{Bt} \int_0^t S_A(p)\exp^{k_Bp}\, d_p \tag{8}$$

and

$$S_C = k_C \exp^{-}{}_{k}{}^{Ct} \int_0^t S_B(p)\exp^{k_Cp}\, d_p \tag{9}$$

where P denotes the running variable of the integration.

In the specific case where S_A declines exponentially as a function of time (this is the case when A receives a labeled injection at time zero and no feedback is considered), then:

$$S_A = S_{A_o} \exp^{-k_At} \tag{10}$$

the integration of Eqs. (8) and (9) can be carried out to yield:

$$S_B = S_{A_o} \frac{k_B}{k_B - k_A} (\exp^{-k_A t} - \exp^{-k_B t}) \tag{11}$$

and

$$S_C = S_{A_o} \frac{k_B k_C}{k_B - k_A} \left[\frac{\exp^{-k_A t} - \exp^{-k_C t})}{k_C - k_A} - \frac{\exp^{-k_B t} - \exp^{-k_C t})}{k_C - k_B} \right] \tag{12}$$

S_{A0} is now calculated by extrapolation from the data points estimated at various times after injection of the radioactive precursor.

If steady-state is preserved while labeling the plasma choline and if the radioactivity of A (A*) declines exponentially, then:

$$A^* = A_o^* \exp^{-k_A t} \tag{13}$$

A^* and A_0^* are the radioactivities associated with choline present in A at times *t* and *0*, respectively. The terms of the equation can become dimensionless if one defines:

$$\frac{A^*}{A_o^*} = A^*_N \tag{13a}$$

where A^*_N is the transformed radioactivity of A at time *t* and

$$\frac{1}{k_A} = \tau \tag{13b}$$

where (tau) is the transformation unit that defines time.

By incorporating (13a) and (13b) into Eq. 13, one obtains

$$A^*_N = \exp^{-t}/\tau \tag{14}$$

where A^*_N and *t*/tau are pure, dimensionless numbers.

$$\text{When } t = \text{tau, then } A^*_N = e^{-1} = \frac{1}{2.718281}$$

Therefore, tau, the turnover times of plasma choline, can further be defined as the time necessary for the plasma choline radioactivity to decline by 63.3%. An example is shown in Fig. 6. Mice have been injected intravenously with phsophoryl (^{14}C)choline (250 μc/kg). The tau value is calculated from k_A = 1/tau. From the equations, tau is defined as the time in which S_A declines to 36.7%. The experimental data then is transformed using the time rate of A* decline in order to facilitate analysis. Moreover, this transformation is important for pharmacological studies of brain ACh turn-

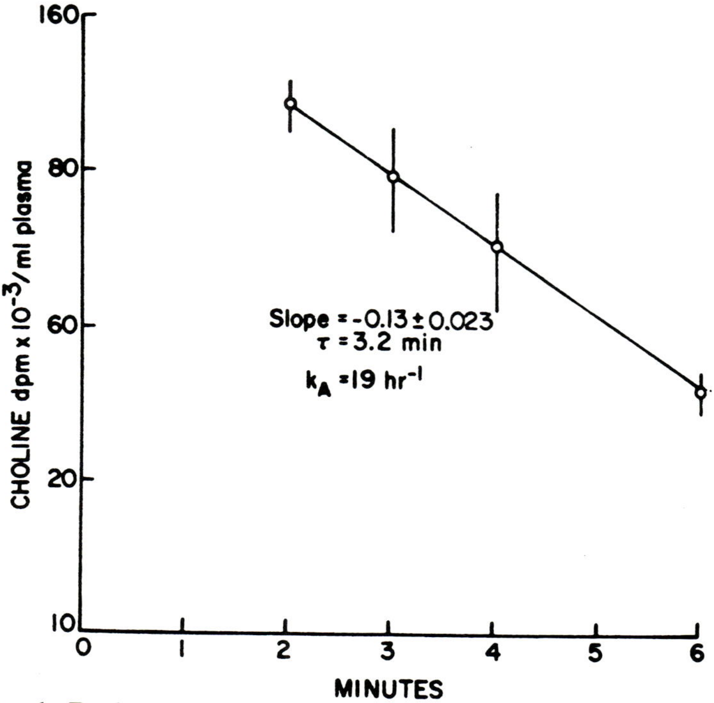

Fig. 6. Decline of plasma choline radioactivity in mice injected intravenously with phosphoryl [Me^{14}C]choline (250 μCi/kg, iv). The tau value is calculated from $k_A = 1/\text{tau}$, where (tau is defined as the time in which S_A declines to 36.7%. (From Cheney et al., 1975c.)

over rate, because it minimizes the consequences of drug effects on the rate of change of plasma choline specific radioactivity.

Where S_B reaches its maximum, the time rate of increase of S_B is equal to zero. Mathematically, this is expressed as:

$$\left[\frac{dS_B}{dt}\right]_{t = t_{max}} = 0 \tag{15}$$

Applying this relation to Eq. (12), t_{max} is related to the fractional rate constant by the equation

$$t_{max}/\tau = \frac{1}{k_{NB} - 1} \ln k_{NB} \tag{16}$$

where tau $= 1/k_A$ and $k_{NB} = k_B k_A \cdot k_{NB}$ is the normalized fractional

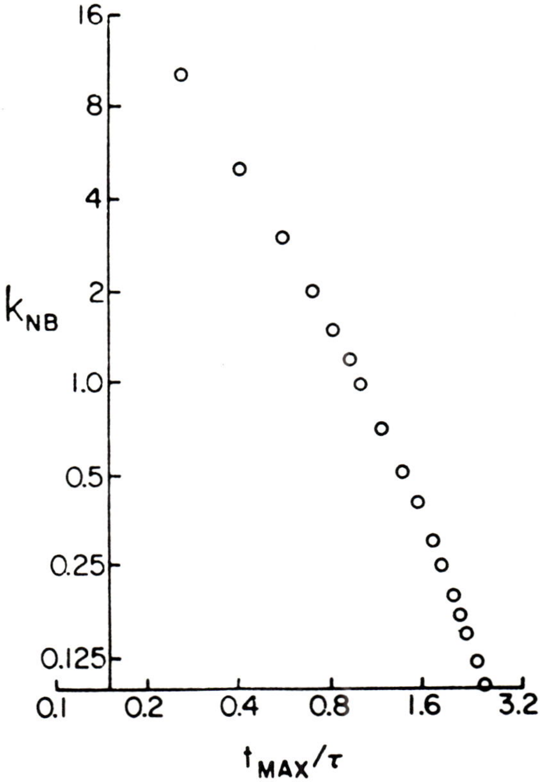

Fig. 7. Plot of t_{max}/tau calculated as a function of k_{NB} according, to Eq. (16). (From Cheney et al., 1975c.)

rate constant of *B*. t_{max}/tau in Eq. (16) has been calculated as a function of k_{NB} and is plotted in Fig. 7. Tau and t_{max}/tau are experimentally determined. Thus, k_{NB} can be calculated from Eq. (16).

Thus, S_B and S_C have been plotted as a function of *t*/tau, and the transformation procedure has been defined as follows:

$$S_{NB} = \frac{S_B}{S_{A_o}} \text{ and } k_{NB} = \frac{k_B}{k_A} \tag{17}$$

Thus, Eq. (11) becomes

$$S_{NB} = \frac{k_{NB}}{k_{NB} - 1} (\exp^{-t/\tau} - \exp^{-k_{NB}t/\tau}) \tag{18}$$

where S_{NB} and k_{NB} are dimensionless, pure numbers. When S_{NB} reaches its maximum value (t_{max}/tau), its rate of change is equal to zero. Therefore, Eq. (16) allows estimation of k_{nB} from experimental data points S_{NB}. This equation has been plotted in Fig. 7 for a selected sequence of theoretical values of t_{max}/tau. From this plot, the experimental value t_{max}/tau allows estimation of the value of k_{NB} for a given set of experimental data points. The validity of this estimation can be verified by comparing the theoretical values of S_{NB} for the assumed k_{NB} with the experimental data. To further validate these estimations, the value of k_{NB} has been evaluated statistically by the least mean squares method described by Wessel et al. (1965).

Transformation of S_C and k_C into S_{NC} has been preformed using definitions similar to those used for S_{NB} and k_{NB} in Equation XVII. Therefore, Eq. (12) can be formulated

$$S_{NC} = \frac{k_{NB} \cdot k_{NC}}{k_{NB} - 1} - \left[\frac{\exp^{-t/\tau} - \exp^{-k_{NC}t/\tau}}{k_{NC} - 1} - \frac{\exp^{-k_{NB}t/\tau} - \exp^{-k_{NC}t/\tau}}{k_{NC} - k_{NB}}\right] \tag{19}$$

A family of curves can be generated from Eq. (19) by assuming a set of theoretical values for k_{NB} and k_{NC}. These curves are graphically compared with the experimental values of S_C that are normalized and plotted as a function of t/tau. The best fitting curve is then verified statistically (Wessel et al., 1965). From k_{NC} and k_{NB}, k_B and k_c can be calculated as shown in Eq. (20).

$$k_B = \frac{k_{NB}}{\tau} \text{ and } k_C = \frac{k_{NC}}{\tau} \tag{20}$$

The turnover rate of ACh in the brain (TR_C) can be calculated from the equation

$$TR_C = k_C C \tag{21}$$

3.2.2.3. Treatment of the Animals and Preparation of the Tissue. Mice may be injected rapidly with an intravenous injection of 250 μCi/kg of phosphoryl (^{14}C)choline (Cheney et al., 1975a) or of 135 μmol/kg deuterated phosphorylcholine (Wood and Cheney, 1978) in physiological saline. The animals must be sacrificed by microwave irradiation at various times between 1–10 min after the injection of label. As indicated earlier, the microwave irradiation inactivates the enzymes responsible for the synthesis and degradation of ACh and choline. The microwave beam may be generated by a source operating with 2.5 kW at a frequency of 2.45 gHz, and focused for 1 s on the head of the mouse. Both brain choline and ACh concentrations are stabilized for at least 10 min by this radiation treatment (Cheney et al., 1975a).

The whole brain is then removed and homogenized in 3 mL of ice-cold perchloric acid (0.4*N*) with a Polytron homogenizer (20 s at full speed). The homogenate is centrifuged for 20 min at 35,000*g* and the supernatant transferred to a 13 mL centrifuge tube. An internal standard of 10 μmol of pentanoylcholine iodide, and 2 μmol of pentanoyl-1-^{14}C choline iodide is added to the supernatant (Hanin et al., 1973). Potassium acetate (7.5*N*) in water is then added to each tube (0.19 mL) to adjust the pH of the solution to between 4.2–4.4. After centrifugation at 35,000*g* for 20 min at 0°C, 2.5 mL of the supernatant is transferred to prechilled glass, 15-mL screw-cap centrifuge tubes. Endogenous quaternary ammonium compounds are precipitated as the Reinecke salts in the presence of tetraethylammonium chloride (10 m*M*; 0.25 mL) as a precipitant and 2.5 mL 2% ammonium Reineckate solution. After 45 min in ice, the mixture is centrifuged at 2500 rpm and 0°C and 10 min, and the supernatant discarded. The precipitate is freeze-dried and the quaternary ammonium reineckates in the precipitate are converted to their corresponding tosylates by batch treatment with 5 m*M* silver tosylate in acetonitrile (300–600 μL). The supernatant containing the quaternary ammonium toxylates is transferred to a 1 mL Kontes Microflex tube equipped with a screw cap and Teflon-faced liner, and the organic phase is evaporated to dryness under a stream of dry nitrogen gas (Hanin et al., 1972).

Esterification of choline to hexanoylcholine is achieved by adding 0.3 mL hexanoyl chloride in methyl ethyl ketone to each tube, and incubating for 45 min at 80°C with intermittent shaking. Methyl ethyl ketone and excess hexanoyl chloride are next evaporated under a stream of dry nitrogen. The contents of the reaction vessel are washed twice with 0.2 mL of trimethylamine in pentane (0.1*M*; 0.1 mL of 25% trimethylamine in methanol/4.12 mL of pentane). This approach quantitatively and selectively removes all tertiary amine from the tube. Traces of trimethylamine and pentane are removed under a stream of dry nitrogen, and the quaternary ammonium compounds are demethylated in the presence of sodium benzenethiolate in methyl ethyl ketone (50 m*M* benzenethiolate, 25 m*M* benzenethiol, 0.2 mL; 45 min, 80°C). The resultant tertiary amines are next separated from excess solvent and other nonbasic compounds, and concentrated into a small volume by a series of extractions. These extractions consist of the addition of 20 μL of 0.5*M* citric acid, mixing for 2 min, centrifuging the sample, and aspirating the organic phase; two washes with 0.2 mL pentane including mixing for 2 min, centrifuging, and aspirating the pentane; addition of 20 μL of base [ammonium citrate (2*M*) in ammonium hydroxide (7.5*N*)] and 5 μL of chloroform, mixing for 2 min, centrifuging and aspirating the chloroform pellet for injecting into the gas chromatograph.

3.2.2.4. An Example (Taken from Cheney et al., 1975a). Following injection of radiolabeled phosphorylcholine at various times between 1–10 min, the animals are sacrificed by microwave irradiation, the whole brain is removed and homogenized, and samples are prepared for injection into the gas chromatograph. From the areas of the peaks and the radioactivities of the respective peaks, the specific activities of choline (S_B) and ACh (S_C) can be determined and plotted as shown in Fig. 8.

From the value at the first time point of measurement, the S_B increases to reach its maximum at 1 min and then declines during 25 min. The curves S_B and S_C intersect at about 4 min. At this time point, the S_C has reached its maximal value. This is in line with the postulates defining a precursor–product relationship.

Using the value of tau calculated from the initial exponential decline of plasma choline radioactivity (Fig. 6), the data obtained between 0–6 min from Fig. 8 is transformed. The data is replotted in the form of S_{NB} and S_{NC} as a function of t/tau (Fig. 9). The S_{NB} and S_{NC} are plotted on the y-axis the t/tau on the x-axis. Since k_{NB} relates to S_{NB} as indicated in Eq (18), k_{NB} defines how values of S_{NB} change as a function of t/tau. The t_{max}/tau is 0.31 as read from Fig. 9. Using the graph from Fig. 4, the k_{NB} can be estimated to be 6.5. The data reported in Fig. 9 show that the experimental values of S_{NB} fit the calculated values S_{NB} when a $k_{NB} = 6.5$ is used in Eq. (18). Calculation of the value of k_{NB} with the lowest standard error according to the least squares method (Wessel et al., 1965) indicates that 6.5 is the best value for the data of Fig. 9.

Using $k_{NB} = 6.5$ and applying Eq. (19) to the experimental values of S_{NC}, it is found that they can represent a curve similar to that characterized by a k_{NC} of 1.05. Verification of this analysis by the least mean squares method reveals that $k_{NC} = 1.05$ is the one with the lowest standard error. When the value of this k_{NC} is transformed into the corresponding k_C, the a value of 20/h is

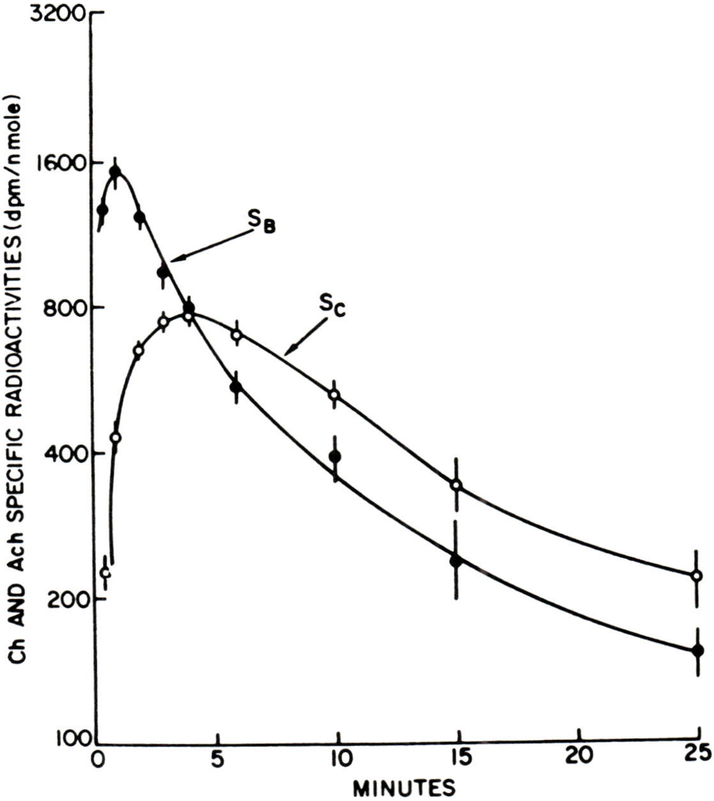

Fig. 8. Time course of choline and ACh specific radioactivities (S_B and S_C) in brains of mice receiving an iv bolus of radioactive choline (250 μCi/kg).

obtained. The average turnover rate of brain ACh is calculated to be 0.36 μmol/g/h [Eq. (21)].

3.2.3. Constant Rate Infusion of Isotopic Label

3.2.3.1. Application of the Technique. This approach utilizes intravenous infusion of isotopic labeled choline or phosphorylcholine at a constant rate for several minutes to label sufficiently the endogenous stores of choline and ACh. When radiolabeled choline or phosphorylcholine is used, the concentrations of ACh and choline as well as the specific activities of choline and ACh can be measured using a gas chromatograph with an attached fraction collector. ACh turnover rate can be measured in various regions of

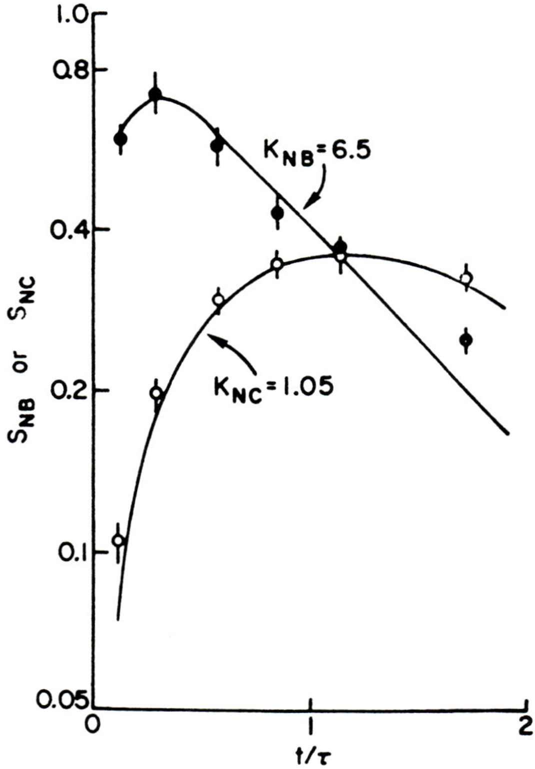

Fig. 9. S_{NB} and S_{NC} as a function of t/tau. Data points are the transformed S_B and S_C from Fig. 5.

rat brain weighing 100–200 mg (Cheney et al., 1974; Racagni et al., 1974; Saelens et al., 1973). In brain samples of 0.5–10 mg, the radiolabeled method cannot be used to measure the turnover rate (Racagni et al., 1975b). In such small tissue samples, a constant rate infusion with deuterated phosphorylcholine can label the choline and ACh content without changing the steady-state content of either ACh or choline (Costa et al., 1975; Zsilla et al., 1976, 1977).

3.2.3.2. Description of the Model. Figure 10 shows a simple model of neurotransmitter function. The rate of entrance of a label into compartment B (the precursor compartment) is indicated by J. The fractional rate constants for the efflux of products from compartment B and compartment C (the neurotransmitter compartment) are $k_{B1} + k_{B2}$ and $k_C = k_{C1} + k_{C2}$. Neurotransmitter from compartment C is constantly being released and/or metabolized

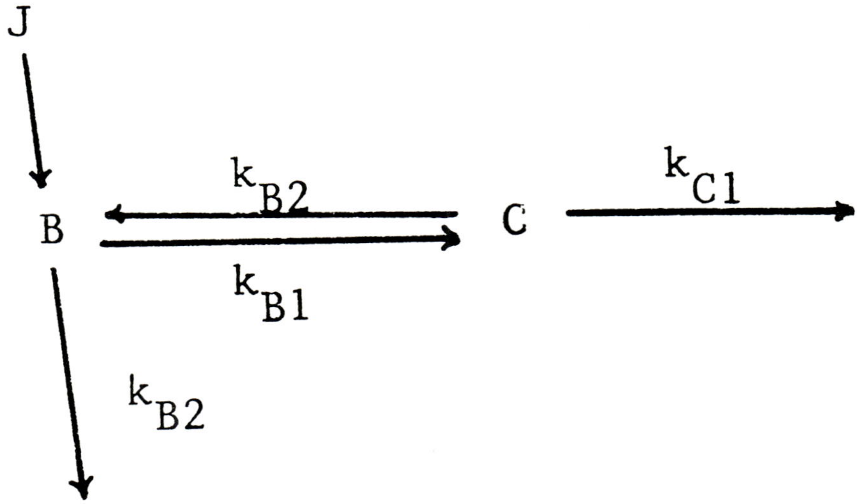

Fig. 10. Model for studying ACh and choline dynamics by constant infusion labeling. C, neurotransmitter compartment; J, rate of entry of the label into compartment B, the precursor compartment; k_B and k_C are the fractional rate constants for compartments B and C, respectively; $k_B = k_{B1} + k_{B2}$; $k_C = k_{C1} + k_{C2}$.

either intraneuronally or extraneuronally. Since the neurotransmitter compartment C is maintained at steady-state, metabolism and/or release represents a net loss of neurotransmitter from compartment C, regardless of whether the enzymatic alteration occurs intraneuronally or extraneuronally. The rate of this loss is an important kinetic parameter and is described by the product k_C[C]. Although the transmitter turnover rate fluctuates rapidly and continuously, it is assumed that, within the time constant of the sampling, the transmitter compartment remains at steady-state. The rate of neurotransmitter loss, k_C[C], is balanced by the rate of neurotransmitter formation, which is described by k_B[B]. Neither k_B nor compartment B can be measured directly, since in brain and other tissues the precursors for the neurotransmitters are also precursors for many other metabolic processes. Since there is no way to physically separate the various metabolic pathways involving the neurotransmitter precursors, the changes with time of the precursor and neurotransmitter specific radioactivities (when radioactivity is administered) and percent incorporation (when stable isotopes are administered) have been related. Equations have been derived, using this relationship, from which k_B and k_C can be estimated (Racagni et al., 1974) (*see* Appendices 1 and 2 for computer programs).

The S_B in compartment B and the S_C in compartment C would be expected to change with time as shown in Eqs. (22) and (23).

$$\frac{dS_B}{dt} = J - k_B S_B \tag{22}$$

$$\frac{dS_C}{dt} = k_C (S_B - S_C) \tag{23}$$

Imposing the condition that $S_B = 0$ at $t = 0$ where t is the duration of infusion, Eq. (22) can be integrated.

$$S_B(t) = \frac{J}{k_B} (1 - \exp^{-k_B t}) \tag{24}$$

$$k_B = \frac{-1}{t_2} \ln \left[1 - \frac{S_{B(t_2)}}{S_{B(t_1)}} (1 - \exp^{-k_B t_1}) \right] \tag{25}$$

Under the condition where $S_C = 0$ at $t = 0$, integration of Eq. (23) yields:

$$S_{C(t)} = \frac{J}{k_B} \left(1 - \frac{k_C \exp^{-k_B t}}{k_C - k_B} + \frac{k_B \exp^{-k_C t}}{k_C - k_B} \right) \tag{26}$$

Since J can be controlled and S_B and S_C can be experimentally measured, k_B can be calculated from Eq. (25). Following the calculation of k_B, k_C can be determined from Eq. (26). A solution can be found to Eq. (26), so that the turnover rate of ACh can be evaluated using a single time point of phosphorylcholine infusion. However, the k_B must be first determined from multiple points. Equation (24) is solved for J/k_B, substituted into Equation (17), and the natural log taken. Thus

$$k_c = \frac{-1}{t_1} \ln \left[1 - \frac{k_C}{k_B} + \frac{k_C}{k_B} \exp^{-k_B t_1} + \frac{}{m} \left(\frac{k_C}{k_B} - 1 \right) (1 - \exp^{-k_B t}) \right] \tag{27}$$

where $m = S_{B(t_1)}/S_{C(t_1)}$. Since t_1, $S_{B(t_1)}$ and $S_{C(t_1)}$ can be experimentally determined, Eq. (25) can be solved for k_B and Eq. (27) can be solved for k_C. Once the k_C is calculated, the turnover rate for ACh can be determined according to Equation (21).

4. Conclusions

Various techniques have been discussed that have been employed to study how drugs affect the dynamic regulation of cholinergic function. The numerous studies carried out with these techniques have contributed greatly to our understanding of the interactions of drugs and neurotransmitters with central cholinergic neurons (Cheney and Costa, 1977; Cheney, 1984). The choice of the technique will depend on the question being asked. Each of the available methodologies is based on a set of assumtions; and each has specific limitations. Nevertheless, all approaches allow for the in-depth study of drug actions on cholinergic neurons. Many of the discussed approaches complement each other in the study of complex systems. For example, in vitro studies of rat striatal slices (Lehmann and Scatton, 1982) have clearly defined the characteristics of the excitatory amino acid receptors regulating the activity of the cholinergic interneurons in this preparation. Studies of ACh turnover in decorticate rats also have demonstrated the in vivo significance of this positive modulatory input to these cholinergic neurons. Such studies clearly demonstrate how different techniques can be utilized to solve complex problems in the study of cholinergic neurons.

Appendices

Appendices 1 and 2 follow on pp. 482-485.

```
list
10 REM  NONLINEAR REGESSSION OF 2 VARIABLES
15 BT=0
20 PRINT
25 PRINT "NONLINEAR REGRESSION OF FIRST ORDER INCREASE
30 PRINT
35 PRINT "              -K*T"
40 PRINT "Y=A*(1 - e      )"
45 DIM X(50), Y(50), YY(50), W(50), WW(50)
50 PRINT
55 PRINT "TYPE OF WEIGHTING?"
60 PRINT "                          : 1 = CONSTANT S.D."
65 PRINT "                          : 2 = PROPORTIONAL S.D."
70 PRINT "                          : 3 = BETWEEN THE ABOVE"
75 PRINT "                          : 4 = S.D. SUPPLIED"
80 PRINT
85 INPUT "CHOICE=",N2
90 PRINT
95 INPUT "BISQUARE WEIGHTING (Y or N)=",Y$
100 IF Y$="N" THEN 110
105 BT=1
110 B=0 : C=0 : D=0 : E=0 : N1=0 : I=0
115 PRINT
120 INPUT "# OF X/Y PAIRS=",NP
125 PRINT
130 FOR N1=1 TO NP
135 INPUT "X=",X(N1)
140 INPUT "Y=",Y(N1)
145 YY(N1)=LOG(Y(N1))
150 IF N2<4 THEN 170
155 INPUT "S.D.=",SD
160 W(N1)=1/SD^2
165 GOTO 195
170 W(N1)=1
175 IF N2=1 THEN 195
180 W(N1)=1/Y(N1)
185 IF N2=3 THEN 195
190 W(N1)=W(N1)^2
195 PRINT
200 B=B+X(N1) : C=C+X(N1)^2 : D=D+YY(N1) : E=E+YY(N1)^2
205 I=I+(X(N1)*YY(N1))
210 NEXT N1
215 PRINT
220 G=NP
225 H1=C-(B^2/G) : H2=E-(D^2/G) : H3=I-((B*D)/G) : J1=H3/H1
230 J2=(D-(J1*B))/G : J3=ABS(H3^2/(H1*H2)) : J3=SQR(J3)
235 J4=ABS((1/(G-2))*(H2-J1*H3)) : J4=SQR(J4)
240 PRINT
245 B(1)=EXP(J2) : B(2)=J1
250 PRINT "INITIAL ESTIMATES : A=";B(1)
255 PRINT "                  : B=";B(2)
260 PRINT
265 PRINT "ITERATION","A","K","SSQ"
270 PRINT
275 I1=0
280 I1=I1+1
285 IF I1>10 THEN 510
290 S1=0 : S2=0 : S3=0 : S4=0 : S5=0 : S6=0 : R1=0
295 FOR I=1 TO NP
300 X=X(I)
```

```
2080 PRINT
2090 FOR I=1 TO NP
2100 A(I)=20-(Y(I)-NY)*YS
2110 IF (A(I)-INT(A(I)))>.5 THEN A(I)=A(I)+1
2120 WW(I)=(X(I)-NX)*XS :IF (WW(I)-INT(WW(I)))>.5 THEN WW(I)=WW(I)+1
2130 NEXT I
2140 PRINT
2150 PRINT
2160 DEF FNHOME$=ESC$+"h"+ESC$+"J"
2170 PRINT FNHOME$;
2180 FOR I=1 TO 20
2190 PRINT "      I"
2200 NEXT I
2210 PRINT "      "; :FOR I=1 TO 61 :PRINT "-"; :NEXT I
2220 DEF FNHOME$=ESC$+"H"
2230 FOR I=MX/50 TO MX STEP MX/50
2240 BB=(I-NX)*XS : IF (BB-INT(BB))>.5 THEN BB=BB+1
2250 BB=BB+4 : YD=B(1)*(1-EXP(-B(2)*I))
2260 CC=20-(YD-NY)*YS : IF(CC-INT(CC))>.5 THEN CC=CC+1
2270 PRINT FNHOME$;
2280 IF CC<0 OR CC>22 THEN 2310
2290 FOR J= 1 TO CC : PRINT " " : NEXT J
2300 PRINT TAB(BB);"."
2310 NEXT I
2320 FOR I=1 TO NP
2330 PRINT FNHOME$;
2340 IF A(I)<0 OR A(I)>22 THEN 2380
2350 FOR J=1 TO A(I) : PRINT " " : NEXT J
2360 WW(I)=(X(I)-NX)*XS :IF (WW(I)-INT(WW(I)))>.5 THEN WW(I)=WW(I)+1
2370 WW(I)=WW(I)+4 :PRINT TAB(WW(I));"*"
2380 NEXT I
2390 PRINT FNHOME$;
2400 FOR I=1 TO 20
2410 PRINT "      I"
2420 NEXT I
2430 PRINT FNHOME$;
2440 PRINT MY
2450 DEF FNHOME$=ESC$+"F"
2460 PRINT FNHOME$;
2470 PRINT "    ";NX; TAB(65);MX
2480 DEF FNHOME$=ESC$+"A"
2490 PRINT FNHOME$;
2500 PRINT FNHOME$;
2510 PRINT "  ";NY
2520 PRINT " "
2530 PRINT " "
2540 INPUT "CHANGE AXES ( Y or N )=",B$
2550 IF B$="Y" THEN 2050
2560 PRINT
2565 RT=0
2570 INPUT "ANOTHER CALCULATION ( Y or N )=",B$
2580 IF B$="Y" THEN 50
2590 END
Ok

 run

 NONLINEAR REGRESSION OF FIRST ORDER INCREASE
```

```
305 GG=B(1)*(1-EXP(-B(2)*X))
310 Z=Y(I)-GG : R5=SQR(W(I))*Z : R1=R1+ABS(R5)
315 FOR J=1 TO 2
320 B(J)=1.02*B(J)
325 GG=B(1)*(1-EXP(-B(2)*X))
330 U=GG: B(J)=B(J)*.98/1.02
335 GG=B(1)*(1-EXP(-B(2)*X))
340 B(J)=B(J)/.98
345 P(J)=(U-GG)/(.04*B(J))
350 NEXT J
355 R3=1
360 IF BT=0 THEN 385
365 IF I1=1 THEN 385
370 R3=0 : R4=(R5/R2)^2
375 IF R4>=1 THEN 385
380 R3=(1-R4)^2
385 S1=S1+R3*W(I)*P(1)^2 : S2=S2+R3*W(I)*P(1)*P(2) : S3=S3+R3*W(I)*P(2)^2
390 S4=S4+R3*W(I)*P(1)*Z : S5=S5+R3*W(I)*P(2)*Z : S6=S6+R3*W(I)*Z^2
395 NEXT I
400 R2=6*R1/N1 : D=S1*S3-S2^2 : Q(1)=(S3*S4-S2*S5)/D
405 Q(2)=(S1*S5-S2*S4)/D : C=ABS(Q(1)/B(1))+ABS(Q(2)/B(2))
410 B(1)=B(1)+Q(1) : B(2)=B(2)+Q(2)
415 PRINT I1,B(1), B(2), S6
420 IF C>.00001 THEN 280
425 V=S6/(N1-2) : Q(1)=SQR(V*S3/D) : Q(2)=SQR(V*S1/D)
430 PRINT
435 PRINT "FINAL VALUES : A=";B(1)
440 PRINT "              S.D.(A)=";Q(1)
445 PRINT
450 PRINT "                    K=";B(2)
455 PRINT "              S.D.(K)=";Q(2)
460 PRINT
465 PRINT "X","Y","Y(CALC)","DIFF"
470 PRINT
475 FOR I=1 TO NP
480 X=X(I)
485 GG=B(1)*(1-EXP(-B(2)*X))
490 PRINT X(I),Y(I),GG,Y(I)-GG
495 NEXT I
500 PRINT
505 GOTO 520
510 PRINT "TERMINATED AFTER 10 ITERATIONS"
515 GOTO 425
520 PRINT
2000 PRINT
2010 REM X - Y PLOTS OF EXPERIMENTAL DATA
2020 INPUT " PLOT ( Y OR N )=?",L$
2030 IF L$="N" THEN 2560
2040 ESC$=CHR$(27)
2050 INPUT "X - RANGE VALUES (MIN X, MAX X)=", NX,MX
2060 INPUT "Y - RANGE VALUES (MIN Y, MAX Y)=", NY,MY
2070 XS=60/(MX-NX) : YS=20/(MY-NY)
```

```
         -K*T
Y=A*(1 - e    )

TYPE OF WEIGHTING?
                          : 1 = CONSTANT S.D.
                          : 2 = PROPORTIONAL S.D.
                          : 3 = BETWEEN THE ABOVE
                          : 4 = S.D. SUPPLIED

CHOICE=4

BISQUARE WEIGHTING (Y or N)=n

# OF X/Y PAIRS=4

X=.05
Y=14.91
S.D.=1.65

X=.1
Y=23.93
S.D.=.89

X=.15
Y=32.15
S.D.=1.54

X=.2
Y=37.22
S.D.=1.36
```

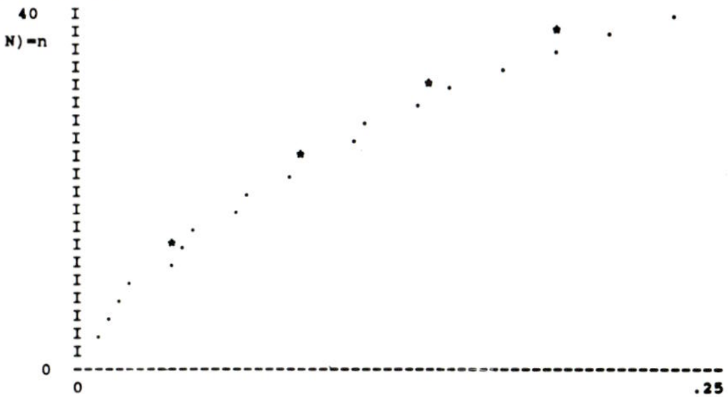


```
INITIAL ESTIMATES : A= 11.95529
                  : B= 6.079453

ITERATION      A           K           SSQ

 1          52.11247    6.871327    1194.476
 2          51.8748     6.290645    7.274831
 3          52.1445     6.256665    .4994283
 4          52.36212    6.21247     .4055348
 5          52.44351    6.195472    .3905971
 6          52.45989    6.191884    .3906459
 7          52.46376    6.191022    .390652
 8          52.465      6.190752    .3905709
 9          52.46539    6.190665    .3905402
10          52.46551    6.190638    .3905324

FINAL VALUES : A= 52.46551
          S.D.(A)= 2.666147

                K= 6.190638
          S.D.(K)= .5053762

 X          Y          Y(CALC)       DIFF
 .05        14.91      13.96682      .9431782
 .1         23.93      24.21554      -.2855415
 .15        32.15      31.73596      .4140473
 .2         37.22      37.25436      -3.435898E-02
```

Appendix 1. Computer program to calculate the $k_{PRECURSOR}$ constant for constant infusion experiments. The program is in basic (HP 150) and is adapted from the program of Duggleby (1981). The data are regressed to an exponential function by weighted nonlinear analysis. The example provided involves the incorporation of D_9-phosphorylcholine into the choline pool of the rat parietal cortex. The *x* values are time of infusion (h), whereas the y values are % incorporation of the deuterium label.

```
LIST
10 PRINT "ITERATIVE CALCULATION OF K(PRODUCT)"
20 PRINT "FOR AN M OF 1.2 TO 7.0"
30 PRINT
40 RR=1000
50 INPUT "INFUSION TIME (T1) IN MIN=",T1
60 INPUT "               K(PRECURSOR)=",KA
70 QZ=0 : X=40 : A=T1/60 : B=-1/A : C1=EXP(-KA*A) : C2=(1/KA)*C1-(1/KA)
80 PRINT
90 PRINT "M","K(PRODUCT)"
100 PRINT
110 FOR M=1.2 TO 7.1 STEP .1
120 D1=(1-C1)*(1/M) : D2=1-D1 : D3=(1/KA)*D1 : D4=D3+C2 : IT=1
130 IF QZ=0 THEN 150
140 GOTO 160
150 GOSUB 660
160 IF QZ=1 THEN 180
170 GOTO 200
180 X=1
190 GOSUB 760
200 IF QZ=1 THEN 220
210 GOTO 240
220 X=X-1
230 GOTO 250
240 X=X+1
250 IT=.1
260 IF QZ=0 THEN 280
270 GOTO 290
280 GOSUB 660
290 IF QZ=1 THEN 310
300 GOTO 320
310 GOSUB 760
320 IF QZ=1 THEN 340
330 GOTO 360
340 X=X-.1
350 GOTO 370
360 X=X+.1
370 IT=.01
380 IF QZ=0 THEN 400
390 GOTO 410
400 GOSUB 660
410 IF QZ=1 THEN 430
420 GOTO 440
430 GOSUB 760
790 IF P3<0 THEN 792
791 GOTO 800
792 RETURN
800 X=X+IT
810 IF X>KA-.1 THEN 830
820 GOTO 760
830 P2=0
840 RETURN
Ok

RUN
ITERATIVE CALCULATION OF K(PRODUCT)
FOR AN M OF 1.2 TO 7.0

INFUSION TIME (T1) IN MIN=9
              K(PRECURSOR)=6.26

M              K(PRODUCT)

 1.2            29.66
 1.3            22.05
 1.4            17.86
 1.5            15.13
 1.6            13.18
 1.7            11.71
 1.8            10.55
 1.9            9.609999
 2              8.83
 2.1            8.18
 2.2            7.61
 2.3            7.13
 2.39           6.16
 2.49           6.16
 2.59           5.98
 2.69           5.68
 2.79           5.41
 2.89           5.16
 2.99           4.94
 3.09           4.73
 3.19           4.54
 3.29           4.37
 3.39           4.21
 3.49           4.06
 3.59           3.92
```

```
440 IF QZ=1 THEN 460
450 GOTO 480
460 X=X-.01
470 GOTO 490
480 X=X+.01
490 IT=.001
500 IF QZ=0 THEN 520
510 GOTO 530
520 GOSUB 660
530 IF QZ=1 THEN 550
540 GOTO 560
550 GOSUB 760
560 IF QZ=1 THEN 580
570 GOTO 600
580 X=X-.001
590 GOTO 610
600 X=X+.001
610 P1=D2+(D4*X) : P2=B*LOG(P1) : P3=P2-X
620 PRINT (INT(M*100))/100,(INT(P2*100))/100
630 NEXT M
640 PRINT
650 END
660 P1=D2+(D4*X)
670 IF P1<=0 THEN 710
680 P2=B*LOG(P1) : P3=P2-X
690 IF P3<0 THEN 692
691 GOTO 700
692 RETURN
700 IF X<=KA/2 THEN 720
710 X=X-IT : GOTO 660
720 P2=0
730 X=1 : QZ=1
750 RETURN
760 P1=D2+(D4*X)
770 IF P1<=0 THEN 800
780 P2=B*LOG(P1) : P3=P2-X
```

```
3.69      3.79
3.79      3.66
3.89      3.55
3.99      3.44
4.09      3.34
4.19      3.25
4.29      3.16
4.39      3.07
4.49      2.99
4.59      2.92
4.69      2.84
4.79      2.77
4.89      2.71
4.99      2.65
5.09      2.59
5.19      2.53
5.29      2.47
5.39      2.42
5.49      2.37
5.59      2.32
5.69      2.28
5.79      2.23
5.89      2.19
5.99      2.15
6.09      2.11
6.19      2.07
6.29      2.04
6.39      2
6.49      1.97
6.59      1.93
6.69      1.9
6.79      1.87
6.89      1.84
6.99      1.81
7.09      1.78

Ok
```

Appendix 2. A program written in basic to calculate the $k_{PRODUCT}$ for a given $k_{PRECURSOR}$ and varying m value. In the example, the experimental infusion time (in this case 9 min) and the $k_{PRECURSOR}$ are provided, with the program subsequently calculating the $k_{PRODUCT}$ for m values from 1.2–7. The relationship of $k_{PRODUCT}$ to m can be best fit to a third order polynominal, for computerization of the standard curves.

References

Aguilonius, S. M., Lundholm, B., and Windbladh, B. (1972) Effects of some anticholinergic drugs on cortical acetylcholine release and motor activity in rats. *Eur. J. Pharmacol.* **20,** 224–230.

Ansell, G. B. and Spanner, S. (1971) Studies on the origin of choline in the brain of the rat. *Biochem. J.* **122,** 741–750.

Ansell, G. B. and Spanner, S. (1975) The origin and metabolism of brain choline, in *Cholinergic Mechanisms,* (Waser, P. G. ed.) 117–129. New York, Raven.

Barker, L. A. and Mittag, T. W. (1975) Comparative studies of substrates and inhibitors of choline transport and choline acetyltransferase. *J. Pharmacol. Exp. Ther.* **192,** 86–94.

Bartholini, G., Stadler, H., Gadea-Ciria, M., and Lloyd, K. G. (1976). The use of the push–pull cannula to estimate the dynamics of acetylcholine and catecholamines within various brain areas. *Neuropharmacology* **15,** 515–519

Beani, L. and Bianchi, C. (1970) Effects of adrenergic blocking and antiadrenergic drugs on the acetylcholine release from the exposed cerebral cortex of the conscious animal, in *Drugs and Cholinergic Mechanisms in CNS* (Heilbronn, E. and Winter, A. eds.) 369–386. Res. Inst. of Natl. Defense, Stockholm.

Beani, L., Bianchi, C., Santinoceto, L., and Marchetti, P. (1968) The cerebral acetylcholine release in conscious rabbits with semipermanently implanted epidural cups. *Int. J. Neuropharmacol.* **7,** 469–481.

Beani L., Bianchi C., Siniscalchi A., Sivilotti L., Tanganelli S., and Veratti E. (1984) Different approaches to study acetylcholine release: Endogenous ACh versus tritium efflux. *Naunyn-Schmiedebergs Arch. Pharmacol.* **328,** 119–126.

Bergman, M. (1969) Kinetic modeling in physiology. *Fed. Eur. Biochem. Soc. Lett.* **2,** 556–557.

Bymaster, F. P., Perry, K. W., and Wong, D. T. (1985) Measurement of acetylcholine and choline in brain by HPLC with electrochemical detection. *Life Sci.* **37,** 1775–1781.

Casamenti, F., Corradetti, R., Loffelholz, K., Mantovani, P. and Pepeu, G. (1982) Effects of 4-aminopyridine on acetylcholine output from the cerebral cortex of rat in vivo. *Br. J. Pharmacol.* **76,** 438–445.

Casamenti, F., Deffenu, G., Abbamondi, A. N., and Pepeu. G. (1986) Changes in cortical acetylcholine output induced by modulation of nucleus basalis. *Brain Res. Bull.* **16,** 689–695.

Casamenti, F., Mantovani, P., Amaducci, L., and Pepeu, G. (1979) Effect of phosphatidylserine on acetylcholine output from the cerebral cortex of the rat. *J. Neurochem.* **32,** 529–533.

Casamenti, F., Pedata, F., Corradetti, R., and Pepeu, G. (1980) Acetylcholine output from the cerebral cortex, choline uptake and muscarinic receptors in morphine-dependent freely moving rats. *Neuropharmacology* **19,** 587–605.

Celesia, G. G. and Jasper, H. H. (1966) Acetylcholine released from cerebral cortex in relation to state of activation. *Neurology* **16,** 1053–1063.

Cheney, D. L. (1984) Drug effects on transmitter dynamics: An overview, in *Dynamics of Neurotransmitter Function* (Hanin, I., ed.) 63–80 New York, Raven.

Cheney, D. L. and Costa, E. (1977) Pharmacological implication of brain acetylcholine turnover. *Ann. Rev. Pharmacol. Toxicol.* **17,** 369–386.

Cheney, D. L., Costa, E., Hanin, I., Trabucchi, M., and Wang, C. T. (1975a) Application of principles of steady state kinetics to the in vivo estimation of acetylcholine turnover rate in mouse brain. *J. Pharmacol. Exp. Ther.* **192,** 288–296.

Cheney, D. L., LeFevre, H. F., and Racagni, G. (1975b) Choline acetyltransferase activity and mass fragmentography measurement of acetylcholine in specific nuclei and tracts of rat brain. *Neuropharmacology* **14,** 801–809.

Cheney, D. L., Trabucchi, M., Hanin, I., and Costa, E. (1975c) Morphine dependence and in vivo turnover rate in whole mouse brain. *J. Pharmacol. Exp. Ther.* **195,** 288–295.

Cheney, D. L., Racagni, G., Zsilla, G., and Costa, E. (1976) Differences in the action of various drugs on striatal acetylcholine and choline content in rats killed by decapitation or microwave radiation. *J. Pharm. Pharmacol.* **28,** 75–77.

Cheney, D. L., Trabucchi, M., Racagni, G., and Costa, E. (1974) Effects of acute and chronic morphine on regional rat brain acetylcholine turnover rate. *Life Sci.* **15,** 1977–1990.

Choi, R. L., Roch, M., and Jenden, D. J. (1973) A regional study of acetylcholine turnover in rat brain and the effect of oxotremorine. *Proc. West. Pharmacol. Soc.* **16,** 188–190.

Collier, B. and Mitchell, J. F. (1966) The central release of acetylcholine during stimulation of visual pathways. *J. Physiol.* (Lond) **184,** 239–254.

Consolo, S., Ladinsky, H., and Garattini, S. (1974) Effect of several dopaminergic drugs and trihexyphenidyl on cholinergic parameters in rat striatum. *J. Pharm. Pharmacol.* **26,** 275–277.

Costa, E. (1970) Simple neuronal models to estimate turnover rate of noradrenergic transmitters in vivo, in *Advances in Biochemical Psychopharmacology,* (Costa, E. and Giacobini, E. eds.) 169–204. New York, Raven.

Costa, E., Boullin, D. J., Hammer, W., Vogel, W., and Brodie, B. B. (1966) Interactions of drugs with adrenergic neurons. *Pharmacol. Rev.* **18,** 577–597.

Costa, E., Cheney, D. L., Racagni, G., and Zsilla, G. (1975) An analysis at synaptic level of the morphine action in striatum and n. accumbens: Dopamine and acetylcholine interactions. *Life Sci.* **17,** 1–8.

Costa, E. and Neff, N. H. (1966) Isotopic and nonisotopic measurement of the rate of catecholamine biosynthesis, in *Biochemistry and Pharmacology of the Basal Ganglion* (Costa, E., Cote, L., and Yaks, M. eds.) 141–156. New York, Raven.

Costa, E. and Neff, N. H. (1970) Estimation of turnover rates to study the metabolic regulation of the steady-state level of neuronal monoamines, in *Handbook of Neurochemistry* (Lajtha, A., ed.) 45–90. New York, Plenum.

Damsma, G., Westerink, B. H. C., and Horn, A. Horn. (1985) A simple, sensitive, and economic assay for choline and acetylcholine using HPLC, an enzyme reactor, and an electrochemical detector. *J. Neurochem.* **45,** 1649–1652.

Domino, E. F., Donelson, A. C., and Tuttle, T. (1977) Effects of delta-9-tetrahydrocannabinol on regional brain acetylcholine. *Adv. Behav. Biol.* **24,** 673–678.

Domino, E. F., Matthews, J. D., and Labrecque, G. (1973) Effects of morphine, nalorphine and naloxone on rat brain release of acetylcholine. *Fed. Proc.* **32,** 757.

Domino, E. F. and Wilson, A. (1973) Effects of narcotic analgesic agonists and antagonists on rat brain acetylcholine. *J. Pharmacol. Exp. Ther.* **184,** 18–32.

Domino, E. F. and Wilson, A. E. (1975a) Brain acetylcholine in morphine pellet-implanted rats given naloxone. *Psychopharmacology* **41,** 19–22.

Domino, E. F. and Wilson, A. E. (1975b) Decreased rat brain acetylcholine utilization after heroin and cross tolerance to I-methadone. *Biochem. Pharmacol.* **24,** 927–928.

Dross, K. and Kewitz, H. (1966) Der einbau von i.v. zugefuhrtem cholin in das acetylcholin des gehirns. *Naunyn-Schmiedebergs Arch. Phamacol.* **255,** 10–11.

Dross, K. and Kewitz, H. (1972) Concentration and origin of choline in the rat brain. *Naunyn-Schmiedebergs Arch. Pharmacol.* **274,** 91–106.

Dudar, J. D. (1975) The effect of septal nuclei stimulation on the release of acetylcholine from the rabbit hippocampus. *Brain Res.* **83,** 123–133.

Duggleby, R. G. (1981) A nonlinear regression program for small computers. *Anal. Biochem.* **110,** 9–18.

Eva, C., Hadjiconstantinou, M., Neff, N. H., and Meek, J. L. (1984) Acetylcholine measurement by high-performance liquid chromatography using an enzyme-loaded postcolumn reactor. *Anal. Biochem.* **143,** 320–324.

Goldberg, A. M. and McCaman, R. E. (1973) The determination of picomole amounts of acetylcholine in mammalian brain. *J. Neurochem.* **20,** 1–8.

Guidotti, A., Cheney, D. L., Trabucchi, M., Doteuchi, M., Wang, C., and Hawkins, R. A. (1974) Focussed microwave radiotion: A technique to minimize postmortem changes in cyclic nucleotides, DOPA and choline and to preserve brain morphology. *Neuropharmacology* **13,** 1115–1122.

Guyenet, P. G., Agid, Y., Javoy, F., Beaujouan, J. C., Bossier, J., and Glowinski, J. (1975) Effects of dopaminergic receptor agonists and antagonists on the activity of the neostriatal cholinergic system. *Brain Res.* **84,** 227–244.

Guyenet, P., Agid, Y., Javoy, F., Beaujouan, J. C., and Glowinski, J. (1974) Action selective des neuroleptiques sur les neurones cholinergiques du neostriatum chez le rat: Antagonisme vis-a-vis de l'apomorphine. *C.R. Acad. Sci. Paris* **278,** 2679–2682.

Guyenet, P., Lefresne, P., Rossier, J., Beaujouan, J. C., and Glowinski, J. (1973) Inhibition by hemicholinium-3 of C14-acetylcholine synthesis and H3-choline high affinity uptake in rat striatal synaptosomes. *Mol. Pharmacol.* **9,** 630–739.

Hanin, I. (1974) *Choline and Acetylcholine: Handbook of Chemical Assay Methods* New York, Raven.

Hanin, I., Cheney, D. L., Trabucchi, M., Massarelli, R., Wang, C. T., and Costa, E. (1973) Applications of principles of steady state kinetics to measure acetylcholine turnover rate in rat salivary glands: Effects of differentiation and duct ligation. *J. Pharmacol. Exp. Ther.* **187,** 68–77.

Hanin, I., Massarelli, R., and Costa, E. (1972) An approach to the in vivo study of acetylcholine turnover in rat salivary glands by radio gas chromatography. *J. Pharmacol. Exp. Ther.* **181,** 10–18.

Hemsworth, B. A. and Mitchell, J. F. (1969) The characteristics of acetylcholine mechanisms in the auditory cortex. *Br. J. Pharamcol.* **36,** 161–170.

Hemsworth, B. A. and Neal, M. J. (1968) The effect of central stimulant drugs on acetylcholine release from rat cerebral cortex. *Br. J. Pharmacol.* **34,** 543–550.

Higgins, A. J. and Neal, M. J. (1982a) Potassium activation of [^{3}H]choline accumulation by isolated sympathetic ganglia of the Rat. *Br. J. Pharmacol.* **77,** 573–580.

Higgins, A. J. and Neal, M. J. (1982b) Effect of potassium depolarization and preganglionic nerve stimulation on the metabolism of [^{3}H]choline in at isolated sympathetic ganglia. *Br. J. Pharmacol.* **77,** 581–590.

Israel, M. and Lesbats, B. (1981) Continuous determination by a chemilumine scent method of acetylcholine release and compartmentation in Torpedo electric organ synaptosomes. *J. Neurochem.* **37,** 1475–1483.

Jasper, H. H. and Tessier, J. (1971) Acetylcholine liberation from the cerebral cortex during paradoxical (REM) sleep. *Science* 601–602.

Jenden, D. J., Choi, L., Silverman, R. W., Steinborn, J. A., Roch, M., and Booth, R. A. (1974) Acetylcholine turnover estimation in brain by gas chromatography-mass spectrometry. *Life Sci.* **14,** 55–63.

Jenden, D. J., Jope, R. S., and Weiler, M. H. (1976) Regulation of acetylcholine synthesis: Does cytoplasmic acethylcholine control high affinity choline up take? *Science* **194,** 635–637.

Johnston, M. V., McKinney, M., and Coyle, J. T. (1979) Evidence of a cholinergic projection to neocortex from neurons in the basal forebrain. *Proc. Natl. Acad. Sci.* **76,** 5392–5396

Kelly, P. H. and Moore, K. E. (1978) Decrease of neocortical choline acetyltransferase after lesion of the globus pallidus in the rat. *Exp. Neurol.* **61,** 479–484.

Kaneda, N., Noro, Y., and Nagatsu, T. (1985) Highly sensitive assay for acetylcholinesterase activity by high-performance liquid chromatography with electrochemical detection. *J. Chromatogr.* **344,** 93–100.

Koelle G. B., Davis R., Smyrl E. G., and Fine A. V. (1974) Refinement of the bis-(thioacetoxy)aurate (I) method for the electron microscopic localizati on of acetylcholinesterase and nonspecific cholinesterases. *J. Histochem. Cytochem.* **22,** 252–259.

Ladinsky, H., Consolo, S., and Garattini, S. (1974) Increase in striatal acetylcholine levels in vivo by Piribedil, a new dopamine receptor stimulant. *Life Sci.* **14,** 1251–1260.

Lehmann, J., Nagy, J. I., Atmodia, S., and Fibiger, H. C. (1980) The nucleus basalis magnocellularis: The origin of a cholinergic projection to the neocortex of rat. *Neuroscience* **5,** 1161–1174.

Lehmann J. and Scatton B. (1982) Excitatory amino acid receptor mediated release of [^{3}H]acetylcholine from striatal slices of the rat: effect of magnesium and pharmacolgical characteristics. *Brain Res.* **252,** 77–89.

Lehmann, J., Schneider, J., McPherson, S., Murphy, D. E., Bernard, P., Tsai, C., Bennett, D. A., Pastor, G., Steel, D. J., Boehm, C., Cheney, D. L., Liebman, J. M., Williams, M., and Wood, P. L. (1987) CPP, a selective NMDA-type receptor antagonist: Characterization in vitro and in vivo. *J. Pharmacol. Exp. Ther.* **240,** 737–746.

Liberato, D. J. and Yergey, A. L. (1986) Solvent selection for thermospray liquid chromatography/mass spectrometry. *Anal. Chem.* **58,** 6–9.

Lo Conte, G., Bartolini, L., Casamenti, F., Marconcini-Pepeu, I., and Pepeu, G. (1982a) Lesions of cholinergic forebrain nuclei: Changes in avoidance behavior and scopolamine actions. *Pharmacol. Biochem. Behav.* **17,** 933–937.

Lo Conte, G., Casamenti, F., Bigl, V., Milaneschi, E., and Pepeu, G. (1982b) Effects of magnocellular forebrain lesions on acetylcholine output from the cerebral cortex, electrocorticogram and behaiour. *Arch. Ital. Biol.* **120,** 176–188.

McGeer, P. L., Grewaal, D. S., and McGeer, E. G. (1974) Influence of non-cholinergic drugs on rat striatal acetylcholine levels. *Brain Res.* **80,** 211–217.

MacIntosh, F. D. (1963) Synthesis and storage of acetylcholine in nervous tissue. *Can. J. Biochem. Physiol.* **41,** 2555–2571.

MacIntosh, F. D. and Oborin, P. E. (1953) Release of acetylcholine from the intact cerebral cortex. *Abstr. XIX Int. Physiol. Congr.* 580–581.

Magnani, M., Mantovani, P., and Pepeu, G. (1984) Effect of cholecystokinin octapeptide and ceruletide on release of acetylcholine from cerebral cortex of rat in vivo. *Neuropharmacology* **23,** 1305–1309.

Marchi M., Paudice P., and Raiteri M. (1981) Autoregulation of acetylcholine release in isolated hippocampal nerve endings. *Eur. J. Pharmacol.* **73,** 75–79.

Mitchell, J. F. (1960) Release of acetylcholine from the cerebral cortex and the cerebellum. *Proc. Physiol. Soc.* **155,** 22P.

Mitchell, J. F. (1961) Acetylcholine release from the cerebral cortex during stimulation. *J. Physiol.* (Lond) **158,** 20–21.

Mitchell, J. F. (1963) The spontaneous and evoched release of acetylcholine from the cerebral cortex. *J. Physiol.* (Lond) **165,** 98–116.

Moroni, F., Corradetti, R., Casamenti, F., Moneti, G., and Pepeu, G. (1981) The release of endogenous GABA and glutamate from the cerebral cortex in the rat. *Naunyn-Schmiedebergs Arch. Pharmacol.* **316,** 235–239.

Mulas, A., Mulas, M. L., and Pepeu, G. (1974) Effect of limbic system lesions on acetylcholine release from the cerebral cortex of rats. *Psychopharmacologia* (Berlin) **39,** 223–230.

Mullin, W. J. (1974) Central release of acetylcholine following administration of morphine to unanesthetized rabbits. *Can. J. Physiol. Pharmacol.* **52,** 369–374.

Murrin L. C., DeHaven R. N., and Kuhar M. J. (1977) On the relationship between [^{3}H]choline uptake activation and [^{3}H]acetylcholine release. *J. Neurochem.* **29,** 681–687.

Myers, R. D. and Beleslin, D. B. (1970) The spontaneous release of 5-hydroxytryptamine and acetylcholine within the diencephalon of the unanestherized rhesus monkey. *Brain Res.* **11,** 539–552.

Neff, N. H., Ngai, S. H., Wang, C. T., and Costa, E. (1969) Calculation of the rate of catecholamine synthesis from the rate of conversion of [^{14}C]tyrosine to catecholamines: Effect of adrenal demedullation on synthesis rate. *Mol. Pharmacol.* **5,** 90–99.

Neff, N. H., Spano, P. F., Groppetti, A., Wang, C. T., and Costa, E. (1971) A simple procedure for calculating the synthesis of norepinephrine, dopamine, and serotonin in rat brain. *J. Pharmacol. Exp. Ther.* **176,** 701–710.

Oelszner, W., Scherber, A., Nehring, K., and Staib, H. H. (1973) Spontane und evozierte Freisetzung von Azetylcholn in verschiedenen Hirnregionen der Ratte. *Acta Biol. Med. Ger.* **30,** 407–414.

Pepeu, G. (1973) The release of acetylcholine from the brain: An approach to the study of central cholinergic mechanisms, in *Progress in Neurobiology* (Kerkut, G. A. and Phillis, J. W., eds.) 257–288. Oxford, Pergamon.

Pepeu, G., Garau, L., Mulas, M. L., and Marconcini-Pepeu, I. (1975) Stimulation by morphine of acetylcholine output from the cerebral cortex of septal rats. *Brain Res.* **100,** 677–680.

Phillis, J. W. (1968) Acetylcholine release from the cerebral cortex: Its role in cortical arousal. *Brain. Res.* **7,** 378–389.

Phillis, J. W. and Chong, G. C. (1965) Acetylcholine release from the cerebral and cerebellar cortices: Its role in cortical arousal. *Nature* **207,** 1253–1255.

Potter, P. E., Hadjiconstantinou, M., Meek, J. L., and Neff, N. H. (1984) Measurement of acetylcholine turnover rate in brain: An adjunct to simple HPLC method for choline and acetylcholine. *J. Neurochem.* **43,** 288–290.

Potter, P. E., Meek, J. L., and Neff, N. H. (1983) Acetyhcholine and choline in neuronal tissue measured by HPLC with electrochemical detection. *J. Neurochem.* **41,** 188–194.

Quastel J. H., Tennenbaum M., and Wheatley A. H. M. (1936) Choline ester formation in and cholinersterase activities of tissues in vitro. *Biochem. J.* **30,** 1668–1681.

Racagni, G., Cheney, D. L., Trabucchi, M., and Costa, E. (1975a) In vivo actions of clozapine and haloperidol on the turnover rate of acetylcholine in rat striatum. *J. Pharmacol. Exp. Ther.* **196,** 323–332.

Racagni, G., Trabucchi, M., and Cheney, D. L. (1975b) Steady-state concentrations of choline and acetylcholine in rat brain parts during a constant rate infusion of deuterated choline. *Naunyn-Schmiedebergs Arch. Pharmacol.* **290,** 99–105.

Racagni, G., Cheney, D. L., Trabucchi, M., Wang, C., and Costa, E. (1974) Measurement of acetylcholine turnover rate in discrete areas of the rat brain. *Life Sci.* **15,** 961–975.

Rommelspacher H. and Kuhar M. J. (1975a) Effects of drugs and axotomy on acetylcholine levels in central cholinergic neurons. *Naunyn-Schmiedebergs Arch. Pharmacol.* **291,** 17–21.

Rommelspacher, H. and Kuhar, M. (1975b) Effects of dopaminergic drugs and acute medial forebrain bundle lesions on striatal acetylcholine levels. *Life Sci.* **16,** 65–70.

Saelens, J. K., Simke, J. P., Allen, M. P., and Conroy, C. A. (1973) Some of the dynamics of choline and acetylcholine metabolism in rat brain. *Arch. Int. Pharmacodyn. Ther.* **203,** 305–312.

Sanfacon, G. and Labreque, G. (1982) Ketamine is the anesthetic of choice in studies on neocortical acetylcholine release. *Eur. J. Pharmacol.* **85,** 347–350.

Schmidt, D. E. and Buxbaum, D. M. (1978) Effect of acute morphine administration on regional acetylcholine turnover in the rat brain. *Brain Res.* **147,** 194–200.

Schuberth, J., Sparf, B., and Sundwall, A. (1969) A technique for the study of acetylcholine turnover in mouse brain in vivo. *J. Neurochem.* **16,** 695–700.

Sedvall, G. C., Weise, V. K., and Kopin, I. J. (1968) The rate of norepinephrine synthesis measured in vivo during short intervals: Influence of adrenergic nerve impuse activity. *J. Pharmacol. Exp. Ther.* **159,** 274–279.

Sethy, V. H. and VanWoert, M. H. (1974) Modification of striatal acetylcholine concentration by dopamine receptor agonist and antagonists. *Res. Commun. Chem. Pathol. Pharmacol.* **8,** 13–28.

Simon, J. R., Simon, J. R., Atweh, S., and Kuhar, M. J. (1976) Sodium-dependent high affinity choline uptake: A regulatory step in the synthesis of acetylcholine. *J. Neurochem.* **26,** 909–922.

Slater, P. (1969) The effects of intraventricular injections of quarternary ammonium compounds on the acetylcholine content of rat brain. *Arch. Int. Pharmacodyn. Ther.* **181,** 253–262.

Smith, C. M. (1972) The release of acetylcholine from the rabbit hippocampus. *Br. J. Pharmacol.* **45,** 172P.

Somogyi G. T. and Szerb J. C. (1972) Demonstration of acetylcholine release by measuring efflux of labelled choline from cerebral cortical slices. *J. Neurochem.* **19,** 2667–2677.

Sparf, B. (1973) On the turnover of acetylcholine in the brain. *Acta Physiol. Scand.* [Suppl.] **397,** 1–47.

Stadler, H., Lloyd, K. G., and Gadea-Ciria, M. (1973) Enhanced striatal acetylcholine release by chlorpromazine and its reversal by apomorphine. *Brain Res.* **55,** 476–480.

Stavinoha, W. B., Weintraub, S. T., and Modak, A. T. (1973) The use of microwave heating to inactivate cholinesterase in the rat brain prior to analysis for acetylcholine. *J. Neurochem.* **20,** 361–371.

Szerb, J. C. (1975) The release of acetylcholine from central cortical slices in the presence or absence of an anticholinesterase, in *Cholinergic Mechanisms* (Waser, P. G., ed.) 213–216. New York, Raven.

Trabucchi, M., Cheney, D. L., Hanin, I., and Costa, E. (1975) Application of principles of steady-state kinetics to the estimation of brain acetylcholine turnover rate: Effects of oxotremorine and physostigmine. *J. Pharmacol. Exp. Ther.* **194,** 57–64.

Vorhees, C. V., Schmidt, D. E., and Barrett, R. J. (1978) Effects of pyrithiamin and oxythiamin on acetylcholine levels and utilization in rat brain. *Brain Res. Bull.* **3,** 493–496.

Wenk, H., Bigl, V., and Meyer, U. (1980) Cholinergic projectsions from magnocellular nuclei of the basal forebrain to cortical areas in the rat. *Brain Res.* **2,** 295–316.

Wessel, R. H., Willett, E. R., and Simone, A. J. (1965) *Statistics as Applied to Economics and Business.* Chap. 10, (New York, Holt, Rinehart and Winston, Inc.).

Westfall T. C., Besson M. J., Giorguieff M. F., and Glowinski J. (1976) The role of presynaptic receptors in the release and synthesis of 3H-dopamine by slices of rat striatum. *Naunyn-Schmiedebergs Arch. Pharmacol.* **292,** 279–287.

Wilson, A. E. and Domino, E. F. (1973) Verification of rat brain acetylcholine antidepletion by morphine using gas chromatography. *Biochem. Pharmacol.* **22,** 1943–1944.

Wood, P. L. and Cheney, D. L. (1978) The effects of muscarinic receptor blockers on the turnover rate of actylcholine in various regions of the rat brain. *Can. J. Physiol. Pharmacol.* **57,** 404–411.

Wood, P. L. and Peloguin, A. (1982) Increases in choline levels in rat brain elicited by meclofenoxate. *Neuropharmacology* **21,** 487–497.

Wood, P. L. and Stotland, L. M. (1980) Actions of enkephalin, mu and partial agonist analgesics on acetylcholine turnover in rat brain. *Neuropharmacology* **19,** 975–982.

Yaksh, T. L. and Yamamura, H. I. (1974) Factors affecting performance of the push–pull cannula in brain. *J. Appl. Physiol.* **37,** 428–434.

Yaksh, T. L. and Yamamura, H. I. (1975) Blockade of morphine by acetylcholine release from the caudate nucleus in the midpontine pretrigeminal cat. *Brain Res.* **83,** 520–524.

Yamamura, H. I. and Snyder, S. H. (1973) High affinity transport of choline into synaptosomes of rat brain. *J. Neurochem.* **21,** 1355–1374.

Zilversmit, D. B. (1960) The design and analysis of isotope experiments. *Am. J. Med.* **29,** 832–848.

Zsilla, G., Cheney, D. L., Racagni, G., and Costa, E. (1976) Correlation between analgesia and the decrease of acetylcholine turnover rate in cortex and hippocampus elicited by morphine, meperidine, viminol R2 and azidomorphine. *J. Pharmacol. Exp. Ther.* **199,** 662–668.

Zsilla, G., Racagni, G., Cheney, D. L., and Costa, E. (1977) Constant rate infusion of deuterated phosphorylcholie to measure the effects of mor phine on acetylcholine turnover rate in specific nuclei of rat brain. *Neuropharmacology.* **162,** 25–30.

Index